ADVANCES IN
UNDERWATER TECHNOLOGY,
OCEAN SCIENCE
AND OFFSHORE ENGINEERING

Volume 6

Oceanology

ADVANCES IN UNDERWATER TECHNOLOGY, OCEAN SCIENCE AND OFFSHORE ENGINEERING

Vol. 1 Developments in Diving Technology

Vol. 2 Design and Installation of Subsea Systems

Vol. 3 Offshore Site Investigation

Vol. 4 Evaluation, Comparison and Calibration of Oceanographic Instruments

Vol. 5 Submersible Technology

Vol. 6 Oceanology

CONFERENCE COMMITTEE

D. W. Colvin, SUT
C. J. Antonakis, SUT
D. A. Ardus, SUT
D. Wardle, SUT
R. N. Bray, WODCON committee representative
D. C. C. Stott, Spearhead Exhibitions Limited
L. A. Sandbach, Spearhead Exhibitions Limited

0 860107728 7413

ADVANCES IN UNDERWATER TECHNOLOGY, OCEAN SCIENCE AND OFFSHORE ENGINEERING

Volume 6

OCEANOLOGY

Proceedings of an international conference (Oceanology International '86), sponsored by the Society for Underwater Technology, and held in Brighton, UK, 4–7 March 1986

Published by
Graham & Trotman Ltd.

First published in 1986 by

Graham & Trotman Limited
Sterling House
66 Wilton Road
London SW1V 1DE

Graham & Trotman Inc.
13 Park Avenue
Gaithersburg, MD 20877
USA

ISBN 086010 772 8

PRINTED AND BOUND IN GREAT BRITAIN
AT THE ALDEN PRESS, OXFORD

Contents

PART X
HYDROGRAPHY — INSTRUMENTATION

PART XI
HYDROGRAPHY — DREDGING/MUD/GENERAL

Part I

Environmental Observations

MODERN NAVIGATION AND POSITIONING TECHNIQUES

Prof. V Ashkenazi, Dr M E Napier,
University of Nottingham, UK

Presented on **Tuesday, 4 March 1986** in
Clarence Conference Room at **14.30.**

EMBARGOED AGAINST DELIVERY

1

Modern Navigation and Positioning Techniques

V. Ashkenazi and M. E. Napier, University of Nottingham, UK

INTRODUCTION

During the last two decades, significant developments have taken place in navigation and positioning techniques. These have occurred not only in the more classical radio and acoustic methods, but also in satellite and inertial techniques. Radio navigation is still the most commonly used surface navigation and positioning method. It spans a wide range of applications from the more localized high-accuracy positioning systems (e.g. Trisponder, Maxiran, Syledis) to worldwide navigation (e.g. Omega). The proliferation of systems reflects their relative merits and drawbacks, there being, in particular, a trade-off between range and accuracy. Moreover, the different techniques may not only be based on different co-ordinate systems, but also differ from one another by systematic biases. This can lead to ambiguities, especially in the determination of international and commercial concession boundaries. This is where a satellite-based navigation system comes into its own.

At present Transit (the Navy Navigation Satellite System) can provide positioning accuracies ranging from 50 m for single-pass navigation down to 2–5 m for multi-pass positioning. The Global Positioning System (GPS), which is due to replace Transit at the end of the current decade, is capable of quasi-instantaneous positioning to an accuracy of 10–20 m, with higher accuracies obtainable from observations over a longer interval. The US Department of Defense estimates that by the beginning of the 21st century there will be nearly half-a-million GPS receivers in operation throughout the world. Equally important, but covering a more limited objective than GPS, is the development of inertial navigation and surveying systems.

Radio, acoustic and inertial positioning methods are described in the next section. The two main satellite positioning systems, Transit and GPS, are reviewed in the following section. This is followed by a discussion of the various scientific and commercial applications, notably in surface navigation, surveying and geodesy and underwater positioning.

RADIO, ACOUSTIC AND INERTIAL POSITIONING

Radio Positioning Systems

Radio positioning systems comprise a chain of radio beacons (transmitters or transponders) located at points of known geodetic co-ordinates, and one or more mobile units on vessels whose positions are required. The fixed beacons are usually on land, but may also be on fixed platforms out at sea. There are two main positioning methods in common use, the circular mode and the hyperbolic mode.

In circular mode, the mobile unit transmits either a pulsed or a continuous-wave radio signal (depending on the system), which the transponders receive and return in a suitably coded form. In the case of a pulsed signal (e.g. Syledis) the ranges to each transponder are calculated from the precisely measured time differences between the transmitted pulse and each returned pulse. The position is then computed by trilateration. In the case of a continuous-wave signal (e.g. Argo) the phase difference between the transmitted signal and each return signal is measured. For each transponder, the loci of zero phase differences are a series of concentric circles centred on the transponder and separated by half the wavelength of the modulated signal. The phase measurements give the portion between two such circles but not the lane in which the mobile unit is located (i.e. between which two circles). The resulting ambiguity may be resolved by including one or more additional modulation of longer wavelength in the measuring process. In the circular mode, the mobile unit is active, and therefore most systems are restricted to a single user, although some can support a few users by time-sharing (e.g. Syledis: 10 users).

In the hyperbolic mode, the mobile unit receives either pulsed or continuous-wave signals from two or more pairs of transmitters within the chain. It is usual to have a single master station, which is a transmitter common to all pairs and with which all the other transmitters (slaves) are synchronized. In the case of pulsed signals (e.g. Pulse/8) the time difference between the arrival of pulses from a pair of transmitters is a measure of the range difference from the mobile unit to the transmitters. Lines of constant range difference define a family of hyperbolae whose foci are the two transmitting stations. Two or more range difference measurements from different pairs of transmitters are required to compute the position of the mobile unit. In the case of continuous-wave transmissions (e.g. Decca Navigator) the phase difference between the signals from a pair of transmitters gives the portion of a lane, but not the lane count. As for pulsed time, the loci of zero phase differences are a family of hyperbolae. Lane ambiguity can be resolved by the inclusion of a longer wavelength modulation, as for the circular mode. The hyperbolic mode systems can support an indefinite number of users, because the mobile units are passive.

There is a proliferation of radio positioning systems using either circular or hyperbolic modes, and offering a variety of working ranges and accuracies. There is a trade-off between range and accuracy, both being functions of the carrier wave frequency. Microwave systems offer the highest accuracy, but can only operate within line of sight (e.g. Tellurometer MRDI, 1 m). Ultra-high frequency systems (e.g. Syledis, 2–10 m) down to medium frequency systems (e.g. Hyperfix, 5–25 m) offer high accuracy and ranges of up to 400 km. Low frequency systems offer slightly greater range, but at much reduced accuracies (e.g. Pulse/8 and Decca Navigator, 50 m) and, lastly, very low frequency systems can increase the ranges up to several thousand kilometres, but with very poor accuracies (e.g. Omega, 1–3 km).

Acoustic Positioning Systems

There are two main acoustic positioning techniques. The first is the long baseline acoustic positioning system (LBL), which is used for the positioning of underwater vehicles, divers and structures, as well as for the positioning of surface vessels, particularly when the site is out of range of a suitable radio positioning system. The second is the super-short baseline acoustic positioning system (SSBL), which has been developed as a dynamic positioning system for holding a drilling ship or semi-submersible over a fixed point on the sea bed. The latter is also used as a system for tracking one or more underwater vehicle and for determining the positions of divers relative to that of the mother ship.

Long baseline acoustic position fixing is based on trilateration. An interrogating acoustic beacon at an unknown point measures ranges to three or more transponders sited over fixed points with known co-ordinates. The transponders are usually deployed close to the sea bed in a triangular or a square array, with baselines varying from as little as 100 m to several kilometres. The interrogating beacon may be located, as required, either on a submersible or a diver or on the underside of a surface vessel. Using the known velocity of the propagation of sound in water and the time delays within the equipment, the ranges are determined by observing the time which an acoustic signal takes to travel from the interrogating beacon to each of the transponders and back.

Calibration of the transponder array requires both the relative positioning of the transponders and the absolute location and orientation of the whole array. Relative positioning may be done by ranging from a surface vessel to all the transponders from a number of unknown positions. However, a higher accuracy is achieved if direct baseline measurements are made, thus avoiding ranging through the water column. This can be done by using an intelligent transponder system (e.g. Sonardyne Compatt). Absolute calibration can then be done by ranging from a surface vessel which is, in turn, positioned relative to another available surface positioning system.

The effect on the path length of refraction due to temperature gradients in the water column, reflections from the sea bed, surface or nearby structures, and the uncertainty in the value of the speed of sound in water are the main limiting factors in the accuracy which can be achieved by acoustic positioning methods. In addition to these environmental effects, the accuracy of LBL is dependent on the frequency used, with the accuracy increasing as the frequency increases. As the attenuation of the acoustic signal also increases with frequency, a balance has to be struck between range and accuracy. In high frequency systems (e.g. HATS) 1–10 cm accuracies in range are obtainable for ranges of up to 100 m, whereas in lower frequency systems (e.g. Oceano Subnav) relative positioning accuracies of 1–3 m can be achieved from a well calibrated network of beacons with 500 m baselines.

In the super-short baseline acoustic systems (e.g. Honeywell RS9), the position of a transponder relative to the interrogating hydrophone is determined by measuring the length, heading and angle of dip of the sound path between the two. The length is calculated from the elapsed time for the acoustic signal to travel the double distance to the transponder. The heading and angle of dip are obtained from the phase comparison of the returned signal at three closely spaced receivers, which are mounted on the same unit as the interrogating hydrophone and are separated by only a few centimetres. Corrections may be applied for the heading, roll and pitch of the ship in order to obtain the relative position of the transponder in the adopted co-ordinate reference system (e.g. UTM).

The errors in range measurements due to refraction, reflection and speed of sound also apply to SSBL positioning. Refraction also causes errors in the measurement of the angle of dip. This can be countered by an independent measure of the depth of the transponder. The relative accuracy of SSBL positioning is estimated to be about 1% of the ranges involved. This compares favourably with LBL, but does not take into account the additional error introduced by the (absolute) surface positioning system, which may in turn be quite large.

Inertial Positioning

The basis of inertial navigation is the computation of position by the measurement and subsequent integration of the 3-D acceleration vector of the vehicle. It is a dead-reckoning method of navigation which gives attitude, velocity and position, starting from a known point. Inertial systems are self-contained, they are free from environmental influences and, unlike radio and satellite systems, they are independent of any radio signal or external organization. However, due to component and alignment errors, the accuracy of the determined velocity and position vectors deteriorate as a function of time. A modern military aircraft system has a position error which increases by about 2000 m per hour. Inertial systems have been used mainly for the navigation of missiles, spacecraft, ships, submarines and aircraft, which may tolerate an error of this order between infrequent position updates.

During the last decade, several high-precision systems have been developed and successfully used in many land surveying (e.g. Ferranti Inertial Land Surveyor) and other applications (e.g. Ferranti FINDS for borehole surveys). These systems achieve greater accuracy by using direct velocity updates. A series of perfect velocity measurements is made by frequently bringing the system to rest ($v = 0$). In some of the systems, a simple curve may be fitted and then used to model the velocity error during the periods between velocity updates. In other systems a dynamic model of the inertial system errors is corrected for by the frequent velocity error measurements used in conjunction with a Kalman filter. For either method of velocity error modelling the zero velocity updating must be made every 3–5 min. Zero velocity updates are practical in land survey applications, where either a road vehicle can be stopped or a helicopter landed at pre-arranged and suitably spaced landing sites. In this way, accuracies of 10–20 cm can be achieved over distances up to 100 km.

Clearly, it is impractical to bring a ship to rest for a zero velocity update. However, with a dynamic error model and a Kalman filter, the inertial system may be updated by using independent measurements of either velocity (as may be provided by a Doppler speed log) or position (from a radio, acoustic or satellite navigation system). The high relative accuracy of an inertial system complements well the high absolute, but low short-term, accuracy of modern radio positioning systems. The integration of these two provides real-time filtered and post-mission smoothed position and velocity vectors, with an accuracy superior to that which could be achieved by either system alone. Moreover, with this combination of systems, the (Kalman filter) computed track of the vessel will be very close to its actual track, irrespective of the dynamic behaviour of the vessel.

SATELLITE POSITIONING SYSTEMS

Transit Doppler System

The US Navy Navigation Satellite System (NNSS), also known as Transit Doppler, has now been in existence for over 20 years and has been generally available to non-military users since 1967. It consists of five or six operational satellites with circular orbits at an altitude of 1100 km. Each satellite transmits two continuous signals using the same highly stable oscillator. One frequency is nominally 400 MHz and the other always exactly 3/8 of that frequency. These frequencies are modulated to provide specific time marks and to carry the satellite ephemeris, from which the position of the satellite at this epoch may be calculated. The basic observable in the Transit system is the Doppler count, between two time marks, of the "beat frequency", which is obtained by subtracting the received frequency from one generated by the receiver. The observed Doppler counts are corrected from atmospheric effects and then used to derive "range difference" measurements between the receiver and the two corresponding successive positions of the satellite in its orbit. These "range difference" observations, together with the corresponding satellite positions, are adjusted by "variation-of-co-ordinates" to obtain the 3-D geocentric Cartesian co-ordinates of the receiver, X, Y, Z.

Repeated tests have shown that the Transit Doppler system is capable of an independent point positioning accuracy of 1–1.5 m in each direction by using the "Precise Ephemeris" and 3–5 m with the "Broadcast Ephemeris". These would require 30–50 good satellite passes taken over a period of 2–3 days, depending on the latitude of the observing station. Relative positioning accuracies can be substantially improved if observations are made in the "translocation mode", i.e. simultaneously at two or more observing stations. This leads to a significant reduction of errors due to satellite ephemeris and atmospheric correction models. A further improvement can be achieved if the translocation data are processed in the so-called "orbit relaxation" mode, when the satellite orbits are allowed translations and accelerations following predetermined mathematical bias models. Repeated field tests have established that the Transit system, used in orbit relaxation mode, is capable of relative positioning accuracies of the order of 30 cm for baselines up to 200 km, and 60 cm for baselines up to 600 km. These accuracies can be achieved by using the broadcast ephemeris with observations taken over 2–3 days.

Global Positioning System

The NAVSTAR Global Positioning System (GPS) was conceived in the early 1970s as a highly accurate satellite based navigation system. When the system reaches its full configuration (in or soon after 1988), it will consist of 18 satellites arranged in six orbital planes of inclination 55°. The near-circular orbits will have an altitude of 20 200 km and a period of 12 h. The configuration will make it possible to have a minimum of four satellites always in view, each transmitting a unique data set for identification and navigational positioning through two L-band (r.f.) frequencies centred on 1575.42 MHz (L1) and 1227.60 MHz (L2).

Range between receiver and satellite is one of the two basic observables in the GPS system. Ideally, one could obtain the position of a navigator from three measured ranges to three satellites. However, these ranges are measured by "delay time" which is contaminated by the receiver's clock offset (or bias) in relation to the satellite clocks which are (nominally) set in the GPS time framework (very close to UTC). Hence, the measurements consist of "pseudo-ranges" (i.e. ranges affected by the receiver's clock bias), and a minimum of 4 pseudo-ranges are necessary in order to solve for the three co-ordinates of the receiver and its unknown "clock bias". This constitutes the first basic observational mode in GPS, the pseudo-range mode.

The satellite signal consists of a carrier wavelength which is modulated with a Precision (P) code, a Coarse/Acquisition (C/A) code and the formatted data message. At present (1985), only the L1 carrier frequency has both the P and the C/A modulations, whereas the L2 frequency has only the P code. The two carrier frequencies are made up, respectively, of 154 and 120 times the fundamental GPS frequency of 10.23 MHz, and enables some correction to be made for ionospheric delays. It is with the two timing codes that the GPS receiver measures the pseudo-ranges to the satellites.

GPS Navigation and Positioning

Pseudo-ranging is the basic mode of observation in GPS. The pseudo-range method, which involves four simultaneous pseudo-ranges to four different satellites and leads to the complete solution of the receiver's 3-D position and time bias, is the only real-time GPS navigational technique. Several multi-channel receivers have been built and successfully tested in high dynamic environments for instantaneous 3-D position and velocity. Other single-channel instruments have also been developed for low dynamic and static applications, notably surveying and geodesy. Such a receiver (e.g. TI 4100) may typically multiplex four satellites, in turn, each for 5 ms and receive all four data messages (including the broadcast ephemeris).

In 1985, only five of the eight launched GPS satellites were operational, and simultaneous four-satellite cover is only available for very short periods outside continental USA where the system is being tested. This does not generally affect static positioning techniques, although some interferometric processing algorithms require data from several simultaneously observed satellites. Extensive field tests have repeatedly confirmed earlier predictions based on simulations that the GPS system used in the pseudo-range mode is capable of (quasi-)instantaneous positioning accuracies of about 10–20 m, depending on satellite geometry, clock errors, etc. Moreover, it appears that this very high "instantaneous" navigational accuracy can be achieved with either the C/A code or the P code, making the latter look unexpectedly redundant.

These high navigational accuracies have led to thoughts of denial of access to the P code (and by implication the L2 carrier frequency), down-grading the C/A code (through jitter) and/or down-grading the broadcast ephemeris to civilian users. However, present indications are that bona fide users will have access to a down-graded (100 m accuracy) C/A code (and by implication only the L1 frequency). A system of annual charges on a national basis, depending on the amount of usage, is being considered but the US Department of Defense is opposed to this.

GPS Surveying and Geodesy

Only an interferometric mode of observation, similar to that used in VLBI (radio astronomy), appears capable of providing the breakthrough in accuracy which is necessary for surveying and geodetic applications. The basic observable in GPS interferometry is the single phase difference, which corresponds to the range difference between a satellite and two receivers. As the observable is strictly a phase measurement, it contains a so-called 2π-ambiguity or an unknown number of integer wavelengths of the carrier frequency. Successive phase-difference observation equations all contain the same unknown initial ambiguity (provided the receivers keep track of the phase without interruption), and this can then be solved as part of the least-squares solution.

Tests were carried out early in 1983 by the US Federal Geodetic Control Committee (FGCC) with a pair of Macrometer V-1000 receivers on a small geodetic network consisting of eight stations. For the short baselines (up to 2 km), differences between the GPS derived and the terrestrial distances ranged from 1:200 000 to 1:500 000. The azimuth differences averaged 7 s, but in terms of co-ordinates the differences amounted to a few centimetres in the three co-ordinate components. For the longer baselines (8–42 km), the results obtained by processing the data with the precise ephemeris were significantly better. Scale differences were reduced to an average of 1:500 000 and azimuth differences to 0.5 s. In terms of co-ordinates, the differences ranged from 2 to 10 cm in each of the three directions, with the differences in height largely attributable to uncertainties in the terrestrial geoid. It should be emphasized that in the interferometric GPS mode, the results obtained are always in terms of "co-ordinate differences", and absolute co-ordinates can only be obtained if these are known for at least one of the observing stations.

GPS in the 1990s

The Global Positioning System will be commissioned some time during or after 1988. In all likelihood, bona fide civilian users will have access to a down-graded C/A code and an appropriate broadcast ephemeris which will limit instantaneous positioning accuracies to somewhere between 100 and 250 m. Although the so-called "differential GPS" approach may help with some real-time navigation situations, only interferometric techniques will be able to achieve the level of relative positioning accuracies required in surveying and geodesy. It is also likely (but not certain) that by 1988 the L2 carrier frequency will have the C/A code (as well as the current P code) timing modulation, in order to access this second frequency and correct for first-order ionospheric effects.

The high initial cost of early geodetic GPS receivers (typically US $150 000 for a pair of receivers and processing unit) would restrict their use to national survey organizations and large companies involved in geodetic control and positioning work for major engineering projects (e.g. oil companies). As the cost of GPS receivers begins to drop, the number of users will increase. According to a Canadian study, at around US $40 000 (in 1985 prices) per unit, a small surveying firm may consider acquiring one and using it in interferometric mode with another base receiver, operated by a service company. At this stage, not only the purchase price (and the size) of the receivers may come down, but also the length of time taken to achieve high-accuracy positioning. The threshold, according to that same Canadian study, is 3–5 cm within an observing time of 10–15 min. It is then that GPS becomes competitive with existing technology. This was thought likely to happen in the late-1990s, but it may now happen in a few years' time.

SCIENTIFIC AND COMMERCIAL APPLICATIONS

Navigation

There is no worldwide navigation system which can fulfil the maritime needs for accurate position-fixing in all the seas of the world. Traditionally, visual aids (e.g. lighthouses, buoys) and radar have been used for harbour approach and channel navigation, and astronomical observations have been used for ocean navigation. The potential of worldwide electronic position fixing has only been realized since the advent of Transit and Omega. Neither system is wholly satisfactory. Transit receivers need accurate course and speed inputs if they are to maintain the required degree of accuracy between satellite passes, and Omega is not sufficiently accurate. As a result, hybrid navigation systems are being used (e.g. Transit with Doppler-log, Transit with Omega). However, these hybrid systems, which are suitable for ocean-going vessels, are not accurate enough for coastal and estuarine navigation, harbour approach or docking. Radio positioning systems, such as Loran-C (e.g. the USA) and Decca Navigation (e.g. the North Sea), are now the main techniques for coastal navigation. Microwave systems have been demonstrated of being capable of handling harbour approach and docking, but are nevertheless not generally used for this purpose.

Clearly, there is a need for a single worldwide accurate and universal navigation system for the safe and economical operation of marine transport. The Global Positioning System is confidently expected to fulfil this requirement from the 1990s onward, but there is some concern about being entirely dependent upon a military-controlled navigation system, which may be affected during periods of political tension or worse. This is one of the main reasons for initiating studies for civilian global satellite positioning systems (e.g. the ESA's NAVSAT).

Positioning and Surveying

Surveying at sea is concerned primarily with the determination of the position of a stationary or slowly moving structure or vessel. Successive positions may be required for bathymetry, for seismic surveys, for the positioning of an oil platform or similar structure, or for the laying of a pipeline. At present, bathymetric and seismic surveys, and pipelaying would involve the more accurate radio positioning systems (e.g. Syledis or Argo), where such systems are available (e.g. in the North Sea). Elsewhere, the two options are either to set up a temporary radio positioning chain or to use a long baseline acoustic system whose absolute calibration would be effected by using the Transit system.

The positioning of large offshore structures, such as oil platforms, is preceded by a detailed site survey and requires the establishment of an array of acoustic transponders. These are then used for the continuous low dynamic positioning of the platform as it is towed to its designated position. Once erected, Transit receivers are used, either in independent point or translocation mode (with respect to a known control point on land), to verify the final location of the platform and determine the positioning accuracy.

This procedure is likely to change substantially with the inauguration and commissioning of the full GPS system in 1989. The latter system will eventually not only replace the functions of Transit, but will also be used in a continuous low dynamic navigation mode when the platform is towed to its final destination. However, even if bona fide civilian users are allowed complete access to the timing codes, the GPS system in its full configuration will lead to (quasi-instantaneous) positioning accuracies of the order of 15–25 m. Platform towing, pipeline laying, bathymetry or seismic surveys may require accuracies of the order of 1–5 m. This could only be achieved in one of the following three ways:

(a) a Kalman filter based on primary GPS position determinations coupled with a secondary dynamic model of the vessel;

(b) an inertial surveying system with GPS position updates;

(c) a "differential GPS" system, based on one or more close permanent land-based stations, which continuously monitor the performance and accuracy of the GPS signals and, in turn, broadcast appropriate corrections.

Underwater Surveying

There are several applications of underwater positioning. These include the location of deep submergence vehicles (for the exploration of the ocean bed), divers, manned submersibles, remotely controlled vehicles (for survey and inspection tasks) and (the installation of) pipelines and structures. In the offshore industry, the accuracy requirements for these tasks range from 0.1 m to 10 m. Generally speaking, relative positioning to nearby features is required to a far greater degree of accuracy than absolute positioning.

Long baseline acoustic positioning is the most widely used method of underwater positioning. However, super-short baseline has become very popular in the last few years for the tracking of vehicles during, for example, pipeline surveys and inspections, because of the saving in time by not having to deploy and calibrate successive arrays of transponders. Inertial systems (e.g. HASINS) have also been considered for the positioning of submersibles, but only moderate success has been achieved due to the need to stop the vehicle every few minutes. However, in and around structures conventional acoustic methods are prone to distortion from multipath rays reflected off various surfaces. The integration of an inertial system with other sensors should help with the shortcomings of the acoustic system on its own and continuously provide the position, velocity and attitude vectors required to locate and maintain the vehicle on station.

Geodetic Applications

Geometric geodesy is concerned with very high positioning accuracies, with a view to determining minute periodical changes in the co-ordinate values. Traditionally, geodetic co-ordinates are established by using classical triangulation, trilateration and precise levelling. With the advent of the Transit system, it has been possible not only to connect two disjoint trangulation systems, but also to extend these land-based networks to include reference marks established on offshore structures. To achieve the orders of accuracy required (i.e. a few decimetres) it has been necessary to carry out Transit observations in translocation mode over several days. GPS will change this situtation significantly. It is expected that by the 1990s it will be possible to use the system in interferometric relative positioning mode and achieve accuracies of the order of a few centimetres over several hundred kilometre long baselines in about 15–30 min. Even higher (sub-centimetre) accuracies in all three dimensions are predicted for relatively short baselines. This may lead to a whole new range of geodetic and geophysical applications.

The slow and laborious process of spirit levelling on land may also be replaced by GPS used in combination with a high-accuracy geoidal model. It may even be possible to carry out this levelling process to platforms at sea. Indeed, *changes* in the absolute height of a platform above an arbitrary geodetic reference surface can already be obtained by using GPS interferometric phase measurements. This, in turn, can lead to the monitoring of the local crustal dynamics of the sea bed in an area such as the North Sea. It is confidently expected that GPS coupled with VLBI and Satellite Laser Ranging will usher a new era of continuous geodetic measurements in support of plate tectonic motion and earthquake studies. Eventually, the very need for a triangulation control network will be questioned. The current hierarchical and dense system of triangulation stations (300 *primary* stations in a country the size of Great Britain) may be replaced by a few base stations used in conjunction with any number of GPS units, operating throughout the country and the seas around it.

CONCLUSION

With the advent of GPS, surface navigation and positioning at sea will be largely carried out by satellite. In the interim period, which is likely to extend some years after the proposed inauguration of the GPS system in 1988, some of the existing radio positioning systems will continue to function, and may even be used in conjunction with GPS updates in a Kalman filter mode. However, it has already been announced that two of the long-range radio positioning systems, Omega and Loran-C, are due to be phased out some years after the introduction of GPS. Clearly, the scene is now set for GPS to become the single universal worldwide surface navigation system. Underwater positioning will still be carried by acoustic methods, with much improved accuracies resulting from the more precise positioning of transponders in combination with GPS observations.

GPS is also likely to affect significantly traditional surveying and geodetic positioning methods. It is confidently expected that GPS will help with the effective towing of structures to predetermined positions and the more economical laying of pipelines. It will allow improved positioning of vessels in seismic surveys and the precise monitoring of the vertical movements of production platforms. The very high accuracies which are achieved by interferometric phase techniques will also lead to high-accuracy geodetic measurements capable of supporting geophysical exploration and geodynamic investigations.

REFERENCES

Ashkenazi, V. 1985. Future developments in satellite surveying. *Int. Seminar on Surveying by Satellite for the Developing World, Dublin.*

Ashkenazi, V. and Diederich, P. 1984. Positioning by second generation satellites: GPS and NAVSAT. Proc. 2nd Int. Hydrographic Technical Conf., Plymouth

Ashkenazi, V. and Sykes, R. M. 1980. Doppler translocation and orbit relaxation techniques. *Phil. Trans. R. Soc. A* **294**

Laurila, S. H. 1976. *Electronic Surveying and Navigation*. New York: Wiley.

Milne, P. H. 1980. *Underwater Engineering Surveys*. Spon.

Napier, M. E. 1985. Application of inertial positioning to high accuracy underwater surveys. Ph.D. Thesis, Newcastle University.

Department of Defense and Department of Transportation, Washington, D.C. 1984. *Federal Radionavigation Plant.*

EQUIPMENT FOR DIFFERENTIAL GPS

W F Blanchard, Racal Positioning Systems
Ltd, UK

Presented on **Tuesday, 4 March 1986** at **15.00**
in the Clarence Conference Room.

EMBARGOED AGAINST DELIVERY

2

Equipment for Differential GPS

W. F. Blanchard, Racal, Surrey, UK

It could be claimed that in one sense all navigational systems are differential by nature. The navigator is interested in where he or she is relative to other points of interest (e.g. the destination) and not particularly where he or she is in some arbitrary pattern of grid lines produced by a set of radio transmitters. Such a pattern has to be related to a more universal system that can be used for position definition irrespective of whether that radio system is in use or not. These "universal" systems, such as latitude and longitude, have themselves to be defined, usually by having a specified origin and reference framework. Thus, the radio system acts as a "position transponder", relaying this universal origin and reference via its own grid. Were the radio grid to be absolutely accurate and unchanging, it could become totally "transparent" to the user, who would obtain position in the universal reference without even realizing it had been "transponded" through another system. Unfortunately, this is not so, because the radio system, whatever it is, has its own inaccuracies and although they may be quite well-known and calibrated, they cannot be ignored. There is, for satellite-based systems, the additional problem that although they may nominally produce positions in the same universal grid as is used at the Earth's surface, their definition of this system, as seen from space, is necessarily different from that seen from the Earth's surface, and sometimes the errors this produces are not well-calibrated. One way around these problems is to calibrate them accurately by providing a monitor station whose position is known precisely in the universal grid, that measures the position indicated by the radio system and relates one to the other in such a way that a navigator using the radio system anywhere in an area fairly close to the monitor can apply a correction transmitted to him by the monitor. This method is, of course, of general application and not restricted to any one system, but in the case of GPS it has become known as Differential GPS (DGPS). GPS itself is so well-known now that it is not proposed to describe it here — there are many references available if a description is needed. Superficially it may seem an easy task to set up such a system — all that is needed is a monitor receiver and a method of transmission but, in practice, as usual, it is not. Not only does the actual design and implementation of the monitor need care, but data transmission has to be done within some very stringent requirements. It is the purpose of this paper to outline some of these and illustrate them by reference to measurements on representative equipment. A

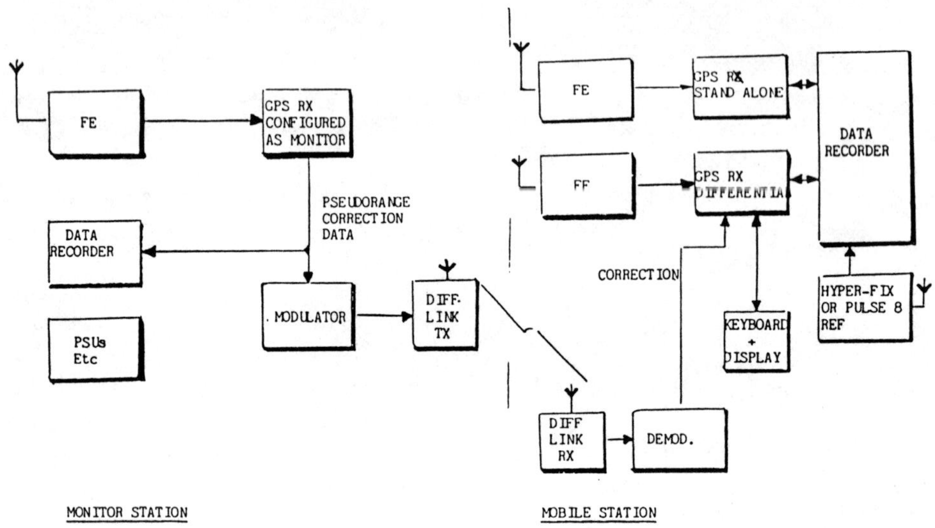

Fig. 1

representative DGPS system block diagram is given in Fig. 1.

CORRECTABLE ERRORS

Table 1 shows the variety of sources that can cause GPS errors. They contribute in two ways:

- they affect either the geometry of the system itself or

TABLE 1

SPS error budget with ephemeris degraded by accuracy denial

Segment	*Description*	*Error**
Space	Clock stability	4.5
	Space perturbation	3.0
	Other	0.5
Control	Ephemeris prediction	2.5
	Accuracy denial corruption	59.5
	Other	0.5
User	Ionospheric delay	10.0
	Tropospheric delay	2.0
	Receiver noise	15.0
	Multipath	1.2
	Other	0.5
RSS ranging error (UERE)		62.5

*Errors are 24-h 1 σ user equivalent range errors.

- the accuracy of the pseudo-range measurements on which positioning is based.

As with other systems, if there were no measurement errors, geometry would be irrelevant except in limiting cases, but clearly they are not not. Geometry is set by the satellite positions and cannot be altered; all that can be done is to select the combination of satellites that provides the best geometry, or, possibly, wait until the movement of the satellites in their orbits has provided a better situation. Provided that all receivers are already making the best selection, a DGPS station cannot help, but it can certainly help in cases where receivers do not do this by telling them which ones to use for best results. It may, in any case, wish to do this to cover the occasions when a satellite may be giving poor results for other reasons, or has recently changed its navigation message. Many receivers do not themselves constantly check this navigation message, after having decoded it once, and will not be aware of a change.

Pseudo-range measurement errors are made up of many factors, some of which are recoverable via DGPS, but others not. Thus, errors in the satellites' own positions, timing errors and some propagation errors can be compensated, but not errors in the receiver itself, or very local propagation disturbances (multipath).

Satellite position errors will clearly produce a 1:1 error along the observer-to-satellite line of sight which may propagate into a much larger fix error on the ground, depending on geometry and whether similar errors are occurring on other satellites. In certain cases, the resultant fix error is critically dependent on geometry and can vary considerably over an area normally considered well within the range of a DGPS monitor station. In this case, a monitor observing only fix errors (i.e. latitude/longitude variations) can produce a correction that is quite invalid for a navigator within its nominal range, whereas if it produced corrections for the basic range measurement only, this would not happen.

Timing errors in the satellite can produce much the same result, and are treated the same way, but are different in character from satellite position errors in that ephemeris errors tend to occur for all satellites in view at the same time and in the same sense, since they are usually produced by orbit forecast errors rather than space perturbations (as was the case with Transit, for instance), but timing errors are usually particular to one satellite.

Rate of change of these errors is also quite important, since no data link can transmit complete data on all satellites instantly and continuously. Particularly in cases of rapid change, the monitor station should be able to provide its own forecast of rate of change so that the receiving station can allow for it.

Propagation errors are limited to the delays encountered in the ionosphere and troposphere which become significant if high accuracy is required. Most users of GPS will be tracking only the C/A code signals at 1575 MHz and will be unable to compensate directly for ionospheric effects by using two radio frequencies (as they could with Transit), but these effects are proportional to actual radio frequency and since GPS is operating at a much higher frequency than Transit, the effects will be accordingly smaller. Without compensation, it has been estimated the average effect of ionospheric refraction will be to produce a 10-m error in range, but this can be reduced by using the data on ionospheric effects contained in the GPS navigation message. Provided the ionosphere itself is reasonably homogeneous over the comparatively small area of a DGPS station, almost complete compensation can be made differentially, especially at periods of low ionospheric activity, as at present.

Tropospheric effects are rather larger and are not frequency-dependant, hence the uncompensated 20–30 m errors of Transit are the same for GPS. They can be reduced quite considerably by using one of several algorithms for computing the error from measured values of pressure, temperature and humidity at ground level, while DGPS can almost eliminate them as long as weather conditions are not greatly different between monitor and receiver sites. Even where they are different, transmission of local weather parameters from the monitor can enable the remote to calculate a differential correction using his own local measurements. The net result of using DGPS is shown in Table 2.

TABLE 2
DGPS error budget

Segment	*Description*	*Error**
Space	Clock stability	0
	Space perturbation	0
	Other	0
Control	Ephemeris prediction	0
	Accuracy denial corruption	0
	Other	0
User	Ionospheric delay	0
	Tropospheric delay	0
	Receiver noise	15.0
	Mulipath	1.2
	Other	0.5
RSS ranging error (UERE)		15.0

*Errors are 24-h 1 σ user equivalent range errors.

UNCORRECTABLE ERRORS

What DGPS cannot do anything about are remote receiver characteristics and reception conditions. All GPS receivers need to do a certain amount of time and frequency prediction in order to be able to track the satellites at all, the manner in which they do this is highly variable and, as far as the monitor is concerned, totally unpredictable. The monitor receiver will of necessity be a very accurate machine, probably tracking every visible satellite continuously, but not having to deal with motion or acceleration, nor much concerned with size, weight or even cost. Hence it will be capable of producing second-by-second corrections of actual range measurements to each satellite, and not necessarily fix errors. Fixes are produced in most receivers by using a maximum of four satellites, and are compounded by geometry, so both the monitor and the remote receiver must use the same combination of satellites for fix corrections to mean anything. Were a DGPS system to be implemented on the basis of fix corrections, because up to 11 satellites might be visible at any one time (in the full system) there might be several combinations of four capable of giving reasonable results, and the monitor would have to produce differential fix corrections based on all of these possible combinations. This might not be too bad a task for a high-speed computer, but the problem would be to transmit them — it is very unlikely they could all be transmitted at the necessary speed. Even then, the remote receiver would no doubt be applying filtering to its own fixes due to the need to cope with probably much "noisier" information than at the monitor, and the resultant time-pattern of its own fixes would be quite different from that received from the monitor (see Fig. 2). Predictive filters of the Kalman variety are coming into wide use for mobile work, and unless the data input from the monitor DGPS station is correctly matched the results might be disastrous. One could even envisage a situation where in order to get over this problem the remote receiver used no filtering at all, making a DGPS input absolutely necessary to produce anything intelligible.

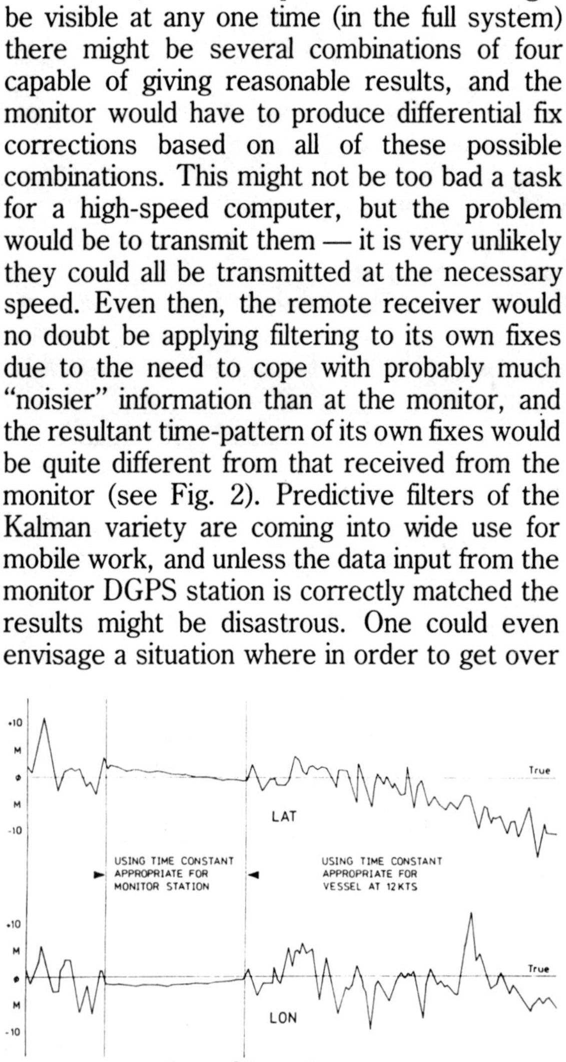

Fig. 2

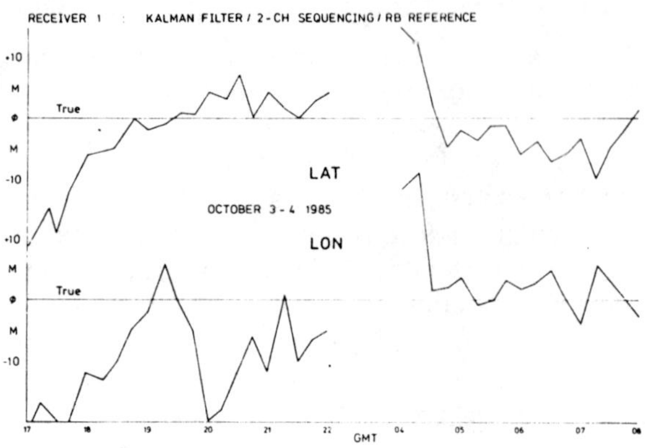

Fig. 3

Figure 3 is a plot of smoothed fix output from one type of receiver over the full period for which GPS satellites were available, some 9 h. In this period, it tracked two, three or four satellites as they appeared or disappeared, and in the two-satellite case was controlled by a rubidium frequency standard providing the timing reference. As plotted, no differentiation is made between 2-, 3- or 4-satellite operation, on the basis that a DGPS system will have to correct in all situations and not merely in carefully selected circumstances.

The first thing to note is that for 7.5 h all fixes were within 10 m of the true position, which was established not merely from standard survey methods but also by Transit translocation. Since fixes were being produced every 10 s but the plot is only every 15 min, it is legitimate to wonder what the noisiness would be like if every 10-s point were added, but the answer is not much more than can already be seen.

The pattern is what is important, major deviations (at the start, at 2000 h and at 0400 h) being the result of poor geometry and/or limited satellite availability. The rest of the pattern is typical of that receiver's response, and to check that it was not a totally random response, a repeat run made 24 h later is shown in Fig. 4.

Satellite geometry does not change very much from day to day, and the fact that Figs 3

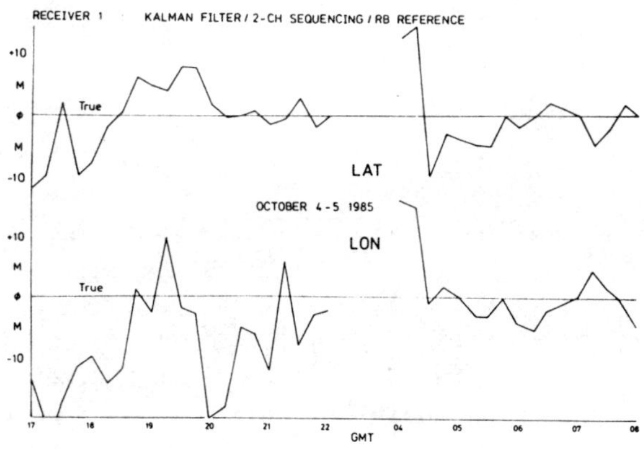

Fig. 4

and 4 are remarkably similar shows that receiver response to given conditions does not vary much, either.

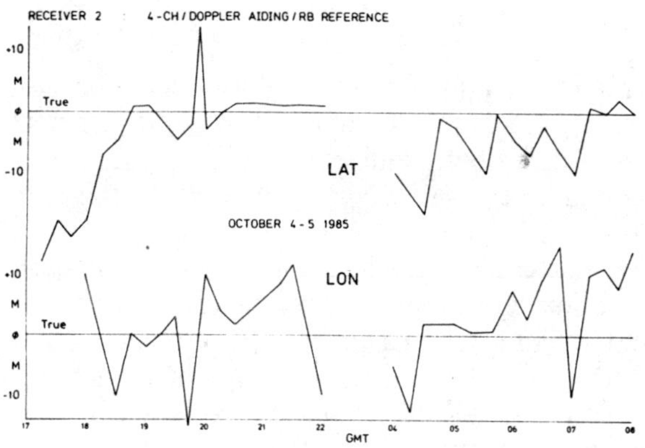

Fig. 5

Figure 5 data taken at the same time as that of Fig. 4, but with a totally different type of receiver, shows that this receiver, too, stayed within 10 m almost all of the time, but apart from one or two similarities, its pattern is quite different.

Both receivers were run under, as far as possible, identical conditions, sharing the same frequency standard, with smoothing appropriate to a 12-knot vessel, and with their aerials only 5 m apart, yet it is obvious that any attempt to use receiver 1 to correct receiver 2, or vice versa, would not have helped overall accuracy at all.

It might be argued that had 10-s corrections from one set been applied to the other, then its filter parameters would have changed to accommodate the more-precise information, and this would indeed have some truth if the corrections were in the form of delta-pseudo-ranges, but not if they were fix positions--which are determined *after* the filter, not before it.

It is quite clear from these results that each manufacturer's receiver has its own characteristic pattern of errors, and that any attempt to use a different type of receiver for correction will not work. The obvious conclusion is that any DGPS system based on fix corrections must be limited one particular type of receiver, and would not then have universal applicability. This difficulty, together with the problem of satellite selection, makes it highly questionable whether fix-based DGPS systems are worth developing.

Another error that cannot be corrected by DGPS is that caused by multipath reception. There is already considerable evidence that, even with Transit, some of the anomalous results obtained on-board offshore platforms are attributable to multipath, or assymetric, reception. What the effects are at the L-band frequencies of GPS has not been evaluated, but although it has been stated that in military trails multipath has not been as big a problem as was once thought, it would be incorrect to assume that these results will read across into operation on-board platforms occupied by many large metal structures.

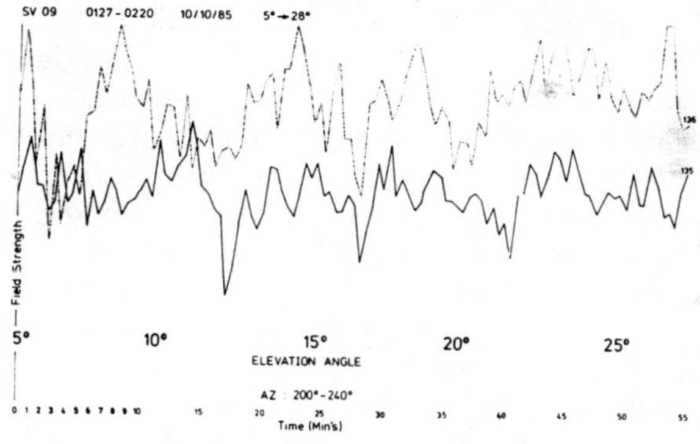

Fig. 6

Figure 6 shows the signal strengths of a single satellite as received at two GPS receivers with aerials only 5 m apart. It can be seen that there are systematic fades present that are uncorrelated between the two aerials, and although there are factors other than multipath present, some of these fades can only be due to local mulitpath effects. Detailed analysis shows that systematic fix errors were associated with some of these fades, although not all. Since the monitor site was reasonably clear of obstructions and certainly had no large metal objects near it (such as cranes), it would be only prudent to allow for bigger effects on platforms and ships.

One last point is that DGPS cannot, of course, do anything about sheer loss of signal at the receiver, which may occur through local noise, local transmitters or a poor aerial site. In particular, it should be noted that INMARSAT terminals, when operated close to GPS receiver aerials, cause severe reception problems. This is a matter of receiver design, and DGPS cannot help.

RECEIVERS FOR DGPS

For the reasons already given, simple lat/lon position correction is unlikely to be of wide application, and the best method is to correct measured pseudo-ranges, before they are converted to fixes. To do this the receiver must be able to accept the input of corrections from the data-link and relate them to what it is itself measuring. There is evidence to show that when selective accuracy is applied an update rate of no more than 15 s will be necessary, and that this is therefore the maximum time allowable between successive measurements. It follows that no sequencing receiver with a cycle time greater than 15 s will be usable. However, if the remote receiver tracks all satellites in parallel and continuously, this problem does not occur and it is only necessary to time-tag the corrections. Parallel tracking also eliminates another problem–that monitor and remote may not be tracking the same satellites at the same time. Although some types of sequencial receivers have facilities for being steered onto a particular set of satellites, none has facilities for enforcing any particular sequence of tracking, and indeed this may change from time to time even without change of satellite, due to noise, etc.

Having obtained the correction, the receiver must calculate position based on corrected ranges, but it is also important that it accepts the other data carried on the link and acts on it as necessary, if the DGPS system is to be properly effective. Unfortunately, no receiver currently on the market can do this. Uncritical correction of pseudo-ranges carried out without reference to the editing information contained in the rest of the link message is likely to create more problems than it solves.

A possible way around this is for the receiver simply to output time-tagged measurements of pseudo-range and let an exterior computer deal with both them and the DGPS inputs. The data-link receiver would then produce the final corrected DGPS co-ordinates using known correction algorithms taking into account both data-link and monitor receiver characteristics, and making all allowances for editing as required. The task of the receiver manufacturer could be simplified, in that a receiver could be produced without having to consider the vagaries of, possibly, several different DGPS systems.

THE DATA-LINK

Corrections at a rate of once every 15 s, 24 h per day, requires a link of very high reliability. Technical considerations of the link itself limit the data rate to about 50 baud unless a short-range link in the VHF/UHF/microwave spectrum is used. Such a line-of-sight link is not likely to use DGPS to its greatest effectiveness (although it may have some very local applications) and is not considered further here. The necessary data for a full system are shown in Fig. 7 (as recommended by the RTCM, USA) and it is clear that this cannot be accommodated in full every 15 s. Since the priority is for survey high-accuracy work, this format is rearranged as shown in Fig. 8, giving the survey data as a foreground message and relegating the general information message to the background. In this way, the essential information is repeated in 12 s for eight satellites and the general information is repeated about once every 4 min (Table 3).

Outages of more than a few repetitions of the

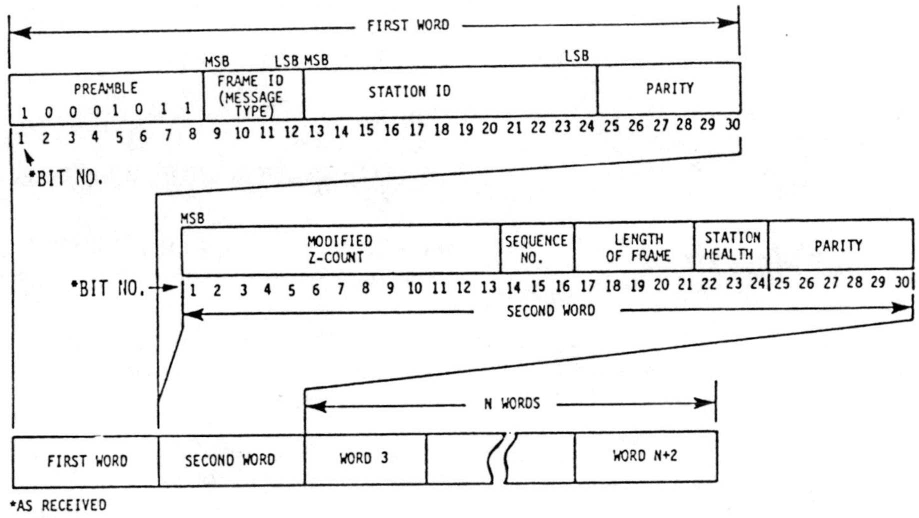

Fig. 7

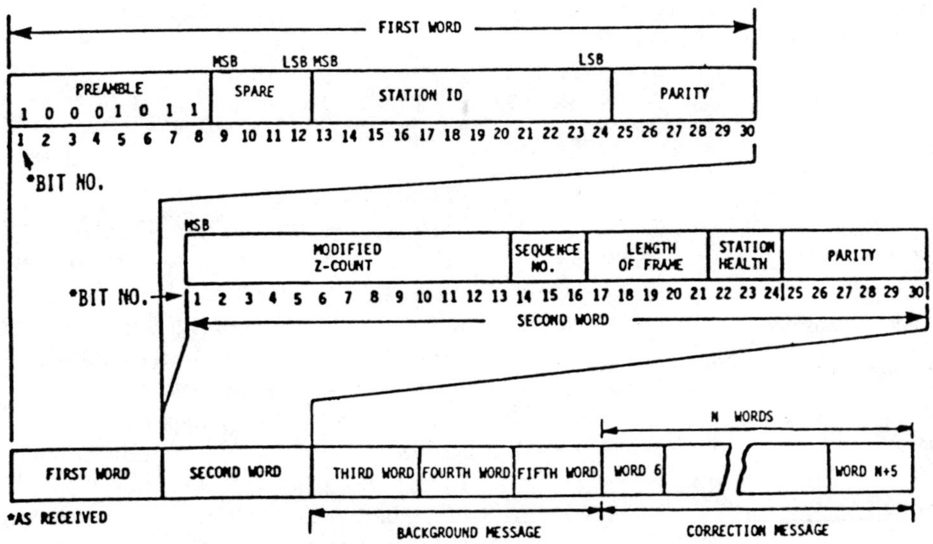

Fig. 8

foreground message cannot be tolerated in this system and the link itself will have to be of very high integrity to satisfy this requirement 24 h per day.

Existing navigational transmitters are designed to provide this high reliability, and with their known areas of coverage and unique frequency allocations appear at first to be a good choice. However, almost all use phase information as part of their navigational function, prohibiting phase or frequency modulation, and amplitude modulation is notoriously difficult to resolve except under very good conditions. The best way of using them would be to add DGPS information in a format that does not affect their basic navigational functions. For instance, Pulse/8, the Racal 100 kHz system, uses a series of eight pulses to provide basic timing for navigation, and while no interference with these eight can be allowed, there is nothing to stop the addition of further pulses that can be modulated in various ways which

TABLE 3
Table of transmission rates (times in seconds)

		No. of satellites in foreground				
		6	*7*	*8*	*9*	*10*
Message 1	No. of bits	—	—	—	—	—
	Time to complete	—	—	—	—	—
Message 2	No. of bits	150	180	210	210	240
	Time to complete	16	23	28	31	37
Message 3	No. of bits	450	450	450	450	450
	Time to complete	48	57	60	66	69
Message 4	No. of bits	390	450	510	570	630
	Time to complete	42	57	68	84	97
Message 5	No. of bits	150	180	180	210	210
	Time to complete	16	23	24	31	32
Message 6	No. of bits	30	30	30	30	30
	Time to complete	3.2	3.8	4.0	4.4	4.6
Message 7	No. of bits	270	270	270	270	270
	Time to complete	29	34	36	40	41
Foreground message	Background baud rate	9.38	7.89	7.5	6.82	6.52
	Complete message length	480	570	600	660	690
	Time to transfer in seconds	9.6	11.4	12	13.2	13.8

will not be "seen" by a navigation receiver. However, if more than one or two are added, engineering problems arise and, unfortunately, even with two extra pulses the resultant maximum data rate is still below optimum. Continuous-wave systems such as Hyper-Fix are even worse off, with maximum theoretical data rates for this purpose of less than 2 baud.

Tests carried out over representative paths of data transmission reliability using a modulated 9th pulse on a Pulse/8 transmitter have shown that even at a slow data rate (5 baud) complete reliability is not achieved except at rather short ranges (100 km) (Fig. 9). In part, this is due to the large number of Loran-C and Pulse/8 transmitters operating in the North Sea area, where the tests were done, and it could be that results would be better in less-congested areas of the world.

It does not seem feasible to use any of the other navigation systems, although tests are under way to determine whether this is so.

Failing the use of these systems, a specialized link becomes necessary, which could take one of several forms.

- a pulse signal at 100 kHz specially designed for DGPS data link use;
- a 2 MHz link using essentially the same hardware as is already used by the 2 MHz positioning systems;
- frequency-multiplexed HF;
- satellite links.

All of these have problems of their own, and while it is not intended to give an exhaustive description of them here, the following points should be noted.

Pulse systems, in the 100 kHz band, even if non-navigational, will always be subject to cross-rate interference from navigation transmitters and considering the very high power used by some Loran-C transmitters will be a problem not easily solved.

2 MHz transmissions really work only over sea–water paths and if a requirement arose for DGPS to work over land they would be almost unusable. Also, skywave problems rule out their use at ranges over about 250 km in any case. Skywave resolution techniques used in positioning systems cannot be used in data links. A possible exception to this is a system based on spread-spectrum techniques, such as Geoloc, which is, however, a rather expensive solution.

Frequency-multiplexed HF is not reliable enough for the purpose if it is limited to, say, one set of three frequencies. Racal has had such a system in operation in several parts of the world for some time (for non-DGPS purposes) and a considerable amount of data is available to show that even three carefully-chosen frequencies cannot provide 24-h reliability at the ranges DGPS requires. Figure 10 shows the results of monitoring data transmissions on three frequencies in the North Sea area over a 24-h period. If multiple sets of frequencies were used, the problem would lessen, but the frequency allocations may be unobtainable.

Satellite links can provide an almost totally

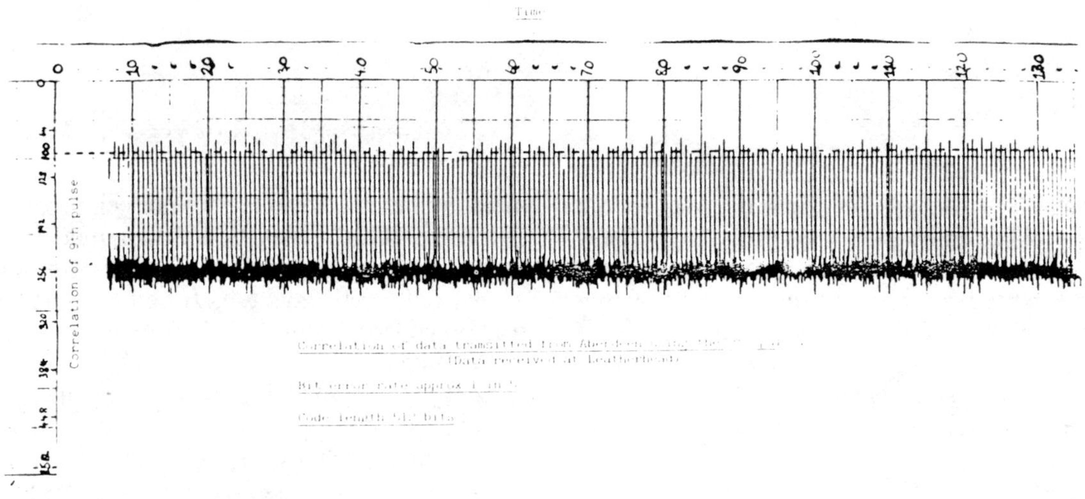
Correlation of 9th pulse

Fig. 9

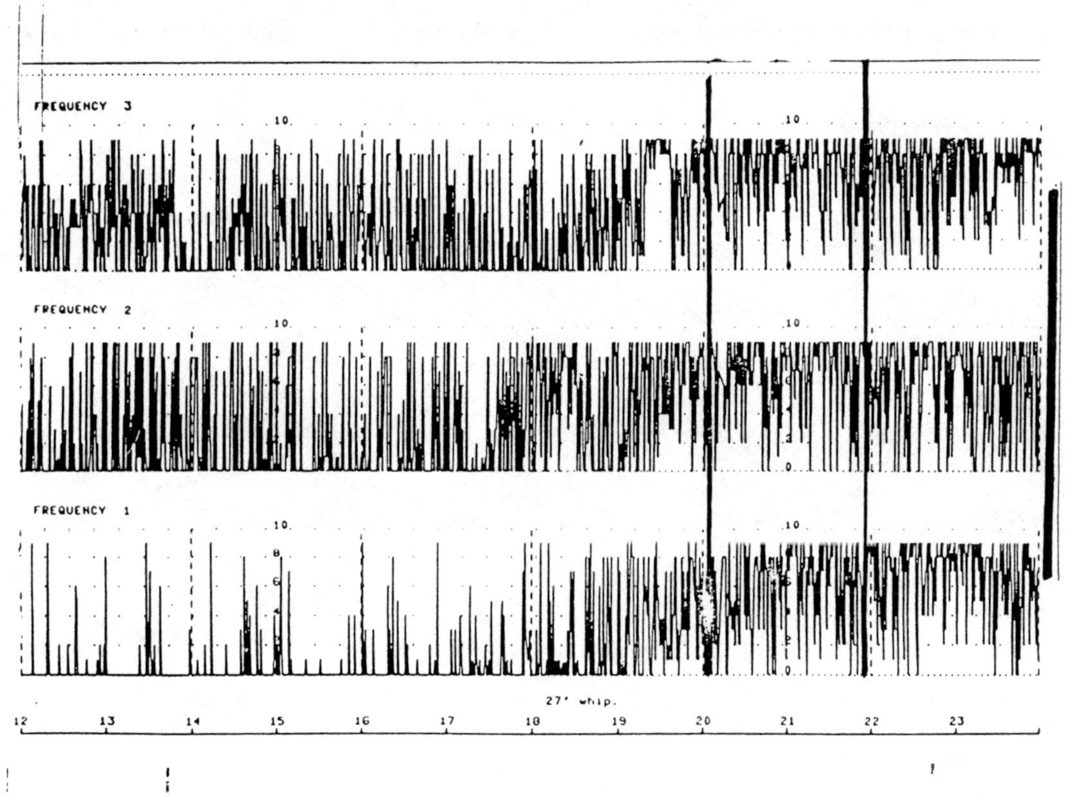
FREQUENCY 3
FREQUENCY 2
FREQUENCY 1
27' whip.
12
13
14
15
16
17
18
19
20
21
22
23

Fig. 10

reliable 24-h service of very high-speed data and could certainly solve the problem technically, but the cost would be high, not only the cost of the rented channel, but the ground transmission and reception equipment. Against that is the fact that one satellite channel could carry all the necessary information for quite a large number of DGPS systems simultaneously.

SUMMARY

It is obvious from the foregoing, necessarily rather brief, narrative that the implementation and operation of a DGPS system is not as easy as is sometimes thought, and that many areas of doubt still exist. Racal is continuing its tests and development on a realistic basis engendered by many years of hard experience of attempting similar differential systems based on the more-traditional land-based radio positioning devices. The outstanding problem is undoubtedly that of accurate error-matching between monitor and remote. Receivers capable of accepting DGPS information in full are not with us yet and may not be for some time. Having said that, the caveat must be entered that this is being written some 6 months before the conference, and in such a very fast-moving field as this it is almost inevitable that some of the statements made may have been proved wrong in the meantime.

THE APPLICATION OF FAST PARALLEL PROCESSING TO ENCHANCE THE PERFORMANCE OF NAVIGATION SYSTEMS

J G Boulton, A N Cockcroft, Dr V J Hughes,
Dr M F Richardson, Cambridge Consultants Ltd, UK

Presented on **Tuesday, 4 March 1986 at 16.00**
in the Clarence Conference Room.

EMBARGOED AGAINST DELIVERY

3

The Application of Fast Parallel Processing to Enhance the Performance of Navigation Systems

J. G. Boulton, A. N. Cockcroft, V. J. Hughes and *M. F. Richardson,* Cambridge Consultants Limited, UK

INTRODUCTION

In this paper, the application of transputers to improve the performance of navigation and positioning systems is considered. The development of increasingly complicated navigation and positioning systems is examined to highlight common features such as high data capture rates, sensor diversity and high processing loads. It is these factors which make transputer technology particularly appropriate for such systems.

Transputers are a new generation of microprocessors which extend the speed and flexibility of parallel processing beyond the capacity of currently available chips. An overview of their capabilities in relationship to two commonly used microprocessors is presented.

Finally, in order to demonstrate how transputer technology can, indeed, handle the requirements of navigation systems, a specific example of their appropriateness for an advanced Doppler sonar navigation system is considered. Their performance is compared with two alternative systems built up from conventional microprocessors. It is shown that the transputer-based design can indeed satisfy the requirements for fast parallel processing coupled with fast data read-in, whilst offering low power, low development costs and flexibility in design.

AN OVERVIEW OF NAVIGATION SYSTEMS; THE NEED FOR ENHANCED SYSTEM PERFORMANCE

In the offshore industry, there is a continuing demand to improve the performance of navigation and positioning systems. These systems provide the operator with the information necessary to minimize the risk of the operation in hand. Risks involved in offshore operations range from an expensive failure to complete the task to encountering unforeseen problems which inevitably cause time-scales and costs to increase rapidly.

As the demands made throughout the industry continue to become more exacting, these systems must perform more reliably, with improved flexibility and increased accuracy. For example, positioning installations (whether they are cables, pipelines or production platforms) to a given degree of accuracy is less time-

enable the determination of N-year return values of significant *wave height*, H_s, and of the most probable value of the height of the highest individual wave in a 3-h interval, $H_{max\ (3\ hours)}$, for N from 1 to 100. Figure 31 gives the cumulative distribution of significant wave height for the whole year, plotted at half-metre intervals on the log-normal scale; Fig. 32 presents the same data on the Weibull scale; and Fig. 33 on the Fisher-Tippett type I scale. Figure 34 gives the cumulative distribution of $H_{max\ (3\ hours)}$ for the whole year plotted at half-metre intervals on the log-normal scale. Figure 35 presents the same data on the Weibull scale; and Fig. 36 on the Fisher–Tippett type I scale. There is no theoretical justification for these distributions, but they have been widely and satisfactorily used elsewhere. The best straight lines have been fitted by eye.

TABLE 4
Results of extreme value analyses (return values: heights are in metres)

Year	*Log-normal*	*F–T I*	*Weibull*
H_s			
1	16.5	16	15.5
10	22	19.5	19
20	24	20.5	20
50	26	22	21
100	28.5	23.5	22
$H_{max\ (3\ hours)}$			
1	31	29.5	31
10	41	36.5	37
20	44	39	39
50	47	42	41
100	52	44	43

There is no objective way of choosing from these three methods, but it will be observed that the data are more nearly in a straight line on the Fisher–Tippett type I scale.

Wave periods are assessed by examining the value of T_z associated with the 50-year wave height. Extrapolations of zero-up-crossing periods are unreliable for determining extreme-condition design parameters because many of the longer-period values relate to swell when wave height is not large. High waves generally have a significant-height steepness of between 1:15 and 1:20, and this applies in this data set (Fig. 29). Accordingly, the zero-up-crossing period to be associated with the 50-year wave height can be deemed to lie between about 13 and 18 s.

The period of the individual 50-year return wave has a broad probability distribution and is difficult to assess. According to Longuet–Higgins (1983) the median period of high individual waves corresponds to the mean spectral frequency, and so is $1.07T_z$ assuming a JONSWAP spectrum (giving about 14–19 s for these data), but Bell (1972) suggests, from an analysis of North Sea instrumental data, a value of $1.28T_z$. Clearly, for design purposes it is wise to consider a range of values.

Persistence

Because the records are discontinuous, it is not possible to produce persistence diagrams for these stations, although recent studies by Kuwashima and Hogben (1984) indicate that it might be possible to produce an empirically derived distribution based on the scatter diagram, such as Fig. 29.

DISCUSSION ON PRESENT DATA AND THEIR COMPARISONS WITH EARLIER DATA

In the earlier analyses it was difficult to distinguish between the *India* and *Juliett* (*India 1* and *Juliett 1*) data sets, but this is not so with the newer measurements. In comparison with the earlier data, those in this report (*India 2* and *Juliett 2*) suggest a markedly more severe climate. At *Juliett* the all-year average increase in median wave height since the 1960s is 11% and at *India* it is 27%, with the latter's "winter" of January, February and March showing an increase of 36%.

The actual ratios of median wave height exceedance are as follows:

	India 2/India 1	*Juliett 2/Juliett 1*
"Year"	1.27	1.11
Winter	1.36	1.16
Spring	1.25	1.00
Summer	1.31	1.21
Autumn	1.16	1.06

carefully which data channels should be considered at any given time. This flexibility is appropriate when data on any channel is not being updated or when the signal-to-noise ratio on any sensor becomes unsatisfactory. This approach prevents the unnecessary degradation of the navigation calculation.

The radical design of transputers along with the features of OCCAM, their programming language, is extremely useful in fulfilling these types of requirements. Furthermore, since the complexity of all of these systems continues to increase, the ease with which transputer-based systems can be expanded should be invaluable. The attributes of transputers will now be considered.

THE ATTRIBUTES OF TRANSPUTERS

The INMOS transputer, which became available in October 1985, is an innovative microprocessor which incorporates a number of attributes not found in other microprocessor chips. Like the Motorola 68000 series processors, it is a fast, 32-bit orientated processor. However, unlike the 68000, it incorporates a limited amount of fast on-chip memory and, more importantly, considerable support for external interfaces. The chip provides four high-speed communication channels which can be used to communicate with other transputers or peripherals. These can operate independently of the main processor, without significantly degrading the performance of the processor.

Although it will run conventional programming languages, the transputer has been designed in conjunction with the language OCCAM. This maximizes the performance of the processor, and incorporates a number of features which allow a system to be designed largely independently of the number of transputers to be used in the target hardware.

The transputer is physically small, and consumes only a modest amount of power. Also, it incorporates a memory interface which drastically reduces the number of support chips needed to connect it to additional memory and peripherals. These points make it suitable for use in environments where space and power are limited.

As a conventional processor, the transputer is one of the fastest microprocessors available, and outperforms, for example, the Motorola 68000 by a factor of 4 to 5 (see Table 1(a)). It can be used as a "number cruncher" of respectable power in its own right, and is currently being considered as a building block for fast, inexpensive array processors.

One of the major strengths of the transputer lies in its ability to handle high bandwidth streams of data. It can accept incoming data, and pass results on, via the on-chip communication channels, at high speed with almost no degradation in the speed at which programs, running concurrently, are executed. This compares favourably with other microprocessors, where the requirement for high communication rates imposes significant overheads, at the expense of program execution.

The communication facilities allow easy construction of multi-processor systems. For instance, in a system where there are several streams of data arriving from various sensors, a separate transputer could be used on each stream. The outputs from each stream could then be passed on to further transputers to combine the data, in an optimal way. If sensor data require a number of operations to be performed one after the other, as is the case with a sequence of filtering functions, then a "pipeline" of transputers could be used, each performing one operation, and passing its results onto the next. Hence, the transputer is well-suited to systems which require high performance in both data transmission and processing, and can give a degree of hardware uniformity not previously possible.

The transputer, and the OCCAM programming language are designed in a way which allows considerable flexibility in the final hardware implementation of a system. It is possible to write software which can be run on different numbers of transputers without change. A minimal system may only have a single transputer, but a computationally more powerful one would use several. It is therefore easy to produce a range of systems with increasing performance with a minimal redesign overhead. In the parallel and pipeline examples above, each individual operation is coded as an

TABLE 1

(a) General specifications and attributes of two popular microprocessors and the transputer

	Microprocessor		*Transputer*
	TMS320–5	*68000–12*	*T414–20*
Register–register Multiply	0.2 μs/16-bit —	7 μs/16-bit 22 μs/32-bit	— 2 μs/32-bit
Register-register Divide	10 μs/16-bit —	12 μs/16-bit —	— 2 μs/32-bit
Register–register Add	— 0.2 μs 32-bit	0.3 μs/16-bit 0.5 μs/32-bit	— 0.05 μs/32-bit
FFT butterfly	5–15 μs	75 μs	15–30 μs
Internal RAM	288 bytes	None	2048 bytes
Input/output configuration	16-bit ports	Memory mapped	Memory mapped + 4 × 10 MHz channels
Programming language	Assembler	Assembler (most languages available	OCCAM (C, Pascal soon)
OCCAM style process switch	N/A	60 μs typical	0.6 μs typical

(b) Performance and cost estimates for two microprocessor- and one transputer-based configurations for the Doppler speed log specification. These performance figures include data I/O as well as FFT tasks

	Microprocessor		*Transputer*
Configuration	*4 × TMS320 + 68000*	*5 × 68000*	*5 × T414*
Relative performance (%)	100	38	77
Power (W)	35	70	10
Relative board area	2½	5	1½
Relative cost (%): hardware*	100	38	100
software	167	106	100
TOTAL (%)	146	85	100

*These figures are for custom TMS320 and T414 boards bought-in 68000 boards.

OCCAM "process". These processes can then be distributed over however many transputers are used in the target hardware.

A further possibility is for systems which can run in a number of modes, by physically turning particular transputers on or off. In a minimal power mode, a single transputer would support the entire computation. The system could then increase throughput, at the expense of power consumption, by turning on further transputers and distributing the workload. This technique can also be used to give a degree of fault-tolerance in systems where high reliability is important.

THE APPLICATION OF TRANSPUTER TECHNOLOGY TO DOPPLER SONAR

So far, only the general requirements of navigation systems have been considered. In a detailed system design, factors such as speed, power consumption and physical size become crucial considerations. We have undertaken a detailed investigation into the use of transputers for the processing and control requirements of a Doppler sonar speed log.

For the Doppler log to give a three-dimensional velocity vector, three or four transducers are required, in the Janus arrangement, as shown in Fig. 2. Monochromatic pulses are transmitted from each transducer. These are backscattered from the sea bed, and are received by the transducers, where their frequency is estimated. The received frequency is Doppler shifted in proportion to the velocity

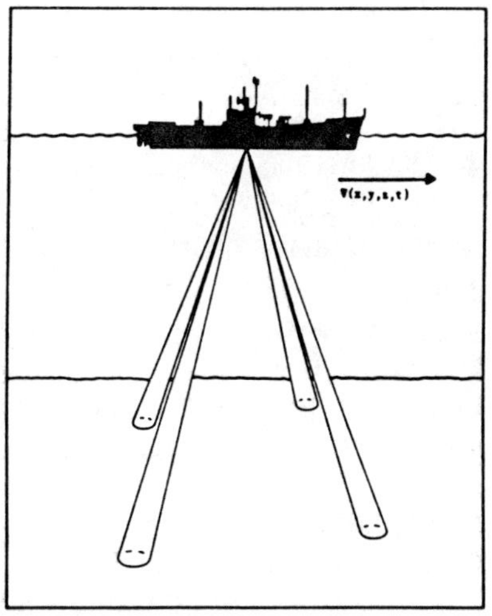

Fig. 2. Janus arrangement of four single frequency acoustic beams. Doppler shifts measured on each beam are combined to calculate the three-dimensional velocity vector for the vessel.

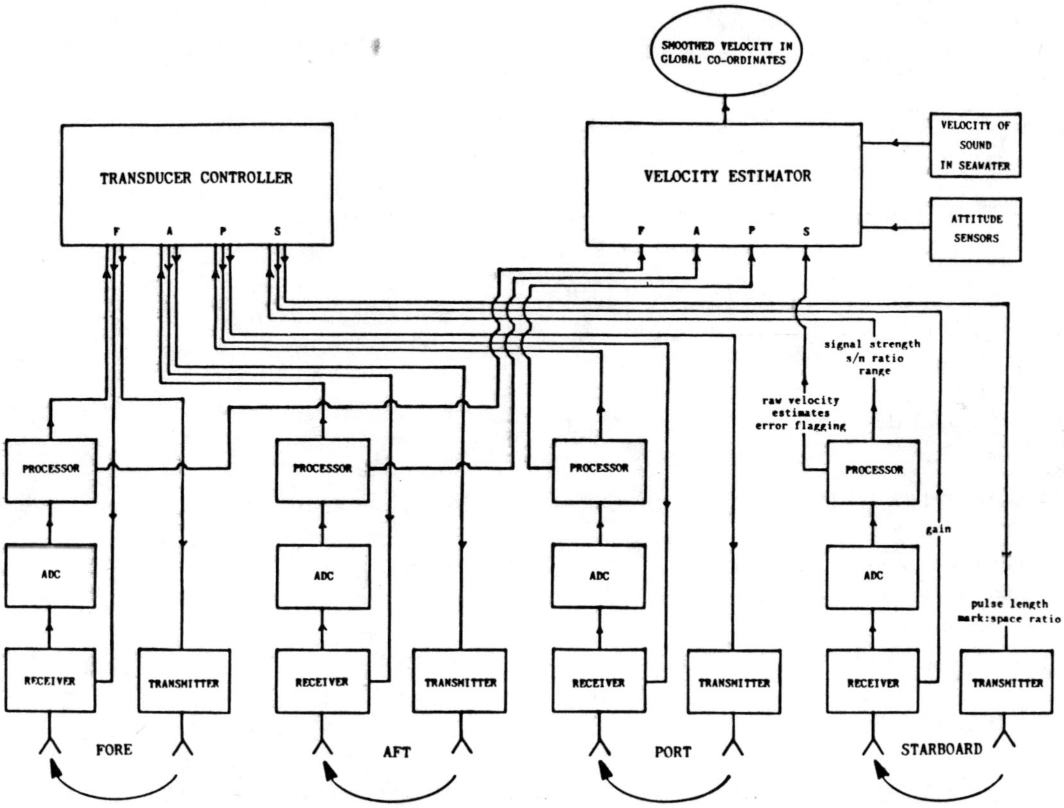

Fig. 3. Functional block diagram of a four-beam Doppler speed log. The transducer controller sets on identical pulse length and mark : space ratio for all of the transducers. It also receives individual range, signal strength and signal-to-noise ratio values from each transducer. This information is used to set individual receiver gains. The velocity estimator receives raw velocity estimates and error flags from each transducer channel. These values are combined with attitude sensors and the velocity of sound in sea water inputs to calculate a smoothed velocity in global co-ordinates.

of the ship. Hence, a measure of the frequency of the return signals gives a measure of the speed of the ship.

Several frequency discrimination techniques to measure the observed frequency shift have been investigated by comparing their performance using trials data. It was established that a Fourier transform based method gave the greatest accuracy and robustness, compared with the zero-crossing technique and the maximum entropy method, both techniques which are commonly used for this purpose. In view of this, the system has been designed around the Fourier transform method, so that one of the major design considerations has been the speed with which fast Fourier transforms could be performed.

Figure 3 shows the functional block diagram of the Doppler speed log. The four beams require independent pre-processing before handling information to both a central transducer controller and velocity estimator. The major processing and control requirements of a Doppler speed log can be summarized as follows.

Data Processing to Calculate the Velocity

The pre-processor for each channel is required to

(a) digitize the return signals at high speed (20kHz),

(b) Fourier transform the correct part of the returns and

(c) find the centre frequency of the Fourier peak.

Raw velocity estimates, together with any error coding from each channel are then combined in an optimal way, in the central velocity estimator, together with pitch, roll and yaw information, to give smoothed velocity and position updates.

Feedback

Feedback is required:

(a) which responds to range variations on each channel and calculates the correct pulse length, and repetition rate to be used, and

(b) which monitors signal strength and signal-to-noise ratio variations from each receiver and sets an individual gain for each channel.

The critical factors of the navigation system which will limit performance are:

(i) the need for fast digitization of an almost continuous data input,

(ii) the ability to perform fast Fourier transforms (FFTs), which are variable in length, depending on range and hence pulse length, and

(iii) the need for parallel processing of three or four data streams, and then to combine their outputs, in an optimal way, together with other sensor information.

In the detailed system design of the Doppler sonar speed log, a variety of processors were compared in order to establish their suitability and cost for these tasks. The configurations considered, two based on microprocessors and one based on transputers were:

(i) four TMS320's, plus a Motorola 68000 as a controller,

(ii) five Motorola 68000's and

(iii) five INMOS T424 transputers.

It was found that the TMS320's performed worst-case FFTs twice as fast as the transputer did. The 68000 was an order of magnitude slower. However, it is the combination of the time taken to read the ADC data into the processor together with the FFT processing time which limit the performance. In the case of the TMS320 and the 68000, times for the two stages must be added to give the total since no useful work can be performed on the 50 μs gaps between samples (Table 1(a)). Using a concurrent FFT on a transputer, however, each sample is read from the ADC, and it is immediately fed into the FFT. Thus, half of the FFT has been performed by the time the last sample is read from the ADC. This is a special property of the concurrent FFT where each FFT butterfly is a concurrent OCCAM process. This is only possible because of the very fast (0.6 μs) process swap time for OCCAM on the transputer. A concurrent FFT on a 68000 would be too sluggish to work because the 68000 takes 60 μs to perform an OCCAM process swap — 100 times slower than a transputer. It is because half of the FFT is performed during the ADC input operation on a transputer that the system performance approaches that of the TMS320 while the 68000 remains unacceptably slow.

For this application, comparisons were made between commercially available 68000 VME bus processor boards and custom-built TMS320 and transputer boards. The 68000 and TMS320 are NMOS microprocessors requiring significant support circuitry which increases board area and power consumption figures (Table 1(b)). In contrast, the transputer is built in CMOS and additionally requires very little support circuitry, thus allowing compact and relatively low-power designs to be implemented.

The board area requirements are exacerbated by the need for inter-processor communications in this application. For the TMS320, a custom-designed interface is required, but in the case of the 68000 the VME bus interface can be used. This leads to further increases in board area and power consumption. The transputer has communication facilities built into the chip. Two pins on one transputer are simply connected to two pins on another and a 1 Mbyte s^{-1} bidirectional link is available. Four such links are available on each transputer, and software support for the links is provided directly in OCCAM. From the point of view of adequate processing capability, coupled with low power consumption, and minimal design of custom hardware, and inter-processor links, the transputer system was the most appropriate of the three considered. Furthermore, the use of OCCAM for design purposes is likely to minimize design costs.

OCCAM is an innovative programming language designed to run on the INMOS transputer. The language supports parallel processes, inter-process communications and multiple processors The instruction set of the transputer has been designed to execute OCCAM very efficiently, ensuring that instruction dead time is minimized. In particular, the ability to implement a concurrent FFT algorithm, in contrast to the TMS320, allows the FFT to be overlapped with the ADC input so that half of the FFT has been performed by the time the last value is read. OCCAM has a further advantage, in that its use is estimated to reduce the software development costs, compared with the TMS320 option, by a factor of 10. The TMS320 must be programmed in assembler, which, although efficient, is unproductive in terms of software development.

In summary, the use of transputers for Doppler sonar systems requiring fast parallel processing and fast data transfer, gives competitive processing power for the lowest power consumption, and the fastest development time compared with other options; this minimizes design and manufacture costs.

DISCUSSION

Transputers are a new technology which has opened up more possibilities for system designers in navigation and positioning systems. However, they are not a panacea for all system design problems. The transputer does not allow the implementation of systems which could not be built using conventional microprocessors. However, it can make the construction of such systems simpler, with reduced hardware and software development cost, and viable in situations where current configurations are limited by space and power.

The programming language of the transputer, OCCAM, allows flexibility in design since it makes no distinction between multiple processes running on the same transputer, and the same processes running on an array of transputers. Therefore, it would be possible to manufacture a multipurpose system, which uses more or fewer transputers, depending on the system option chosen.

This enables a unique flexibility in product specifications. There are requirements for one-axis Doppler logs, for docking purposes, lower performance, low-power systems, using three transducers, and high performance logs using four transducers. Similarly, Doppler sonar current meters are required which are very low power and can be mounted on the sea bed. In addition to these, ship or platform systems with the highest performance are also in demand for use in commercial and academic research projects.

The flexibility of transputers and OCCAM could minimize software development, hardware design and manufacturing costs for development of three systems meeting three different requirements (in the case of the Doppler logs) without compromising performance in any single mode. Transputers could be added to, or not included on, the board, as required.

In this paper, the applicability of transputers to navigation systems, and Doppler sonar speed logs in particular, has been considered.

Incorporating transputers into such systems which require fast parallel processing can give benefits due to their:

(i) increased speed,
(ii) high data transfer rates,
(iii) low power consumption,
(iv) small physical size,
(v) flexibility in configuration and reconfiguration, and
(vi) versatility in handling a variety of data inputs.

These features should enable existing system performance to be enhanced, as well as offering an opportunity to implement novel processing algorithms which have so far been considered impracticable.

REFERENCE

Newsfocus 1985. New fibre optic cable route. *First Break* **3** (9), 2.

DYNAMIC COMPARISONS OF SURFACE POSITIONING SYSTEMS

J P Kovach, Del Norte Technology Inc, USA

Presented on **Tuesday 4 March at 16.30** in the Clarence Conference Room

EMBARGOED AGAINST DELIVERY

4

Comparison of Dynamic Positioning Systems

John P. Kovach, Del Norte Technology, Inc.

INTRODUCTION

In early 1985, Del Norte Technology, Inc. (DNTI) deployed a new ranging system in the Gulf of Mexico with the objective of testing the system's performance. The first question was "How can we compare the new non-line-of-sight (non-LOS) system to anything but a known static calibration checkpoint?" This proved to be an interesting challenge since February in the Gulf of Mexico is not conductive to parking a vessel exactly at a known position for any length of time without roll, pitch, yaw, and heave becoming completely uncontrollable. The answer was to use standard X-band equipment to track the non-LOS system's gyrations at the calibration points. By clustering X-band within LOS of strategic calibration points and placing non-LOS equipment elsewhere so that it would not be in line-of-sight to the same point, the systems could be compared within the same reference grid. Standard summing techniques show the accuracy of the non-LOS system (for brevity, let us call the non-LOS system UHFB) in comparison with the X-band system.

Such a simple answer instigated the second question: "Can we track the UHFB performance dynamically with X-band in a similar fashion?" Certainly! An Apple IIth computer was set up with three serial ports so that measurements from the UHFB system would enter the computer regularly at a three-second update, after which a trigger would be sent to the X-band system so that a second set of ranges could be acquired. Upon reception of the second data stream the entire data block would be saved on a digital cassette recorder. Calculations showed that at 9600 Baud, the data sets would be separated by about 50 ms, which was an acceptable time differential in the error budget (see Appendix A). It was thought that positions and residuals from each type of system could be compared as if they were static. This proved to be untrue.

After data were colected in the Gulf, laboratory tests showed that the UHFB and X-band data were approximately 800 ms apart rather than 50 ms apart due to the Apple II's I/O handling. Obviously, at 6 kn the time offset error was significant. Conventional centroid-fix analysis proved to be useless for comparison purposes since UHFB and X-band fixes occurred at different times. Therefore a second approach was necessary.

THE LSA APPROACH

Matrix reduction of the data seemed to be the proper approach, particularly as presented by Cross (1981). Linear algebra solutions of positional fixes are common, particularly since many small computers have matrix manipulation embedded into their operational repertoire. Essentially, the equation

$$\mathbf{Ax} = \mathbf{b} + \mathbf{v} \qquad (1)$$

is solved recursively for a best **x** fix based upon range measurements. One must have *a priori* information (a guess) estimating a proposed position of the ranging system under scrutiny in order to ascertain the difference between ranges as measured and ranges as calculated from the guess. The **b** term in (1) is that difference of opinion:

b = observed range – calculated range

The **A** term in (1) is a transition term which relates ranges calculated from the guess to grid positions calculated from that same guess. Deployment geometry is introduced by this **A** term, yet no accuracy-destabilizing transcendental functions are used. A typical entry in the **A** matrix would be

easting guess – station easting/calculated range

As the **b** terms are adjusted during recursion they might not be nulled completely, giving rise to the **v** terms which represent residual components of the minimization effort. Residuals **v** have traditionally been principal figures of merit of a positioning system, partially because each input measurement possesses a unique residual. The RHS of equation (1) is normally solved as a single quantity, and residuals are extracted after the recursion is optimized.

Consider now that each measured range has an inherent standard error, which can be deduced during static calibration by performing elementry statistical analysis. Standard error relates to the stability of an individual measurement and is expressed in a covariance matrix **W**, so that now

$$\mathbf{A}^{\mathrm{T}}\mathbf{WAx} = \mathbf{A}^{\mathrm{T}}\mathbf{Wb} \qquad (2)$$

is the weighted LSA matrix system to be iterated, and

$$\mathbf{v} = \mathbf{Ax} - \mathbf{b} \qquad (3)$$

extracts the residuals (Cross, 1981; Brogan, 1985; LeMay *et al.*, 1985).

The above equations are a condensation of a few hundred man-years' labor in matrix algebra and linear system theory by various investigators, and are not meant to be rigorous by any means. Those mathematically curious would invest time wisely by perusing the included references. The procedures followed in this paper are outlined in Appendix A.

INDEPENDENT SYSTEMS

An analysis of two "time independent" range-measurement systems must begin with an assessment of individual quality. Fix quality is most important in the system assumed to be the reference, i.e. if the reference system is of questionable value, then nothing is gained by comparing it to any other system. Alternately, if the system under test fails any reasonable independent quality examination, then no reason exists for comparing it against a reference.

(1) The reference system, X-band, is examined first. Two main results provided by the LSA are the residuals and the error ellipse. Table 1 shows typical data from a run, including ranges in meters, position in the South-Central Texas zone (Lambert projection, in feet), residuals of each range in meters, and the error ellipse orientation and dimension. The fix number is included for comparison later with UHFB fixes. The format is thus:

Fix No.	easting	northing	
Range 1	Range 2	Range 3	Range 4
Residual 1	Residual 2	Residual 3	Residual 4

[(ellipse orientation)(error)] orthogonal error

The ellipse information in brackets '[]' is a pair; 'orientation' is the angle from grid North of

TABLE 1

27	3489011	609643.6	
40522.55	21307.24	16935.01	26910.83
−.034	−.047	.085	−.073
167.302	.213	.262	
28	3488991	609685.3	
40518.45	21297.44	16920.71	26903.23
−.303	−.081	.372	−.39
167.29	.213	.262	
29	3488972	609722.8	
40514.05	21288.04	16907.71	26896.53
−.418	−.08	.476	−.512
167.364	.213	.262	
30	3488953	609761.2	
40509.55	21278.64	16895.21	26888.83
−.102	.108	−.03	−.026
167.431	.213	.262	
31	3488933	609802.9	
40504.95	21269.05	16880.81	26881.13
.01	−.246	.27	−.179
167.5	.213	.262	
32	3488913	609843.1	
40499.95	21259.45	16867.71	26873.03
.609	−.296	−.222	.427
167.574	.213	.262	
33	3488895	609880.3	
40496.65	21250.15	16855.12	26866.34
−.038	.185	−.175	.102
167.646	.213	.262	
34	3488875	609921.4	
40491.75	21241.05	16840.72	26858.84
.366	−.576	.32	−.053
167.712	.213	.262	
35	3488854	609962.2	
40487.65	21230.15	16826.72	26851.74
−.418	.107	.261	−.366
167.785	.213	.262	
36	3488835	609999.6	
40482.75	21220.55	16814.82	26844.04
.121	.474	−.649	.498
167.859	.213	.262	
37	3488814	610039.9	
40478.65	21211.05	16800.32	26837.34
−.33	.16	.12	−.231
167.926	.213	.262	
38	3488792	610078.4	
40472.86	21200.76	16786.62	26830.74
−.067	−.049	.117	−.111
167.998	.213	.262	
39	3488771	610116.6	
40468.86	21190.66	16773.73	26823.94
−.596	.555	−.085	−.211
168.068	.213	.262	
40	3488748	610155.4	
40463.05	21180.16	16760.13	26817.14
−.325	.507	−.279	.044
168.136	.213	.262	

either the semi-major or semi-minor axis of the ellipse, and 'error' is the one sigma error along that axis. 'Orthogonal error' is the one sigma error perpendicular to the orientation. As this paper progresses the semi-minor and semi-major axes will swap roles occasionally; therefore some visualization is required.

Now, for this straight-line X-band run the error ellipse is oriented with the semi-major axis about 77° from grid North. The maximum/minimum error is approximately 0.24/0.19 m and slowly changing as the ship proceeds northwest towards the coast. Residuals are small and variable, with a maximum near 0.5 m. In LSA analysis it is important that the stability of the ranges be comparable to the entries of the weighting matrix lest unreliability and unbelievability creep in. Recursive filter theory (*à la* Kalman) teaches that a running average of 'residuals' keeps a good account of system stability. In fact most recursive filters constantly update the weighting matrix with such a running average.

This paper deals with raw unfiltered data. Ascertaining the quality of a typical evaluation run requires that analysis of stability be accomplished to ensure that the assumed 0.33 m one sigma error of the weighting matrix (see Appendix A) is not exceeded or else the confidence in the X-band 'standard' becomes low. Averaging ten range residuals per traveling group for the five groups representing 14 combined X-band and UHFB fixes defines the X-band set as:

Fix span	*Avge R1*	*Avge R2*	*Avge R3*	*Avge R4*
27–36	−0.0207	−0.0431	0.0706	−0.0574
28–37	−0.0503	−0.0223	0.0740	−0.0731
29–38	−0.0268	−0.0191	0.0486	−0.0453
30–49	−0.0448	0.0444	−0.0074	−0.0152
31–40	−0.0671	0.0738	0.0176	−0.0082

'Fix span' indicates the ten fixes used to calculate the residual averages. Clearly the averages are less than the assumed weighting; therefore the X-band fixes are of sufficient quality and reliability to use as a standard for comparison.

(2) Similar examination of UHFB reveals whether or not the data are of sufficient quality to bear comparison against the standard. Table

TABLE 2

27	3489005	609651
40521.09	16932.24	53602.33
–.045	.025	–.043
196.388	.242	.308
28	3488990	609697.4
40517.49	16918.24	53595.73
.858	–.481	.813
196.388	.242	.308
29	3488963	609729.5
40511.09	16905.34	53596.13
.316	–.177	.299
196.389	.242	.309
30	3488949	609768.9
40508.89	16892.24	53592.53
–.463	.259	–.438
196.407	.242	.309
31	3488925	609807.1
40503.19	16878.64	53590.33
–.384	.215	–.363
196.41	.242	.309
32	3488909	609848.7
40499.59	16865.35	53585.53
–.057	.032	–.054
196.423	.242	.309
33	3488890	609888.8
40494.99	16852.05	53581.73
.294	–.164	.278
196.426	.242	.309
34	3488869	609928.8
40489.49	16838.55	53578.23
.878	–.489	.831
196.434	.242	.309
35	3488851	609969.4
40486.59	16824.65	53575.13
.017	–.009	.016
196.444	.242	.309
36	3488828	610006.2
40480.7	16811.65	53572.83
.397	–.221	.376
196.449	.242	.309
37	3488806	610046
40474.8	16798.05	53569.63
1.02	–.567	.966
196.462	.242	.309
38	3488786	610082.7
40471.8	16784.46	53568.53
–.574	.319	–.544
196.473	.242	.309
39	3488765	610124.7
40466.8	16770.46	53564.63
–.154	.085	–.145
196.482	.242	.309
40	3488742	610161.1
40461.39	16757.16	53563.13
–.403	.223	–.382
196.49	.242	.309

2 shows (in the same format as Table 1) the necessary ingredients for analysis, except that only three stations were available for the test, one of which was about twice line-of-sight.

The semi-major axis is alligned ≈106° from grid North and the maximum/minimum position error of 0.31/0.24 m implies that the ellipse is approaching a circular shape. Residuals are quite variable, so the 'running average technique' is a required analysis tool. In this case the averages are shown below using the criterion of 0.33 m for a GO/NO-GO decision as determined during equipment calibration. The UHFB ranges are numbered 5 through 7 to avoid confusion with X-band, and ten residuals per group are used as above.

Fix span	*Avge R5*	*Avge R6*	*Avge R7*
27–36	0.1805	–0.1006	0.1710
28–37	0.2876	–0.1602	0.2724
29–38	0.1444	–0.0802	0.1367
30–39	0.974	–0.540	0.0923
31–40	0.1034	–0.714	0.0979

For this typical straight-line UHFB run, the ranges are of sufficient quality to go forward with the system comparison.

COMBINED SYSTEMS

With regard to the anticipated 50 ms time skew, the first joining of the two systems was to expand the matrices and treat all incoming ranges as simultaneous. Table 3 attests that this attempt was fruitless since the error ellipse semi-major axis rotated to ≈90° as geometry dictated for all seven stations, but the fix was only slightly removed from the X-band coordinates shown in Table 1. Even by inspection, the residuals exceeded any quality limits imaginable. In all honesty, however, as melodramatic as Table 3 appears, the comparison of Lambert coordinates of each independent system's solution (see Fig. 1) reveals a consistent disagreement of a few meters at common fix reference marks. This could not be possible at 6 kn with only 50 ms of time skew. A variable 800 ms skew was the culprit, as explained above.

It is possible to extend the **A** matrix to include such ambiguities as propagation velocity variances by assigning an overall 'scale factor'

TABLE 3

28			
3488991			
609690.3			
40518.45	21297.44	16920.71	26903.23
–.057	–.533	–.946	–1.74
40517.49	16918.24	53595.73	
.9	1.523	1.965	
180.793	.162	.196	
29	3488968	609725.1	
40514.05	21288.04	16907.71	26896.53
–1.303	–1.246	–.628	–.624
40511.09	16905.34	53596.13	
1.65	1.739	.029	
180.838	.162	.196	
30	3488951	609764.2	
40509.55	21278.64	16895.21	26888.83
–.451	–.628	–1.045	–.565
40508.89	16892.24	53592.53	
.209	1.922	–.161	
180.88	.162	.196	
31	3488930	609804.2	
40504.95	21269.05	16880.81	26881.13
–.881	–1.263	–.524	–.023
40503.19	16878.64	53590.33	
.877	1.644	–.843	
180.928	.162	.196	
32	3488912	609845.2	
40499.95	21259.45	16867.71	26873.03
.232	–.943	–1.016	.103
40499.59	16865.35	53585.53	
.588	1.352	.044	
180.972	.162	.196	
33	3488894	609883.5	
40496.65	21250.15	16855.12	26866.34
–.51	–.704	–1.33	–.435
40494.99	16852.05	53581.73	
1.146	1.738	.505	
181.015	.162	.196	
34	3488873	609924.1	
40491.75	21241.05	16840.72	26858.84
–.304	–1.586	–.787	–.366
40489.49	16838.55	53578.23	
1.954	1.381	.803	
181.062	.162	.196	
35	3488853	609965.1	
40487.65	21230.15	16826.72	26851.74
–.599	–.45	–.638	–.958
40486.59	16824.65	53575.13	
.456	1.431	.401	
181.108	.162	.196	
36	3488832	610002	
40482.75	21220.55	16814.82	26844.04
–.576	–.511	–1.662	.293
40480.7	16811.65	53572.83	
1.482	1.508	.251	
181.151	.162	.196	
37	3488811	610041.9	
40478.65	21211.05	16800.32	26837.34
–1.351	–1.086	–.949	–.184
40474.8	16798.05	53569.63	
2.504	1.321	.512	
181.198	.161	.196	
38	3488789	610079.9	
40472.86	21200.76	16786.62	26830.74
–.699	–.845	–.599	–.125
40471.8	16784.46	53568.53	
.359	1.569	–.789	
181.243	.161	.196	
39	3488769	610119.6	
40468.86	21190.66	16773.73	26823.94
–1.202	–.434	–1.244	–.624
40466.8	16770.46	53564.63	
.856	2.023	–.075	
181.288	.161	.196	
40	3488746	610157.4	
40463.05	21180.16	16760.13	26817.14
–1.015	–.421	–1.185	–.079
40461.39	16757.16	53563.13	
.645	1.784	–.569	
181.334	.161	.196	

FIGURE I

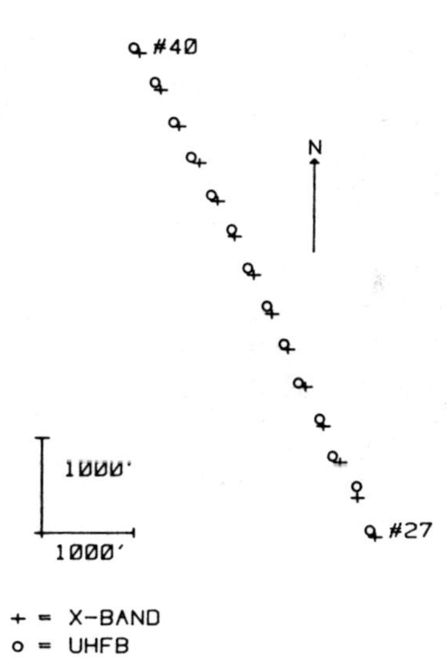

to an individual system. Cross (1981) indicates that such an approach may be of use when mixing two different measuring systems. Since both systems under scrutiny in this paper are of acceptable quality, except for time skew, the **A** matrix was modified such that each range from the X-band system has its own scale factor error. Each X-band measurement has a portion of the ≈800 ms skew according to the ship's heading. The LSA is now solved for six unknowns — two positional coordinates and four proportional time skews associated with the four X-band ranges. With seven measurements, the degrees of freedom criterion is not violated.

Table 4 shows the results of this approach. It is immediately obvious that the residuals from the X-band stations are zero (as far as 64 bit double precision is concerned), and the positional fix is shifted close to the previous UHFB fixes of Table 2. Effectively, the influence and stability of the X-band system has been time-shifted to the UHFB system's solution, causing a slight yet noticeable change in error ellipse parameters and residuals. An accounting of running average residuals of the UHFB system is:

Fix span	*Avge R5*	*Avge R6*	*Avge R7*
27–36	0.1841	−0.1027	0.1743
28–37	0.2907	−0.1620	0.2753
29–38	0.1473	−0.0820	0.1395
30–39	0.1003	−0.0557	0.0950
31–40	0.1062	−0.0592	0.1006

Minor changes of the mixed systems' results demonstrate that the two are equivalent in quality, despite time, geometry, and philosophical differences. Both systems, while operating dynamically, fell within the projected limits as defined during static calibration.

TABLE 4

28	3488990	609697.4	
40518.45	21297.44	16920.71	26903.23
0	0	0	0
40517.49	16918.24	53595.73	
.862	−.483	.816	
196.389	.242	.309	
29	3488963	609729.6	
40514.05	21288.04	16907.71	26896.53
0	0	0	0
40511.09	16905.34	53596.13	
.317	−.177	.3	
196.407	.242	.309	
30	3488949	609768.9	
40509.55	21278.64	16895.21	26888.83
0	0	0	0
40508.89	16892.24	53592.53	
−.464	.259	−.439	
196.41	.242	.309	
31	3488925	609807.1	
40504.95	21269.05	16880.81	26881.13
0	0	0	0
40503.19	16878.64	53590.33	
−.38	.212	−.359	
196.432	.242	.309	
32	3488909	609848.7	
40499.95	21259.45	16867.71	26873.03
0	0	0	0
40499.59	16865.35	53585.53	
−.058	.032	−.055	
196.426	.242	.309	
33	3488890	609888.8	
40496.65	21250.15	16855.12	26866.34
0	0	0	0
40494.99	16852.05	53581.73	
295	−.165	.28	
196.434	.242	.309	
34	3488869	609928.8	
40491.75	21241.05	16840.72	26858.84
0	0	0	0
40489.49	16838.55	53578.23	
.882	−.491	.835	
196.444	.242	.309	
35	3488851	609969.4	
40487.65	21230.15	16826.72	26851.74
0	0	0	0
40486.59	16824.65	53575.13	
.018	−.01	.017	
196.449	.242	.309	
36	3488828	610006.2	
40482.75	21220.55	16814.82	26844.04
0	0	0	0
40480.7	16811.65	53572.83	
.4	−.222	.379	
196.462	.242	.309	
37	3488806	610046	
40478.65	21211.05	16800.32	26837.34
0	0	0	0

40474.8	16798.05	53569.63	
1.02	-.567	.966	
196.473	.242	.309	
38	3488786	610082.7	
40472.86	21200.76	16786.62	26830.74
0	0	0	0
40471.8	16784.46	53568.53	
-.571	.317	-.541	
196.482	.242	.309	
39	3488765	610124.7	
40468.86	21190.66	16773.73	26823.94
0	0	0	0
40466.8	16770.46	53564.63	
-.152	.084	-.144	
196.49	.242	.309	
40	3488742	610161.1	
40463.05	21180.16	16760.13	26817.14
0	0	0	0
40461.39	16757.16	53563.13	
-.401	.222	-.38	
196.503	.242	.309	

CONCLUSIONS

Conclusions from this exercise are mixed. Constant time offsets may be alleviated by the above method, perhaps automatically in real-time. However, asynchronous data with inconsistent time offsets would be difficult, even with human intervention or an artificial intelligence approach. Many newer ranging systems, including DNTI's, provide for external computer control such that data may be acquired and held until requested by a logging device. Until these types of equipment are more widespread, time synchronization will continue to be an object of concern. Nevertheless, the greatest lesson to be learned from this exercise is that one must be sure that the 'standard' is of high enough quality to produce meaningful results.

APPENDIX A

The computer used for this analysis was a 512K Macintosh TM, running the binary version of Microsoft TM Basic in double precision. Algorithms were taken from Cross (1981, pp. 11–12) and the actual matrix calculations were written as subroutines.

The weighting matrix **W** is a diagonal matrix, the elements of which are the inverse of the static standard deviation squared. In this case, all seven stations displayed one sigma deviation of less than 0.33 m during pre- and post-calibration checks so the diagonal entries were ≈ 9.1827.

The standard A matrix for four ranges is

$$\begin{bmatrix} (E^1-E_1)/C_1 & (N^1-N_1)/C_1 \\ (E^1-E_2)/C_2 & (N^1-N_2)/C_2 \\ (E^1-E_2)/C_3 & (N^1-N_3)/C_3 \\ (E^1-E_4)/C_4 & (N^1-N_4)/C_4 \end{bmatrix}$$

where E^1 and N^1 are the estimated coordinates, E_x and N_x are shore station coordinates, C_x is the distance calculated between an estimated position and a shore station.

An **A** matrix with systematic scale error modifies the standard A matrix in this manner:

$$\begin{bmatrix} (E^1-E_1)/C_1 & (N^1-N_1)/C_1 & O'_1/10^6 \\ (E^1-E_2)/C_2 & (N^1-N_2)/C_2 & O'_2/10^6 \\ (E^1-E_3)/C_3 & (N^1-N_3)/C_3 & O'_3/10^6 \\ (E^1-E_4)/C_4 & (N^1-N_4)/C_4 & O'_4/10^6 \end{bmatrix}$$

where O'_x is the observed (measured) range from station x. Three unknowns are solved for a pair of coordinates and a systematic error. The division by 10^6 ensures a systematic error's value to be in parts per million (ppm).

The variant form of the above matrix is to stagger the observations so that six unknowns

REFERENCES

Brogan, W. L. 1985. *Modern Control Theory.* Prentice-Hall, Englewood Cliffs, New Jersey.

Cross, P. A. 1981. The computation of position at sea. *The Hydrographic Journal,* No. 20, 7-16.

LeMay, J. L., Bierman, G. J. and Brogan, W. L. 1985. Linear algebra for modern control and kalman filtering. *UCLA Short Course E5678R Class Notes,* August 1985.

would be solved for a pair of coordinates and an error factor associated with each observation. This amounts to four rows by six columns in the matrix which violates degree of freedom restrictions. Application of systematic error removal by solving for scale factor is not usually used unless such an error is anticipated. In the case of comparing two systems, each of which is of acceptable quality, the anticipated systematic error is time. A saving grace of system comparisons is the wealth of measurements, and in this case three additional ranges yielding seven equations with six unknowns and degrees of freedom is satisfied. The overall matrix now looks like this:

$$\begin{bmatrix}
(E^1-E_1)/C_1 & (N_1-N_1)/C_1 & O'_1/10^6 & 0 & 0 & 0 \\
(E^1-E_2)/C_2 & (N_1-N_2)/C_2 & 0 & O'_2/10^6 & 0 & 0 \\
(E^1-E_3)/C_3 & (N_1-N_3)/C_3 & 0 & 0 & O'_3/10^6 & 0 \\
(E^1-E_4)/C_4 & (N_1-N_4)/C_4 & 0 & 0 & 0 & O'_4/10^6 \\
(E^1-E_5)/C_5 & (N_1-N_5)/C_5 & 0 & 0 & 0 & 0 \\
(E^1-E_6)/C_6 & (N_1-N_6)/C_6 & 0 & 0 & 0 & 0 \\
(E^1-E_7)/C_7 & (N_1-N_7)/C_7 & 0 & 0 & 0 & 0
\end{bmatrix}$$

The dE and dN limits to escape the recursion loop were 0.05 m each. By using the current fix coordinates as an estimate for the next fix, only one pass through the loop was necessary.

Part II

Navigation

NOT SUBMITTED

A REVIEW OF THE ODAP PROJECT: 1980-1985

Dr T Audmunson, S F Barstow, Oceanographic
Centre, Norway

Presented on **Tuesday 4 March 1986 at 15.00**
in the Norfolk Conference Room

EMBARGOED AGAINST DELIVERY

6

A Review of the ODAP Project: 1980-1985

Tore Audunson and *Stephen F. Barstow*, Oceanographic Centre, Trondheim, Norway

The Oceanographic Data Acquisition Project (ODAP) was started in 1980 as an interinstitutional program within Norway for the collection of wave, current and meteorological data from the Norwegian Shelf and the surrounding waters. The project is sponsored by the oil companies operating north of 62° and the Norwegian Council for Scientific and Industrial Research. The effort put into ODAP has to date produced more than sixty data reports, six summary reports and several articles, conference proceedings and papers based upon data collected within the project. The data collected under the ODAP program are open and are stored at the Environmental Data Centre at the Norwegian Meteorological Institute. The promising oil resources on Haltenbanken established over the last two years have resulted in an increased demand both for the available data and for various types of analysis based upon the data sets from this area. Figure 1 is a map giving an overview of the measurements being carried out under the ODAP program in 1985. The cost of the project has typically been between 12 and 15 million Norwegian kroner per year.

As far as ODAP is concerned 1985 may be considered as a suitable break point for reviewing the measurement program and bringing some of the project results to the attention of the worldwide scientific and engineering community. During 1985 the main data acquisition effort in ODAP shifted northwards into the Barents Sea whilst measurements further south off mid-Norway were scaled down and rationalized. The measurements carried out under ODAP have included wave measurements by Waverider buoy near the coast and metocean directional wave buoys offshore; current metre measurements at various depths, both long term and short term; hydrographic measurements; and drifting buoy measurements both for mapping residual flow and obtaining meteorological data from areas with sparse coverage. These measurements are reviewed and discussed.

The data set from the ODAP project to have attracted most attention is undoubtedly that from Haltenbanken (65° N, 7.5° E), from where a unique data set exists consisting of several years of simultaneous current, meteorological and directional wave data. This information has been extensively utilized in a number of research programs. The paper summarizes some

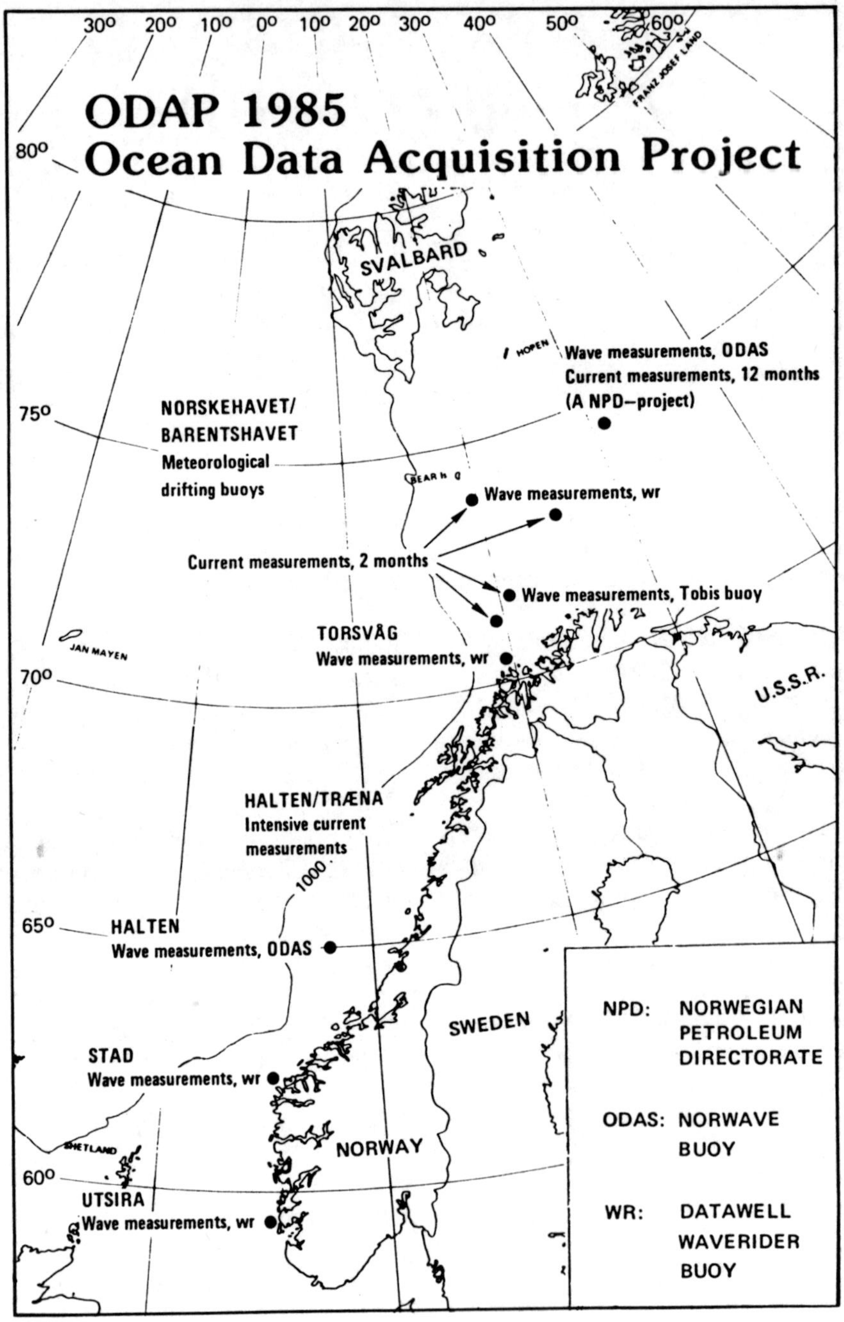

Fig. 1 Overview of 1985 ODAP measurements

of the results from the analysis, including studies of the wind-drift current, the directional wave climate and its presentation for design and operational purposes, extreme oceanographic events, ocean swell, current residuals and so on.

Two results arising from the project are briefly discussed below. Firstly, it has been found that the contribution of the Atlantic swell to the wave climate offshore mid-Norway is more important than that in the North sea. The reason for this is that for waves propagated between Scotland, the Faeroe Islands and Iceland this region is more directly exposed to the North Atlantic. Swells with periods in the region of 20 s and heights over 5 m have been measured. In fact, based on these observations, a research project is now in progress which aims at improving the modelling and forecasting of swell on the shelf. The ODAP data are used for calibration and testing. Secondly, the relationship between wind and current in the near-surface layer has been investigated using ODAP measurements. The measurements include observations of currents from 2 m depth, using an ultrasonic current meter and the tracking of drifters which average the current over the upper 0.35 m. There are also directional wave measurements which permit the correction of the drifter data for Stokes Drift. The results have, among other things, shown that the wind factor for surface current is about 1.5 to 2.0%, depending on the stratification.

Apart from the development of a comprehensive suite of data handling, analysis, and presentation software, the ODAP project has also indirectly led to the development of oceanographic equipment in order to improve the accuracy of the measurements, the data return and, of course, keep the data acquisition costs down to a reasonable level. We briefly discuss some of these developments, which include directional metocean wave buoys, surface drifters, meteorological buoys and equipment for the real-time computation and transmission of wave and current data for forecasting and data surveillance purposes.

WAVE MEASUREMENTS AT NORTH ATLANTIC WEATHER STATIONS 'INDIA' AND 'JULIETT' IN THE NINETEEN SEVENTIES

L Draper, Institute of Oceanographic Sciences, UK

Presented on **Tuesday 4 March 1986 at 15.30** in the Norfolk Conference Room

EMBARGOED AGAINST DELIVERY

7

Wave Measurements at North Atlantic Weather Stations "India" and "Juliett" in the 1970s (India 2 and Juliett 2)

Laurence Draper, Marine Information and Advisory Service, Institute of Oceanographic Sciences, Wormley, Godalming, Surrey GU8 5UB, UK

Measurements of surface waves have been made intermittently from 1953 to 1975 at Ocean Weather Stations *India* (59°N 19°W) and *Juliett* (52° 30′N 20°W) using Ship-borne Wave Recorders. Measurements made during the first decade of the project were analysed, and published in the late 1960s. Measurements made in the second decade have been analysed and the results are published in this report; they may be taken as representative of wave conditions existing over a typical year, divided into the four seasons. Estimates of extreme heights having various return periods are also made.

There are significant increases in wave heights deduced from these analyses compared with those in the previous studies; they were more severe in the 1970s than in the 1960s. All conceivable sources of error have been investigated to try to understand the reasons for the differences, but none of any significance has been found.

It therefore seems necessary to accept that wave conditions can be more severe than had previously been believed.

Because of space limitations, the figures cannot be presented here, but will be available in a report from the Institute of Oceanographic Sciences.

INTRODUCTION

Background to Surface Wave Measurements at Ocean Weather Stations

Measurements started in 1953, using a Ship-borne Wave Recorder (Tucker, 1956) mounted on *OWS Weather Explorer* (formerly a Flower Class Corvette), one of the four British vessels in service at that time. They occupied four stations, *Alpha, India, Juliett* and *Kilo* in rotation with Dutch and French ships. Because of this sparse occupation level by a wave-recorder-carrying vessel, it was not possible to obtain a continuum of records. Indeed, it took many years before the ships had been on station long enough to accumulate sufficient records to produce an acceptable number for every calendar month, from any one station, to allow wave climate data to be published; this stipulation was made to allow a uniform representation throughout the year. The situation is exacerbated by the fact that the recorded wave period is apparently altered by an incalculable amount when the vessel itself moves relative to the water. If the sea waves were all to come from one known direction and the speed and direction of the vessel were known, there would be no problem, but when wave energy comes from a range of directions, as is always the case, and the vessel has to move to keep station, the apparent period, and to a lesser extent the height, is irretrievably modified. The acceptable criteria in these analyses were that the vessels were within about 20 nm of their normal positions and that the station-keeping speed was 2 knots or less; this latter one makes only a 10% modification to the lowest period waves, about 4 s, considered, and proportionately less to longer-period waves. In addition, electronic devices have been known to fail or lose their calibrations, and when any doubt at all existed, a recording was deemed to be unusable for this purpose. The mean depths were 2000 m at *India* and 3000 m at *Juliett*.

In May 1958, *Weather Explorer* was retired and the instrument transferred to *Weather Reporter* (formerly a Castle Class Corvette). By 1964 there were enough records available to permit an analysis of 1440 records from *Juliett* (now referred to as *Juliett 1*) on location 52° 30′N 20°W, and 2400 records from *India* (now referred to as *India 1*) on location 59° N 19°W. The results of these analyses were published in two papers (Draper and Whitaker, 1965; Draper and Squire, 1967). They suggested that the wave climates at the two locations were indistinguishable. These two reports, and especially the *India* one, have become the base references on which many engineering projects have been planned. In particular, *India* data have been the main components of environmental input to the design of wave-energy devices for use in severe conditions.

In November 1966 a second instrument was deployed on *Weather Adviser*, a sister ship to *Weather Reporter*. Because there were then two ships in service, it took less time to find enough records to assemble a composite "year" of wave records correctly distributed throughout the calendar months. The analysis was undertaken starting from the more recent dates and working back in time to fill gaps in the sequence. For *India* and *Juliett*, these records have now been analysed and are referred to as *India 2* and *Juliett 2*.

In mid-1975 *Alpha*, *India* and *Juliett* were abandoned, the latter two being replaced by station *Lima* on location 57°N 20°W, between their old locations. From this time there was always a wave recorder on station giving a continuous sequence of records, interrupted only by the ship steaming and by electronic faults.

Most of the records for both stations were taken in 1975, 1974 and 1973, with a few from 1971 and 1970. At that time the Ship-borne Wave Recorder was calibrated to give wave height in feet, so this measurement unit was used for analysis of individual wave-recorder traces, but before presentation the data were converted to metres.

Analysis of Wave Records

The method of analysing the analogue records is based upon a method developed by Tucker (1961) from theoretical studies of Cartwright and Longuet–Higgins (1956); it is fully explained by Tann (1976). Values for the following parameters were taken from each record:

H_1 = the sum of the distances from the mean water level of the highest crest and the lowest trough;

H_2 = the sum of the distances from the mean water level of the second highest crest and the second lowest trough;

T_z = the mean zero-up crossing period, obtained by dividing the duration of the record (in seconds) by the number of occasions the trace passes through the mean water level in an upward direction;

T_c = the mean crest period.

Values of H_1 and H_2 were corrected for the depth below the surface of the pressure sensors, using the formula, involving T_z, given by Darbyshire (1961) (her equation (1) with k = 2.5). Depths of the pressure sensors were assumed constant; 6.3 ft (1.9 m) for *Weather Adviser* and 7.1 ft (2.2 m) for *Weather Reporter*. These are the mean depths for each vessel. On each voyage the depths were greater on departure and less on return because of the consumption of fuel and water.

From these corrected values of H_1 and H_2, together with T_z, the following were estimated:

H_s = the significant wave height (defined as 4σ, where σ is the root-mean-square surface elevation);

$H_{max\ (3\ hours)}$ = the most probable value of the height of the highest zero-up-crossing wave during a three-hour interval (Draper, 1963).

RESULTS OF THE MEASUREMENT PROGRAMME

India 2

The Most Severe Wave Conditions

The highest value of H_1 in this analysis, actual wave height after correction for instrumental response, measured at *OWS India* was 26.3 m (significant height 15.7 m) with a mean zero-crossing period T_z of 12.9 s, at 00.00 hr on 30 December 1972, although there were two other records with higher values of significant wave height. Not only was it an extreme value of H_1, but it was a real wave with the crest and trough being consecutive. This 26.3 m wave is the highest which we, or to our knowledge anyone else, has ever measured with an instrument. Also, the depth of the trough was about one metre greater than was the height of the crest, conflicting with the common concept of huge waves being largely composed of disproportionately high crests, but giving credibility to stories of "holes in the ocean" (Byles, 1965). This is a relatively unusual wave because with such a value of significant wave height (see Table 1, third line down) in a fairly broad-band sea, only one individual wave in about 1400 can be expected to be above that height, assuming a Rayleigh distribution; this is equivalent to a duration of about 5 h. The second highest wave in the measurements had a H_1 value of 23.0 m springing from the fifth highest significant wave height, of 13.6 m, and with a mean zero-crossing period of 11.5 s, at 09.00 hr on the same day as the highest wave.

Values of significant wave height and the most probable value of the height of the highest wave in 3 h (the recording interval) are normally derived statistically from the two measurements H_1 and H_2 (first and second highest wave) taken from the short analogue record. In general, errors caused by the statistical variations inherent in the short sample and the necessarily rapid analytical techniques can be expected to cancel out for the bulk of the data, but for the very highest recorded waves there is a strong possibility that the highest wave on a record was itself the 3-h (or even longer-interval) wave. Accordingly, it is sensible to determine significant wave height by a more comprehensive method. This has been done for each of the five records showing the highest individual values of H_1, calculating the average height of the highest one-third of the crests, using excursions from the mean, and similarly with the average troughs, and then adding the two values to yield a more widely based estimate of significant wave height. These values have replaced the mass-produced values in the original data set. The interesting, and unexpected, result from this check is that three of the five highest values in the initial analysis underestimated the true value of significant wave height by about one metre; the other two were, as expected, over-estimates. These five most severe (significant height) conditions were as given in Table 1.

Calms

There were no calms (defined as significant height of 0.5 m or less); on only about 10 occasions did the significant height fall below 1 m, the lowest calculated value being 0.58 m

Seasonal and Annual Distributions

The results of the analysis of all the 3-h records are shown in the figures (not presented here), prepared using the IOS forms of presentation recommended for engineering purposes (Draper, 1966). Some of these figures give results for individual seasons defined as follows:

Winter: January–March

TABLE 1

Records having the five largest values of significant wave height (values derived from H_1 and H_2 are given in brackets)

Day no.	Date	Time	T_z (s)	Heights (m) Significant	Highest on record (H_1)
364	30 December 1972	03.00	15.4	17.4 (16.4)	22.1
299	26 October 1974	21.00	12.2	15.8 (14.7)	21.8
364	30 December 1972	00.00	12.9	15.7 (17.0)	26.3
364	30 December 1972	06.00	12.3	15.2 (14.1)	20.4
364	30 December 1972	09.00	11.5	13.6 (14.5)	23.0

Spring: April–June
Summer: July–September
Autumn: October–December

Note that in this analysis the leap year is ignored.

Figures 1–4 show the cumulative distributions of H_2 and of $H_{max\ (3\ hours)}$ for each of the four seasons, and Fig. 5 gives the distribution for the whole year. For example, Fig. 1 shows that during the winter the significant wave height exceeded 5.7 m for 50% of the time, while Fig. 5 shows that a 3.7 m wave height was exceeded for 50% of the entire year.

Figures 6–10 are histograms of the distribution of T_z for each season, and for the year. A comparison of the four seasonal histograms shows that the modal value of T_z was least during the summer, at 7.5–8 s, and greatest during the winter, at 9–9.5 s. The largest zero-crossing periods were around 15 s in the autumn and winter seasons. The histograms show the smallest values of T_z to be about 5 s; the rapid attenuation of short waves with depth causes the Ship-borne Wave Recorder pressure sensors to be insensitive at periods shorter than this.

Figure 11 is a scatter diagram showing the joint distribution throughout the year of significant wave height and zero-up-crossing period. For example, the most frequently occurring combinations were waves with significant wave heights of between 2 and 4 m and periods of between 7.5 and 8.5 s. These occurred for 18.4% of the time, or, since there are 2920 three-hour durations (and samples) in an average year, on about 540 occasions.

Wave steepness is defined as the ratio of wave height to wave length; lines of constant significant wave height steepness 1:7, 1:10, 1:15, 1:20 and 1:40 are drawn on Fig. 11 (wavelength was computed using simple wave theory, i.e. $L = gT^2/2\pi$, where T is deemed to be the zero-crossing wave period T_z).

Figure 12 is a more detailed version of Fig. 11, but it excludes the most severe conditions. Note that rounding can cause apparent differences of one in the totals within particular height and period ranges.

Extreme Conditions

Figures 13–18 present the N–year return values of significant *wave height*, H_s, and the most probable value of the height of the highest individual wave in a 3-h interval, $H_{max\ (3\ hours)}$, for N from 1 to 100. Figure 13 gives the cumulative distribution of significant wave height for the whole year, plotted at half-metre intervals on the log-normal scale; Fig. 14 presents the same data on the Weibull scale; and Fig. 15 on the Fisher–Tippett type I scale. Figure 16 gives the cumulative distribution of $H_{max\ (3\ hours)}$ for the whole year, plotted at half-metre intervals on the log-normal scale. Figure 17 presents the same data on the Weibull scale; and Fig. 18 on the Fisher–Tippett type I scale. There is no theoretical justification for any of these distributions, but they have been widely and satisfactorily used elsewhere. The data points above about 12 m significant and 22 m maximum wave height do not lie on a smooth curve, but best straight lines have been fitted by eye.

TABLE 2
Results of extreme value analyses (return values; heights are in metres)

Year	*Log-normal*	*F–T I*	*Weibull*
H_s			
1	17.5	17.5	17.5
10	23	22	22
20	24.5	23.5	23
50	27	25	25
100	29	26.5	26
H_{max} (3 hours)			
1	32	33	32
10	42	41	39
20	45	43	40.5
50	48	46.5	43
100	52	49	46

There is no objective way of choosing from these three methods, but it will be observed that the data are more nearly in a straight line on the Fisher–Tippett type I scale.

Wave periods are assessed by examining the value of T_z associated with the 50–year wave height. Extrapolations of zero-up-crossing periods, T_z, are unreliable for determining extreme-condition design parameters because many of the longer-period values relate to swell when wave height is not large. High waves generally have a significant-height steepness of between 1:15 and 1:20, and this applies in this data set (Fig. 11). Accordingly, the zero-up-crossing period to be associated with the 50-year wave height can be deemed to lie between about 13 and 18 s.

The period of the individual 50-year return wave has a broad probability distribution and is difficult to assess; according to Longuet–Higgins (1983) the median period of high individual waves corresponds to the mean spectral frequency, and so is $1.07T_z$, assuming a JONSWAP spectrum (giving about 14–19 s for these data), but Bell (1972) suggests, from an analysis of North Sea instrumental data, a value of $1.28T_z$.

Clearly, for design purposes it is wise to consider a range of values.

Persistence

Because the records are discontinuous, it is not possible to produce persistence diagrams for these stations, although recent studies by Kuwashima and Hogben (1984) indicate that it might be possible to produce an empirically derived distribution based on the scatter diagram, such as Fig. 11.

Juliett 2

The Most Severe Wave Conditions

The highest value of H_1 in this analysis, actual wave height after correction for instrumental response, measured at *OWS Juliett* was 20.3 m (significant height 11.3 m) with a mean zero-crossing period T_z of 11.6 s, at 18.00 hr on 12 February 1973, although there were four other records with higher values of significant wave height. The second highest wave (H_1) was 20.1 m with a mean zero-crossing period T_z of 11.4 s at 09.00 hr on 14 February 1973.

Values of significant wave height and the most probable value of the height of the highest wave in 3 h (the recording interval) are normally derived statistically from the two measurements H_1 and H_2 (first and second highest wave) taken from the short analogue record. In general, errors caused by the statistical variations inherent in the short sample and the necessarily rapid analytical techniques can be expected to cancel out for the bulk of the data, but for the very highest recorded waves there is a strong possibility that the highest wave on a record was itself the 3-h (or even longer-interval) wave. Accordingly, it is sensible to determine significant wave height by a more comprehensive method. This had been done for each of the five records showing the highest individual values of H_1, calculating the average height of the highest one-third of the crests, using excursions from the mean, and similarly with the average troughs, and then adding the two values to yield a more widely based estimate of significant wave height. These values have replaced the mass-produced values in the original data set. As expected, the new values are slightly lower than the mass-produced ones. These five most severe (significant height) conditions were as shown in Table 3.

TABLE 3

Records having the five largest values of significant wave height (values derived from H_1 and H_2 are given in brackets)

Day no.	Date	Time	T_z (s)	Heights (m) Significant	Highest on record (H_1)
43	12 February 1973	21.00	12.3	12.4 (12.8)	18.6
45	14 February 1973	09.00	11.4	12.3 (13.3)	20.1
23	23 January 1975	09.00	11.2	11.7 (12.8)	19.4
19	19 January 1975	12.00	10.3	11.6 (12.6)	20.0
43	12 February 1973	18.00	11.6	11.3 (13.1)	20.3

Calms

There were no calms (defined as significant height of 0.5 m or less); on only about 20 ocassions did the significant height fall below 1 m, the lowest calculated value being 0.78 m. There was one memorable occasion in 1972 when it did not exceed 1 m for 10 consecutive measurements and this was, rather unexpectedly, in mid-September close to the equinox.

Seasonal and Annual Distributions

The results of the analysis of all the 3-h records are shown in the figures (not presented here), prepared using the IOS forms of presentation recommended for engineering purposes (Draper, 1966). Some of these figures give results for individual seasons defined as follows:

Winter:	January–March
Spring:	April–June
Summer:	July–September
Autumn:	October–December

Note that in this analysis the leap year is ignored.

Figures 19–22 show the cumulative distribution of H_2 and of $H_{max\ (3\ hours)}$ for each of the four seasons, and Fig. 23 gives the distribution for the whole year. For example, Fig. 19 shows that during the winter the significant wave height exceeded 5.0 m for 50% of the time, while Fig. 23 shows that a 3.2 m wave height was exceeded for 50% of the entire year.

Figures 24–28 are histograms of the distribution of T_z for each season, and for the year. A comparison of the four seasonal histograms shows that the modal value of T_z was least during the summer, at 7.5–8 s, the greatest during the winter at 9–9.5 s. The largest zero-crossing periods in the winter season approached 13.5 s, although swell in the summer slightly exceeded this value. The histograms show the smallest values of T_z to be about 5 s; the rapid attenuation of short waves with depth causes the Ship-borne Wave Recorder pressure sensors to be insensitive at periods shorter than this.

Figure 29 is a scatter diagram showing the joint distribution throughout the year of significant wave height and zero-up-crossing period. For example, the most frequently occurring combinations were waves with significant wave heights of between 2 and 3 m and periods of between 7.5 and 8.5 s. These occurred for 11% of the time, or, since there are 2920 three-hour durations (and samples) in an average year, on about 320 occasions (if the range is extended to 4 m the occurrence is 18% and about 530 occasions).

Wave steepness is defined as the ratio of wave height to wave length; lines of constant significant wave height steepness 1:7, 1:10, 1:15, 1:20 and 1:40 are drawn on Fig. 29 (wavelength was computed using simple wave theory, i.e. $L = gT^2/\pi$ where T is deemed to be the zero-crossing wave period T_z.

Figure 30 is a more detailed version of Fig. 29, but it excludes the most severe conditions. Note that rounding can cause apparent differences of one in the totals within particular height and period ranges.

Extreme Conditions

Figures 31–36 present the data in forms which

consuming in continental shelf water depths than if an identical operation were performed in deeper continental slope waters. Indeed, it may well be that entirely new modes of operation have to be invoked in deep water situations where a much higher proportion of the tasks have to be performed by sophisticated remotely operated vessels (ROVs). Furthermore, in these operations, initial detailed survey work may require specialized equipment such as a deep-towed high-resolution side-scan sonar equipment like GLORIA II. Such systems require precise positioning in order to gain the maximum benefit from the increased picture

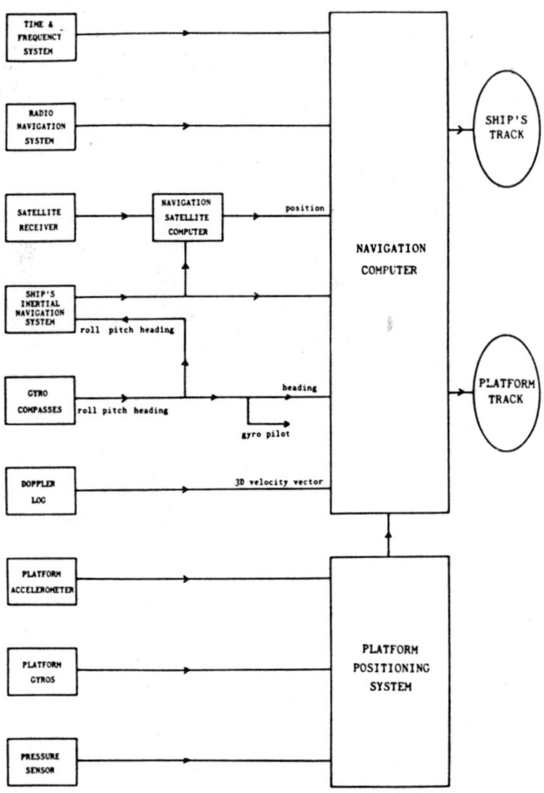

Fig. 1. Function block diagram of a modern integrated navigation system with an additional towed platform navigation system. Although, when depicted at this level, the system appears relatively simple, the ability to process data from a wide range of sensors concurrently in real time imposes severe design constraints on each data channel processor and also in merging these channels. This aspect is particularly well-suited to transputer-based designs.

resolution. This system has been used recently by British Telecom to delineate local troughs and canyons on the continental slope of the Western Approaches. A detailed knowledge of the local seafloor topography is required to assist in determining the exact branching point for the UK link into the new transatlantic fibre-optic communication cable (Newsfocus, 1985).

Manufacturers have responded to these operational demands by devising reliable, sophisticated systems for navigation and positioning. Taking a modern integrated navigation system as an example (Fig. 1), several common system requirements can be identified. One fundamental requirement is to integrate information from a wide range of sensors. The system illustrated includes the following sensor inputs:

(i) a satellite data receiver and navigation computer,
(ii) a ship's inertial navigation system,
(iii) gyro-compasses and
(iv) a speed log which is either an EM- or Doppler-based system.

The integration and processing of information from a diverse range of inputs requires:

(i) fast parallel processing of a diverse range of sensors and inputs,
(ii) fast autonomous preprocessing of individual data channels,
(iii) intelligent, usually optimal, combination of data from the sensors,
(iv) flexibility in the way in which the sensor information is interpreted and
(v) flexibility in the mode of operation.

To satisfy these requirements and, in particular, satisfy the demands of (i) and (ii), leads to the necessity of rapid data transfers between processors on a repetitive basis. In addition to this, "intelligence" in the way in which data channels sampled at different rates are interleaved before being passed on to the next stage of processing is crucial.

Microprocessors used in such applications need the ability to run as both pipeline and parallel processors. In order that the "best" result is calculated by any optimal filtering process, it may well be necessary to select

A difference, but almost certainly an insignificant one, is that the original "winters" started on 23 December and ended on 22 March, whereas now a winter is deemed to start on 1 January and end on 31 March. Wave heights are generally greater at the end of December than in the last week of March, so that, if anything, the series 1 data should have been more severe than in this second series.

It is apparent from a comparison of Figs. 11 and 29 with their equivalents in the *India 1* and *Juliett 1* data publications that the mode of the period distribution has decreased from about 8.75 s in *India 1* and *Juliett 1* to about 8.25 s in *India 2* and *Juliett 2* (a decrease of about 6%). It is difficult to accept that this is a real change, especially as the wave heights seem to have increased, but as the chart speed was governed by the frequency of the ships' generators it is just possible that there was a permanent change in both ships' systems in the same sense between about 1965 and 1970. However, the ships' engineers feel that the frequencies rarely deviated by this amount and that the mean values would have been close to the specified 50 Hz. It is also possible that this change in apparent period is an artefact of the manual method of analysis, but it is difficult to understand how a change as large as 6% could occur from this cause.

The first effect of a slowing of the chart speed is a reduction of the apparent period of the waves, and therefore an increase in the apparent number of waves at the site. This difference can be seen in Table 5. It is of interest that the series 2 data are close in period to *OWS Lima* wave data measured after *India* and *Juliett* were terminated. About two-thirds of the *Lima* data were from records which were taken with an accurately controlled chart speed, so it seems reasonable to conclude that the *India* and *Juliett* series 2 data are more reliable in terms of period than are series 1. From this premise it is reasonable to conclude that height data from the *India* and *Juliett* series 1 data are underestimates by perhaps 2 or 3%, because of the changes in frequency response of the instrument over the frequency range.

The changes in the maxima of the wave height distributions are in good agreement with the changes in the median values; for example, a comparison of the annual scatter diagrams shows that the "summit" of the contours on *India 1* lies at about 7.7 ft (2.3 m), and on *India 2* it lies just below 3 m, an increase of about 30%. Similarly, the maximum of the 20 parts per thousand level contour is at about 12 ft (3.7 m) on *India 1* and about 5 m on *India 2*, an increase of about 35%; the maximum of the 30 parts per thousand level contour is at about 10 ft (3 m) on *India 1* and 4 m on *India 2*, an increase of about 30%.

Checks made on the Climate, Instruments and Analyses

Because of these significant increases in average wave heights at *India* and *Juliett* in the early 1970s compared with the figures previously published of records taken around 1960, it became necessary to investigate all aspects relevant to both sets of measurements and analyses.

Briefly, no discrepancy has been found.

Recordings were made on the same actual instruments and on the same type of chart, and on ships which occupied one station or the other on an apparently random basis; it was not until the mid-1970s, after *India* and *Juliett* had been abandoned, that the recorders were superseded by transistorized versions. Certainly, there is no preponderance of occupation of one ship at either location, so the difference in increase of 27% at *India* and 11% at *Juliett* cannot be attributed to a faulty calibration of one particular instrument. The principal instrumental change between series 1 and series 2 has been an increase in chart scale to accommodate extreme waves which occasionally occur, but this cannot account for any change in sensitivity. There was only a small amount of data from *Weather Explorer* (in series 1 only), and none of these came from the winter.

The first reaction was that the climate had become more severe. A check on the wind conditions at *India* for the winter months when the bulk of the severe-weather data was obtained shows that there was no significant change in the mean wind speeds. This is a simple check, but it seems unlikely that it would not show up some difference in the winter when waves at *India* appear to be 36% higher, which would mean perhaps a 20% increase in wind speed. However, it is just conceivable that by chance the wave measurements in series 1 were made at times when the winds were less severe, or the wave measurements in series 2 were made at times when winds were more severe, or a combination of both, so the wind speeds at the actual times of the wave measurements of both series at both stations were investigated. This shows that the populations of the wind data were virtually the same in both series from each station, whereas for significant wave heights H_s the differences are significant at *Juliett* and highly significant at *India*. The checks were repeated to compare waves with winds six hours previous to each wave measurement, and the results were the same. Detailed results of these investigations are available.

The next aspect to come under scrutiny was the instrument calibration. The very earliest calibrations were not adequately recorded; in particular this concerns the actual settings of the sensitivity controls on the instruments during service. However, all controls were in a very inaccessible position in service and were locked, and all servicing until the mid-1960s was undertaken by the present author personally, who feels that there were no sizeable errors anywhere in the procedure. It was a matter of pride that considerable care was always taken to achieve the best possible calibrations. As time progressed, improved notes were made and kept, and by about 1970 the system was adequately standardized and recorded. The laid-down procedures were followed without problems by subsequent engineers, overall responsibility still being with the present author until 1974. All relevant calibration sheets have recently been checked by present engineering staff and no error has come to light. The bulk of the data in the earlier analyses had come from the 1960s; only a few gap-filling records were of earlier date.

The only other variable has been the change in chart scales from 50 to 60 ft and then to 80 ft, followed by metrication which occurred after these data sets had been completed. There is no evidence that wrong charts were used either in calibration or in service (all checks undertaken show that the correct ones were indeed used at all times). It is difficult to see how a change in overall sensitivity, and therefore in chart scale, can alter the sensitivity to either small or large waves. In the earlier days the larger wave crests and troughs were obviously running more often into the more non-linear parts of the chart recorder, and it might be arguable that it would consequently under-read on big waves, but the chart itself is highly non-linear and almost certainly copes with that problem. The accuracy of the entire system out to the chart limits was one of the routine checks during calibration.

A further check was to calculate all the parameters from the early (*India 1* and *Juliett 1*) analyses, starting with a check on the numbers taken from the chart records, and then using the present computer program. Although occasional errors were shown to have occurred in the early work, the bulk of the early calculations, performed on hand calculators, were close to their proper values; errors which were identified were random and seemed to cancel out, and the overall picture has not changed at all.

There is only one other Ship-borne Wave Recorder station which was in operation around 1960 and also in the early 1970s, and that is Sevenstones, off Land's End; to confuse the issue further, in an analysis of several years' wave data from the late 1960s and early 1970s, Fortnum and Tann (1977) have shown that heights (at the 50% level) show an 11% increase on the 1962 values (Draper and Fricker, 1965) which was an "average" year wind-wise. This is the same increase as at *Juliett.* Challenor (private communication) has significance-tested the 1962 data to see if they are part of the same family as the later data, and cannot identify any fundamental difference.

An independent source of wave data, visual observations published in the *Oceanographic Atlas of the North Atlantic Ocean,* suggested that the average wave height at *India* was 4% lower than that at *Juliett,* although in any particular named month (averaged over many years) sometimes the average wave height was greater at one station and sometimes at the other. In this atlas in each 5° square there were typically 500 to 1000 visual observations per calendar month, so this amount of random scatter can be expected. Previously, it seemed from instrumental measurements and from the atlas that the climates at the two stations were very much the same. The picture now seems to have changed.

In a study of changes in the wind and wave regime of the North Sea, Lamb and Weiss (1979) show that there was a continued increase in wave heights from 1950 to the 1960s, followed by a slight decrease to the 1970s. The largest increase (but not the largest waves) occurred in the southeastern area of the North Sea. Although these changes are in the opposite sense to those at *India* and *Juliett,* it does demonstrate that significant changes can and do occur. In the North Sea it is postulated that the changes might be due to a (demonstrable) increase in the frequency of occurrence of northerly winds at the expense of westerlies. In the middle of the North-east Atlantic it is difficult to understand how a change in the distribution of wind directions, but not speeds, where all fetches are large, can affect the average wave height. It could have been explained if the average duration of high wind speeds had increased, as it would have resulted in more fully developed seas, but an analysis of data provided by the Meteorological Office of winds at *OWS India* for the years 1960–64 and 1970–74 shows that the average duration of high wind speeds decreased from the 1960s to the 1970s, the opposite sense to that required to explain the change in wave conditions.

Lamb and Weiss also present a Gale Index Survey for the British Isles and Eastern North Atlantic from 1880 to 1977, in which they show that the number of days of gale per year from 1965–77 were higher than any other period since 1889, but not quite as high as in 1880–9. It is possible therefore that in the early 1970s the waves at *India* and *Juliett* were also more severe than in the rest of this century. The value given for 1960–5 is markedly the lowest for this century, but the authors have major reservations about its validity, as that was the time when computer-automated map drawing was introduced, and the pressure gradients may have been unrealistically smoothed.

In a hindcast study of the longer-term changes in wave conditions of the North Atlantic from 1970 to 1982, Neu (1984) concluded that there are large variations in wave height from one year to another, ranging from –25% to + 45% of the mean. This is therefore greater than the differences occurring in these two measured data sets. Unfortunately, Neu does not present an analysis of conditions in the early 1960s. What is perhaps more alarming is that he maintains that there was a continuous increase in wave heights (except, mercifully, in the more severe wave conditions in the eastern North Atlantic) from 1970 to 1982. Perhaps that trend had started by the early 1960s, but if so it is difficult to believe that it has not by now made itself more evident to mariners in other ways.

Methods Used in the Calculation of Extremes

In the analysis of the two data sets, extreme conditions were estimated using four different methods, log-normal, Fisher–Tippett type I, Weibull, and Battjes. It was observed that the Battjes method was extremely sensitive to the highest individual records; omission of the one highest measured value in the *India 2* series results in a reduction of 1.5 m in the estimated value of the 50-year return wave height. Accordingly, this method has been excluded from the study. A presentation of the Battjes results is given in Fig. 37; note that *Juliett 1* and *Juliett 2* are indistinguishable on this plot, a somewhat surprising result. This demonstrates the differences between the two stations, and also the effect of removing the one highest value from the *India 2* data.

Table 5 presents the number of waves derived by the Battjes method for each series, and gives the best estimate of the 50-year return value of individual wave height using a visual linear fit to the curves in Fig. 37 (values to the nearest half metre).

TABLE 5

	No of waves/year	*50-year wave height (m)*
India		
11	3.41×10^6	37.0
12	3.76×10^6	41.5
12mod	3.76×10^6	40.0 (the one highest wave omitted)
Juliett		
J1	3.39×10^6	35.0
J2	3.70×10^6	35.0

CONCLUSIONS

The inescapable conclusion is that there is no obvious explanation for the apparent change in wave climate from the 1960s to the 1970s. It might just be within the realms of possibility that the necessarily short samples can have produced the variability. Measurements made subsequently on station *Lima* (nearer to *India* than to *Juliett*) bear more similarity to *India 2* than to the older measurements or, indeed, to *Juliett 2.* On this basis, considering all available evidence, it seems likely that the severity of wave conditions in the North Atlantic at about 20°W and between 52°N and 59°N have changed from the early 1960s to the early 1970s, both in absolute values and in geographic distribution. The only really firm conclusion is that we shall need to obtain many more measurements before we can truly claim to know the wave climate in this area to an acceptable level. Even so, the problem of future climatic variation is one which must be borne in mind with due caution.

ACKNOWLEDGEMENTS

The author wishes to express his appreciation to the Meteorological Office for permitting the fitting of Ship-borne Wave Recorders to Ocean Weather Ships *Adviser* and *Reporter,* and his thanks to those who, over the years, maintained the instruments and obtained the records. This applies especially to Mr E. Binner, Shore Meteorologist, who helped enormously in a multitude of ways throughout the entire period, and Mr J. S. Driver (now of J. S. Driver SEAWAVE Services) who maintained the instruments in tip-top condition. The author is also grateful to colleagues in the Institute of Oceanographic Sciences who helped with the instrumentation, analysed the records, and wrote and ran the computer programs, and also those, D. J. T. Carter in particular, who have commented constructively on this draft, A. R. Tabor who undertook much of the data handling, and P. G. Challenor who made the wind and wave comparisons. The help of Professor H. H. Lamb on the problem of the relative severity of the measurement period, and the Meteorological Office in the provision of comparative environmental data, are very much appreciated.

REFERENCES

Bell, A. O. 1972. *North Sea Wave Spectra Report.* London: BP, for North Sea Environmental Study Group.

Byles, W. S. 1965. The one from nowhere. *The British and Commonwealth Review.* Also published in *The Marine Observer* **XXXV** (210), 195–197, October 1965.

Cartwright, D. E. and Longuet-Higgins, M. S. 1956. The statistical distribution of the maxima of a random function. *Proc. R. Soc. (Lond.) A* **237**, 212–237.

Darbyshire, M. 1961. A method of calibration of shipborne wave recorders. *Deutsche Hydrograph. Z.* **14** (2), 56–63.

Draper, L. 1963. Derivation of a design wave from instrumental records of sea waves. *Proc. Instn Civ. Engrs* **26**, 291–304.

Draper, L. and Whitaker, M. A. B. 1965. Waves at *Ocean Weather Ship Station "Juliett". Deutsche Hydrograph. Z.* **18**, 25–30.

Draper, L. and Fricker, H. S. 1965. Waves off Land's End (Jan. 1962–Jan. 1963). *J. Inst. Navig.* **18**, 180–187.

Draper, L. 1966. The analysis and presentation of wave data—a plea for uniformity. *Proc. 10th Conf. on Costal Engng, Tokyo* **1**, 1–11.

Draper, L. and Squire, E. M. 1967. Waves at *Ocean Weather Ship Station "India". Trans. R. Instn Nav. Archit.* **109**, 85–93.

Fortnum, B. C. H. and Tann, H. M. 1977. Waves at Sevenstones Light Vessel (Jan. 1968–June 1974). *IOS Report 66.*

Gleason, R. (unattributed) 1985. Waves recorded at *Ocean Weather Station Lima*, December 1975–November 1981. *Offshore Technology Report OTH 84 204*. Report prepared by the Institute of Oceanographic Sciences, Taunton, for the Department of Energy. London: HMSO.

Kuwashima, S. and Hogben, N. 1984. The estimation of persistence statistics from cumulative probabilities of wave height. *NMI Report R183*. May.

Lamb, H. H. and Weiss, I. 1979. On recent changes of the wind and wave regime of the North Sea and the outlook. *Fach. Mitt Wehrgeophys*, no. 194.

Longuet-Higgins, M. S. 1983. On the joint distribution of wave periods and amplitudes in a random wave field. *Proc. R. Soc. (Lond.)* A **389**, 241–258.

Neu, H. J. A. 1984. Interannual variations and longer-term changes in the sea state of the North Atlantic from 1970–1982. *J. Geophys. Res.* **89** (C4), 6397–6402, 20 July.

Oceanographic Atlas of the North Atlantic Ocean 1963. US Naval Oceanographic Office Publication 700; Section IV: sea and swell.

Tann, H. M. 1976. The estimation of wave parameters for the design of offshore structures. *IOS Report 23*.

Tucker, M. J. 1956. A shipborne wave recorder. *Trans. Inst. Nav. Archit.* **98**, 236–250.

Tucker, M. J. 1961. Simple measurement of wave records. *Proc. Conf. on Wave Recording for Civil Engineers* (at NIO—now IOS), L. Draper, ed., 22–23.

THE ADVANTAGES AND PROBLEMS OF USING ACOUSTIC DOPPLER CURRENT PROFILERSIN SUPPORT OF EXPLORATION DRILLING ON THE WEST SHETLAND CONTINENTAL SHELF

Dr S Archer, Miros, UK

Presented on **Tuesday 4 March 1986 at 16.30**
in the Norfolk Conference Room

EMBARGOED AGAINST DELIVERY

8

The Advantages and Problems of using Acoustic Doppler Current Profilers in Support of Exploration Drilling on the West Shetland Continental Slope

S. J. Archer, MIROS Ltd*
** Formerly with Offshore Environmental Systems Ltd.*

INTRODUCTION

Acoustic instrumentation for obtaining water current velocities over a depth profile evolved by applying range-gating techniques to ship Doppler speed logs (Joyce *et al.*, 1982). However, the errors in navigational position and pitch/roll motions of the vessel limit the accuracy that can be achieved unless compensated for by other sensors (Crocker, 1983). This spurred the development of self-contained acoustic Doppler current profilers (ADCPs), so that by mounting the ADCP on the seabed in an upward-looking configuration such sources of error could be avoided (Pettigrew and Irish, 1983). This development also allowed mounting such instruments on offshore production platforms (Gordon, 1982), and with oblique beam axes as in the CUMEX study (Gordon and Skorstad, 1985).

This paper describes applications where ADCPs have been used in a downward-looking configuration, mounted from semisubmersible vessels being used for exploration drilling. Since a semisubmersible is subject to some motion, not all errors due to motion of the ADCP transducer can be avoided. However, the deployment methods described here allowed some decoupling from the semisubmersible motion, and the remaining motion-induced errors were generally within acceptable limits, as discussed later. Also, used in the direct reading mode, the instrument is freed from constraints on power supply and data-recording capacity.

Immediate and recent historical data on current profiles can be of great benefit while conducting exploration drilling, especially on the West Shetland Continental Slope (see Fig. 1). Here, the upper Atlantic water, above about 500 m, flows generally parallel with the shelf break towards the north east, modulated by the shelf tidal currents with major axes also roughly aligned with the shelf break, so that during weak residual flows reversal of the total flow can occur. Below this, in the Faroe-Shetland Channel, cold water from the Norwegian Sea flows out to the Atlantic.

Changes in the depth of the boundary between these flows can cause interaction of the currents on the Continental Slope. Also, perturbations in response to storms can cause inertial currents with semi-diurnal oscillations or quasi-steady residual currents (Heaps,

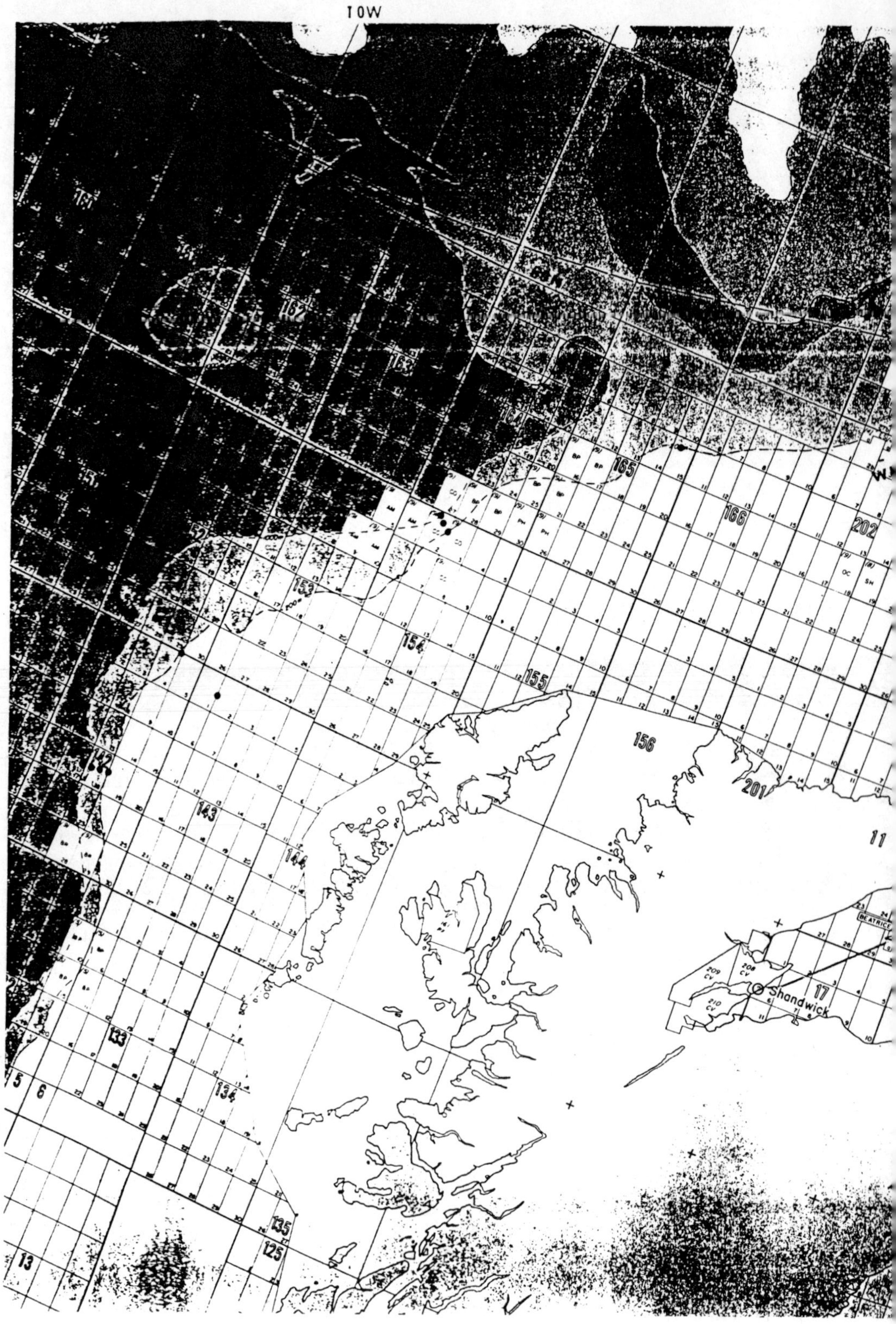
10W
Shandwick
BEATRICE

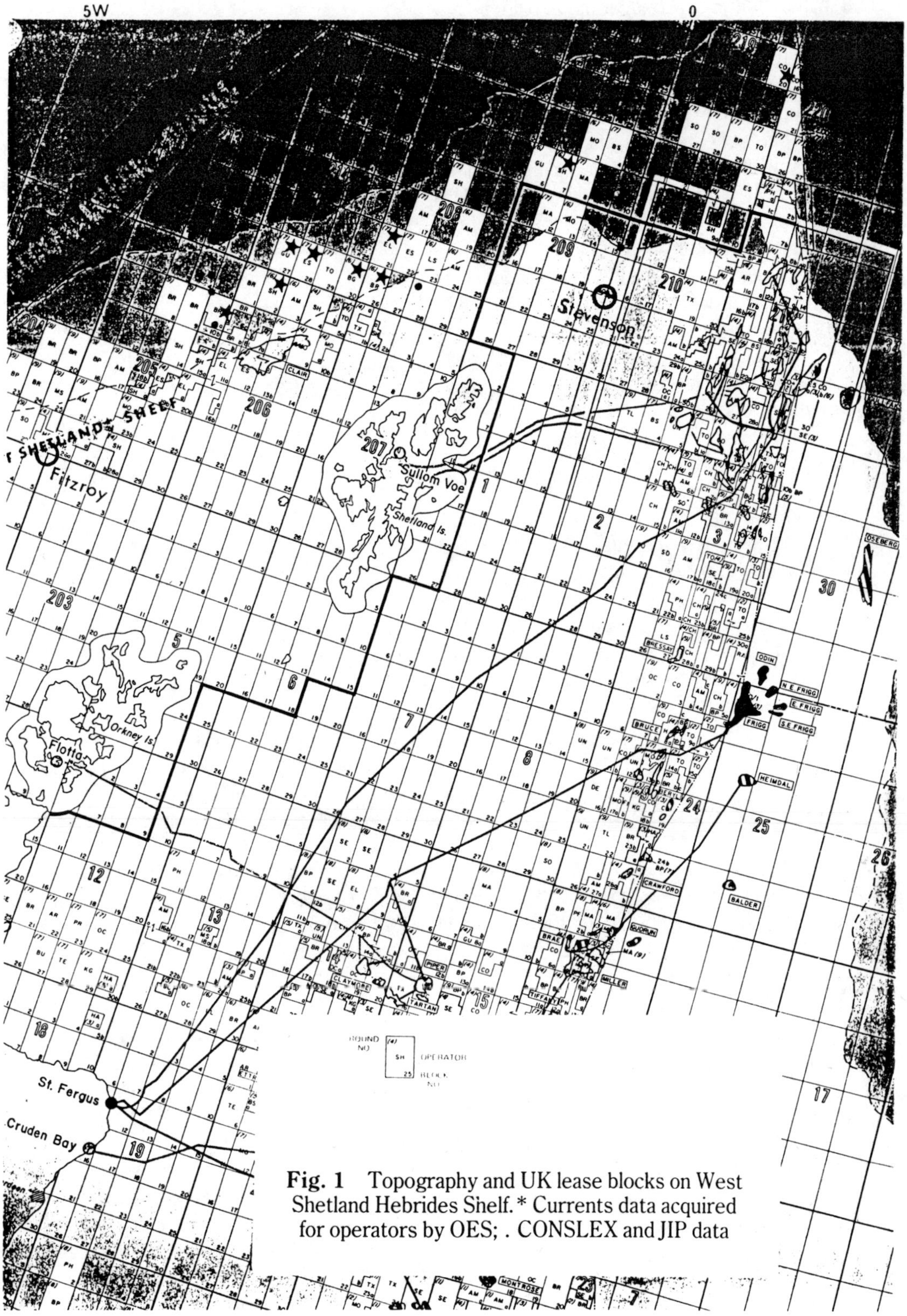

Fig. 1 Topography and UK lease blocks on West Shetland Hebrides Shelf. * Currents data acquired for operators by OES; . CONSLEX and JIP data

1985). Interactions of these currents can set up shelf wave motions propagating to the north-east which can develop into eddies, causing current reversals to the south-west (Gould, 1984).

Therefore, while the general current conditions to be expected in the area can be reasonably well predicted, the actual conditions experienced at a particular site may be quite complex and unpredictable; hence the benefit in monitoring the current profile during the drilling operation. Unexpected problems with the riser catenary, with station-keeping of the vessel and mooring line tensions, or with ROV operations can be related to the known current conditions, and appropriate decisions made. Also, trends in the current can be deduced to assist in scheduling critical operations such as 'spudding in' the drill string, and pulling off or reconnecting the riser to the wellhead.

While vibrations of the riser induced by depth-coherent vortex-shedding can be avoided or minimized by careful selection and design of the riser, a knowledge of the history of the current profile during the drilling operation to date, combined with engineering calculations, can provide confidence that fatigue stresses have not reached critical values, so avoiding costly riser failures.

DEPLOYMENT METHODS

The configuration of the four acoustic beams of the RD Instruments RD-SC/DR Acoustic Doppler current profiler is shown in Fig. 2. The

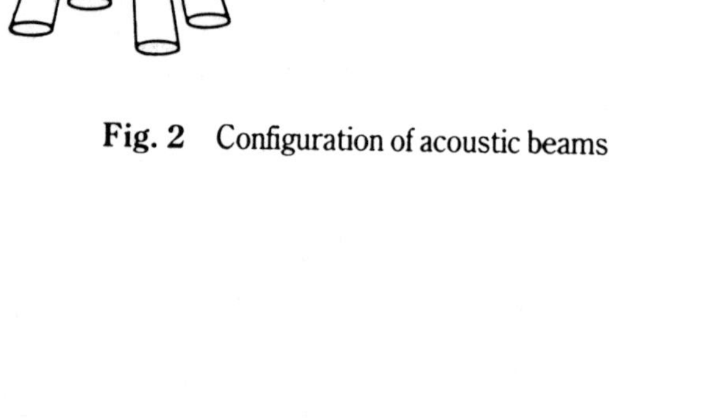

Fig. 2 Configuration of acoustic beams

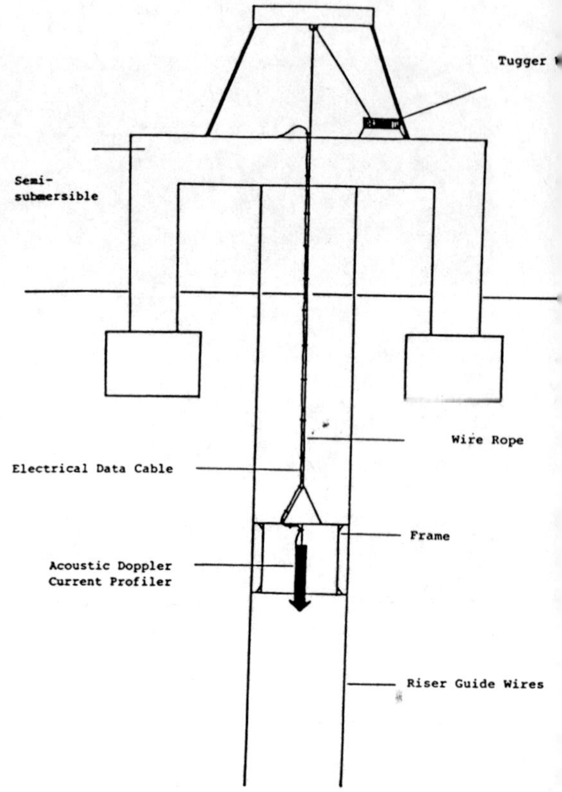

Fig. 3 Mounting configuration

ADCP was connected by an underwater cabl carrying power to the ADCP and allowin command and data exchange between th ADCP and a topside computer system.

In each case, the ADCP was mounted on frame with its axis vertical in a downward looking direction (see Fig. 3). In the firs deployment, the frame used was a secondary camera frame. This was a satisfactory expedient, although no severe conditions were experienced over the 30-day deployment period, and the ADCP would have had to be recovered if the primary camera had needed to be replaced. The axes of the acoustic beams were carefully orientated so that no beam intersected the riser. In fact, one beam intersected one of the riser guidewires but no effect was seen in the data.

For the second deployment, which lasted for six months, two ADCPs were used to cover the full depth range. Two special taut guidewires were deployed away from the riser with a bottom ballast weight. The 150 kHz ADCP was mounted on a frame which was lowered to a depth of 250 m, while a 300 kHz ADCP was mounted on a second frame on the same guidewires at a depth of 20 m. A Suber SLS 21 vector-averaging current meter (VACM) was also mounted on this frame, with its reading interrogated every 20 minutes. This provided a check on the velocity values from the ADCPs. Although a variety of conditions were experienced during the deployment, no mechanical problems were encountered. The heading of the semisubmersible was fixed throughout the deployment, otherwise alteration of the heading would have required partial or full recovery of the system. The guidewire system took just over three hours to deploy, excluding preparation time.

Upon recovery, extensive marine fouling was observed on the upper frame. The ADCP had remained operational, but the rotor and vane of the Suber VACM had ceased to rotate. Some pitting corrosion had occurred in the aluminium bodies of the ADCPs.

The third deployment was over a one month period. As in the first deployment, a frame was used on the riser guidewires. However, problems were experienced with mechanical fouling of the frame assembly against some hydraulic control lines. Also, during occasions of severe weather the electrical cable was sheared by wear against the riser and mechanical damage was sustained by the ADCP and frame.

These experiences indicate the general suitability of the guidewire method of deployment. However, care in planning and design of the arrangement is necessary to ensure trouble-free operation, and this requires co-operation between the oceanographic contractor, the drilling contractor, and the operator. One problem is that often the operator may have to charter an alternative rig at short notice, and different operating practices of the new drilling contractor may require a revised arrangement from that originally planned.

SAMPLING INTERVALS AND ACCURACY

The radial current velocity component vr along a single acoustic beam is determined from the measured Doppler shift ΔF in frequency

$$vr = \frac{C\Delta F}{2F}$$

where C is the sound velocity at the transducer (nominal value 1500 m/s and F is the transmitted acoustic frequency.

The horizontal and vertical components of velocity are obtained by solving three equations involving the radial velocity components along any three beams.

The standard deviation Δv of the random horizontal velocity error is given by

$$\Delta v = \frac{C \tan \theta}{FL(TNP)^{1/2}}$$

where θ is the acoustic beam angle relative to the velocity vector (60 degrees), L is the depth cell length (m), T is the averaging period (s), N is the number of beams (four in this case), and P is the ping rate (Hz).

The maximum ping rate is determined by the maximum range of penetration, since it is undesirable for acoustic energy from the previous ping to be received during an opening of the range gate for the current ping, although a time-variable threshold allows discrimination against reverberation. A ping rate of 1 Hz allows a round trip of 750 m, which is adequate for any of the applications considered here, and this ping rate was selected in each case.

Table 1 shows the standard deviation of the velocity error for the various configurations described here. It can be seen that for a single ping the error would be unacceptably high (of the order of 10 cm/s), but that by averaging over 20 or 30 minutes this random error is reduced to an insignificant amount (less than 1 cm/s).

However, this analysis assumes that the transducer is stationary. In practise, the methods of mounting used here allowed some movement of the transducer. The guidewires and lowering wire were not heave-compensated, so that the mounting frame was

TABLE 1
Typical set-up configuration

	No. 1	No. 2		No. 3
Transducer frequency (kHz)	150	150	300	150
Deployment depth (m)	25	250	20	27
Averaging time (min)	30	20	20	20
Ping repetition interval (s)	1.0	2.0	2.0	2.0
Cell length (m)	8	4	4	4
Number of cells	54	104	64	124
Maximum profile range (m)	432	416	256	496
Total depth penetration (m)	460	669	279	526
Depth at site (m)	548	650	650	530
Short-term error of single ping (cm/s)[b]	11	22	8	22
Error after averaging (cm/s)[c]	0.26	0.89	0.44	0.89

a The first depth cell was at 3 m range (minimum allowed).
b The standard deviation of the short-term error is calculated from

$$\Delta v = \frac{15000 \tan\theta}{FL(TNP)^{1/2}} \text{ cm/s}$$

F acoustic frequency (kHz)
L depth cell length (m)
T averaging time (s)
N number of beams
θ beam angle relative to velocity vector (60 degrees)
P acoustic ping rate (Hz)
c Not allowing for transducer motion

subject to the same heave motion as the semisubmersible drilling vessel.

The vertical heave motion will cause a maximum measurement error in the radial component of current velocity per ping, given approximately by

$$\Delta vh = \tfrac{1}{2}H\omega \sin\theta$$

$$= \frac{\tfrac{1}{2}H2\pi \sin\theta}{Th}$$

where H is the peak-to-peak heave (m), and Th is the heave period (s).

The peak-to-peak heave motion is typically less than 2 m, with a period of less than 10 s, but can reach 5 m with a period around 16 s in limiting conditions. This results in a per pi velocity error of 0.85 m/s. However, since th motion is largely periodic, with a period muc shorter than the averaging interval, this erro will be filtered out to a worst-case error of

$$\frac{\Delta vh}{\omega T}$$

$$= \frac{H \sin\theta}{T}$$

$$= 0.4 \text{ cm/s}$$

which is sufficiently small.

Yaw, pitch, and roll of the transducer woul have more serious effects on the velocity erro since returning pings would be interpreted i terms of the nominal geometry of the configura tion. In all the applications described here individual pings were not corrected for yaw pitch, or roll. However, the time-average velocities were corrected for bias in yaw, pitch and roll using the output from the compass an tilt sensors within the instrument. The mount ing configuration provides a considerable de gree of decoupling from the vessel yaw, pitch and roll motion, although some periodic til motion could occur in the mounting frame due to variations in tension (and therefore in caten ary) of the taut-wire guidelines. Also, the watch circle of the vessel in maintaining position over the well would allow slight variation in catenary. A watch circle diameter of 100 m in a water depth of 500 m would allow a negligible (0.003 degree) change in tilt, but dynamic drag of the guidewires may degrade this. For a one degree increase in tilt, the ADCP will in fact be sampling the next nearest range cell of 4 m length at a range of 400 m, which is near the maximum range of the 150 kHz transducer.

In practice, scatter in velocities was seen to increase at the extreme depth ranges in severe weather, although velocities up to mid ranges remained consistent and useful for all practical purposes.

This situation could be improved if individual pings were corrected for yaw, pitch, and roll, which could be implemented in the ADCP intenval software. Alternatively, the selection of longer-range cells (up to 32 m) should improve the situation.

Concurrent measurements were made with two types of direct reading current meter and with the Suber VACM. Only the VACM results allowed reasonable inter-comparisons, although the VACM was on the same frame as the upper 300 kHz ADCP, so that it sampled at a level about 5 m above the first depth cell of this transducer. The data indicated that confidence could be placed in the ADCP results within an absolute accuracy of ±5 cm/s in speed, and ± 5 degrees in direction, subject to the degradation at longer ranges in rough weather.

DATA ANALYSIS AND PRESENTATION

The data, after processing either by the ADCP internal microprocessor or by the computer system, comprises a set of average velocities for the measuring period for each depth 'cell'.

For each depth level, and for each of the four beams, the data can comprise averages of the radial velocity component, the spectral width, the echo amplitude, the percentage of good 'pings' (where the amplitude of the received signal exceeded the gate threshold), and status information. The four beams allow solution of two estimates of the vertical velocity, which should be equal; so disagreement outside certain limits together with the percentage of good pings can be used to flag the data quality. In addition, the date and time, two components of compass reading, two axes of tilt, water temperature and system information are recorded.

The horizontal components are resolved into

RESOLVED VESSEL CURRENTS AT 10:40:39 ON 23/08/85

		SPEED (CM/S)	DIRECTION (DEG TRUE)
DEPTH	30	!	!
DEPTH	70	29.51	14.79
DEPTH	110	28.76	16.32
DEPTH	150	28.42	15.31
DEPTH	190	27.25	11.63
DEPTH	230	26.15	12.07
DEPTH	270	24.33	13.98
DEPTH	310	24.85	13.23
DEPTH	350	21.11	18.66
DEPTH	390	18.47	17.01
DEPTH	430	16.89	21.21
DEPTH	470	11.64	8.501
DEPTH	510	7.94	7.837

MAXIMUM CURRENT
SPEED 29.61 (cm/s)
DIRECTION 16.18 (degrees)
DEPTH 62 (metres)

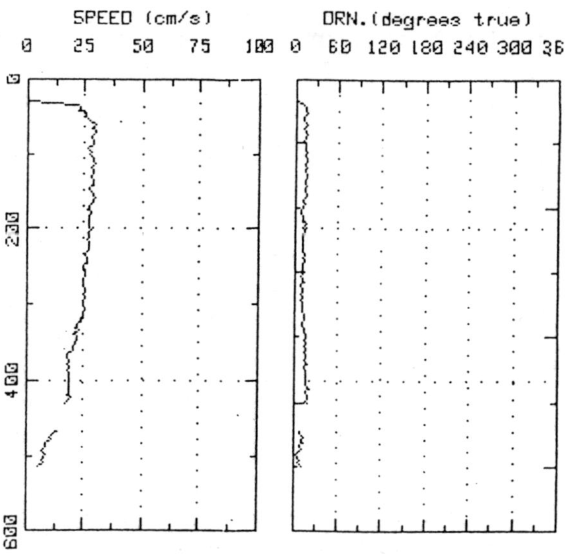

RESOLVED VESSEL CURRENTS AT 00:30:48 ON

		SPEED (cm/s)	DIRECTION (deg rel to ships head)
DEPTH	24	13.17	157.4
DEPTH		22.83	113.3
DEPTH		39.42	121.9
DEPTH		34.8	117.7
DEPTH		34.7	117.2
DEPTH		9999	9999
DEPTH		9999	9999
DEPTH		27.59	91.86
DEPTH		19.36	107.7
DEPTH		14.86	72.49
DEPTH		13.47	109.8
DEPTH		9.272	97.28
DEPTH		24.76	84.94
DEPTH		25.94	108.6
DEPTH		20.41	137.4
DEPTH		20.44	130.8
DEPTH		5.991	137.6
DEPTH		9.769	58.11

MAXIMUM CURRENT
SPEED
DIRECTION -89.83 (deg rel to ships heading)
DEPTH 232 (metres)

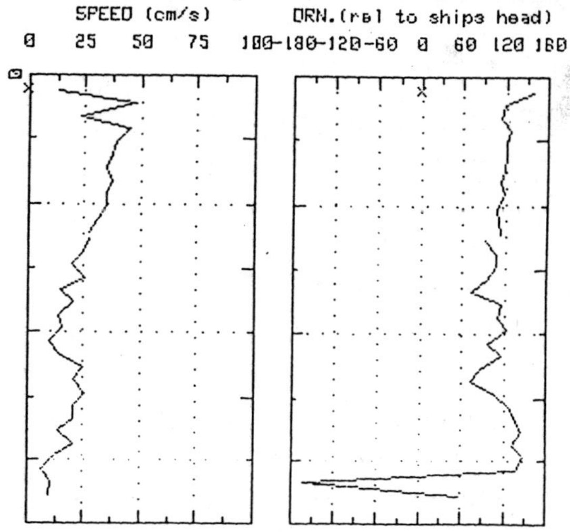

Fig. 4 Examples of operational data presentation

North and East components (or relative to rig heading) using the ADCP compass reading corrected for magnetic deviation. Since there are generally too many depth cells (up to 128) to tabulate usefully, the velocities at selected depths (for example, every 40 m) can be output to a printer. In addition, the horizontal velocity for each depth cell, usually expressed as a speed and direction, can be plotted as a depth profile on the screen and on the printer (see Figs 4a and 4b for examples). This form of presentation can be easily assimilated by operations personnel. In addition, if it required to assess trends, a time series plot of speed and direction for a selected depth can be plotted on the screen and printer (see Fig. 5).

The full data set, including quality flags, is recorded for post-analysis.

The amount of data is considerably more than that obtained by a conventional array of discrete current meters. For example, at the coarsest depth resolution of 32 m, 14 cells are needed to obtain a maximum range of 448 m, which is near the upper range limit for a 150 kHz transducer.

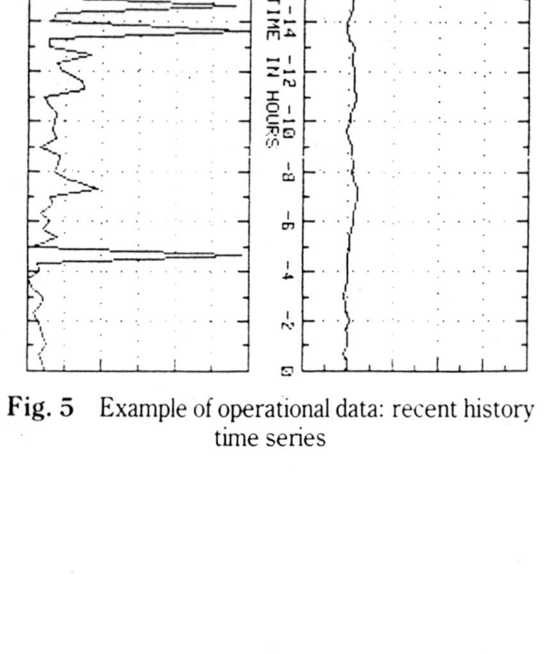

Fig. 5 Example of operational data: recent history time series

For 104 cells, each with 12 data values pl status flags, there are 1248 values (express as 4368 ASCII characters) per sampling inte val, plus compass, tilt, date, and time (48 ASC characters). For a sampling interval of minutes, a 30-day record will contain 9.3 Mby of data.

In order to assess data quality, it is conv nient to plot data as time series of speed a direction (or orthogonal components) in a stac ed form for selected depths. Again, it impractical to produce such a plot for eve depth cell, but a selection covering the f depth range can be shown (see Fig. 6 Therefore, automated methods of checking tl data quality are necessary, so that only da which fails such automated checking need l inspected and edited if necessary.

Persistence of Uniform Currents

Methods of synthesis of the data will deper upon the application. However, for riser analy sis work, where the degree of uniformity of th current velocity profile with depth is of interes a method of sorting the data within selecte tolerance bands has been devised.

The uniformity of currents was considere over a total of 36 vertical sections of minimu length 100 m, through the water column to depth of 460 m. The criteria used to judge th uniformity of currents in a particular section (the water column was as follows:

- For each 30-minute time interval the data a each of the depths lying within the selecte vertical section were scanned for the highes current speed and corresponding direction.
- The current speeds at the other depths i the section were then resolved along th directional axis of this maximum speed.
- The mean of these resolved speeds was the computed. The currents were then consi dered to be uniform over the vertical sectio if each of these speeds fell within a specifie tolerance (±5%, ±10% and ±20%) of th mean.

Having carried out the above tests for uniformi ty, the data were presented in the form of a percentage exceedance of persistence an(

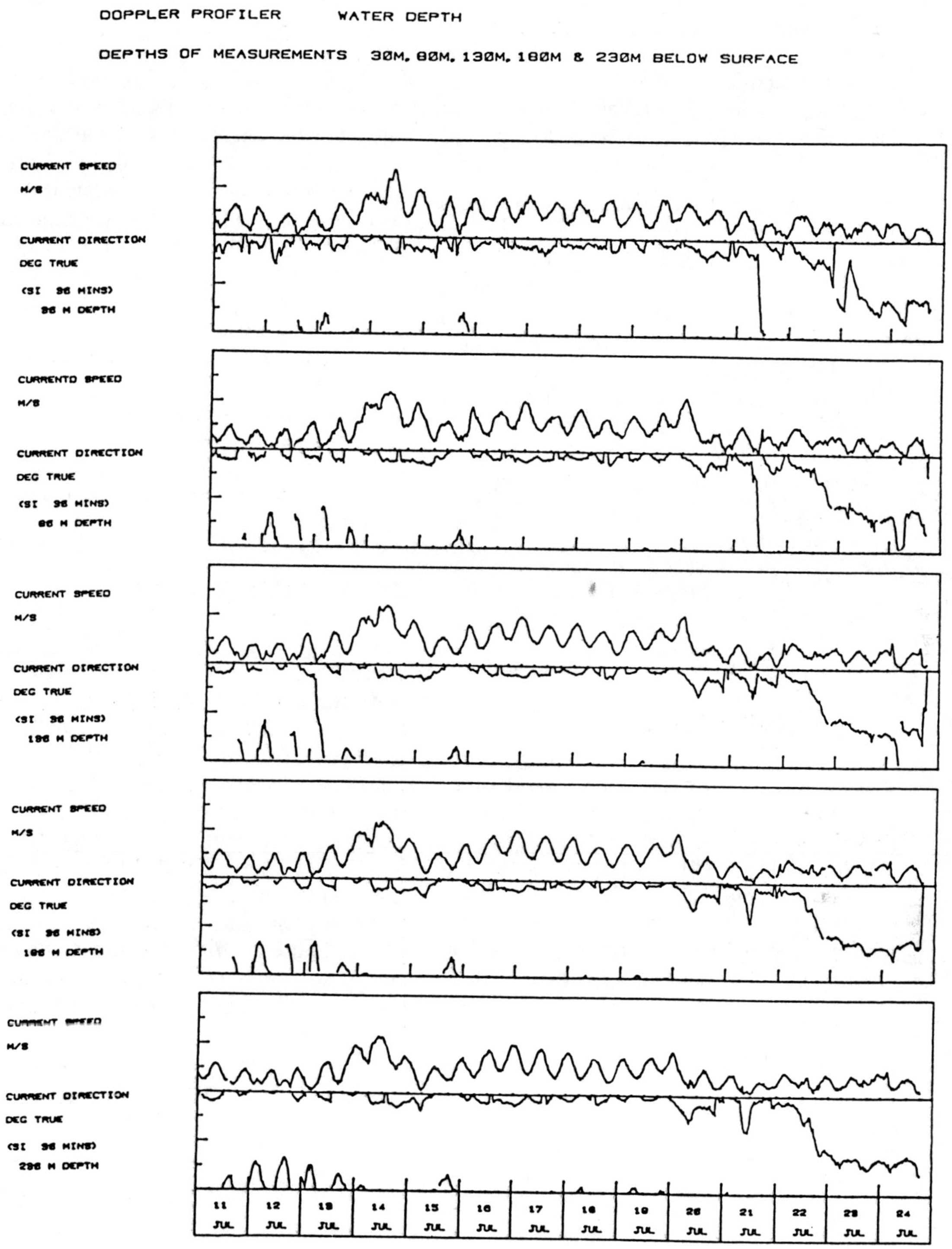

Fig. 6 Example of post-analysis presentation: stacked time series

duration plot for various threshold speeds (0.1–0.7 m/s at intervals of 0.1 m/s). In all, a total number of 108 runs of the computer were made (i.e. three tolerance levels for each of 36 vertical sections).

For each selected vertical section and for each tolerance level the following were presented on a single page (see Fig. 7):

- A plot depicting the selection in relation to the total depth.
- A plot of the instantaneous profile through the entire water column which corresponds to the time at which the highest uniform currents over the selected vertical section actually occurred.
- For the particular section, a plot of percentage exceedance of persistence. From this may be deduced the percentage of time and minimum duration that uniform currents over the specified section of the water column persisted above a given speed.
- A histogram showing the percentage of time, irrespective of duration, for which uniform currents occurred over the vertical section for various speed classes.

CONCLUDING REMARKS

It has been shown that reliable and accurate depth profiles of the current can be obtained by ADCPs in shelf and slope depths during exploration drilling, with considerable benefit to the operator. Automated displays and recording require little intervention and provide immediate data for planning operations, diagnosing problem currents, and calculating riser dynamics.

The 150 kHz transducers have demonstrated ranges up to 450 m, and 75 kHz transducers show promise of ranges to 700 m, although the transducer size will be less convenient.

The data obtained by this method are also of considerable value to the scientific community. Although the data are site specific and often measured over a comparatively short period, the excellent depth resolution and the measurement also of mean vertical currents (indicative of upwelling), taken together with concurrent or historical measurements, should allow greater insight into the processes occurring along the West Shetland and Hebrides Shelf and Slope. The co-ordination and co-operation that would be afforded by the prospective North West Approaches Group, who are planning to drill exploration wells in the UK 9th Round lease blocks in this region, should allow more effective collection and use of such data for both scientific and engineering purposes.

ACKNOWLEDGEMENTS

This work was performed by Offshore Environmental Systems Ltd, and credit goes to all the staff who worked hard to solve software, hardware interfacing, logistic, and deployment problems to ensure a successful outcome. R. Lailey and A. Smith developed the method for treating persistence of depth uniformity of currents, after an initial suggestion by J. Mitchell of Gulf Oil UK Ltd (a subsidiary of Chevron Petroleum (UK) Ltd). Thanks are due to the British Gas Corporation, Gulf Oil UK Ltd, and Elf UK Ltd for the opportunity to carry out the work and for permission to describe the operations and show some of the results.

REFERENCES

Crocker, T. R. 1983. Near-surface Doppler Sonar measurements in the Indian Ocean. *Deep-Sea Research,* **30**, 449–467.

Gordon, R. L. 1982. Measurement of low-frequency current from a fixed platform. *In Proceedings of the Second Working Conference on Current Measurements,* Hilton Head, South Carolina.

Gordon, R. L. and Skorstad, H. E. 1985. North Sea current data will upgrade platform designs. *Ocean Industry,* **20** (7), July 1985, 35–39.

Gould, W. J. 1984. The current regime on the continental slope north and west of the United Kingdom. *Current Measurements Offshore, SUT Seminar, 17 May.* (See also *MIAS News Bulletin* No. 7, 12–13, 1984, and *MIAS News Bulletin* No. 5, 2–5, 1982.)

Heaps, N. S. 1985. Storm-current events on the Scottish Continental Shelf *MIAS News Bulletin* No. 8, 8–9, 1985.

Joyce, T.M., Bitterman, D.S. and Prada, K.E. 1982. Shipboard acoustic profiling of upper ocean currents, *Deep-Sea Research,* **29** (7A), 903–913.

Pettigrew, N.R. and Irish, J.D. 1983. An evaluation of the bottom-mounted Doppler Acoustic Profiling Current meter. In *Proceedings of Oceans 83,* San Francisco.

UTILIZATION OF 'SERVICE ARGOS' FOR THE REMOTE MONITORING OF OCEANOGRAPHIC DATA ACQUISITION SYSTEMS

I Bellamy, J Rigler, A Greig, Hunting Surveys Ltd, UK

Presented on **Tuesday 4 March 1986 at 17.00** in the Norfolk Conference Room.

EMBARGOED AGAINST DELIVERY

9

Utilization of "Service ARGOS" for the Remote Monitoring of Oceanographic Data Acquisition Systems

Ian Bellamy and John Rigler, Senior Oceanographers
and Alan Greig, Chief Oceanographer,
Hunting Surveys Limited

Self-recording oceanographic and meteorological installations in remote areas cannot generally be checked for correct operation, except during routine service visits. In an effort to increase both the time between service visits and reduce the risk of undetected equipment failure or malfunction, a system utilizing the ARGOS data transmission service has recently been implemented for a recording installation in North Africa. This chapter describes the use of "Service ARGOS" in providing daily control data for wave, tide and meteorological recording systems.

The chapter describes the operational aspects of the survey and satellite systems, together with a brief analysis of the cost implications of using the ARGOS service.

INTRODUCTION

Self-recording oceanographic and meteorological installations in remote areas cannot generally be checked for correct operation, except during routine service visits. These visits may be at intervals of several weeks or months depending upon the site location, local logistics and client requirements.

This chapter presents an example of a recently commissioned system for daily monitoring of a remote installation using the ARGOS system. The installation comprised several oceanographic and meteorological sensors at two sites in North Africa. Local logistics required that the system be capable of continuous operation with a minimum of on-site maintenance. This requirement was solved by the implementation of microcomputer controlled processing, recording and data transmission via the ARGOS system.

The following presents a brief summary of the ARGOS satellite system, and a description of the recording installation deployed in the field. The chapter concludes with a discussion of the operational experiences with the ARGOS system, and the relative cost benefits provided by its use.

OPERATION OF ARGOS

ARGOS utilizes the TIROS-N NOAA meteorological satellite and is the result of a joint venture between the Central National d'Etudes Spatiales (CNES, France), the National Aeronautics and Space Administration (NASA, USA) and the National Oceanic and Atmospheric Administration NOAA, USA). The ARGOS system comprises:

- fixed or mobile platforms (i.e. boats, buoys, balloons etc.) fitted with Platform Terminal Transmitters (PTTs);
- two TIROS-N satellites;
- a ground telemetry acquisition and processing network;
- a data distribution system.

Satellite Orbit

The TIROS-N satellites have the following orbit characteristics:

- circular orbit configuration;
- satellite altitude: 830 km (± 18 km) and 870 km (± 18 km);
- orbital period: approximately 101 minutes;
- number of orbits per day: approximately 14 for each satellite.

Each satellite crosses the equatorial plane at a fixed time each day. Consequently any PTT comes within a satellite's visibility at the same local solar time each day. Table 1 shows the frequency of passes for various latitudes. The figures apply for both satellites. The mean duration for a pass is about 10 minutes, while that of a zenith pass is 13 minutes.

TABLE 1

Latitude (degrees)	*Cumulative Visibility Time over 24 Hours (min)*	*Number of Passes in 24 Hours*		
		Min.	*Mean*	*Max.*
±0	80	6	7	8
±15	88	8	8	9
±30	100	8	9	12
±55	170	16	16	18
±65	246	21	22	23
±75	322	28	28	28
±90	384	28	28	28

PTT Operation

The ARGOS system can accommodate PTT transmissions to provide geographic location and/or sensor data.

The PTT unit operates on a nominal frequency of 410.65 MHz. Each message transmitted by the PTT conforms to the format illustrated in Table 2. The maximum amount of usable data which can be transmitted is 256 bits comprising 8 × 32 bit words. Each 32-bit word normally consists of 4 bytes of data.

The whole PTT message is transmitted in less than one second and includes 160 ms of unmodulated carrier, to allow the ARGOS onboard receiver to lock on to the carrier.

Argos Data Collection

The ARGOS Data Collection and Location System (DCLS) receives the message transmitted by PTTs within view of the satellite. Since the system operates on a random access basis, data acquisition is limited by the availability of a vacant channel. Only four channels are available for data reception and consequently there is a possibility of the satellite "missing" data transmission from a visible PTT.

In order to maximize the probability of accurate message reception, the message should be transmitted several times. ARGOS (1984) state that the probability of message reception is 0.9920 for a message repeated three times, and 0.9999 for a message repeated six times.

Bit error probability is fairly low at 10^{-4} (1 in 10 000), and is further reduced by message repetition.

The DCLS system sorts the received data on a time basis, and checks the bit pattern for each message. Only the most recent of equivalent messages are then retained on the ARGOS central processing centre at Toulouse, France. Sensor and/or location processing results for each PTT are transferred to the so-called "DISPOSE" file. Files are stored daily on tape, and are added to the ARGOS data block, which contains three months of data. The "DISPOSE" file data then undergo data compaction and updating to produce the "TELEX" and "AJOUR" files. Users may access all of the three file types, and these contain the following information:

TABLE 2
PTT message structure and format

Unmodulated Carrier Length (ms)	*Modulated Carrier*					
	Preamble	*Format Sync.*	*Initialization*	*Number of 32-Bit Groups*	*ID No. (+ Check Bits)*	*Sensor Data*
160 ±2.5	15 bits (=1)	8 bits (00010111)	1 bit (=1)	4 bits	20 bits	$N \times 32$ bits (1<N<8)

- "AJOUR" file: contains the most recent location and sensor message for each PTT.
- "TELEX" file: contains the most recent location and message for each PTT.
- "DISPOSE" file: contains in chronological order all location data and all sensor messages for each satellite pass and each PTT.

A schematic illustration of the data flow through the DCLS system is shown in Fig. 1.

ARGOS

Data Dissemination

Messages received at the ARGOS centre in Toulouse are sorted according to the user identification number, and coded with the DCLS reception time. Data processing, according to the user specified formats, may be carried out at this stage. Standard data processing includes

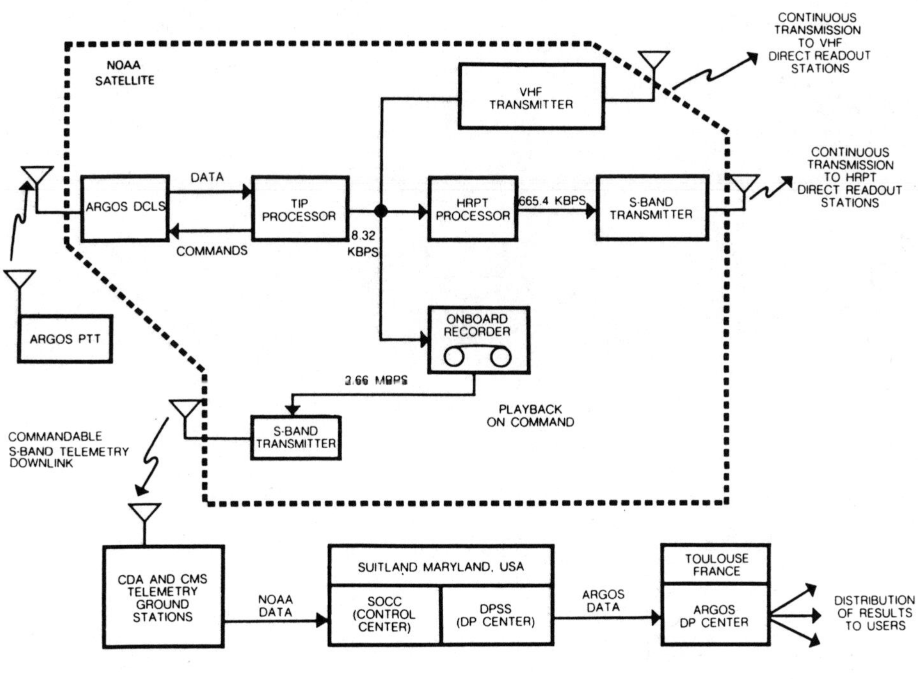

Fig. 1 ARGOS data flow diagram (schematic)

binary to decimal, hexadecimal or octal, and BCD to decimal.

The rate of data flow through the ARGOS system is summarized in Table 3. ARGOS quotes an average time delay of 4–6 hours between data receipt at the TIROS-N satellite and availability on the distribution computer.

TABLE 3
Mean percentages for the number of messages received with given throughput times

Time (hr)	*Percentage*
≤1	3
≤2	8
≤3	28
≤4	52
≤6	74
≤8	86
>10	8

Data retrieval by the user from the ARGOS centre may be accomplished via the French TRANSPAC data network system. Four methods are in use:

- telex at 50 baud;
- telephone/modem at 300 baud;
- telephone/modem at 1200 baud;
- dedicated data line up to 4800 baud.

Our method of access employed during the course of this work is a "dial-up" line at 300 baud, and a full duplex modem interfaced to a microcomputer.

SURVEY DATA ACQUISITION SYSTEM

Two independent data acquisition systems were developed to meet the clients' requirements. One system, installed at a coastal location, was required for the measurement of the following parameters:

- wave height and period;
- tidal elevation;
- wind speed and direction;
- atmospheric pressure;
- air temperature;
- humidity and rainfall.

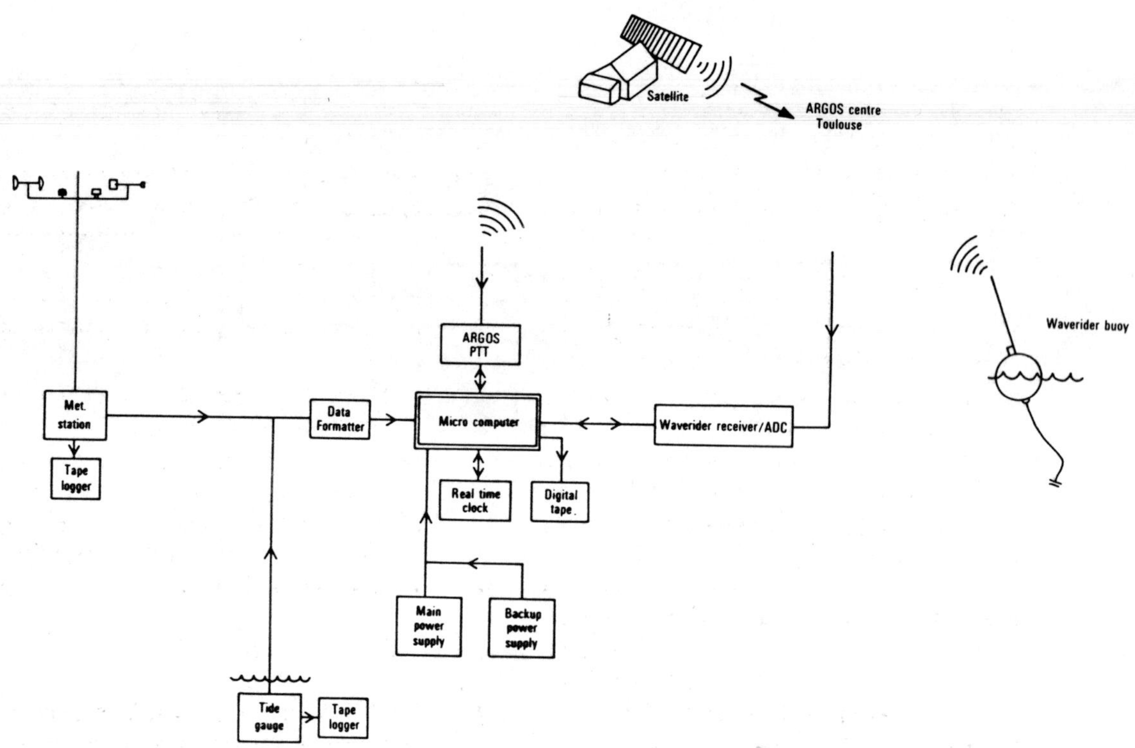

Fig. 2 Flow diagram for data acquisition system (coastal site)

The second system was installed at an inland site and included only the meteorological sensors. As both systems have similar characteristics, only the coastal equipment suite will be described in any detail.

A basic requirement of the survey program was to ensure continuity of data recording and consequently a high data return had to be ensured. For various logistical reasons, service visits were required at extended intervals (up to 3 months), and so the risk of undetected instrument loss or failure could not be permitted. In order to reduce some of the risk element in the acquisition program, the ARGOS data transmission facility was incorporated in the recording system.

Wave Data Collection

A standard Datawell Series 6000 Waverider buoy was deployed about 2 km offshore, and wave height continuously transmitted to the shore base. The shore receiver comprises a digital waverider receiver which demodulates the FM buoy transmission. The unit then undertakes an analog-to-digital conversion (ADC) at 2 Hz on the received data (when required by the system software described below). Control of the receiver and initialization of the ADC process is carried by a Hewlett Packard HP85 microcomputer over a dedicated RS232 link; 2048 wave data samples are collected (representing 17 minutes and 4 seconds of data) every hour. This wave height data is then checked for spurious errors caused by transmission or equipment problems, as detailed in Table 4.

The spectral moments are then computed and used to calculate significant wave height, zero crossing period, peak period, crest period and the spectral width parameter (epsilon). Results of the wave analysis including the wave data error flags (1–10), the spectral density estimates (0.04–0.6 Hz), the spectral moments, and the wave parameters are stored on the HP85 internal tape recorder. In order to reduce data storage requirements, data compression techniques are employed prior to writing the spectral density estimates to tape.

TABLE 4
Wave data checks

Error Flag No.	*Description*
1	Number of 10 consecutive points of equal value
2	Number of zero crossings >25 seconds
3	Number of linear interpolations using rate of change criteria
4	Number of consecutive errors of error 3
5	Number of values greater than 4 times the standard deviation
6	Number of consecutive errors of error 5
7	$E_l/E_t > 5\%$
8	$E_h/E_t < 2\%$
9	Number of samples greater than 10 metres
10	Indicates receiver in "unlocked" condition

E_l = spectral density component < 0.04 Hz.
E_t = total spectral density (0–1 Hz).
E_h = spectral density component > 0.6 Hz.

The wave height data is processed spectrally using a Hewlett Packard Fast Fourier Transform (FFT) package to produce spectral density estimates for the wave record in the bandwidth 0.04–0.6 Hz.

Meteorological Data Collection

Aanderraa weather stations equipped with wind speed, wind direction, atmospheric pressure, air temperature, humidity and rainfall sensors were erected at both recording sites. All data is logged on the standard Aanderraa digital tape recorder. Sampling of all sensors is at hourly intervals.

The output parameters, in the form of 10-bit words, are reformatted as single bytes using a WSOS ARS-1 formatter, and transferred to the computer serial interface port. Meteorological data are converted to engineering units by the computer, stored on the internal tape recorder and then copied to the printer.

Tidal Data Collection

An Aanderraa WLR5 tide recorder is used to measure hydrostatic pressure and sea temperature. Data is stored internally on a data logger, and data for each recording cycle are output via a data formatter to the HP85. The HP85 then corrects hydrostatic pressure for the current measured atmospheric pressure, and computes total tidal height. Both tidal height and sea temperature are stored on the computer's internal tape recorder, and copied to the printer.

COMPUTER SOFTWARE AND TIMING CONTROL

The HP85 microcomputer acts as a controller for wave, tide and meteorological data acquisition and processing. Sampling cycles of all sensors is one hour, which requires a high degree of timing accuracy to avoid loss of data due to clock errors. A stand-alone real-time clock, with an independent failsafe power supply, was incorporated in the system to provide a standard time-base. The real-time clock provides all timing control for the software, and current system time/date are available on demand for coding each record written to tape or transmitted. Consequently, each record is unique and possible timing errors caused by power failure are avoided.

All the recording systems are required to provide data on the hour. However, a small tolerance of several minutes either side of the hour is allowed to accommodate any clock drift on the Aanderraa systems during the course of the survey. Once tide and meteorological data have been acquired, or if no data are available after a preset wait period, then the waverider receiver is prompted to send digitized wave data. This occupies the computer for approximately 20 minutes. On completion of the sampling period, the data is checked and processed as described previously. Meteorological, tide and wave parameters for the current recording cycle are then transferred to an output buffer for transmission to the PTT. Each record is uniquely identified by the record time and date.

A total of 32 bytes of integer data are sent to the PTT for transmission. A full list of the 32 bytes of data transmitted from the coastal location is shown in Table 5. The same data transmission is repeated at time intervals specified for the user by ARGOS. Each hourly record transmission is repeated 24 times to maximize data reception by the satellite.

The whole hourly data acquisition, processing, storage and transmission cycle occupies approximately 50 minutes out of every hour.

TABLE 5
List of data available via ARGOS transmissions (coastal location)

Array Element	*Description*
D(1)	Record time: hours
D(2)	Record time: minutes
D(3)	Record time: day
D(4)	Record time: month
D(5)	Wave record error flag (1)
D(6)	Wave record error flag (2)
D(7)	Wave record error flag (3)
D(8)	Wave record error flag (4)
D(9)	Wave record error flag (5)
D(10)	Wave record error flag (6)
D(11)	Wave record error flag (7)
D(12)	Wave record error flag (8)
D(13)	Wave record error flag (9)
D(14)	Wave record error flag (10)
D(15)	Significant wave height
D(16)	Zero crossing period
D(17)	Peak wave period (from spectral density analysis)
D(18)	Met. station identification no. (DIV 256)
D(19)	Met. station identification no. (MOD 256)
D(20)	Wind speed
D(21)	Wind gust
D(22)	Wind direction (DIV 256)
D(23)	Wind direction (MOD 256)
D(24)	Air temperature
D(25)	Humidity
D(26)	Air pressure (actual pressure − 920)
D(27)	Rainfall
D(28)	Tide recorder identification no.
D(29)	Tide recorder identification no.
D(30)	Tidal height
D(31)	Sea temperature
D(32)	End of file marker

PROGRAM SUPPLY AND PROGRAM E-START

The HP85 computer requires a 240 V supply, which is currently supplied by a diesel generator. However, the supply is prone to interruption and is supplemented by a stand-by battery supply and static inverter. This provides sufficient reserve power for up to 6 hours' operation.

In the event of power interruption, the HP85 software will be lost from the memory and the recording cycle ended. Upon restoration of power, the software is loaded automatically from the computer's internal data cassette. As mentioned previously, the real-time clock is provided with an independent power supply and can run on internal batteries for up to 500 hours. The software, following autostart, requests the current time and data from the clock, and prepares for a new recording cycle. As each record is uniquely identified, the power break is clearly identified and no confusion can result during processing.

Local Logisitic Support

The system has been designed for minimum intervention by local personnel. Local staff duties are restricted to maintenance of the power supply and replacing the HP85 internal cartridge at 14-day intervals.

DATA RECEPTION AND OPERATIONAL PERFORMANCE

The recording installations described above were operational from May 1985. Interfacing problems delayed the full use of the ARGOS system until July 1985. Since then, daily checks on sensor data have been made at the ARGOS centre, via the UK data network (Packet Switch Stream).

A comparison of received messages with those transmitted (and recorded on tape in the field) shows very good agreement. Occasional problems with bit transmission have been experienced, although the frequency of the events is similar to ARGOS' stated bit error rate of 1 in 10 000. However, these problems have not resulted in any loss of information, as the same message is repeated several times.

Although data is transmitted on an hourly basis by the recording systems in North Africa, satellite passes are restricted to an average of eight per day. During August and September 1985, the actual number of observed readings corresponded to the quoted average. During early October (the time of writing) only one satellite is operational, but an average of seven passes per day are being obtained.

ARGOS IMPLEMENTATION COSTS

The implementation of the ARGOS system as a method of data quality control has required some financial investment in addition to the standard survey spread. It is instructive to summarize the hardware and operating costs for this additional element of the recording system.

As a guide, the following costs (at October 1985 prices) are indicative of the scale of investment required.

Hardware Costs

Typical purchase price of an ARGOS transmitter fitted with an RS232 converter	£1800
Typical cost of supplying d.c. power to transmitter	£200
On-site dedicated computer for data processing and formatting data	£2000
Typical purchase price of a full duplex 300/1200 baud modem	£500
Current UK installation fee for telephone	£80
Apportioned use of in-house terminal or computer	£500
Total hardware investment	£5080

Operational Costs

These include telephone charges, line rental, packet switching line rental, charges for transmission of data and ARGOS direct charges. All costs shown below are given on an annual basis:

Telephone line rental (UK)	£72
Telephone call (2 minutes per day at peak rate)	£22

Packet switching rental	£50
Packet switching on-line charge	£15
Packet switching data transmission charge	£80
ARGOS charge per PTT (data collection only)	£646
ARGOS charge for data supplied on magnetic tape (total 12)	£712
Total operating cost (for one year of operation)	£1597

Total Costs

Therefore the total cost including hardware, and one year's operating costs is approximately £6680, excluding any software development and personnel time.

OPERATIONAL SUMMARY

The system described above has been operating on a routine basis for several months (as of October 1985), and has provided a daily check on sensor data. The usefulness of this facility becomes apparent at an early stage of the survey, and its value has been acknowledged on several occasions. A number of recording/hardware problems has been identified by use of the daily updates.

Operational experience of this system to date has been quite satisfactory. The advantage of having a daily check on sensor data acquisition is apparent and outweighs the additional equipment costs. The relative merits of using the ARGOS system must be judged on a job-specific basis but, in a situation where long-term unattended operation is required, it provides a welcome indication of system operation (or failure, as the case may be).

The system described above does not use the ARGOS system as the primary source of data recording, as all field systems have independent recording facilities. However, the use of ARGOS as a means of near real-time transmission has been considered as a viable possibility. This is discussed briefly below.

WAVE DATA PROCESSING ON BOARD A WAVERIDER BUOY

A recent implementation of the ARGOS system by Hunting Surveys has been to equip a Waverider buoy with onboard processing and data-transmission facilities. The modified Waverider comprises a dedicated data logging system which performs simple data checks, and an FFT. This provides the moments of the record spectrum and these are sent, via a formatter, to the PTT. Data transmitted, in addition to the moments, includes a record counter mark and data quality flags. Wave data is acquired every three hours, and the processed results for the record are transmitted by the PTT continuously between recording and processing cycles. This provides the maximum change of data reception by a satellite. Given that there are up to nine passes per day at the latitude of deployment, a high probability of receiving the eight daily transmissions is anticipated.

The buoy is also equipped with an internal digital tape logger to provide 100% back-up of all data transmitted.

An additional bonus of the ARGOS system is that geographic location of the PTT is calculated during each satellite pass. This provides a useful indication that the buoy is still on location and that the mooring is intact.

SUMMARY

The use of the ARGOS data transmission service has been discussed in conjunction with an operating field survey system. The value of "on-line" data on a routine daily basis is considerable for remote recording installations. Operational experience with ARGOS to date has indicated that real benefits can be expected in the monitoring of data acquisition and in extending routine servicing intervals.

REFERENCE

ARGOS 1984. *ARGOS User Guide, Centre* National D'Etudes Spatiales, Toulouse, France: Service ARGOS.

Part III

Underwater Vehicle Applications

AN ADVANCED UNDERWATER TOWED VEHICLE FOR OCEANOGRAPHIC MEASUREMENTS

M Nomoto, Y Tsuji, A Misumi, T Emura, Japan
Marine Science & Technology Centre, Japan

Presented on **Wednesday 5 March 1986 at 10.00**
in Norfolk Conference Room

EMBARGOED AGAINST DELIVERY

10

An Advanced Underwater Towed Vehicle for Oceanographic Measurements

Masao Nomoto, Yoshito Tsuji, Atsunobu Misumi and Tomio Emura, Japan Marine Science and Technology Center (JAMSTEC), Japan

INTRODUCTION

JAMSTEC has developed a new underwater towed vehicle for oceanographic measurements, the Underwater Sliding Vehicle (USV). Unlike other conventional towed vehicles, the USV is not linked to the end of the tow cable but slides along it, while the tow cable catenary is stabilized by a hydrodynamic depressor. This method allows the USV to have good manoeuvrability although the vehicle is small (1 m long) and of light weight (about 10 kg in air, with sensors). The vehicle is equipped with conductivity, salinity and depth sensors. The vehicle motion is controlled from a deck control unit. The USV scans the vertical plane in which the tow cable advances horizontally, and measures the distributions of those parameters in this plane with high spatial resolution. The measured data are digitized, multiplexed and Frequency Shift Keying (FSK) modulated, and are transmitted up to the deck unit using an inductive coupling data communication system which utilizes an electric current loop formed by a tow cable and the sea water. A theoretical analysis of the vehicle motion was also performed before the vehicle was designed, which indicated that no restriction would be imposed on the undulation amplitude of the USV. Field tests were carried out in the Sagami Bay. The USV was towed from a tugboat by a 220 m hydrodynamically faired cable at towing speeds from 4 to 10 knots, and it measured the distributions of water temperature and conductivity down to a depth of 175 m. Data are monitored on a CRT as vertical colour tomograms of the ocean space with respect to one selected parameter (conductivity, temperature or salinity), in which differences in colour correspond to differences of the parameter values.

PRINCIPLES OF THE USV

Conventional towed vehicles used for oceanographic data collection are usually linked to the end of the tow cable. The vehicle varies its depth or undulates in the ocean space to measure the distributions of oceanographic parameters such as temperature and salinity. Towed vehicles of this type must overcome the hydrodynamic forces which are accompanied by the changes of the tow cable catenary, as shown in Fig. 1. As the magnitude of the

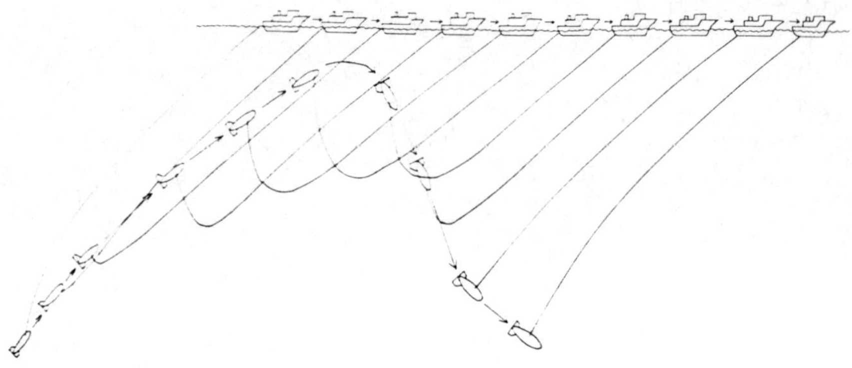

Fig. 1 Undulating motion of a conventional towed vehicle

hydrodynamic forces necessary for the undulation increases with undulation amplitude, a large vehicle is needed for a large undulation amplitude. Hydrodynamic drag of the tow cable also resists the fast undulating motion through the sea water. Figure 2 shows the configuration of the USV system. The undulating motion of the USV is schematically shown in Fig. 3. As the USV need not change the catenary of the tow cable during the undulating motions, a large undulation force is not necessary for the USV. Thus, the USV can be small and of light weight without any affect on the vehicle performance. The rolling motion of the vehicle which, in many cases, causes serious stability problems in the operation of the conventional towed vehicles, is restricted by the towing cable. Thus, no problem exists in the lateral vehicle motions. The basis configuration of the USV is that of an aeroplane with a fixed forward wing and tails with elevators, as shown in Fig. 4. The elevators assume two positions, corresponding to surfacing and diving. These two positions, which are related to the surfacing and diving speeds, are varied from the deck unit. Two depth values between which the USV undulates

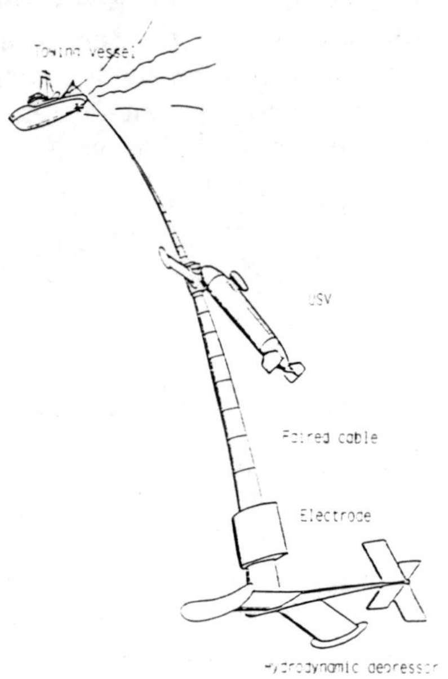

Fig. 2 The USV system

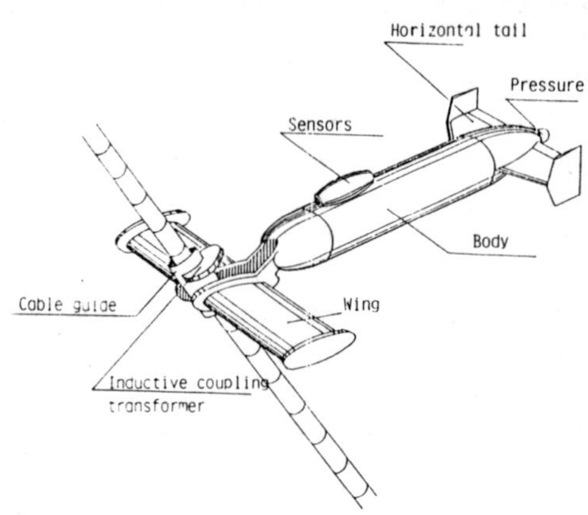

Fig. 3 Undulating motion of the USV

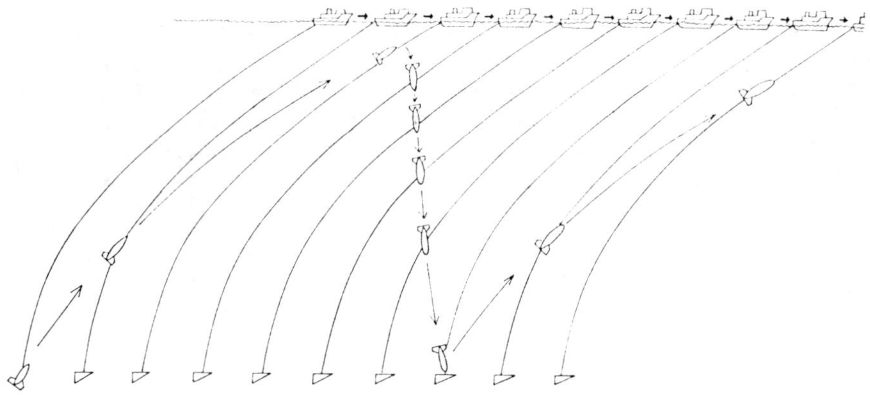

Fig. 4 Configuration of the USV

are also selected from the deck unit. The vehicle depth data from the USV are transmitted five times per second and are monitored by the CPU on the deck unit. When the USV reaches its undulation limit the CPU changes the elevator command to let the USV change its motion from diving to surfacing, or vice versa. As is described in the following section, the parameters which influence the performance of the vehicle are the inclination angle of the tow cable, the lift-to-drag ratio and the weight of the vehicle in water. The first model was designed to meet the following specifications:

towing speed	3–12 knots
operation depth	surface to 200 m
vehicle size	1 m × 0.5 m ×0.15 m (L × W × H)
weight in air	less than 10 kg
sensors	conductivity, temperature, depth

DESIGN AND DESCRIPTIONS

The Body

The body has a cylindrical shape with a streamlined nose and tail cone. Sensor electronics, a data communication circuit board, an elevator control circuit board and batteries are housed in the body. A servo-motor and elevator control mechanisms are mounted in the oil-filled tail cone where the pressure is balanced to the surrounding sea water.

The Wing

The wing is made of fibre-reinforced plastic. It has a rectangular shape, and symmetrical section (NACA 0012). A high aspect ratio is chosen to obtain a high lift-to-drag ratio, which is the most essential parameter to the vehicle performance. End plates are used to increase the effective aspect ratio and to protect the wing from damage during operations. The wing is rigidly mounted on the body through struts with the incident angle of 4° with respect to the body longitudinal axis. The dihedral angle of the wing is zero because there is no stability problem on the lateral motion, as previously mentioned.

The Cable Guide

The cable guide is pivoted on the wing spar. Four plastic rollers on the cable guide minimize the frictioned force between the vehicle and the tow vehicle. A toroidal core for the inductive coupling system which generates alternating electric current in the tow cable conductor is mounted on the cable guide.

The Sensors

Conductivity, temperature and depth sensors are mounted on the body. A thermistor probe is used as the temperature sensor. The accuracy is 0.1°C and the resolution is 0.01°C. The time constant is 0.1 s. The conductivity sensor is of inductive type, and its time constant is matched electrically to that of the temperature sensor to avoid salinity spikes in salinity data calculated from conductivity, temperature and pressure. The pressure sensor is mounted inside of the body nose. The effect of the dynamic pressure due to the motion of the vehicle is avoided by the selection of the layout of the penetration hole.

The Data Communication System

As the USV moves along the tow cable conductive coupling of the vehicle and the tow cable must be abandoned. Several kinds of data communication schemes, including optical and acoustical links, were considered. An inductive coupling system was proved to be the most advantageous for the USV system for the following reasons.

1. The tow cable can be used as a medium of signal transmissions.
2. Long-distance communication is possible with small output power so long as the vehicle is on the tow cable.
3. A high signal-to-noise ratio is easily attained because no strong electric current field exists in the sea.

The inductive coupling system is shown in Fig. 5. This system uses an electric current loop formed by the isolated tow cable conductor and the sea water. Oceanographic data collected by the sensors are converted to 12-bit digital data, and are multiplexed and FSK modulated using 9.6 and 19.2 kHz, and are transmitted from the transformer. The changes of the magnetic flux density in the toroidal core, caused by the changes of the current in the coil of the transformer, generate the alternating current in the towing cable conductor. The communication is bidirectional. The control command, composed of a 3-bit call address and a 5-bit elevator position data are also sent down to the USV. The USV receives the control commands, and if the address datum coincides with that assigned to the vehicle, then the vehicle recognizes that the command is to itself and it sends the collected data back to the vessel as an answer. Thus, eight vehicles can be operated simultaneously with single tow cable.

The Deck Unit

Oceanographic data transmitted from the vehicle are received by the deck unit. The deck

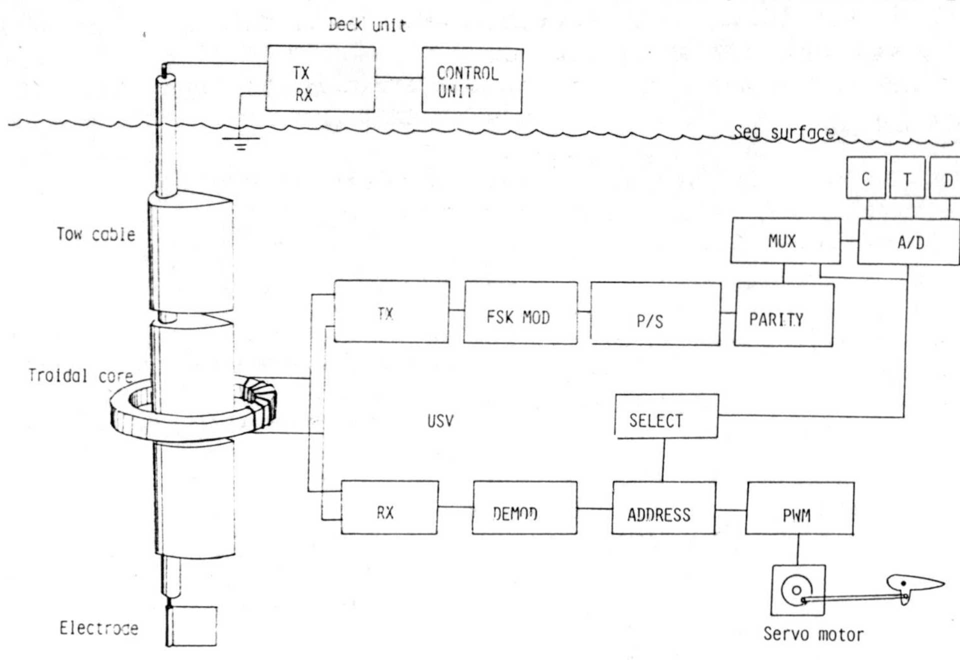

Fig. 5 The inductive coupling data-communication system

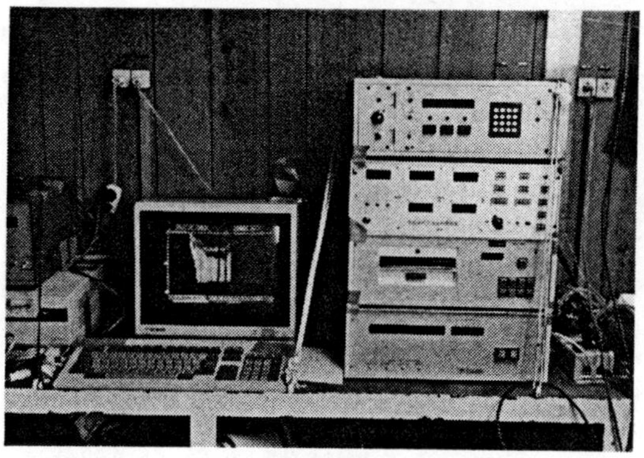

Fig. 6 Deck unit and CRT monitor

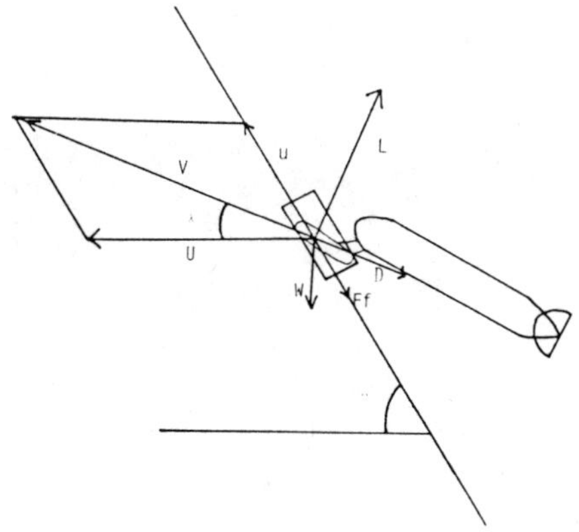

Fig. 7 Forces on a surfacing USV

unit, shown in Fig. 6, consists of a receiver unit, a control unit, a data recorder and an interface unit for data input from external instruments such as a LORAN receiver. The data from the USV are confined to the data from external instruments such as date, time and the position of the vessel, and are recorded on magnetic tape. The data collected are also displayed on a colour CRT as colour graphs. The graph shows the variation of vehicle depth with time. If the speed of the towing vessel is constant, abscissa corresponds to the horizontal distance and the graph approximates the trajectory of the USV. The graph is drawn with eight colours corresponding to the values of the selected parameter (conductivity, temperature or salinity). Thus, the graph offers a colour tomogram of the vertical plane of the ocean space with respect to one selected oceanographic parameter. The full scale of the abscissa is chosen from 4 min, 30 min, 1 h, 2 h, 4 h and 6 h: if auto-mode is selected, 4 min is initially set and when the time exceeds the full-scale value, the full scale is automatically changed to 30 min and the graph is rewritten from the beginning. The data from the USV in the previous 3 h are stored on the RAM of the display computer.

THE PERFORMANCE OF THE USV

The performance of the USV is evaluated by the undulation speed along the tow cable. Since the USV moves along the tow cable, the cable inclination strongly influences the undulation speed. Here we calculate the undulation speed of the USV along the tow cable at various cable inclination angles. Figure 7 shows the USV surfacing with velocity u along the tow cable which is moving horizontally with velocity U. The cable inclination angle θ is measured from the horizontal. The velocity V of the USV relative to the sea water is then obtained as the vector sum of u and U. From geometrical relationships the following equation is derived from the non-dimensional velocity, u^*:

$$u^* = u/U = \frac{\sin\lambda}{\sin(\theta - \lambda)}, \qquad (1)$$

where λ is the angle that V makes with horizontal. The hydrodynamic lift and drag of the vehicle act perpendicular and opposite to V. Then the components of the hydrodynamic force normal and tangential to the tow cable are

$F_n = L\cos(\theta - \lambda) + 78F_f$ caused by the sliding motion of the vehicle along the tow cable. As the frictional force is considered to be proportional to the normal component of the total force acting on the vehicle, the tangential component of the total force which pushes the vehicle upwards along the tow cable is

$$f = F_t - W\sin\theta - \nu(F_n - W\cos\theta), \qquad (4)$$

where ν is the coefficient of friction. Substituting eqns (2) and (3) into (4),

$$F = L\cos(\theta - \lambda) + D\sin(\theta - \lambda) - W\sin\theta$$
$$- \nu[L\cos(\theta - \lambda) + D\sin(\theta - \lambda) - W\sin\theta]. \qquad (5)$$

Equation (5) is divided by the lift L to obtain the non-dimensional tangential force C_t (= F/L):

$$C_t = [\sin\theta - \nu\cos\theta - (\cos\theta + \nu\sin\theta)/r]\cos\lambda - [\cos\theta - \nu\sin\theta - (\sin\theta + \nu\cos\theta)/r]\sin\lambda - f(\sin\theta - \nu\cos\theta), \tag{6}$$

where r (= L/D) is the lift-to-drag ratio and f (= W/L) is the ratio of the weight in water to the lift of the vehicle. If now we assume that the influence of the vehicle motion upon the cable catenary is negligible, the equation of motion of the vehicle along the tow cable can be written as

$$M\frac{du}{dt} = C_t L, \tag{7}$$

where M is the sum of mass and added mass of the vehicle in the direction of motion. If the exact values of u are required, (7) must be integrated. However, as the curvature of the tow cable is small, du/dt is considered small except just after the vehicle switches the direction of motion at undulation limit depths. Thus, we assume that du/dt is negligible; that is, the forces on the vehicle are in equilibrium. This condition is written from eqns (6) and (7) as

$$[\sin\theta - \nu\cos\theta - (\cos\theta + \nu\sin\theta)/r]\cos\lambda - [\cos\theta - \nu\sin\theta - (\sin\theta + \nu\cos\theta)/r]\sin\lambda = f(\sin\theta - \nu\cos\theta). \tag{8}$$

Equation (8) is solved for λ. The solution is obtained as an explicit function of θ with parameters ν, r and f:

Finally, the non-dimensional surfacing velocity along the tow cable is obtained by substituting eqn (9) into (1)

As can be seen from eqn (10), u^* does not depend on the absolute value of L but on L/D and W/L, which are independent of the vehicle size. From this fact, the important characteristics of the vehicle performance are derived, that is:

> The performance of the USV is independent of the vehicle size.

The diving velocity along the towing cable is obtained in the similar manner. The results are shown in Figs 8–13 for different ν and f. In Figs 8, 10 and 12, which represent the surfacing velocities, negative values of u^* mean that surfacing motion is not possible under those

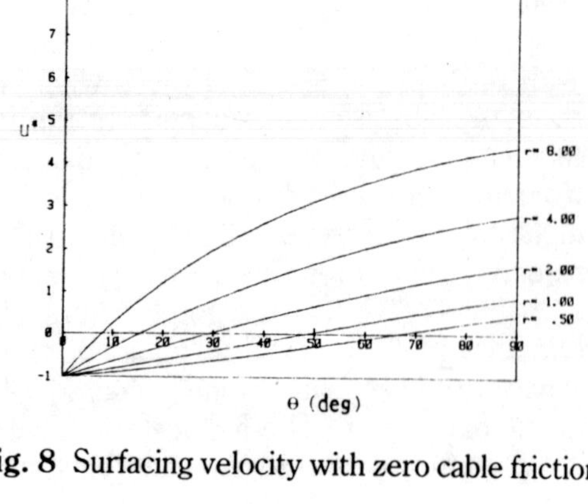

Fig. 8 Surfacing velocity with zero cable friction

$$\lambda = \tan^{-1}\left(\frac{\sin\theta - \nu\cos\theta - (\cos\theta + \nu\sin\theta)/r}{\cos\theta + \nu\sin\theta + (\sin\theta - \nu\cos\theta)/r}\right) - \sin^{-1}\left(\frac{f(\sin\theta - \nu\cos\theta)}{\sqrt{(1+\nu^2)(1+1/r^2)}}\right) \tag{9}$$

EQUATION 10

$$u^* = \frac{\sin\left[\tan^{-1}\left(\frac{(1-\nu/r)\sin\theta - (\nu + 1/r)\cos\theta}{(\nu + 1/r)\sin\theta + (1 + \nu/r\cos\theta}\right)\right] + \sin^{-1}\left(\frac{f(\sin\theta - \nu\cos\theta)}{\sqrt{(1+\nu)(1+1/r)}}\right)}{\sin\left[\theta - \tan^{-1}\left(\frac{(1-\nu/r)\sin\theta - (\nu + 1/r)\cos\theta}{(\nu + 1/r)\sin\theta + (1 + \nu/r\cos\theta}\right)\right] + \sin^{-1}\left(\frac{f(\sin\theta - \nu\cos\theta)}{\sqrt{(1+\nu^2)(1+1/r^2)}}\right)} \tag{10}$$

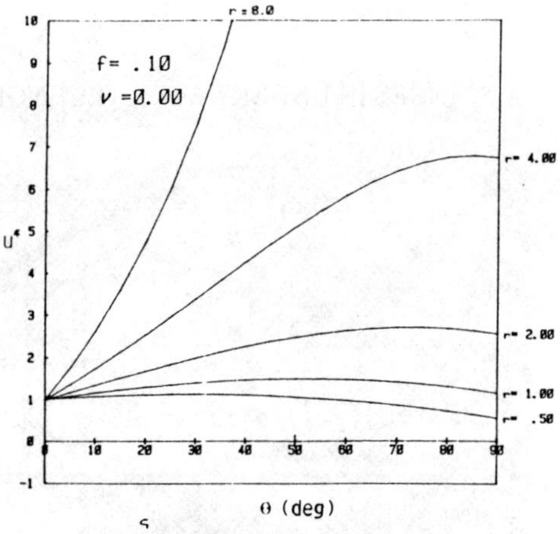

Fig. 9 Diving velocity with zero cable friction

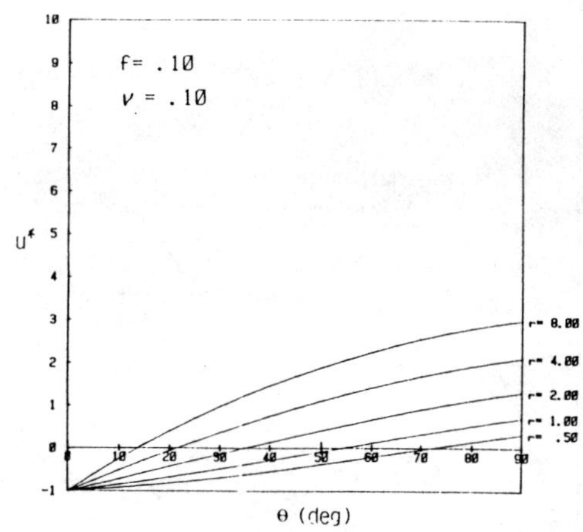

Fig. 12 Surfacing velocity with friction and weight

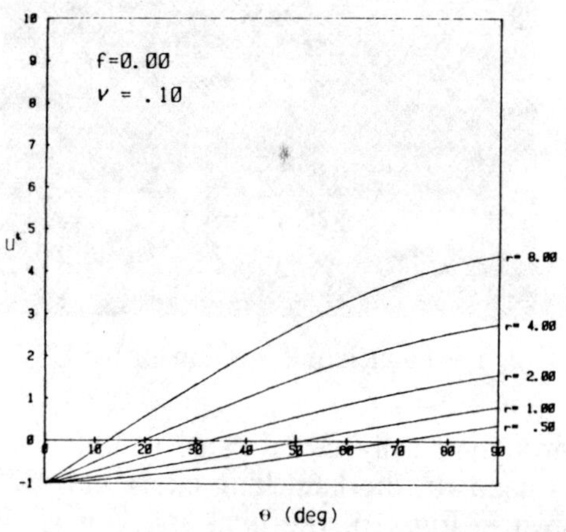

Fig. 10 Surfacing velocity (neutral weight in water)

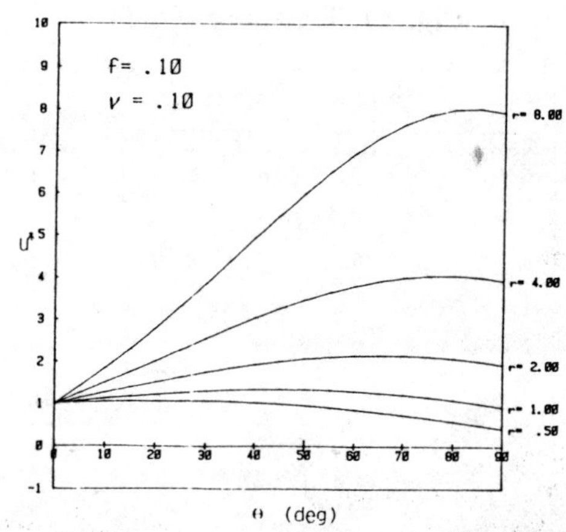

Fig. 13 Diving velocity with friction and weight

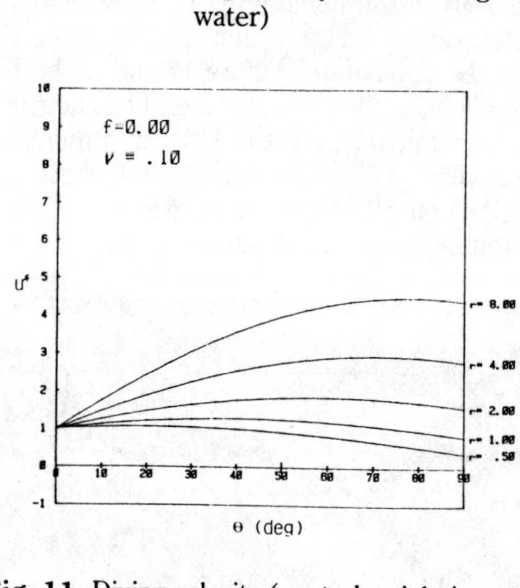

Fig. 11 Diving velocity (neutral weight in water)

conditions. If the catenary of the tow cable is given, the trajectory of the USV is calculated approximately by integrating u with respect to time. Figure 14 shows an example of the vehicle trajectory calculated for the following conditions:

tow cable	8 mm hydrodynamically faired stainless steel wire with polyurethane coating. —drag coefficient is 0.2 based on the frontal area
cable length	250 m
towing speed	8 knots
lift-to-drag ratio	5 for surfacing and 1 for diving
frictional coefficient	0.2

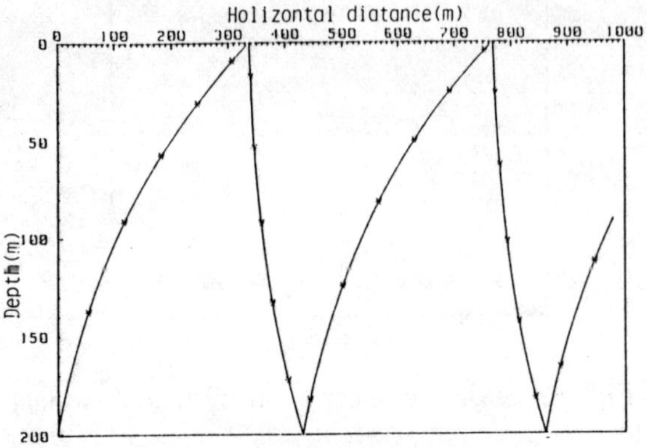

Fig. 14 Trajectory of the USV

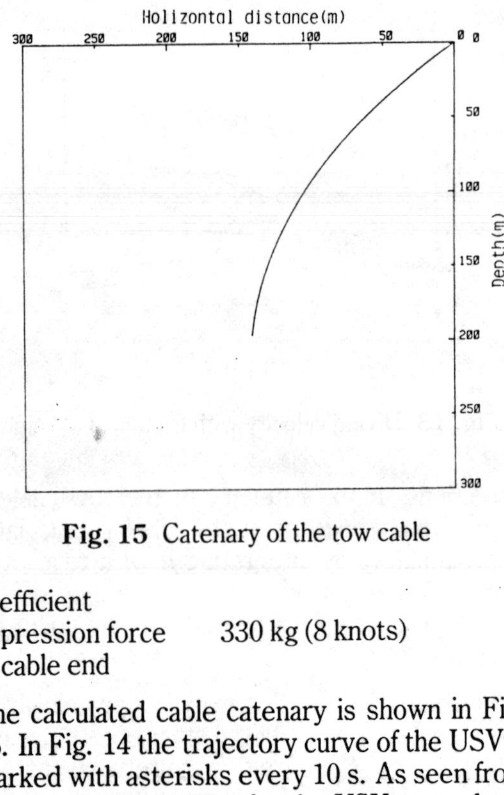

Fig. 15 Catenary of the tow cable

coefficient
depression force at cable end 330 kg (8 knots)

The calculated cable catenary is shown in Fig. 15. In Fig. 14 the trajectory curve of the USV is marked with asterisks every 10 s. As seen from Fig. 14, it takes 105 s for the USV to go down to and come up from 200 m depth, while the vessel advances 440 m. The mean vertical speeds of surfacing and diving are read to be 3.6 m s^{-1} and 4 m s^{-1}, respectively.

FIELD TESTS

The USV was tested in the Sagami Bay. The vessel used was a 500 tonne tugboat. A hydrodynamically faired cable of length 220 m was used. In the launching operation, which is shown in Fig. 16, the boat speed was slowed down to about 2 knots and the tow cable was payed out to the full length. The boat was then accelerated to the operation speed and the USV began the undulation. Figure 17 shows the USV just reaching the sea surface. Oceanographic data transmitted from the USV was monitored on a CRT colour monitor, and were also recorded on magnetic tapes. Figure 18 shows the temperature distributions in the 0–200 m

Fig. 16 Launching operation for the USV

Fig. 17 The USV just reaching the surface

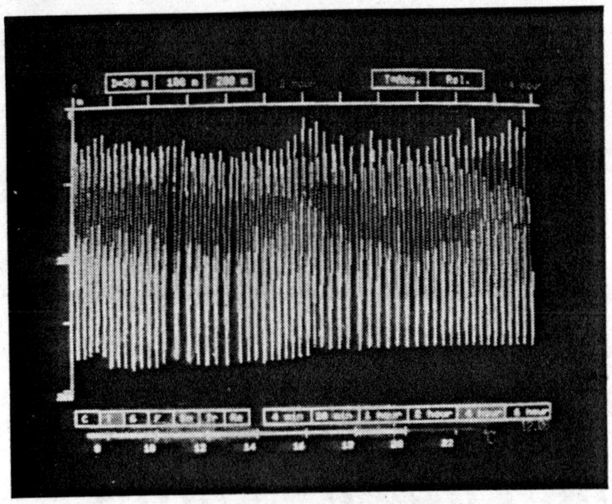

Fig. 18 Water temperature distribution displayed on the CRT colour monitor

layer of the Sagami Bay on 27 September 1984. The graph is plotted using eight colours, corresponding to the water temperature values. The spatial distribution of water

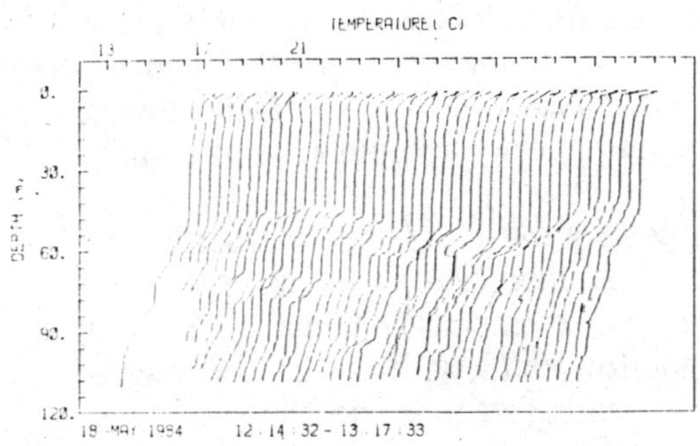

Fig. 19 Waterfall display of temperature distribution

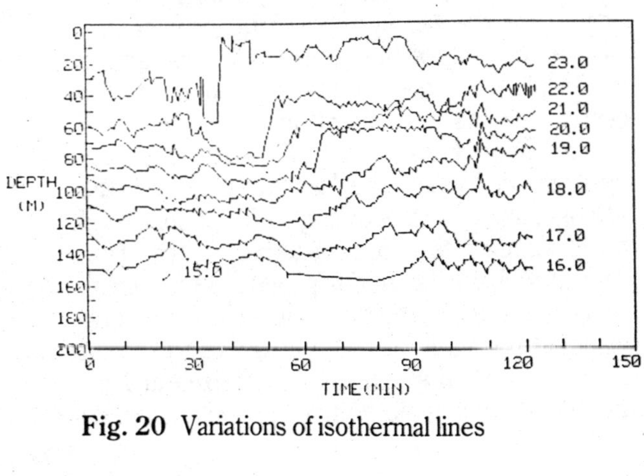

Fig. 20 Variations of isothermal lines

temperature is recognized by the colour distribution. The temperature distribution is shown by other display methods. Figure 19 is the waterfall display of the temperature distribution. The data were collected on 18 May 1984. In this figure the abscissa shifts 0.2°C to the right for each successive trace. The variation of the base depth of the isothermal layer is clearly visualized. Another method is shown in Fig. 20, in which isothermal lines are drawn. The data were collected a few miles off the east coast of Miyake Island on 26 June 1985. The abrupt rise of the 23°C isothermal line shown in this figure is thought to be a local upwelling which is peculiar to this area.

CONCLUSION

The USV, a new towed vehicle for oceangraphic measurements, has been developed. It is small and of light weight. The USV is controlled from a towing vessel. An inductive coupling system is used for the data communications between the USV and the towing vessel. As proved by the theoretical estimations, the USV can undulate at very high speed. Thus, high-resolution scanning of the ocean space is possible. The data collected are monitored on board the vessel as a colour tomogram of the ocean space. The field tests proved the availability of the USV in oceanographic measurements. The USV will be used for the determination of precise physical structure of the ocean.

11

Methods for Reducing Ship-induced Attitude Variations of Passive Towed Bodies

P. J. Wingham, School of Engineering, University of Bath, Bath, UK

INTRODUCTION

It is now possible to simulate, almost in real time, the dynamic response of a towed body to oscillations of the cable top produced by the towing ship's reaction to the waves. It is equally important that one may vary the magnitudes of the physical and hydrodynamic properties of the system to discover which have a dominant influence on the towed body (the "fish").

This technique has been systematically applied to the problem of reducing the angular rotation of passive towed fishes and has shown how far, by careful design, it is possible to reduce the magnitude of ship-induced perturbations in pitch, roll and yaw.

The need for such reductions is well recognized for fishes transmitting or receiving sonar signals, e.g. for seabed surveying.

It may be possible to correct sonar data for changes of body orientation (if the latter is recorded during towing), or to mount equipment on gymbals to isolate it from such changes. However, in both cases it becomes difficult, or expensive, to cope with the horizontal accelerations that may also be present in the body motion. Under these circumstances the precise measurement of attitude requires gyroscopes and accelerometers. In other words, there is no substitute for a steady towed platform experiencing acceptably small angle changes.

The purpose of this chapter is to describe the forms of motion imposed on a single towed fish by the ship, and to emphasize how the choice of fin areas and towstaff length govern the magnitude of the resulting attitude variations.

The properties of the towed fish required to describe its two-dimensional motion in the plane of symmetry may be divided into two classes, the physical and the hydrodynamic. In the former are the mass, the submerged weight, and the location of the centres of mass and of weight-plus-buoyancy forces; in addition the location of the tow point and the radius of gyration in pitch about it. In the latter class are the drag coefficient, the "lift curve slope" (the rate of change of lift with angle of incidence) and the location of the "aerodynamic centre" (the point at which acts the lift produced by a small change of angle of body incidence). In general terms the magnitude of the lift curve slope increases with the size of the horizontal fins, and the aerodynamic centre location depends upon their size and location on the body. Information is also needed about the damping moment, about the tow point, due to a constant angular velocity in pitch. Some of these properties are illustrated in Fig. 1 for the fish configuration used for these studies.

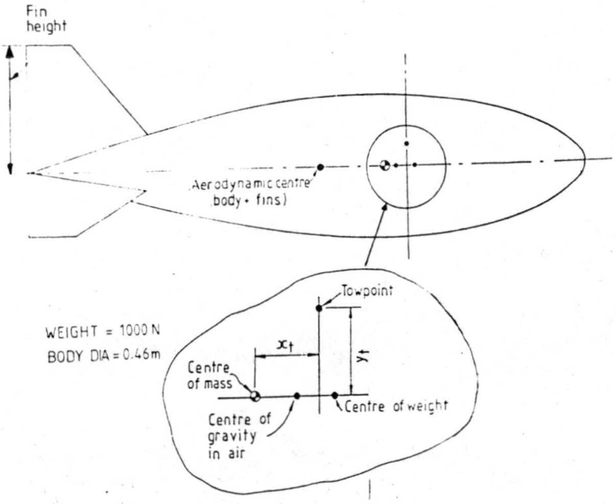

Fig. 1 The fish configuration

The theoretical work on which this chapter is based is contained in Chapman (1982, 1984a, b), which provide much more detail than can be included in this review.

It is useful, initially, to consider the ship-induced motion of a fish in two dimensions when the fish has the (hypothetical) form of a heavy sphere, i.e. non-lifting, requiring no fins, and subjected to only one hydrodynamic force, its drag.

THE RESPONSE OF A TOWED HEAVY SPHERE TO CABLE TOP OSCILLATIONS

Figures 2 and 3 illustrate the motion of a 100 m cable towing a 1000 Newton sphere at 6 knots while a 4 m amplitude simple harmonic motion is applied to the cable top along the direction of its tangent. The period of the SHM is 6 s.

Figure 2 shows how the envelope of the oscillating cable relates to the cable's equilibrium profile, and Fig. 3 shows more clearly the

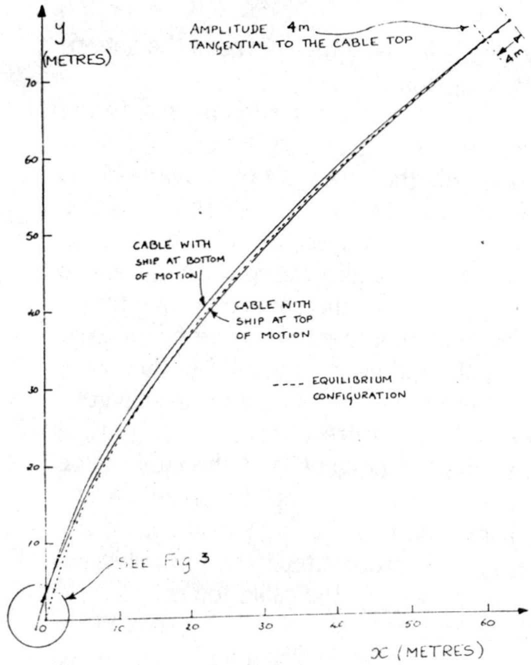

Fig. 2 Equilibrium and extreme position configurations of 100 m of faired cable during 6 s period oscillations of the cable top (tow speed = 6 knots)

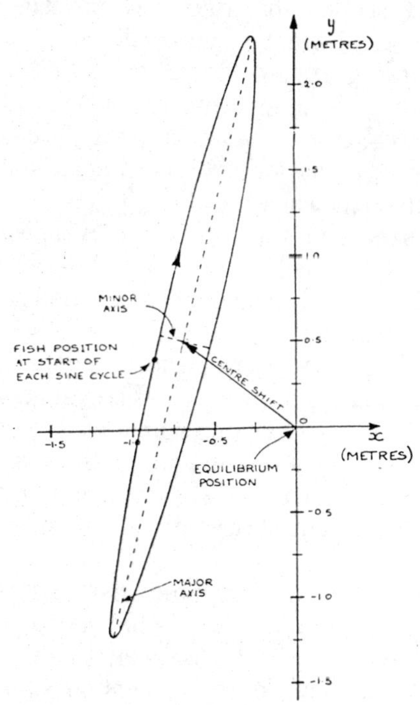

Fig. 3 Locus of the bottom end of the cable in Fig. 2

roughly elliptic locus of the centre of the sphere at the cable bottom.

The major axis of the ellipse is parallel to the cable bottom tangent and its length is 3½ m: thus most of the cable top movement is transmitted directly to the fish by the cable sliding along its curved length. The "lost" ½ m of this amplitude is associated with the much smaller movement of the sphere normal to the bottom tangent represented by the minor axis. This normal motion is caused by the cable flexing and straightening as its tension changes.

If the cable top were oscillated along a line normal to the top tangent (and the cable were not very short) the effect on the sphere would be negligible, due to the high damping of the transverse wave propagated down the cable.

It follows that when the cable top is oscillated along a *vertical* line, assuming the ship to be heaving, the resultant displacement at the cable bottom will depend mainly on the tangential component of the vertical top movement. The ratio of the fish-to-ship amplitudes will therefore vary as the sine of the cable top angle. This is a useful empirical rule.

At short scopes this rule may not apply, owing to the existence of an oscillatory mode capable of being excited. At very short scopes the fish and cable may swing like a pendulum about the cable top. At somewhat greater scopes the fish may swing about a point some fixed length up the cable.

These behavioural patterns are well illustrated in Fig. 4, which shows the sphere's response to a pseudo-random heaving motion of the ship. This motion is based upon a given amplitude-frequency spectrum containing periods ranging from 2 to 20 s. The variations with cable scope and towing speed of the ratio of the r.m.s. fish displacement to the r.m.s. ship heave amplitude are shown in Fig. 4(a). The sine of the equilibrium cable top angle is plotted in Fig. 4(b).

At large scopes the match between Figs 4(a) and (b) is very good. At scopes between 0 and 20 m the amplitude ratio is seen to exceed unity. This is due to the excitation of the pendulum or fixed length modes. Chapman (1982) provides approximate formulae for determining these short scope resonance properties of fish-cable systems).

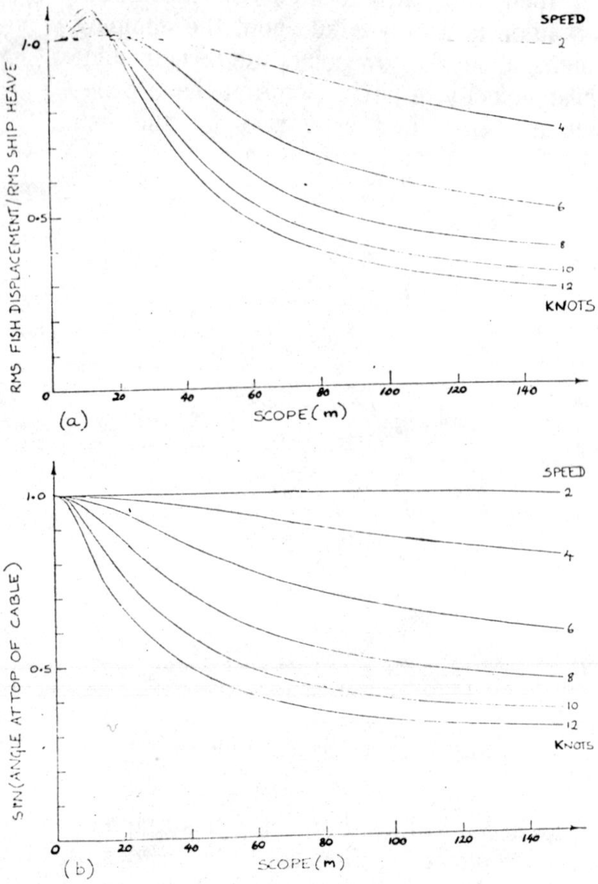

Fig. 4 Variation of the relative r.m.s. fish displacement and the sine of the cable top inclination to the horizontal, with scope

When the r.m.s. displacements are resolved along vertical (heave) and horizontal (surge) axes another feature of the sphere's dynamic response is revealed (see Fig. 5).

It is evident that the variation of the fish heave/ship heave ratio accords with the sine rule but that the surge displacement ratio, albeit of smaller amplitude, peaks at short scopes at high speeds and longer scopes at low speeds. This demonstrates that the short-scope oscillatory modes are predominantly surge-making, as expected. The increase of surge ratio with increase of speed, at long scopes, reflects the increasing curvature of the equilibrium cable profile, which leads to more cable flexing in response to the tangential component of oscillatory motion at the cable top.

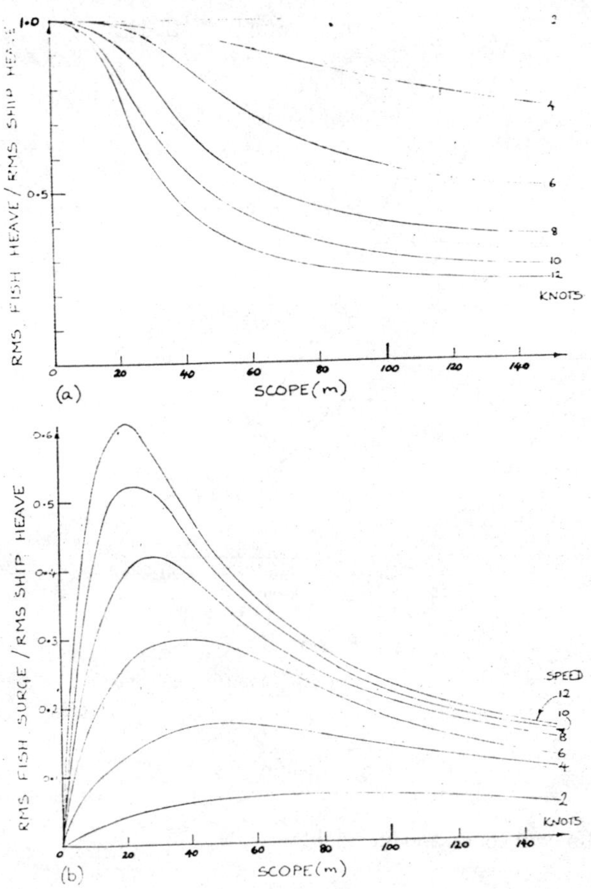

Fig. 5 Variation with scope of the separate r.m.s. heave and surge components of the fish displacement

Summary of Response of a Heavy Sphere to Ship Heave

Simulating random ship heaving motion reveals that the ratio of fish heave to ship heave is equal to the sine of the cable top inclination angle. At very short scopes, the fish surging motion can be significant, but the surge displacement is always less than the heave.

Methods for Reducing Fish Displacement

If one wishes to reduce the ship-induced heaving response of negatively buoyant fishes, as a first step to reducing their attitude variations, the only choice is to reduce the cable top inclination. Some operators prefer to use bare, rather than faired, cable (in spite of its inferior steady state towing performance) to bring this about.

PITCH ATTITUDE RESPONSES OF A FINNED BODY

The motions under discussion are still confined to the plane of symmetry of the cable and body, and are two-dimensional.

The need to wrap a low drag envelope round the payload leads to the use of finned bodies that are pivoted in pitch about the cable bottom. Because closed bodies are hydrodynamically unstable, the fins are needed to provide a degree of "weathercock" stability in pitch. The body will therefore tend to point into the direction from which the flow is approaching the pivot (ignoring non-hydrodynamic forces).

If, due to ship motion, the cable bottom (the pivot) were moving on a sinusoidal path through the water, the body pitch attitude would vary about its equilibrium value. The greater the pitch stability the greater the variation. In the limit, ignoring body pitching inertia, the body attitude would follow the instantaneous slope of the pivot path.

Conversely, if the stability were reduced nearly to zero, the attitude variation would be very small.

It is suggested that many towed fishes have excessive (horizontal) fin areas, causing greater ship-induced pitch attitude changes than necessary.

In qualitative terms, the stability may be reduced by reducing fin area, or moving fins forward, or moving the pivot point aft. The physical constraints applying to existing fishes may prohibit such modification, but there is no excuse in the case of new designs!

It must be noted that theoretical estimates of the pitch stability derivatives are unreliable for any but the simplest fin–body configuration Wind tunnel tests are usually necessary.

Examples of the effects of reducing tail fin size on fish response to ship heave are given in Fig. 6 for the streamlined fish configuration shown in Fig. 1. Three fins are used for this

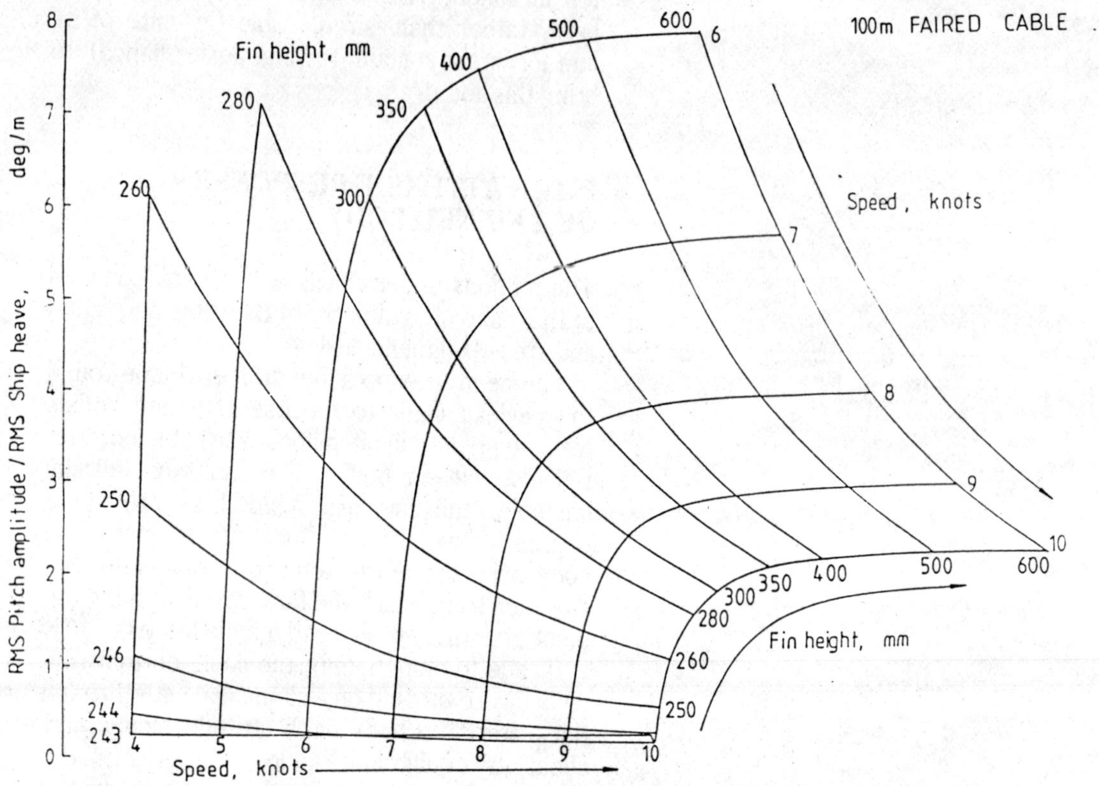

Fig. 6 Variation of r.m.s. pitch amplitude ratio with speed and fin height

fish, and their size varied by changing the height of all three. The effect on the pitch stability of doing so was determined by wind tunnel tests.

Figure 6 illustrates the r.m.s. values of the pitch attitude changes resulting from the pseudo-random ship heaving motion, expressed in degrees per metre ship heave. In this example the cable is 100 m long and the ship speed varied from 6 to 10 knots. The fin height is used as the parameter governing the pitch stability. The dramatic reduction of pitch amplitude ratio as the reducing fin height approaches the value at which the fish becomes neutrally stable is very clear. The reduction with speed increase at each fin height is in accordance, broadly, with the "sine rule" governing the displacement amplitude of the cable bottom as the cable top inclination becomes shallower.

It is emphasized that these examples, which are based upon a "good" hydrodynamic body shape, are typical of what may be expected of any fish shape towed in this manner.

The Effect of Fin Size on the Equilibrium Attitude

Large fins produce large ship-induced attitude perturbations, but only small variations of steady state trimmed attitude when the towing speed varies. Under steady tow, with zero ship heave, the finned body will settle at a particular attitude where the moments about its pivot of the gravity and buoyancy forces balance the hydrodynamic pitching moment created by the fins.

If the fins are large the hydrodynamic moment will balance the weight moment at only a small equilibrium value of the attitude, or angle of incidence. Provided that the hinge (tow point) is close to the body axis, the trimmed attitude will tend to a steady value as the speed increases. It is important to note that if the pivot is located above the body fore and aft axis, the trimmed attitude may become increasingly nose down, as speed increases, due to the additional moment created by the "underslung" drag force on the body.

When the fin size is reduced sufficiently to give nearly neutral stability it is necessary to examine carefully the trimming situation to ensure the avoidance of two possibilities: first, that the variation with speed of trimmed attitude does not become so large as to negate the advantage of having smaller ship-induced variations; secondly, that due to the lower pitching frequency of the body the time required for the fish to settle down following a disturbance in pitch, and the depth changes occurring while it does so, do not become too large.

Under normal circumstances the frequency of the pitching oscillation of the fish is much greater than the frequency of the damped swinging motion of the whole cable; consequently attitude changes following a disturbance are too rapid to influence the cable motion. At very low values of fish pitch stability the two frequencies can become close enough to allow coupling between body pitching and cable swinging, to the detriment of good dynamic behaviour.

Chapman (1984a) has shown that the nearer to neutral stability one "tunes" the fins, the more important it is to consider the geometric relationship between the pivot position, the centres of mass and weight, and the aerodynamic centre of the fins-plus-body. Formulae are offered for determining, in conjunction with chosen values of x_t and y_t (Fig. 1), the smallest pitch stability that will give acceptable dynamic towing behaviour.

One may conclude by noting that rules now exist which allow one to configure a finned fish so that ship-induced pitch attitude changes can be reduced to the order of one or two degrees per metre of ship heave.

If it were necessary to improve upon this, some form of auto-stabilization would be needed.

Additional Effects due to Large Fin Areas

When the horizontal fin area becomes large enough to be described as a "wing", used for the purpose of providing a (depressive) hydrodynamic force, another aspect of the fish response deserves a mention.

The resistance of the winged body to vertical motion becomes significant and has the effect of increasing the surge (horizontal) displacement at the expense of the vertical, in response to an upward pull. An example of this effect is the sliding motion of a large submerged horizontal plate in response to being pulled upwards by a cord attached to its centre of gravity. This is seen in Fig. 7. The upper diagram shows the effect of lift curve slope (proportional to wing area) on the r.m.s. fish displacement/ship heave. The variation is slight. However, the lower diagram shows how the horizontal fish

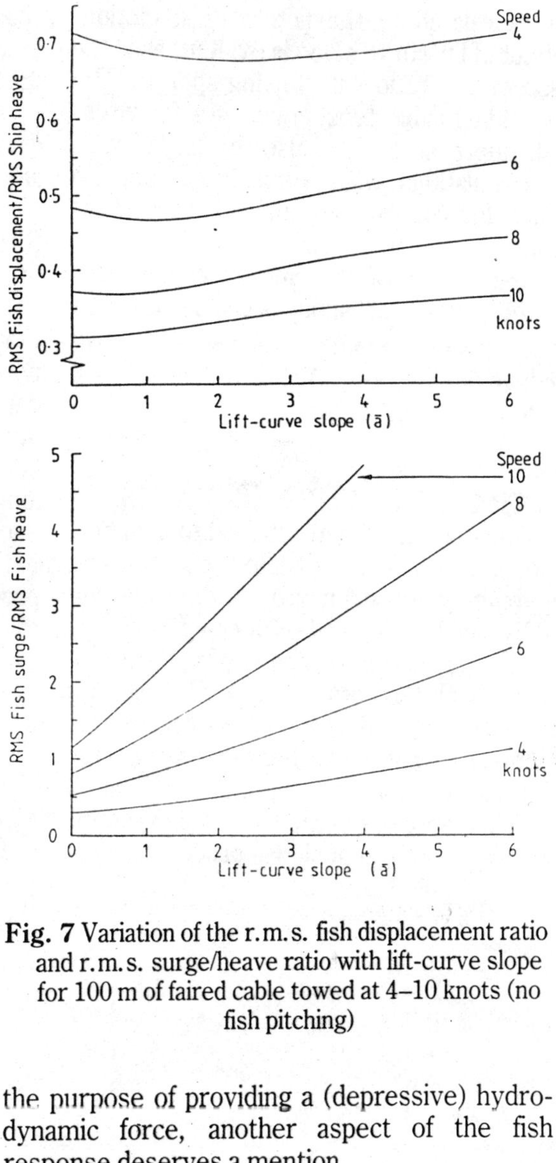

Fig. 7 Variation of the r.m.s. fish displacement ratio and r.m.s. surge/heave ratio with lift-curve slope for 100 m of faired cable towed at 4–10 knots (no fish pitching)

motion is related to the fish vertical motion. The effect of lift-curve slope is evident, as is the increase of this ratio with towing speeds. The corresponding cable tension changes for large-winged depressors can also be significant. These calculations were carried out with no allowance for fish pitching and are thus slight overestimates of the effect.

The magnitude of the surge accelerations associated with such responses can be sufficient to make pendulum type inclinometers give misleading results when used to monitor fish attitude.

LATERAL ATTITUDE RESPONSES

Discussion so far has been confined to symmetric motions of the fish and cable assumed to be excited only by vertical motion of the cable top caused by ship heaving and/or pitching.

If the ship is also rolling and yawing there will be lateral displacements of the cable, causing the fish to roll and yaw. This can be as detrimental to sonar performance as is pitching.

It is possible for lateral ship motion to excite longitudinal responses of the fish–cable system. Fortunately, such coupling produces only second-order longitudinal effects (except in extreme conditions). This allows one to study the lateral behaviour of the system in isolation.

It is useful to consider first the lateral displacement of the cable bottom induced by ship sway.

Ship-induced Lateral Displacement of a Heavy Fish

Forced oscillations of the cable top in directions normal to the top tangent propagate a highly damped wave down the cable. The amplitude decreases exponentially down the cable, to become negligible at the bottom of cables longer than (say) 150 m. For shorter cables, one may estimate the reduction of amplitude between ship and fish. Chapman (1984b) has provided tables to give approximate values of the lateral attenuation, A_L, defined as

$$A_L = 10 \log_{10} \left(\frac{\text{ship lateral amplitude}}{\text{fish lateral amplitude}} \right)$$

If $A_L = 10$, the fish-to-ship amplitude ratio is 10%; if $A_L = 20$ it is 1%; etc.

Figure 8 shows a typical variation of A_L with

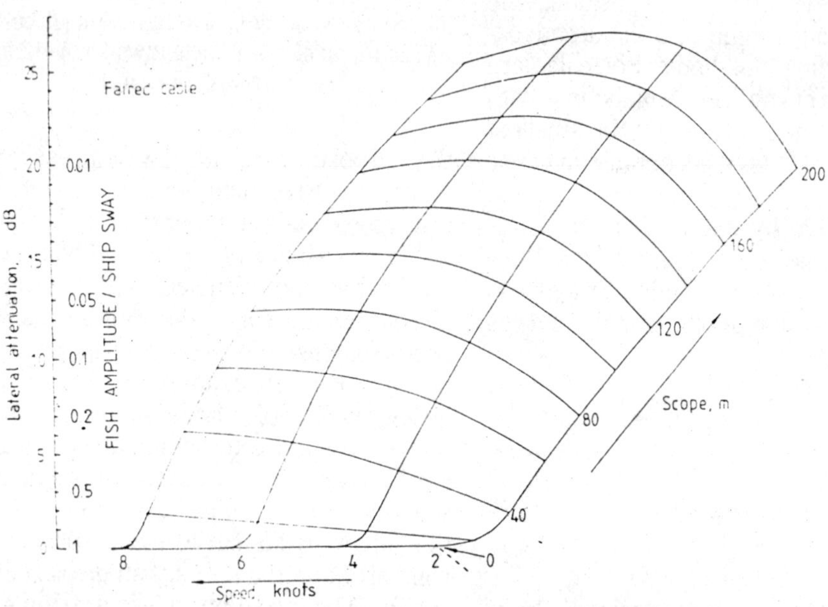

Fig. 8 Variation of lateral attenuation with speed and scope (period 6 s)

towing speed and cable scope for a given towed system, at a period of 6 s, assuming a faired (low C_D) cable.

It is evident that at short scopes the amplitude ratio can be significant; for example, at 2 knots at a scope of 45 m the ratio is 50%. Thus a 1 m lateral sway of the cable top may displace the fish towpoint by ½ m, which could induce significant roll and yaw perturbations.

If the cable fairing were removed, the amplitude ratio at the same speed and depth would only be 3%, due to the longer scope required and the increased damping of the cable due to its higher drag. Thus we have another example of bare cables having the advantage over faired, as noted earlier.

Roll and Yaw Attitude Responses

The displacement ratios of Fig. 8 are estimated without the hydrodynamic properties of the fish. Estimating the roll and yaw angle changes caused by the displacements involves the magnitudes of all the lateral stability and damping derivatives of the fish. These can be so variable, between one fish and another, and the roll–yaw motion so complex that one cannot illustrate typical fish roll–yaw responses. However, Chapman (1984b) has highlighted some of the features of the fish configuration that have significant effects.

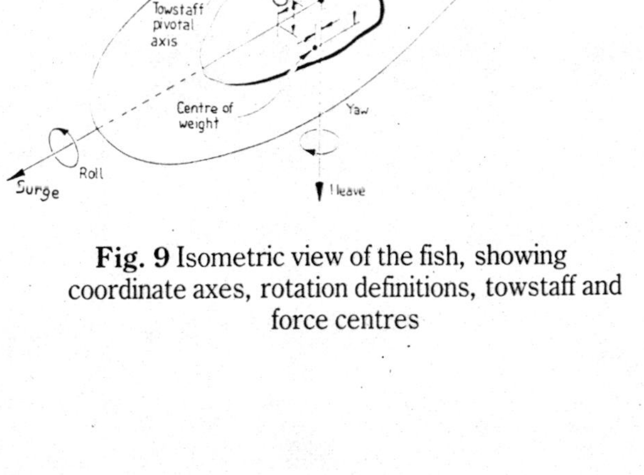

Fig. 9 Isometric view of the fish, showing coordinate axes, rotation definitions, towstaff and force centres

The length of the rigid towstaff (Fig. 9) markedly influences the rolling response of the fish because cable bottom displacements are applied to the fish at the top of the staff. The longer the towstaff the more the fish rolling mode is coupled with the cable, which increases

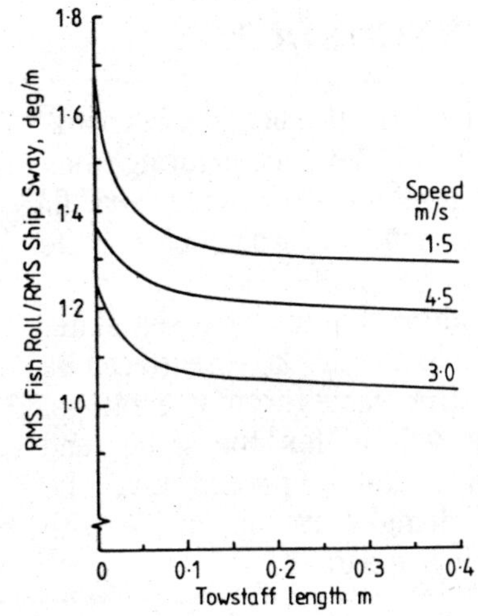

Fig. 10 Variation with towstaff length of the r.m.s. roll of the fish per unit r.m.s. ship sway (50 m faired cable)

the damping; and the greater the roll stiffness, which increases the frequency.

Figure 10 shows the variation with towstaff length of the r.m.s. roll attitude response to pseudo-random ship sway, for this particular fish on 50 m of faired cable. The importance of avoiding too short a towstaff is evident.

A suggested yardstick from Chapman (1984b) is to adopt a towstaff length of order two to four times the fish radius of gyration in roll.

Referring back to Fig. 9 and imagining lateral forces applied to the towstaff top, it is clear that if the projected vertical area of the aft fins is large the yaw attitude changes will also be large.

By analogy with the pitching response, it can be shown that by reducing the fish's directional stability close to neutral one can much reduce excursions in yaw. Detailed response calculations are needed to show how close to neutral stability one may go.

Chapman has also shown that little advantage is to be gained from making the upper (vertical) fin area greater than the lower, or vice versa.

It is concluded that lateral attitude variations are only likely to be significant at scopes less than 60 m (say). In these cases it should be possible, by careful design, to restrict roll and yaw variations to less than 2 degrees per metre of ship sway.

FISH CONFIGURATION DESIGN

Since it is possible to predict the dynamic behaviour of towed bodies, configuring their shape, mass distribution, tow point and fin areas can now be done according to definite principles.

Estimates, based on ship size and sea state, of the expected cable top displacements can be used to provide the cable bottom motions, given preliminary values for the fish's and cables's weight, size, and drag coefficient. The resulting attitude changes are obtainable with more hydrodynamic fish data.

From the desired constraints on pitch and roll and yaw angles under tow, one may judge which aspects of the fish's geometry may be critical.

If the shape is unorthodox it may be necessary to measure the hydrodynamic quantities required to simulate its dynamic responses. This can be costly and time-consuming, but far cheaper than trying to "develop out" adverse features belatedly revealed by sea trials.

CONCLUSIONS

The most important displacements induced by the ship at the cable bottom are those due to ship heave. The ratio of cable bottom amplitudes to fish heave amplitudes is close to the sine of the cable inclination angle at the top. Therefore, at a given speed and fish weight, a low-drag cable will produce larger displacements.

At short scopes this sine law may not estimate the full displacement at the fish, owing to the possible excitation of "pendulum" type modes.

LONG-RANGE UNDERWATER ACOUSTIC IMAGE TRANSMITTING SYSTEM

B Leduc, Institut Francais de Recherche
pour l'Exploitation de la Mer; A Glavieux,
Ecole Nationale Superieure des Telecommunications
de Bretagne, France

Presented on **Wednesday 5 March 1986 at 11.00**
in the Norfolk Conference Room

EMBARGOED AGAINST DELIVERY

12

Long-range Underwater Acoustic Image Transmitting System

B. Leduc and A. Glavieux †*, *Institut Français de Recherche pour l'Exploitation de la Mer, BP 337, 29273 Brest Cédex and †Ecole Nationale Supérieure des Télécommunications de Bretagne, ZI Kernevent, 29263 Plouzane, France

INTRODUCTION

The transmission of data, gathered by different underwater instruments or control systems, was first done using electrical and/or electromechanical cables which linked the source to the receiver. The use of acoustic waves to transmit these data enables the source to become a self-contained instrument and eliminates the drawbacks of the cables. Many scientific institutes and industrial firms tried to design such systems. IFREMER, in the framework of its programmes related to underwater intervention, has designed a self-contained system to transmit pictures from the sea-bottom to a surface vessel.

The use of recently designed components such as CCD cameras, low power consumption microprocessors and high-efficiency wide-band amplifiers led to a compact instrument which can be used in a wide range of applications and which can be installed in manned or unmanned submersibles or in deep sea structures to control underwater operations in real time.

The frequency is the limitation factor on the range. IFREMER chose a low frequency (26 kHz) to get ranges as long as 3000 m. The data

Fig. 1a Fish Epaulard and camera equipment and housing

Fig. 1b

rate depends on the modulation scheme and on the bandwidth of the transmission system, including the transducer. The number of images that can be transmitted per second also depends on the algorithms which are used to reduce the data.

Very good results have been obtained in several configurations with the Epaulard acoustically remote-controlled autonomous submersible (Fig. 1a). The system can be divided into two main parts: the underwater acquisition, processing and transmitting subsystem; and the receiving and decoding station aboard the vessel.

THE UNDERWATER EQUIPMENT

Besides the battery pack and the light, which were existing Epaulard equipment, the underwater equipment consists of a single housing (see Fig. 1b):

diameter	150 mm
length	600 mm
weight	20 kg
wet weight	5 kg

The rated pressure is 600 bars.

Figure 2 presents the block diagram of the equipment which includes:

- a CCD camera and the associated circuits (144 lines × 208 points);
- acquisition and processing circuits based on the IFREMER-designed card set (NSC 800 microprocessor);
- a modulator;
- a wide-band power amplifier; and
- power supplies.

The transducer is linked to this housing by electrical cable and watertight connectors.

The CCD camera

Manufactured by I2S (Bordeaux, France), the CCD camera has been designed to withstand a hard environment. The sensing head is removable and can be separated from the two specific printed circuit boards. The low-weight compact camera is a cylinder of length 80 mm and diameter 60 mm.

The CCD sensor is the TH 7852 manufactured by Thomson, and gives a 144 lines × 208 points image. It is fitted with an anti-blooming circuit. Its opto-electronic characteristics are:

saturation light	17 lux
sensitivity	20 mV lux^{-1}
signal/noise ratio	2000
linearity	+5%
darkness signal	2 mV

As it was to be used in sea water, the CCD camera had to be placed behind a special optic

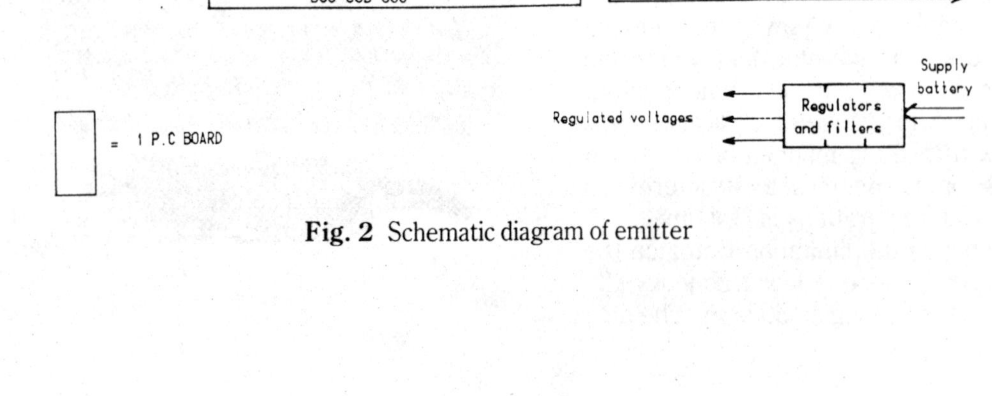

Fig. 2 Schematic diagram of emitter

system. This was designed by CERCO (Paris, France) and is made of removable corrective lenses to take the characteristics of the sea water into account. Its overall specifications are:

focal length	6.3 mm
aperture	F/1.5
angular field in sea-water	50° × 38° (60° diagonal
focus	from 1 m to ∞.

In the air, but with no corrective system, the optical system performance is unchanged.

Two printed circuit boards are used, one to generate synchronization and control signals for the CCD sensor, and the other to digitize the signal (6 bits per pixel) and to enable the direct memory access (DMA) to the microprocessor.

The Acquisition and Processing Circuits

Based on a set of cards called the IFREMER 800 card set, the acquisition and processing circuits are composed of one CPU card (NSC 800), two 16k memory cards and one interface card. These circuits control the image acquisition and the exposure time, and enables the flashing light to be synchronized with the shot when a strobed light is used. They are also used to make some "compression" of the data to reduce the number of data to be transmitted. External commands are used to select the type of image: raw or compressed data.

Before being serially sent to the modulator, the data are formatted to an RS232C standard.

The Modulator

This part has been studied in co-operation with the Ecole Nationale Supérieure des Télécommunications de Bretagne (ENSTB). The "phase shift keying" modulation is used with a choice between PSK 2 and PSK 4. The data rate is then 5300 or 8800 bits s^{-1}. The modulator is connected to the wide-band power amplifier.

The Wide-band Amplifier

This amplifies the 26 kHz signal and drives the transducer. This element must have a very good efficiency so as not to dissipate too much energy. A linear class D amplifier has been studied; the efficiency is about 70%. The nominal electric power applied to the transducer is 20 W r.m.s.

The Transducer

This has been studied by CIT Alcatel; its Sv is 165 dB (re. Pa, 1 m, 1 V). The overall power consumption is 800 mA/48 V, 20 W being absorbed in the transducer.

THE SURFACE EQUIPMENT

The on-board equipment has to receive the acoustic wave, demodulate it and display the transmitted picture on a standard video screen. The equipment comprises (see Fig. 3):

The hydrophone. The main part is a Brüel and Kjaer 8104 hydrophone associated to a low-noise pre-amplifier. Its Sh is –158 dB (re. Pa, V).

The amplifier and the filters. The acoustic signal is amplified and band-pass filtered; the signal level is made compatible with the demodulator.

The demodulator. This detects the phase of the signal and transforms it into a numerical

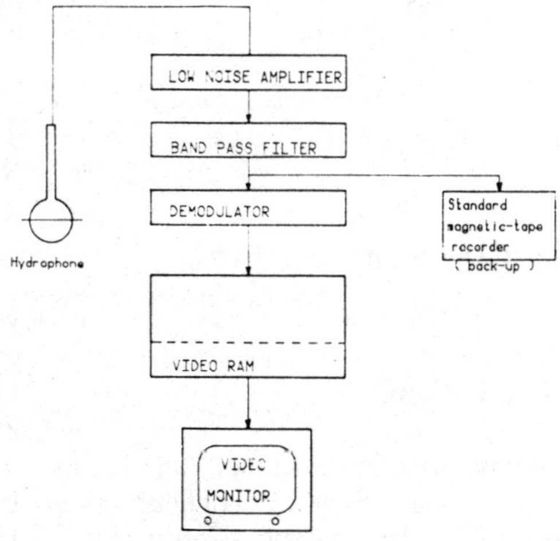

Fig. 3 Schematic diagram of receiver

signal which will then be processed by the data processing unit.

The data processing unit. According to the transmitted picture, different algorithms are used to retrieve the original picture; a RAM is then filled with the data and read 50 times per second to be displayed on a standard video screen.

After testing the whole system in the laboratory and in the testing pool of the Centre de Brest, the underwater equipment was fitted on the Epaulard, the acoustically remote-controlled unmanned submersible.

The camera container was placed in the Epaulard, looking downard. The power supply was the same as that of the submersible and the flashing light was synchronized with the shot (see Fig. 4).

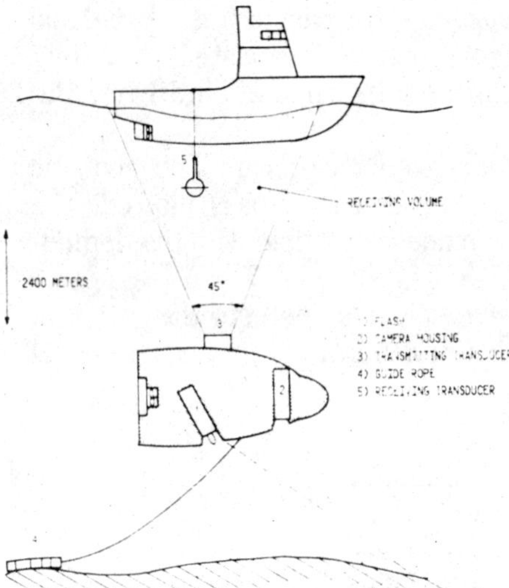

Fig. 4. Operational configuration

AT-SEA TEST RESULTS

A campaign took place in December 1984 in the Mediterranean Sea. Several configurations were tested: PSK 2 or 4, and several algorithms to reduce the number of data to be transmitted.

Transmission Characteristics

Range: More than 3000 m in a tormented environment (canyons and underwater cliffs).
Receiving cone: 45°.

Thermic and salinity gradients can influence the acoustic propagation and create multiple channels, so the receiving cone may vary with the hydrologic conditions, the seasons and the depth.

Results

Whenever the surface vessel is inside the receiving cone, the error rate is very low ($<10^{-3}$), even under very severe propagation conditions (multiple paths, reflection, etc.).

The pictures received are exactly the same as those which have been transmitted. Besides the attenuation (propagation loss), no degradation of the signal was detected. Thus, the quality of the pictures only depends on the shooting conditions (light, subject, light reverberation, etc.) and on the type of the algorithm used to decrease the number of data.

The alphanumeric character-string is generated with each picture and transmitted with the whole message to provide reference signals. The image data are "integrated" by the human eye, which has a tendency to correct the single errors. With a known string of characters it has been easy to transmit known light levels to be able to detect errors more easily.

The most white grey level: JOYEUX NOEL.
The most black grey level: BONNE ANNEE.

The text also shows the size of a pixel. Most of the received messages had no errors. This proves that any kind of data can be sent with a low error rate.

Figure 5 (a + b) is an actual received image with no data compression. It is coded with 6 bits (64 grey levels). Under these conditions, one image is transmitted in 40 s at 5300 baud (PSK 2) or in 26 s at 8800 bauds (PSK 4).

Figure 6 is another actual received image, the data having been "compressed" before transmission. In this case one picture has been transmitted in 15 s at 5300 baud or 9 s at 8800 baud.

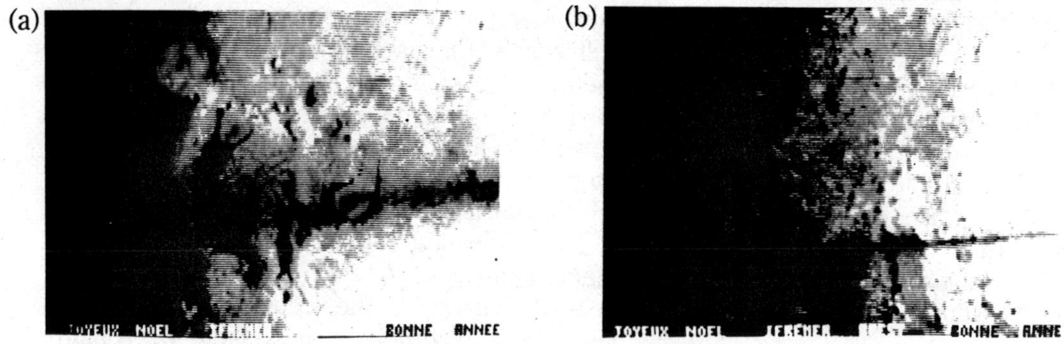

Fig. 5. Original picture (64 grey levels) and compressed picture (16 grey levels)

Unfortunately, the zone which has been explored has a sedimentary bottom with no life, so the images were not very interesting and did not permit an objective judgement as far as the compression algorithms are concerned. However, other trials (at-sea test or simulation with an image processing system) have showed that the used algorithms were often good when the subject was easy to recognize, but it was very often difficult to recognize a picture when there was no particularly interesting subject.

Spectrum Analysis

The spectrum of the received signal is shown in Fig. 6 (26 kHz, 5 kHz bandwith).

FUTURE DEVELOPMENTS

The tests of the equipment were successful, but it could still be optimized. IFREMER and ENSTB will continue the research in several directions:

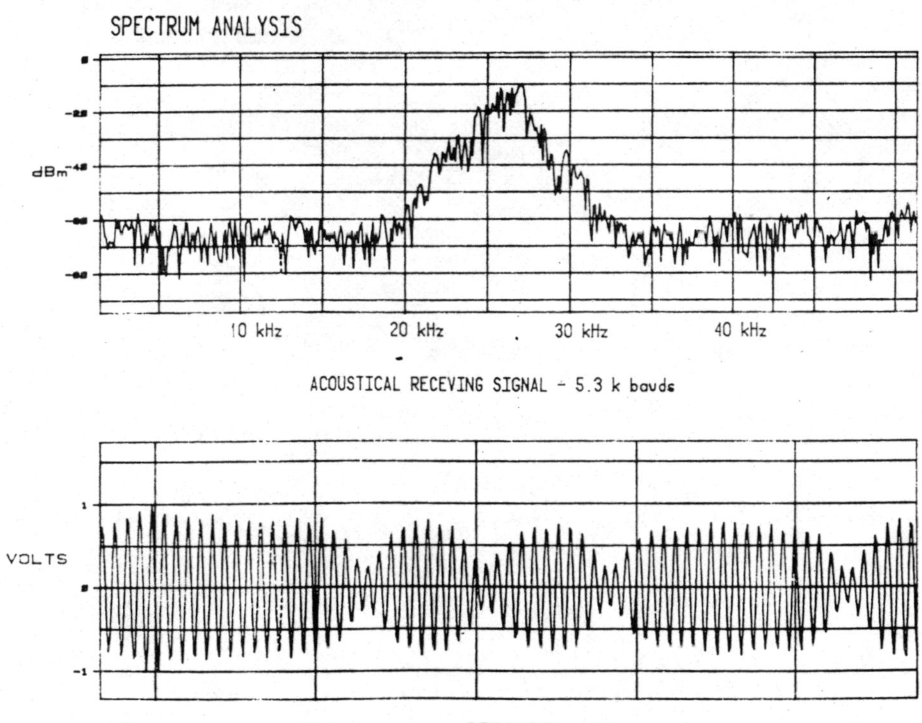

Fig. 6 A received image

- studies of new compression algorithms to try to decrease the number of data to be transmitted;
- changing the technology of the modulator to decrease the number of printed circuit boards—a new modulator has been designed and will soon be available in a DIL package;
- optimizing the surface equipment with an "egalizator" which will correct for the distortion due to the limited bandwidth of the transmitting transducer.

CONCLUSIONS

The equipment is now available and will be used whenever real-time pictures are needed. It can equip any unmanned submersible to get a real-time view of a structure or of the sea bottom.

It can also be used as a self-contained equipment with a flashlight and a battery pack. The frequency can be changed to increase the bandwidth and to increase the data rate.

IFREMER will still be working on this equipment; a two-way communication system will enable the user to change several parameters and to drive or to control external devices such as motors or vanes by using the down-link and to transmit pictures on the up-link (sonar image for example), or other data which need a high-capacity channel.

The equipment can also be used to complement high-resolution picture cameras and to help the user in the choice of the most interesting pictures to be taken with high resolution.

REFERENCES

- Rapport d'essais campagne Transcar Février 83: B. LEDUC
- Rapport d'essais campagne Epautrans – Decembre 85: B. LEDUC
- Transmission d'images sous–marines par ultrasons (thèse doctorat IFREMER – INSA Lyon) 85: A.MAGUER; published in *Rapport Technique de la Direction à l'Ingénierie et à la Technologie*.

OPERATIONAL ASPECTS OF SWIMMING AND SEABED CRAWLING ROV'S FOR SEARCH AND SALVAGE OF LOST DANGEROUS CARGO

F A van Dongen, Rijkswaterstaat, The Netherlands

Presented on **Wednesday 5 March 1986 at 12.00**
in the Norfolk Conference Room

EMBARGOED AGAINST DELIVERY

13

Operational Aspects of Swimming and Seabed Crawling ROVs for Search and Salvage of Lost Dangerous Cargoes

Frans A. van Dongen, Head, Survey and Pipeline Department, Rijkswaterstaat, North Sea Directorate, The Netherlands

Dangerous cargoes lost in the North Sea must be salvaged before they can cause accidents. Lost cargoes must be found using sonar, and then identified with still and video cameras mounted on a remotely operated vehicle (ROV). Finally, the cargoes must be salvaged or neutralized. The practice of search, location, selection, identification and salvage of cargoes and dangerous goods lost in the Dutch sector of the North Sea is presented in this chapter. Special attention is paid to technical and operational aspects of locating and identifying lost cargoes. The applicability of a diver, a swimming ROV or a seabed crawling ROV for these operations is examined. Two recent cases are elaborated.

INTRODUCTION

Shipping lanes in the North Sea have a very high traffic density, with vessels carrying a diversity of cargoes. During bad weather accidents can occur whereby cargo, stored on the upper deck, is lost overboard and sinks to the sea bed. Some cargo lost in this way, such as poisonous, explosive or radioactive material, can form a threat to users of the sea, and to the environment.

Adequate survey techniques are therefore needed to locate and identify these cargoes. This is made difficult because information on the position where a particular incident occurred is often poor. The problem is further complicated if a cargo is lost over a wide area, or when lost containers become embedded or buried in the sandy bottom.

Containers resting on the sea bed are usually located using side-scan sonar, although the first indications as to the whereabouts of a cargo are often reported by fishermen. Containers covered with sand can be detected with a magnetometer or a sub-bottom profiler once the location where the cargo was lost has been pinpointed.

Until recently, detailed surveys of potential areas and identification of lost cargoes were carried out by divers. Poor visibility and strong currents, together with unsystematic operation procedures, hampered these operations, so that surveys seldom achieved the desirable results. A remote operation vehicle (ROV), whether swimming or crawling along the bottom, used in combination with a dynamically positioned vessel, now makes it possible to carry out systematic and accurate surveys. A ROV equipped with a variety of sensors; still and video cameras, an obstacle avoidance sonar, a profiler, a magnetometer, an acoustic positioning responder and an air lift system is used as a bloodhound to search the sea bed. Using this method, engineers from the North Sea Directorate of Rijkswaterstaat have achieved good results. The ROV is directed along a fixed pattern of survey lines, inspecting every square metre of the sea bed.

LOST CARGO REPORTS

Accidents such as the shipwreck of the *Mont Louis,* with its cargo of radioactive material, are very much in the public eye. In this case, containers containing radioactive uranium hexa-flouride stored in the holds broke loose, and there was a very real danger that some of these containers would be lost. It is easy to imagine that, in such a case, the containers could eventually be unwillingly retrieved in uncertain conditions by fishermen.

Regular cargo is also lost at sea. This may form a danger either due to its content or because of its size in relation to the local navigable water depth. Cargoes are most frequently lost during bad weather. Under these circumstances it is not until well after the event that the loss is noticed and reported. This means that the record of where the cargo was lost is vague, sometimes encompassing hundreds of square kilometres. During the past 10 years, cargoes lost in the Dutch sector of the North Sea that can form a threat to either the environment or users of the sea, have been registered: in this period over 100 cases have been reported. Dangerous cargoes lost in the past or simply not reported, as well as illegally dumped cargoes such as toxic chemicals, form an ever-present danger, especially to fishermen.

In addition to dangerous goods, all manner of jetsam and industrial hardware, that forms no immediate threat to the environment, find its way on to the sea floor. Medium-size hardware can lead to critical reductions of water depths in shallow parts of the North Sea, particularly in shipping channels that are dimensioned for optimum use. This hazard is particularly important for very large cargo carriers. Fishermen, using bottom trawl gear, play an important role in providing the first accurate information on the location of many lost cargoes. Their reports effectively reduce the search area within which the lost cargo must be pinpointed.

SEARCH

Two situations can be distinguished for locating lost cargoes. These are:

- locating a known dangerous cargo in a roughly defined area;
- searching a defined area such as a shipping channel, anchoring zone or sand extraction area where an unknown number and assortment of objects can be found.

In the first situation the search starts with a very large area. Attention is focused on locating items belonging to the lost cargo. This involves systematically reducing the search area after each find until the whole cargo is found.

In the second situation, the number, size and properties of the hardware that lies in the search area are unknown, so that many surprises await the searcher. Distinguishing between harmless and dangerous finds is not always an easy task. The hardware can range from jetsam, winches, anchors, rudders, hatch covers, trailers and cables to containers with unknown contents and, just to keep the searcher on his toes, mines and the occasional live torpedo.

LOCATION

Once a lost cargo has been found, the exact location of individual items is pinpointed using sonar systems: side-scan sonar, seafloor mapping or sector scanning sonar. These systems are at the disposal of the Rijkswaterstaat and the Hydrographic Service of the Royal Netherlands Navy. Co-operation between these two departments ensures that the most favourable combination of survey vessel and survey system is selected for search operations.

Within the search area, a dense pattern of tracks is surveyed. During this operation the vessel's speed is kept at between 4 and 6 knots to ensure that the sonar information is recorded with a high discrimination level. From the recordings, the position of the sonar contacts is established. Because the sonar fish is towed behind the vessel on a cable, there exists a degree of uncertainty as to the exact position of each contact. Until now, the position and heading of the sonar fish, with its transducers and hydrophones, were not measured but estimated from the cable length, the speed of the ship and the magnitude and direction of the current. The set-off is used with the recorded ship's position to fix the position of each sonar contact: however, this method is inaccurate. An accurate hydro-acoustic positioning system is now available. It is anticipated that, with this system, this problem will soon be solved.

SELECTION

After the sonar contacts, the size, shape and reflection density of each contact are recorded, interpreted and classified: each contact is then charted. The criteria used for classification of sonar contacts vary depending on the nature or characteristics of the cargo sought. Sometimes the dimensions are important, for example in shipping lanes and channels. Sometimes shape and position are the main selection criteria, when for example containers of a known shape are located along the route that the ship followed when they were lost overboard.

Selection is necessary to reduce the more extensive identification phase. This is particularly important when a large number of contacts is found in a search area. In practice, it is sometimes necessary to adopt appropriate selection criteria depending upon early results.

IDENTIFICATION

The best way to identify objects on the sea floor is by actually going down and visually inspecting the item, using divers, and television or still cameras. To do this, the selected contacts are relocated using a hydro-acoustic positioning system and an obstacle avoidance sonar. Once located, visual inspection of an object reveals whether or not it must be classified as dangerous.

Sensors, be they a diver, a television camera or the obstacle avoidance sonar, must be brought to the vicinity of the object on the sea floor. In practice, there are three ways to bring the "sensor(s)" close to the object:

- lowering a diver, with or without handheld instruments, a camera or magnetometer;
- a swimming vehicle; in general a remotely operated vehicle;
- a seabed crawling vehicle, a bottom-bound vehicle on tracks, in short a crawler.

Both ROVs and crawlers are remotely operated, generally through an umbilical cable. These ROVs were originally developed for deep-water applications. With increasing frequency, these robot divers are now also being used in the shallow southern North Sea where poor visibility, high currents and general sea conditions often hamper work operation.

Advantages of a ROV compared with a diver for underwater inspection jobs are as follows:

- it is less sensitive to currents (diver < 0.5 knots; ROV < 2 knots);
- it is less sensitive (and therefore there is less risk) due to bad weather;
- there is less risk during night and twilight operations;
- work can proceed 24 hours a day, so that the endurance of divers does not play a limiting role;
- positioning is more accurate;
- the execution of survey and inspection operations is more systematic;
- objective observations can be made by an expert on board the ROV support vessel, instead of interpretation of diver reports;
- the simultaneous use of a variety of sensors is possible;
- the ROV can be deployed from a dynamically positioned vessel in autotrack mode,

which means that one has an unlimited range of operation.

This means that, although the day-rate of a ROV is three to four times that of a diving team, the price per object inspected with a ROV is half to one third of that when employing divers.

THE USE OF A SWIMMING OR CRAWLING ROV DURING IDENTIFICATION

For more than 7 years, Rijkswaterstaat, the North Sea Directorate, has employed ROVs for underwater inspections. Experience has been gained over the years with a number of ROVs, from *Recon III*, *Filippo*, *CETUS* and *Sub 1000*, to *Scorpi* and *Duplus II*. The *Duplus II* system (Fig. 1) has been used for more than 3 years now with satisfactory results. Minor disadvantages of a swimming ROV are still the restricted use in currents with a speed greater than 2–3 knots. Particularly, when a straight line must be surveyed or the optimum positioning must be achieved, the manoeuvrability and power of the system is sometimes too small. Furthermore, the payload of a swimming ROV is limited. Another factor is that trim should always be taken into account.

These minor disadvantages can be overcome by using a seabed crawling vehicle. This was the reason why the engineers from the North Sea Directorate investigated the applicability of the seabed crawler *Portunus* for North Sea work. The underwater inspection system *Portunus* was originally designed for underwater inspection work during the construction of the Storm Search Barrier in the Eastern Scheldt, in the south of the Netherlands. Now this work is completed, the system is available for other work.

Comparison of the features of a swimming remote operated vehicle with a seabed crawler reveals that a crawler has certain advantages over a swimming ROV for inspections of objects *on* or *in* the sea bed. The results of the comparison are given in Table 1.

Fig. 1 The *Duplus II* ROV used by Rijkswaterstaat, North Sea Directorate, since 1982

TABLE 1
Comparison of operational aspects: crawler (Portunus) and ROV (Duplus)

	ROV	*Crawler*
Operation		
Launch and recovery	+	+
Reliability	+	+
Maintenance	+	°(...)
Mobilization	+	+
Performance		
Manoeuvrability		
horizontal	+	+
vertical	+	–
constant heading	+	++
bearing capacity of the sea bed	++	°(...)
straight track	–	+
Speed	+	+
Payload	°	++
Execute actions	+	++
Systematic survey	+	++
Environmental conditions		
waves	+	+
current	+	++
Potential	°	+

° = Neutral; – = moderate; + = good; ++ = very good.

A swimming ROV will still be preferred for:

- the inspection of objects with large vertical dimensions;
- areas where bottom sediments have a very low bearing capacity (unconsolidated silt);
- some simple spot-checks which do not justify the complex mobilization of the *Portunus* system.

The *Portunus* system was specially designed for the complex inspection tasks and exceptional conditions during the Eastern Scheldt works. For North Sea applications, the system needs to be simplified. The special sensor carriage will be taken off; the wheels for rotating the vehicle within its own length are no longer needed, and a number of steering options can be left out. The modified system will be installed on board the dynamically positioned survey vessel *Mitra* (Fig. 2). The tentative layout of the modified system and the operation procedure are described in Appendix 1.

In the first half of 1986 the *Portunus* system will be modified for North Sea operations. Then, together with the advanced D.P. survey vessel *Mitra,* a strong team will be available for locating and identification on the sea bed.

Fig. 2 The dynamically positioned survey vessel *Mitra.* The moonpool crane for the launch and *recovery of Portunus* is amidships

TWO RECENT SEARCH AND SALVAGE JOBS USING A ROV

Two examples illustrating the approach of Rijkswaterstaat, the North Sea Directorate, are the location, neutralization and salvage of a cargo of cylinders containing liquid chlorine gas, and the investigation and clearance of the 72 ft shipping channel for very large cargo carriers to Rotterdam.

The Chlorine Cylinders Case

During a heavy storm in December 1979, an Iraqi freighter lost 51 cylinders each containing 800 l of poisonous liquid chlorine gas near the Dutch coast. Due to the bad weather, the sailors on board were not immediately aware of the loss, so when the loss was eventually reported the information available on where the cargo was lost was limited. The track of the ship was known within certain margins. However, it was not clear when the cylinders were washed overboard, and if all the cylinders went overboard at the same time. The cylinders could be spread over a large area.

In the following months, an area of 2 × 14 km along the track was searched by survey vessels of Rijkswaterstaat and minehunters of the Royal Dutch Navy. After charting, the sonar contact locations were investigated by divers. The first cylinder was found with the minehunting ROV PAP. On basis of the first finds the area was reduced. Divers identified cylinders at five locations, and these cylinders were subsequently salvaged. Similar sonar contacts were found in the neighbourhood of the first ones.

Despite several attempts, divers could not find more cylinders in the troubled and dark waters. In the period between 1980 and 1984, 10 cylinders were caught by fishermen. Inspection of the cylinders found in 1984 revealed that corrosion had started and that the seals would not hold for long. This could prove fatal for the crew of a fishing vessel, should a container be lifted from the sea bed in a net.

Four years after the accident, a second search and salvage programme started. The area was very systematically surveyed by sonar and ROV. The circumstances at that time were not favourable for the survey. Some cylinders had been spread over a wide area by fishermen, while others had become partly or completely buried in the sandy sea bed. In spite of this, 23 more cylinders were found with a ROV equipped with video, a magnetometer and an obstacle avoidance sonar. The condition of the cylinders after five years on the sea bed made salvage of the filled cylinders too dangerous (Fig. 3). Therefore, after the sea area was cleared of shipping, the cylinders were blown up one by one, and the ripped up cylinders were then recovered. Now, five years after the accident, 13 cylinders still remain to be found. Each cylinder contains enough liquid chlorine gas to form a 300 m^3 cloud of pure gas. As soon as the *Portunus* is available (Fig. 4), another search will be made to find the remaining cylinders.

Fig. 3 One of the 51 lost cylinders containing liquid chlorine gas

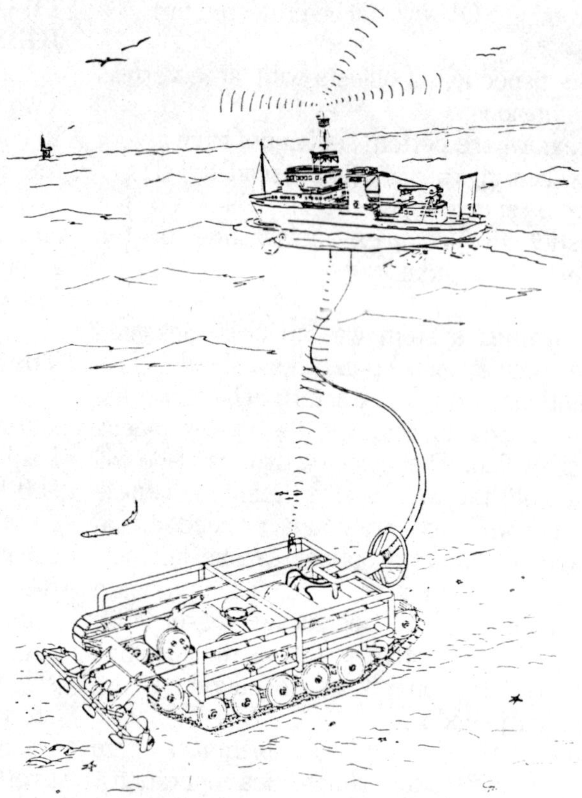

Fig. 4 The modified *Portunus* system operated from the *m.v. Mitra*

Cleaning the Deep-water Shipping Channel to Rotterdam

Between 1982 and 1984, the shipping channel to Rotterdam was dredged to accommodate 72 ft ships. A probabilistic approach was used to determine the underkeel clearance along the route, which runs from the English Channel up through the central southern North Sea to the Hook of Holland.

The actual dredged channel is over 60 miles long. In practice, the underkeel clearance differs along the route depending on the tide, the type of ship and the surge and wave conditions. Under favourable conditions, the underkeel clearance is not more than 2 m. In this respect, it is important that all hardware above the guaranteed maintained depth is detected and removed. In this critical situation not only are small wrecks or standard containers important, but smaller objects like anchors, winches and all kinds of debris also have to be salvaged.

It is not possible to determine from sonar recordings whether the contact is caused by a boulder, an anchor winch or a mine. This problem is solved using a video camera mounted on a ROV in combination with a D.P. vessel. This approach is much more attractive from an operational and economical viewpoint than employing divers from an anchored vessel. Not only the time needed but also flexibility and safety are important aspects to be taken into account. After the contacts are identified and selected, the unacceptable ones are removed in a conventional way by divers.

APPENDIX 1

MODIFIED PORTUNUS SYSTEM

Components

The four main components are:

- the bottom crawler with sensor carriage and sensors;
- the umbilical;
- the operational unit;
- the moonpool crane/umbilical winch.

Bottom crawler (wheels and sensor carriage removed!):

Dimensions: 4.8 × 2.8 × 1.5 m.
Weight in air: 6.5 t.
Weight in water: 4 t.
Maximum velocity: 0.7 m/s.
Pull force: 2.4 t.
Moveable sensor frame; vertical reach 0–2 m; frame rotatable from horizontal to vertical.
Moveable sensor arms on both sides; vertical reach 0–1.5 m; front to alongside horizontal (120°).
Hydraulic cylinder in centre (3 tf); vertical reach 1.5 m.
Active nitrogen pressure system for operational depths up to 200 m.
Two hydraulic motors.
Electronic pods.
Nitrogren filled frame.
Rotating umbilical termination (90°).

Sensors:

In front
video cameras (110°) with adjustable lights (9 × 1000 W) mounted on a moveable and rotatable sensor frame;
obstacle avoidance sonar
magnetometer
Sensor arm
profiler
still camera
colour video camera

At the back
three video cameras (110°) with adjustable lights (fixed positioning)
one pilot camera on umbilical termination (overall view)
obstacle avoidance sonar

Further
responder hydro-acoustic positioning system
gyrocompass
pressure sensor
air lift system
inclinometer

Umbilical:

72 mm ∅
over 300 m
breaking strength: 40 tf
weight in air: 4.5 kg/m
in water: 0.5 kg/m
10 power cables
21 coaxial cables
six twinaxes
two 4mm thick Kevlar coats
one 4.25 mm thick outer coating
Umbilical partly used only due to the omission of specific "Oosterschelde" functions

Operational unit:

installed in 30 ft container
two compartments
entrance (airlock)
operation room (airco)
steering console (tracks + sensorframe + sensor arms + vertical hydraulic cylinder)

video monitors + video tower
sonar display
HPR display
magnetometer display
profiler control unit and display
gyro display
hard copy unit
track plotter
Remote
dynamic positioning system
surface navigation system
ship's gyro

Moonpool crane (available on *Mitra*):
4 × 4 m moonpool
50 t traveller crane
15 t winch with heave compensation system
umbilical sheave
guiding device (for umbilical)
catcher (docking device for crawler/umbilical termination)
moonpool crane for hoisting from keel to deck

Umbilical winch frame (integrated with operational container):
velocity: 30 m/min
umbilical winch 10 t with
heave compensator
constant tensioning (200–1000 kg)
spooling device
operator console + housing
diameter winch: 2.6 m ∅
umbilical winch for hoisting from sea bed to keel

Operating procedure

At the start of operations: crawler on deck; tracks parallel to ship axis; position behind moonpool, moonpool crane above crawler; catcher mounted on guiding device is fixed to crawler (to umbilical termination).

Launch:

- Hoist crawler from deck fixed to guiding device/catcher with moonpool crane.
- Move crawler from horizontal (on deck) to vertical (hanging) position.
- Move moonpool crane to above moonpool.
- Lower crawler through moonpool until at keel level.
- Release catcher; take over by umbilical winch.
- Lower crawler by heave-compensated umbilical winch.
- Land crawler on sea bed (crawler from vertical to horizontal position).
- Take over from heave compensator to constant tensioning.

Operation/inspection on the seabed:

- Follow predetermined track with crawler.
- Ship in D.P./autotrack mode at predetermined position relative to crawler (HPR link).

Recovery (reverse of launch), additional action to be taken before activating the catcher:

- Hoist crawler until 0.2 m under keel.
- Rotate ship in D.P. mode until heading is lined up with crawler gyro.
- Hoist crawler to blocks in catcher.
- Take over by moonpool crane with guidance device, etc.

THE ARCS AND DOLPHIN UNTETHERED REMOTELY OPERATED VEHICLES

R Thomas, International Submarine Engineering Ltd, Canada

Presented on **Wednesday 5 March 1986 at 12.30** in the Norfolk Conference Room

EMBARGOED AGAINST DELIVERY

14

The ARCS and DOLPHIN Untethered Remotely Operated Vehicles

Robert Thomas, International Submarine Engineering Limited, Port Coquitlam, British Columbia, Canada

INTRODUCTION

International Submarine Engineering has developed two types of remotely operated vehicles under the sponsorship of the Canadian Hydrographic Service, Department of Fisheries and Oceans, Bedford Institute of Oceanography. The ARCS (Autonomous Remotely Controlled Submersible) is designed for seabed surveys under the Arctic ice cap. It is characterized by a high degree of autonomous operation and on-board intelligence. It is presently undergoing final development testing in preparation for under-ice trials in the Canadian arctic. The DOLPHIN (Deep Ocean Logging Platform with Hydrographic Instrumentation for Navigation) is designed for surveys of the continental shelf. A prototype vehicle was completed in 1983 and has undergone extensive sea trials. An upgraded version is now nearing completion.

Early work on the ARCS has been reviewed by Jackson and Ferguson (1984), C. Thomas (1985a), and R. Thomas (1985), and on the DOLPHIN by Jackson and Ferguson (1983), Gilbert (1984), and C. Thomas (1985b).

Table 1 compares the general characteristics of the two vehicles. The missions performed by the ARCS are fundamentally limited by the requirement of a self-contained power source, in this case batteries. The diesel-powered DOLPHIN is aimed at missions of approximately the same length, but the new version under development will be capable of refueling in the water. Both vehicles are capable of carrying a variety of instruments and payloads.

TABLE 1
General characteristics of DOLPHIN and ARCS

	DOLPHIN	*ARCS*
Power source	212 hp diesel	10 kWh NiCd battery
Speed	16.5 kn	5 knots
Operating environment	Surface	Under ice
Maximum depth	Sea surface	1000 ft
Telemetry	1200 band UHF	50 band acoustic
Operating range	6.5 miles	5 miles
Endurance	12 h	10–20 h

ARC VEHICLE DESCRIPTION

The ARCS is 17 ft long (extendable to 21 ft with an extra battery/payload section), and 27 in in

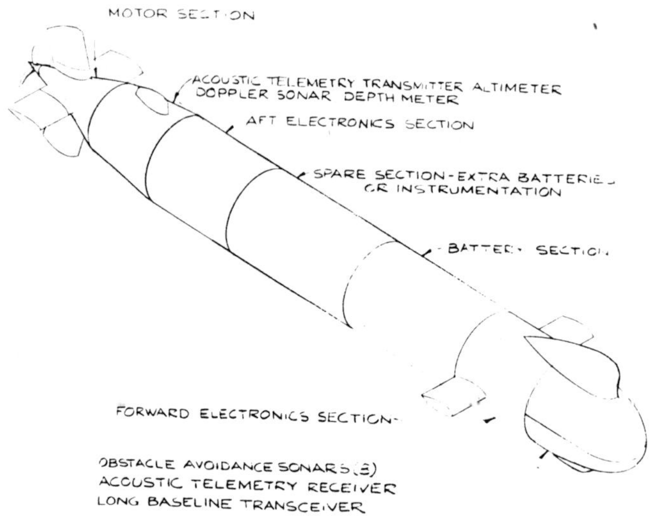

Fig. 1 ARCS: general layout

diameter. The short version weighs 3000 lb, and can operate at depths of up to 1000 ft at 5 knots. Figure 1 illustrates the general arrangement of the vehicle subsystems. They are, from fore to aft:

1. Nose section (flooded)
- 8 Mesotech 980 obstacle avoidance sonars, 107–123 kHz, each with an effective beamwidth of 15–20 degrees;
- Oceano RM101 long baseline transmitter-receiver, 11–14 kHz;
- International Submarine Technology acoustic telemetry receiver, 7.5–10 kHz, 15–17.5 kHz.

2. Forward electronics section (1 atm.)
- main control computers
- acoustic telemetry electronics
- digital tape recorder
- altimeter electronics
- obstacle avoidance sonar electronics
- long baseline electronics
- forward planes controllers and mountings (2)

3. Batteries (1 atm.)
4. Optional second battery or payload (1 atm.)

5. Aft electronics section (1 atm.)
- thruster motor controllers (2)
- doppler sonar electronics
- aft plane controllers (4)
- robertson SKR-80 subsea gyrocompass

6. Aft flooded section
- acoustic telemetry transmitter
- Mesotech 809 digital echo-sounder, 192–208 kHz
- Paroscientific 2400-AS-002 digital pressure transducer
- Ametek MRQ 4015D dual axis doppler sonar, 296–304 kHz

7. Thruster motor section (oil filled)
- brushless DC motors (2)
- aft plane mounts (4)

The ARCS mission requirements call for a high degree of autonomous capability and fault tolerance. Of paramount concern is the ability of the vehicle to detect internal failures and still be able to return to an ice hole for recovery. The reliability of the main control computers is hence of paramount importance. The design of the computer hardware and software has been one of the most demanding tasks of the project.

Figure 2 illustrates the control computer configuration. At the highest level it consists of two identical banks of processors and a Fault Tolerant Monitor (FTM). The FTM monitors the acoustic telemetry link and software-generated 1 Hz square waves output from each processor board in the active bank. The FTM itself consists of three 8051 microcomputers running identical software in independent ROMs. Each is powered by an independent

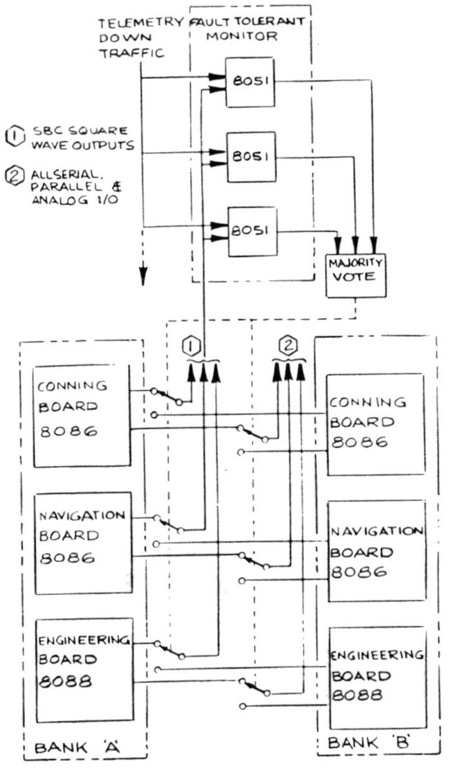

Fig. 2 ARCS control computer configuration

supply, and all outputs are combined in majority vote circuitry.

Each processor bank consists of three Intel Single Board Computers (SBCs). Two of these contain 8086 processors, and one contains an 8088. Two of the three boards also contain 8087 numeric coprocessors. Interfacing to the vehicle subsystems is shared between the boards. The three SBCs share a common back-plane bus and a certain amount of dual-port RAM. The software is written in PLM and runs under the RMX real-time multitasking operating system.

The operator interface with the ARCS is provided by a control panel with on–off switches and planes and throttle levers, and an IBM personal computer.

ARC MISSION PROFILE

Inherent in the overall design of the ARCS design is the assumption that total loss of telemetry and partial loss of acoustic navigational input over moderate periods of time is to be expected in the course of a normal mission.

When the vehicle computers are powered up (with the vehicle out of the water), the control software enters a startup mode in which it expects to receive a series of messages configuring the vehicle for a particular mission. Most of this data consists of navigational information: location of the drop point and long baseline transceivers, a preliminary series of waypoints, parameters used in the navigational algorithms, etc.

The operator then commands the vehicle to enter the manual mode, and proceeds to test the vehicle's response to commands generated at the console. The vehicle can then be detached from the umbilical and lowered through the ice hole, whereupon communication is reestablished via the acoustic telemetry system. At the same time, the vehicle begins to receive acoustic returns from the long baseline transceivers and compute positions which are checked for correspondence to the initial downloaded vehicle coordinates. Next the operator switches the motors on and, if desired, can proceed to drive the vehicle manually from the control panel, either by directly moving the planes or by establishing heading and depth setpoints. Once the operation of the vehicle is seen to be satisfactory, the operator commands a switch to survey mode, whereupon the vehicle proceeds to the first programmed waypoint. Once this point is reached it proceeds in the direction of the second waypoint, and so on.

Whenever an obstacle is detected by the obstacle avoidance sonar, the vehicle attempts to go around, over or under the obstacle. The current approach being taken is a very simple one: vehicle turns, dives, or climbs away from the ostacle, and then continuously attempts to return to the original track, subject to the obstacle no long being within the sonar field. If the time during which this manoeuvre is being carried out exceeds a preset period, of if the distance off track exceeds a preset amount, the waypoint towards which the vehicle had been proceeding is abandoned, and the vehicle attempts to get on track for the following one.

This algorithm may be modified or refined as experience is gained at sea. Once all waypoints have been exhausted, the vehicle heads towards the drop point for recovery in a suspended net.

At any time during a survey, the operator may abort the mission by commanding the vehicle to switch into homing mode. In addition, the vehicle control system initiates an abort under any of the following circumstances:

- the Fault Tolerant Monitor detects failure of the primary processor bank (due to hardware or software failure), and either resets it or switches to the secondary bank;
- the control software determines that too many contiguous waypoints have been abandoned;
- the control software determines that the vehicle has been without surface telemetry for too long a period;
- the control software detects failure of one or more subsystems.

DOLPHIN VEHICLE SUBSYSTEMS

The general configuration of the DOLPHIN is shown is Fig. 3. The hull is divided into the following compartments:

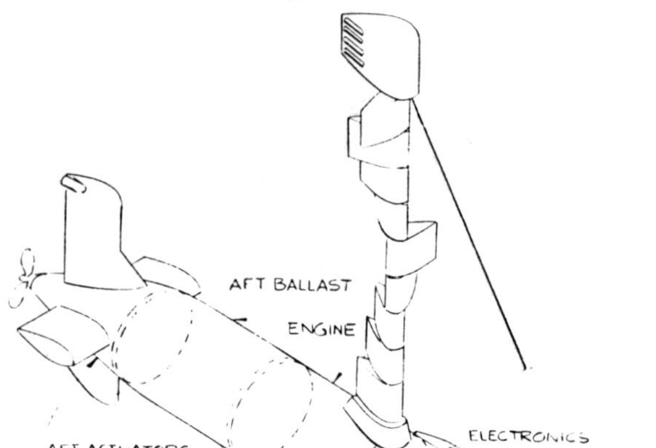

Fig. 3 DOLPHIN: general layout

1. Forward ballast compartment.
2. Forward actuator compartment:
- soft fuel tank
- forward plane actuators
3. Electronics compartment:
- main computers
- power distribution box
- pressure transducer
- radio
- attitude sensors
- compass or gyro
- payload space (12 ft^3)
4. Engine compartment:
- SABRE 212 hp turbo-charged diesel
- bilge pump
- hydraulic pumps
5. Aft ballast compartment:
- aft ballast tank
- air bottles
6. Aft actuator compartment:
- rudder and aft plane actuators
- exhaust fairing

Projecting down 3 ft from the bottom of the hull is a 1500 lb lead keel. Segmented freely rotating fibreglass fairings on the mast are provided for drag reduction. A valve assembly at the top of the mast allows the mast to submerge completely, although the normal average running depth leaves approximately half of the mast above the surface.

The two main vehicle computers, designated Vehicle I/O and Active Controls, are two Siemens SMP systems featuring 8088 microprocessors. The Vehicle I/O computer controls the VHF radio link and the engine. The Active Controls computer provides planes control and most of the interfaces to peripheral devices (gyro or magnetic compass, etc.). They reside on independent backplanes and communicate with each other via a serial channel. Software is written in the C language running under the AMX real-time multitasking operating system. The software algorithms implement second-order linear multivariate control.

Ihherent in the design of the telemetry communications protocol is the capability for controlling multiple DOLPHINs independently over a single time-multiplexed channel. For route surveys the goal is to support simultaneous operation of six vehicles, three on either side of the support ship.

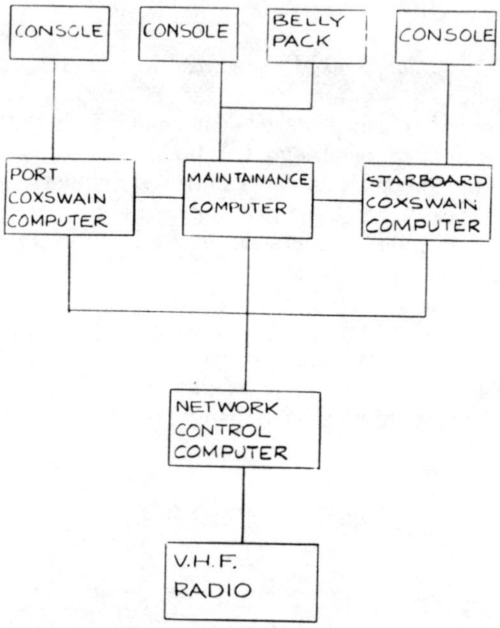

Fig. 4 DOLPHIN surface computer configuration

Figure 4 illustrates the surface control computer configuration provided to support this capability. Three independent control consoles are monitored by IBM personal computers. These are designated the Port Coxswain, Starboard Coxswain, and Maintenance consoles. The Port and Starboard Coxswain consoles each control up to three vehicles, independently or in unison. The Maintenance console is used during launch and recovery. It differs from the other two by the addition of a keyboard and a secondary console configured as a 'belly pack', allowing the operator to move away from the computer. The Maintenance console acts as a master by granting control of a vehicle to either Coxswain console once the vehicle has been moved away from the launch point. When a vehicle is to be retrieved, the Maintenance console resumes control and guides the vehicle back to the recovery point. Communications between any of the control consoles and a vehicle or vehicles over the radio link is managed by the Communications Control Computer, a Siemans SMP system similar to the vehicle units. The four computers communicate via RS-232 serial channels.

CURRENT STATUS AND FUTURE DEVELOPMENTS

The Bedford Institute of Oceanography has ordered three DOLPHIN vehicles with future orders anticipated. Sea trials are expected to begin late in November 1985. A slightly modified version of DOLPHIN, designated Sea Lion, has been ordered by the US Navy, with an option for a second vehicle.

The DOLPHIN is capable of being configured to support a wide variety of civilian and military missions. These include:

- geophysical survey
- parallel hydrographic survey
- mine countermeasures route survey
- torpedo decoy
- communications jamming and deception
- missile countermeasures
- over-horizon targetting
- submarine communications support
- forward communications relay
- coastal surveillance
- beach illumination

In the evolution of the DOLPHIN, we anticipate the development in the near future of an autonomous capability similar to that of the ARCS, possibly incorporating navigation via the Global Positioning System.

REFERENCES

Gilbert, R. L. G. 1984. Field trials of Dolphin. In: ROV 1984 *Conference Proceedings,* Marine Technology Society, San Diego, CA, pp. 348–355.

Jackson, E. and Ferguson, J. 1983. Design and development of a diesel powered semi-submersible ROV. In: ROV 1983 Conference Proceedings, Marine Technology Society, San Diego, CA, pp. 277–281.

Jackson, E. and Ferguson, J. 1984. Design of ARCS — Autonomous Remotely Controlled Submersible. In: *ROV 1984 Conference Proceedings,* Marine Technology Society, San Diego, CA, pp. 365–368.

Thomas, C. 1985a. Sea trials and operation of the ARCS autonomous submersible. In: *ROV 1985 Conference Proceedings,* Marine Technology Society, San Diego, CA, pp. 122–125.

Thomas, C. 1985b. Continuing program of the Dolphin semi-submersible. In: *Proceedings of the Fourth International Symposium on Unmanned Untethered Submersible Technology,* Marine Systems Engineering Laboratory, University of New Hampshire, Durham, N.H., pp. 223–229.

Thomas, R. 1985. Control software in the ARCS vehicle. In: *Proceedings of the Fourth International Symposium on Unmanned Untethered Submersible Technology,* Marine Systems Engineering Laboratory, University of New Hampshire, Durham, N.H., pp. 334–342.

Part IV

Environmental Modelling

15

Activities of the WAM (Wave Modelling) Group

G. J. Komen, Chairman, WAM, KNMI, De Bilt, The Netherlands

The WAM group is an international group of wave researchers, working on the joint development of a so-called third generation ocean wave model. This is a model in which no *a-priori* assumptions are made about the spectral shape and in which source terms are chosen to represent the different physical processes determining the evolution of ocean waves. After a short description of the model, first results from a two-day global test run with realistic winds will be discussed. In the second part of the chapter a short overview of the plans of the WAM group will be given. They comprise growth curve analysis, directional relaxation studies, shallow-water effects and data assimilation.

INTRODUCTION

There exists considerable interest in ocean wave research, and in the development and application of numerical wave models. This interest is strongly stimulated by demand. Worldwide offshore and coastal engineering projects require good knowledge of local wave climates. In addition, shipping, fisheries and offshore operations depend on the availability of accurate wave forecasts. These are usually made with the help of numerical wave prediction models, which are also used to reconstruct wave climates, where observations are lacking.

In recent years, important developments have been made in ocean wave modelling. Our understanding of the dynamics of surface waves has increased significantly as a result of a series of field programs and because of experience with a sequence of first- and second-generation wave models (JONSWAP, Hasselmann *et al.*, 1973; SWAMP, Allender *et al.*, 1984; SWIM, Bouws *et al.*, 1985). The increasing understanding has led to an increasing sophistication of methods which, in turn, has led to increasing collaboration between different wave modelling groups. For example, the SWAMP study involved six European, two American and two Japanese research teams.

The European wave modelling community is currently in the process of developing a new third-generation wave model which may be expected to yield a further significant improvement in wave forecasting skill. The present proposal outlines a plan of action

ACTIVITIES OF THE WAM (WAVE MODELLING) GROUP

G J Komen, Royal Netherlands Meteorological
Institute, The Netherlands

Presented on **Wednesday 5 March 1986 at 10.00**
in the Clarence Conference Room

EMBARGOED AGAINST DELIVERY

directed at further development of this third-generation model, its implementation in a global version on a central efficient vector-processing computer, and the development of regional versions of the model for regional use.

Present wave-predication models are based on the integration of the radiative transfer equation for the two-dimensional wave spectrum. Simple empirical wave-prediction tables relating wind or sea parameters such as the significant wave height and period to the wind speed, fetch and duration are still sometimes used in engineering applications, but they are often inadequate. The transfer equation describes the propagation of the different wave components of the spectrum with different frequencies and propagation directions at their appropriate group velocities, and the change in the energies of these components produced by wind forcing, dissipation and higher-order non-linear wave–wave interaction. The models predict the full two-dimensional wave spectrum at every timestep and grid point. Present second-generation wave models have to make *a-priori* restrictions on the spectral shape, because an accurate representation of the non-linear transfer process was not yet available. The third-generation model will be based on a complete representation of all source terms without any *a-priori* restrictions on the spectral shape. The SWAMP study has clearly indicated that such an approach is essential for a correct description in the important situation of wind-sea to swell transition and in case of turning wind fields when directional relaxation is relevant.

The availability of an operational global wave model will be essential if one wants to make full use of the potential of the first European remote sensing satellite, ERS-1. The high rate of surface wind and wave data to be produced by ERS-1 can be only effectively exploited if one makes use of a large central facility. However, the task of processing and assimilating satellite data in a wave model will require research efforts exceeding the potential of individual countries, and just like the development of the wave model, this will be best made in joint collaboration.

OBJECTIVES OF WAM

The primary objectives of WAM are:

- to jointly develop a third-generation wave model, based on a full description of the physical processes governing wave evolution;
- to develop regional versions of the third-generation model to be nested with the global model — these models could be run by national weather bureaux;
- to implement the global model for operational medium range forecasting (10 days) on the CRAY computer of the European Centre for Medium Range Weather Forecasts (ECMWF) in Reading (or possibly some similar institute);
- to perform physical studies of wave dynamical processes in order to extend our understanding of wave evolution, where needed;
- to develop data assimilation techniques which will make it possible to make full use of the ERS-1 potential for wave forecasting.

FIRST RESULTS WITH THE GLOBAL MODEL

Short Description of the Model

The evolution of the two-dimensional ocean wave spectrum $F(f, \theta, \phi, \lambda, t)$ with respect to frequency f and direction θ (measured clockwise relative to true north) as a function of latitude ϕ and longitude λ on the spherical Earth is governed by the transport equation

$$\frac{\partial F}{\partial t} + (\cos\phi)^{-1}\frac{\partial}{\partial\phi}(\dot{\phi}\cos\phi\, F) + \frac{\partial}{\partial\lambda}(\dot{\lambda}F) + \frac{\partial}{\partial\theta}(\dot{\theta}F) = S \quad (1)$$

where S is the source function and

$$\dot{\phi} = \frac{d\phi}{dt} = vR^{-1}\cos\theta, \quad (2)$$

$$\dot{\lambda} = \frac{d\lambda}{dt} = v\sin\theta\,(R\cos\phi)^{-1}, \quad (3)$$

$$\dot{\theta} = v\sin\theta\tan\phi\, R^{-1} \quad (4)$$

represent the rates of change of the position and direction of a wave packet propagating along a great circle path ($v = g/4\pi f$ is the group velocity and R the radius of the Earth).

Equation (1) follows from the energy conservation equation

$$\frac{\partial \hat{F}}{\partial t} + \frac{\partial}{d\phi}(\dot{\phi}F) + \frac{\partial}{\partial \lambda}(\dot{\lambda}F) + \frac{\partial}{\partial \theta}(\dot{\theta}\hat{F}) = \hat{S} \quad (5)$$

for the four-dimensional spectral density $\hat{F}(f, \theta, \phi, \lambda)$, which is related to the normal spectral density of F with respect to a local cartesian frame through $\hat{F}\, df\, d\theta\, d\phi\, d\lambda = F\, df\, dx\, dy$, *or*

$$\hat{F} = FR^2 \cos \phi \quad (6)$$

Equation (1) applies also in the presence of refraction through bottom topography, if the refraction equation (4) is appropriately modified, but this is not incorporated in the present model.

The source function is represented as a superposition:

$$S \times S_{in} + S_{nl} + S_{ds} \quad (7)$$

of the wind input, non-linear transfer and dissipation source functions.

The wind input source function

$$S_{in} = \max.\ \{0, 0.25 \frac{\rho_a}{\rho_w} (28 \frac{u^*}{c} \cos \theta - 1) \times \omega F(f, \theta, \phi, \lambda)\} \quad (8)$$

(where ρ_a and ρ_w denote the densities of air and water, respectively, and $\omega = 2\pi f$) follows Snyder *et al.* (1981), but is scaled in terms of u^* instead of u_5 following Komen *et al.* (1984).

The dissipation source function

$$S_{ds} = -3.33 \times 10^{-5}\, \bar{\omega}(\omega/\bar{\omega})^2 (\hat{\alpha}/\hat{\alpha}_{PM})^2 F \quad (9)$$

is based on the general white-capping dissipation function proposed by Hasselmann (1974). The coefficient factor and $\hat{\alpha}$ dependence was determined empirically through numerical integrations of (1) using different forms of S_{ds} for given S_{in}, and S_{nl}, which was computed exactly (Komen *et al.*, 1984). In Equation (9)

$$\bar{\omega} = E^{-1} \int\int F(f, \theta)\ \omega\, df d\theta, \quad (10)$$

where

$$E = \int\int F(f, \theta)\, df d\theta,$$

denotes the mean frequency,

$$\hat{\alpha} = E \bar{\omega}^{-4} g^{-2} \quad (11)$$

represents an integral wave steepness parameter, and

$$\hat{\alpha}_{PM} = 4.57 \times 10^{-3} \quad (12)$$

is the theoretical value of $\hat{\alpha}$ for a Pierson-Moskowitz spectrum. The non-linear source function S_{nl} was parameterized according to Hasselmann *et al.* (1984). The model underwent various tests (fetch limited growth and other SWAMP cases) all of which were satisfactory and encouraging. For details, refer to Hasselmann and Hasselmann (1984).

Numerical Implementation on a Global Grid

For the advection term, a first-order upwind difference scheme is used which conserves energy exactly:

$$F(x, t+\Delta t) = F(x,t) + \Delta t S(x,t) + \Delta t \sum_i \left\{ \frac{(\dot{x}_i F)_{x-\Delta x_i} - (\dot{x}_i F_{x_i})}{\Delta x_i} \right\}$$

where $x - \Delta x_i$ is the upstream grid point in the i-direction.

The rate of numerical dispersion of a wave group appears reasonably consistent with the dispersion required theoretically for the finite frequency and angular resolution of the spectrum (Allender *et al.*, 1984).

The theoretical source functions are integrated in the present version of the model with an implicit numerical scheme. The time step is fixed at 20 minutes. The model has 26 frequencies running from 0.04 to 0.42 Hz on a logarithmic scale, with $\Delta f/f = 0.1$, and 12 directional bands (30° resolution). Beyond the high-frequency cut-off, limiting the region of the spectrum which was treated prognostically, an f^{-4} tail is added, with the same directional distribution as the last frequency band of the prognostic spectrum. The model has been implemented and tested for the North Atlantic with a 1° × 1° latitude–longitude and resolution and in a global version with 3° × 3° resolution

In the next section we will give results from a global test run. It should be noted that the model computes two-dimensional spectral distributions without any prior constraints on the spectral shape. This is the principal advantage of the third-generation model over present second-generation models (cf. Allender *et al.*, 1984).

Global Test Run for a Real Wind Case

The global wave model was tested for a two-day forecast using the wind stresses computed by the ECMWF (European Centre for Medium Range Weather Forecasts) operational model for the period 12.00, 13 October–12.00, 15 October 1983. The stress fields were provided every 20 minutes.

As an initial state of the wave field, a uniform JONSWAP spectrum with a mean direction of 90° (eastward travelling waves) and a (high) mean frequency of 0.3 Hz was assumed. The wind-sea adjusts to the wind field direction in about 10 hours. Figure 1 shows the wave field after 44 hours of integration. At that time, deviations between the wind and wave direction have arisen through changes with time in the local wind direction or swell propagation from other grid points.

Of course, a plot such as Fig. 1 hides an enormous amount of information. In every grid point not only the wave energy and the mean wave direction is known, but indeed the full two-dimensional spectrum is available. For special applications one can zoom in on the model. For example, Fig. 2 gives an enlarged representation of the region indicated as A in Fig. 1. Figure 3 gives the full two-dimensional spectrum in the point labelled V. This spectrum clearly shows the complex superposition of changing wind-seas and multiple swells produced by the inhomogeneous, non-stationary wind field. The origin of the various peaks in the spectra can be readily reconstructed in hindsight through inspection of the full wind history, but it would clearly have been difficult to reproduce such spectra with existing second-generation wave models, in which the wind-sea–swell spectra transition is not modelled from first principles through the source function and the shape of the wind-sea spectrum is largely prescribed.

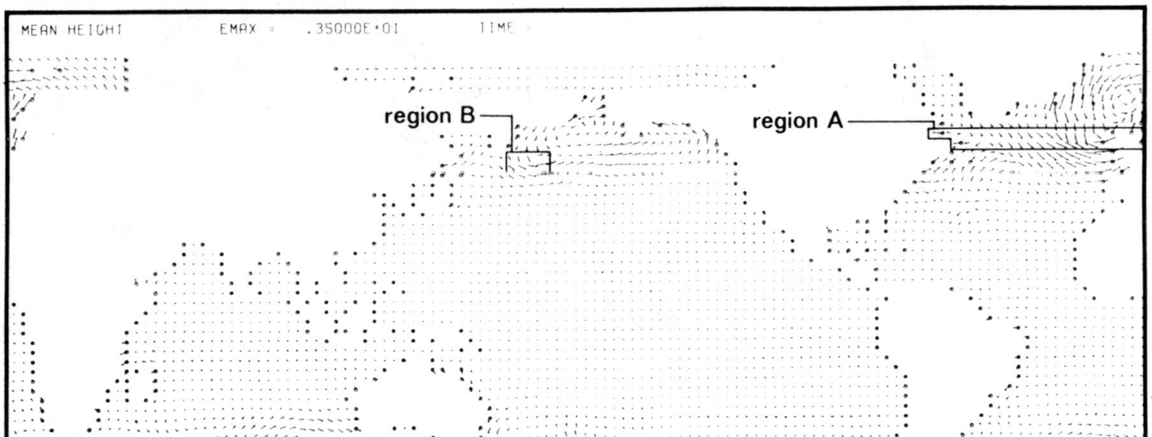

Fig. 1 Significant wave height and direction at 08.00 hrs, 15 October 1983, 44 hours after initial time

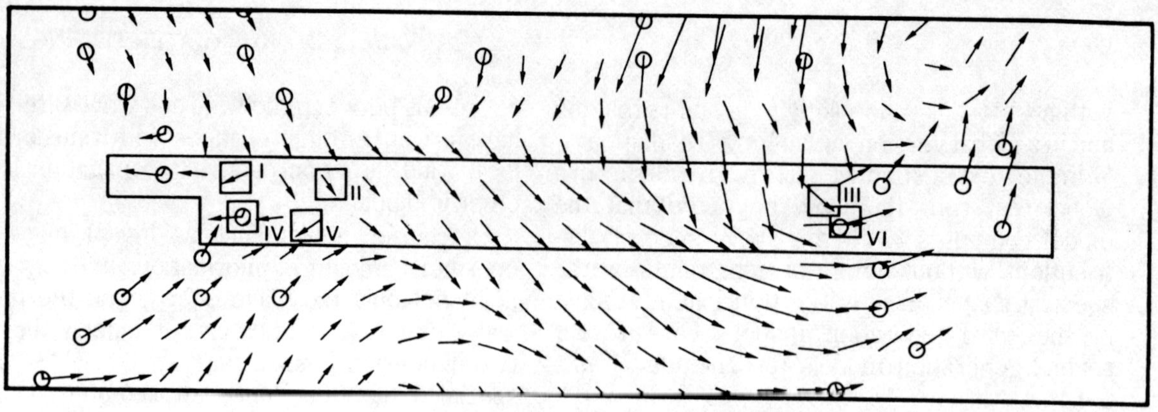

Fig. 2 Significant wave height for region A, same time as Fig. 1

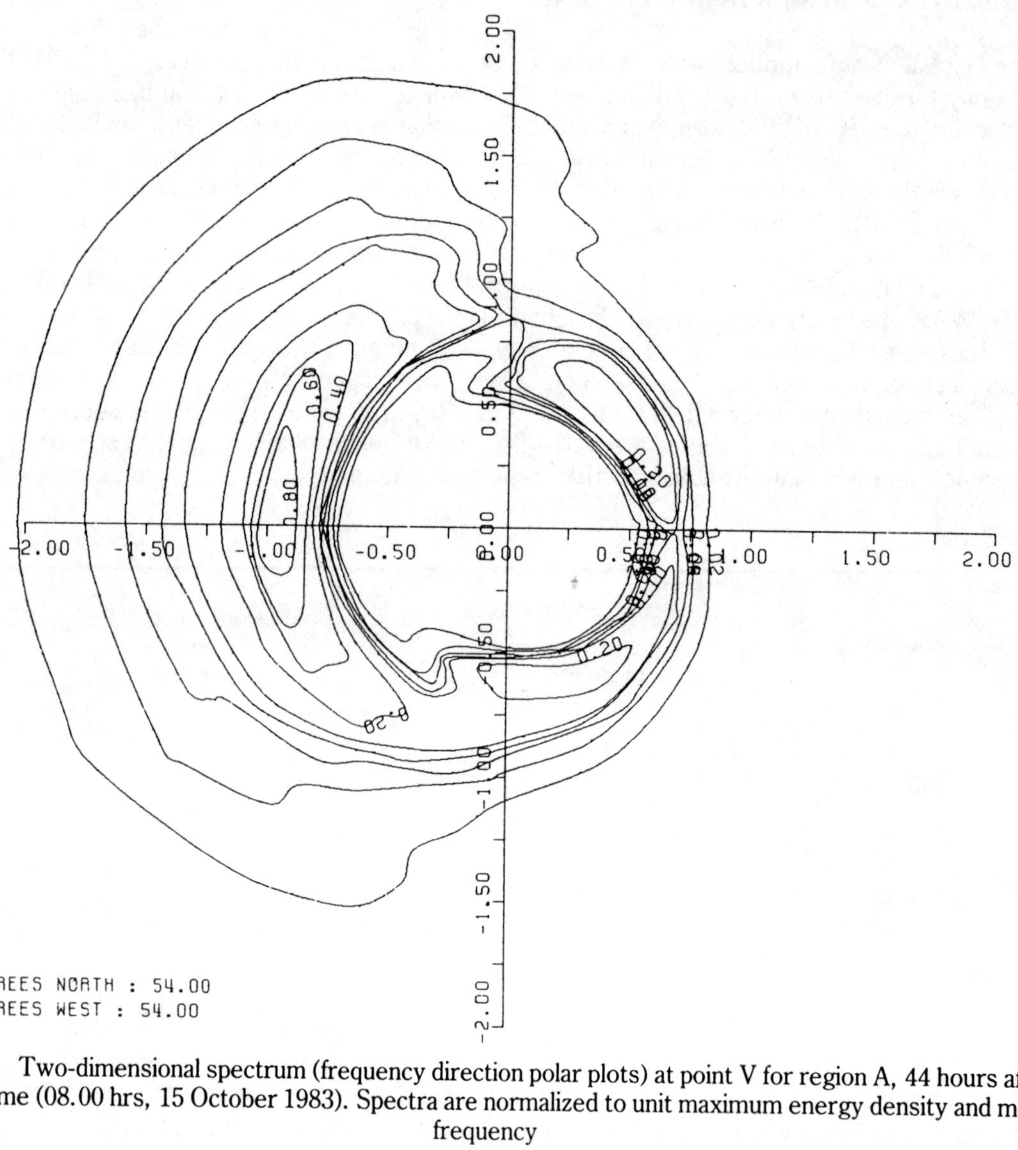

Fig. 3 Two-dimensional spectrum (frequency direction polar plots) at point V for region A, 44 hours after initial time (08.00 hrs, 15 October 1983). Spectra are normalized to unit maximum energy density and mean frequency

STRUCTURE OF THE WAM GROUP

As indicated above, the WAM group consists of more than 30 scientists from 10 different institutes. Therefore, several tasks can be undertaken simultaneously, some of which go beyond the actual modelling effort. To this end, working groups have been formed.

One such group will concentrate mainly on the practical problem of further developing, implementing and validating the new wave model. Their activities comprise such items as documentation, semi-operational implementation (determination of mode of operation) and model validation. They will also consult potential users, in order to further specify the output format and archiving requirements.

Another group is focussing on more general wave dynamical problems. These problems have been clearly isolated in the SWAMP and SWIM studies. In fact, both studies suggested further research, and the time seems ripe to follow these suggestions. The three main topics that have been identified are:

- to re-analyse existing data on fetch limited wave growth;
- to study directional properties of the wave spectrum;
- to investigate the nature of shallow-water limitations to wind-sea.

A third group will consider the wave-data assimilation problem. Here the problems are relatively new, since no experience with wave-data assimilation exists so far. Therefore, it is proposed that, initially, the problem of assimilation from one fixed position will be considered. At a later stage, the more complicated problem of wave-data assimilation from satellites can then be tackled. This exercise will not only improve our ability to make higher-precision global wave forecasts, but it is also expected to play an important rôle in world climate studies. In fact, it is now realized (Working Group on Wind and Wave Data, 1985) that "the effective application of satellite wind and wave data for climate studies requires the implementation of a complete, quasi-real time, operational end-to-end system, starting from satellite sensor algorithms operating on the (calibrated and collocated) physical sensor data on the input side, and producing as output continuous, global wind and wave fields with the space–time resolution appropriate to operational high resolution atmospheric and wave forecast models". The wind fields can then be used to drive coupled ocean–atmostphere models, and to investigate the response of the ocean to global atmospheric forcing, a crucial element in the climate problem.

CONCLUSION

Initial results have been presented from a third-generation global wave model which are promising and encouraging. It is hoped that the model can be put into operational use after a test period in which the model will be validated. The model will be used to perform wave-data assimilation studies, which will both help increase the reliability of forecasts and allow optimal use to be made from future satellite observations. Work on regional versions of the model has not been discussed here at any length. However, this work is well under way; results will be presented elsewhere. Finally, we indicated how the model development activities are embedded in a larger plan for physical wave research.

REFERENCES

Allender, J. H., Barnett, T. P., Bertotti, L., Bruinsma, J., Cardone, V. J., Cavaleri, L., Ephraums, J., Golding, B., Greenwood, A., Guddal, J., Günther, H., Hasselmann, K., Hasselmann, S., Joseph, P., Kawai, S., Komen, G. J., Lawson, L., Linné, H., Long, R. B., Lybanon, M., Maeland, E., Rosenthal, W., Toba, Y., Uji, T. and de Voogt, W. J. P. (the SWAMP group) 1984. *Ocean Wave Modelling, Part I, The Sea Wave Modelling Project (SWAMP). An Intercomparison Study of Wind Wave Prediction Models, Principal Results and Conclusions.* Plenum Press.

Bouws, E., Ephraums, J., Ewing, J. A., Francis, P., Günther, M., Janssen, P. A. E. M., Komen, G. J., Rosenthal, W. and de Voogt, W. J. P. (the

SWIM group) 1985. An intercomparison of shallow-water wave prediction models. *Q. J. R. Met. Soc.* (in press).

Hasselmann, K. 1974. On the spectral dissipation of ocean waves due to white capping. *Boundary Layer Met.* **6**, 107–127.

Hasselmann, K., Barnett, T. P., Bouws, E., Carlson, H., Cartwright, D. E., Enke, K., Ewing, J. A., Gienapp, H., Hasselmann, D. E., Kruseman, P., Meerburg, A., Müller, P., Olbers, D. J. Richter, K., Sell, W. and Walden, H. 1973. Measurements of wind-wave growth andswell decay during the Joint North Sea Wave Project (JONSWAP). *D. Hydrogr. Z.* **A8**, no. 12.

Hasselmann, S. and Hasselmann, K. 1984. A global wave model. Unpublished note.

Hasselmann, S., Hasselmann, K., Allender, J. H. and Barnett, T. P. 1984. Improved methods of computing and parameterizing non-linear energy transfer in a gravity wave spectrum (submitted to *JPO).*

Komen, G. J., Hasselmann, S. and Hasselmann, K. 1984. On the existence of a fully developed wind-sea spectrum. *JPO* **14**, 1271-1285.

Snyder, R. L., Dobson, F. W., Elliott, J. A. and Long, R. B. 1981. Array measurements of atmospheric pressure fluctuations above surface gravity waves. *J. Fluid Mech.* **102**, 1–59.

Working Group on Wind and Wave Data 1985. Report, in: *The Use of Satellite Data in Climate Models.* Proceedings of a Conference held in Alpbach, Austria, 10–12 June. ESA SP–244.

ON THE APPLICATION OF MIROS DATA IN OCEAN WAVE MODELLING

O Gronlie, Miros A/S, J Guddal, The Norwegian Meteorological Institute (DNMI); B Hilmarsen, Esso Norge A/S, Norway

Presented on **Wednesday 5 March 1986 at 10.30** in the Clarence Conference Room

EMBARGOED AGAINST DELIVERY

16

On the Application of MIROS Data in Ocean-Wave Modelling

Øistein Grønlie, MIROS A/S, 1370 Asker Johannes Guddal, DNMI, 5000 Bergen, Norway
and *Bernhard Hilmarsen, Esso Norge A/S, 4000 Stavanger, Norway*

MIROS is a microwave radar system for real-time measurement of directional ocean wave spectra and surface currents. MIROS is the only system of its kind available on a commercial basis. At the present time, several systems have been delivered and will be put into operation in the North Sea in the near future.

During the period from December 1984 until May 1985 one system has been operational from Esso's *Odin* platform in the Norwegian sector of the North Sea under contract with Esso Norge A/S. One important aspect regarding this installation is that other wave and current measurement programs have taken place in the same period; thus data for intercomparison studies will be available.

This chapter starts with a brief description of the MIROS system which includes the most important characteristics of both hardware and software. The principles of operation are also reviewed.

The MIROS directional wave spectra contain amounts of information concerning the directional distribution of sea and swell as well as the unidirectional spectrum and total energy. A new ocean-wave model is being implemented at DNMI (the Norwegian Meteorological Institute). This model simulates the full two-dimensional (direction/frequency) wave spectrum, generated from successive windfields over the N.E. Atlantic, North Sea and Norwegian Sea. An interesting application of MIROS data is in verification and calibration of the wave model. The experience obtained investigating the applicability of MIROS data from the *Odin* installation for this purpose is the main subject of the chapter.

INTRODUCTION

Availability of directional wave information is becoming increasingly more important in several fields related to offshore activities. The present applications involve both the need for statistical information as well as real-time data and forecasts. It is of course impossible to cover all ocean areas of interest with sensors, and therefore a complete maritime weather service system will incorporate both various types of sensors and a wave model, which can be run in both a hindcast and a forecast mode. DNMI (the Norwegian Meteorological Institute) has implemented the WINCH wave model. The windfields used as input are taken from the six-hourly routine wind analysis at DNMI.

A wave model does certainly not eliminate the need for measured data. A wave model must be adapted to a certain geographical area by tuning, and problems with correct windfield estimation render the assimilation of real-time measured wave data in the model necessary, to prevent accumulated errors. Measured data will also be necessary for model verification.

MIROS measures the directional wave spectrum as well as components of the surface current. Wave data from a MIROS system seem to be well suited for the applications described.

The chapter begins with a short review of the MIROS system. Then the application of MIROS data in the maritime weather service is discussed in general terms, followed by a short description of the DNMI wave model. During the winter of 1984/85 one MIROS system was operational from Esso's *Odin* platform in the North Sea. A short description of this installation is given. In a preliminary data analysis, measured data from the MIROS *Odin* installation have been compared with DNMI model hindcast data.

THE MIROS SYSTEM

MIROS is a microwave radar system for real-time directional measurements of ocean wave height frequency spectra and surface currents. MIROS is primarily designed for operation from stationary offshore installations. The MIROS hardware consists an outdoor-mounted microwave sensor head connected by a cable assembly to an indoor-installed central processor cabinet (see Fig. 1). The sensor head is to be installed on the top of the platform building section or at another convenient location giving 50 to 100 m elevation above sea level.

The MIROS software is written in extended PASCAL.

The directional properties of MIROS depend on the actual installation geometry and the sensor head antenna beamwidth. A typical geometry gives resolution better than 30° for waves up to approximately 200 m wavelength. The resolution then gradually decreases for longer waves.

During a typical measurement sequence, observations are taken in six directions with an angular increment of 30°. This represents a 150° rotation of the antenna assembly, which is sufficient for a complete scan since the radar

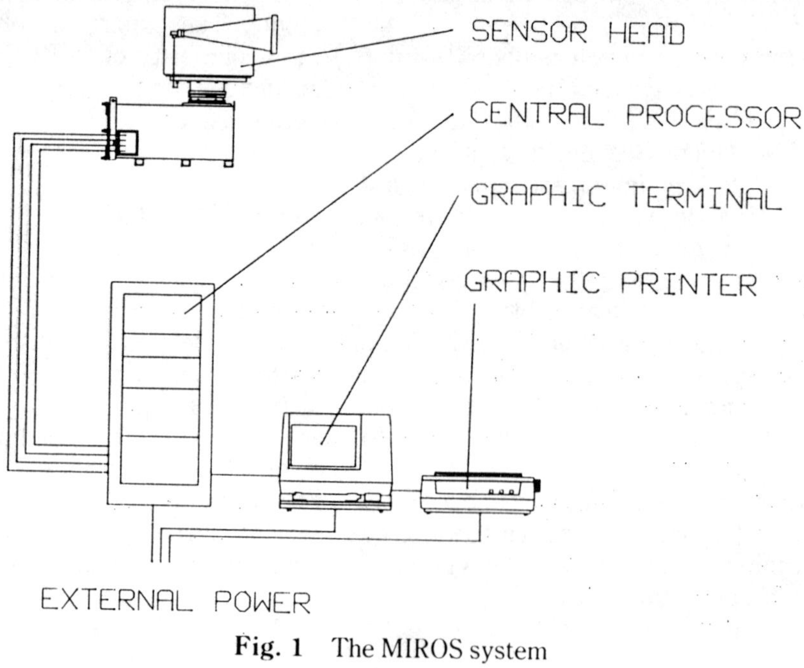

Fig. 1 The MIROS system

can observe both approaching and receding waves. A typical observation time per direction is 12 minutes, giving a total observation time of 72 minutes for a complete directional scan.

MIROS operates in two different modes, a pulse-döppler mode and a dual-frequency mode. Directional wave height–frequency spectra are measured in the pulse-döppler mode, basically obtaining spatially averaged water-particle velocity information. A wave model is applied to transform the velocity spectrum to a wave height spectrum. Surface current information is obtained by measuring the phase velocity of a gravity-wave component in the dual-frequency mode.

The output wave data format is an array with 40 equidistant frequency bands and 12 (6) directions. Presently, additional information is needed to resolve an inherent 180° ambiguity. The frequency bands range from 0.01 Hz to 0.4 Hz, and the resolution is 0.011 Hz.

APPLICATION OF MIROS DATA IN MARITIME WEATHER SERVICE

Hindcasting

For climatological purposes, numerical wave models have been used to produce time-series of wave parameters, such as:

- significant wave height,
- "peak" period, and even
- full directional/frequency spectra

at locations and times where no measurements have been available.

In order to assess model performance, reliable spectral information from measurements is compared to corresponding model spectra, in particular in the case of extremes, i.e. storms with rapidly changing wind. Results from such comparisons might lead to the reconstruction of certain algorithms within the numerical model, i.e. the simulation of wave directional relaxation in connection with rapidly propagating storms.

Nowcasting

For certain wave sensitive operations, such as diving, lifting, etc., special attention will be paid to the onset of long-period swell systems. With a real-time operating MIROS, one will have the opportunity to detect the early stage of long periodic components arriving at the location of interest, and also:

- to identify their generation area by means of their measured direction;
- in co-operation with weather forecasters (with wave models) to decide whether the phenomena are expected to be temporary or to last for a longer period of time;
- to advise operational management on the actual operation.

Forecasting

A numerical wave model used for forecasting is, in principle, the same as the one used for hindcasting, which means that there is the same need for model quality assessment and tuning (reconstruction).

Additionally, a wave model requires the definition of the sea state for a large ocean area at some initial time, t_0. In principle, this should imply the full knowledge of the spectral density function

$$F = F(f, \theta)$$

where f = frequency, and θ = partial direction, for every required grid point in a 75 (or 25) km grid network covering all ocean areas that could generate wave systems to travel towards the North Sea.

In practice, this is impossible. What a model does to define the initial sea-state field, is to let the successive windfields, as input to the model, generate the initial wave fields by a 36 (or 48) hours preceeding integration starting from zero sea state ($F = 0$ for all locations). From this initial (although artificial) sea state, the actual prognostic integration carries on for a 48- or 72-hour forecast period.

However, errors accumulate in the initial field during model run. It has been demonstrated, i.e. for closed basins like the North Sea, that assimilation of even limited amounts of real-time measurement data into the initial fields gives significant benefits to wave-forecast prognosis.

Network Requirement

Certain requirements are stated to the wave-sensoring network supplying the measured data:

- adequate directional information/resolution;
- continuous time-series provision;
- geographical coverage of certain extents, in compliance with operational marine weather service requirements.

A realistic future system will be a blend of buoy and fixed platform based (MIROS) systems. Recommendations for platform instrumentations come partly from national meteorological agencies and joint committees, i.e. the North Sea Panel.

DNMI'S WAVE MODEL

DNMI has implemented the WINCH wave model, which was developed by Vincent Cardone of Oceanweather Inc. The model is of the discrete coupled type and operates with 15 frequency and direction bands, so that the directional resolution is 15°. The frequency bands range from 0.04 Hz to 0.239 Hz, with a ratio between adjacent bands of $3^{1/9}$.

All spectral bins are propagated individually. Growth for each spectral bin is the largest of a parametric increment and an input term of the Miles type, i.e. proportional to the energy in the band.

The parametric growth term is derived from empirical fetch-limited growth data and forces the growing wind-sea to conform with a reference spectrum that possesses the most important characteristics of the JONSWAP spectrum, which can be integrated analytically and reduces to the Pierson–Moskowitz (PM) spectrum as the sea approaches full development. Growth is scaled to the friction velocity, dimensionless energy and the PM spectrum to the wind speed at 10 m above sea level.

The high-frequency tail of the spectrum is restored to an αf^{-5} form, where α is related to the dimensionless energy. Thereafter, a directional redistribution is performed so that frequencies below the PM peak frequency are left unaltered, i.e. they propagate as swell, and frequencies above are subjected to relaxation.

The time step is 2 hours.

The grid is a nested rectangular mesh on a transverse Mercator projection, with an assumed equator along the Greenwich meridian. The grid spacing is 75 km on the fine mesh and 150 km on the coarse mesh. The integration area is shown in Fig. 2.

The windfields used as input are taken from the six-hourly routine wind analysis at DNMI.

MIROS AT ESSO'S *ODIN* PLATFORM

During the period from December 1984 until May 1985, one MIROS system has been operational from Esso's *Odin* platform in the Norwegian sector of the North Sea (Fig. 3), under a contract with Esso Norge A/S. One important aspect regarding this installation is that other wave and current measurement programs have taken place in the same period; thus data for intercomparison studies will be available.

This exercise represents the first real offshore experience with a MIROS system. Apart from some minor technical problems at the beginning, from the manufacturer's point of view the experiment has been a success.

The sensor head was installed on the lifeboat deck approximately 45 m above sea level, looking in the directions 235°, 265°, 295°, 325°, 355° and 25° relative to true north (see Fig. 4). The central processor cabinet was installed in a cable shaft at the same level in the living module. The sensor head was connected to the central processor cabinet with appoximately 50 m of cable. Mechanical installation and cabling were performed by Esso-*Odin* personnel.

Apart from periodical inspection by MIROS personnel, the system was left unattended. Data were logged on 3M ¼ inch data cassettes. One cassette has a storage capacity of 67 Mbytes. Measurements were taken every 3 hours using radar mode 3, wave spectrum and

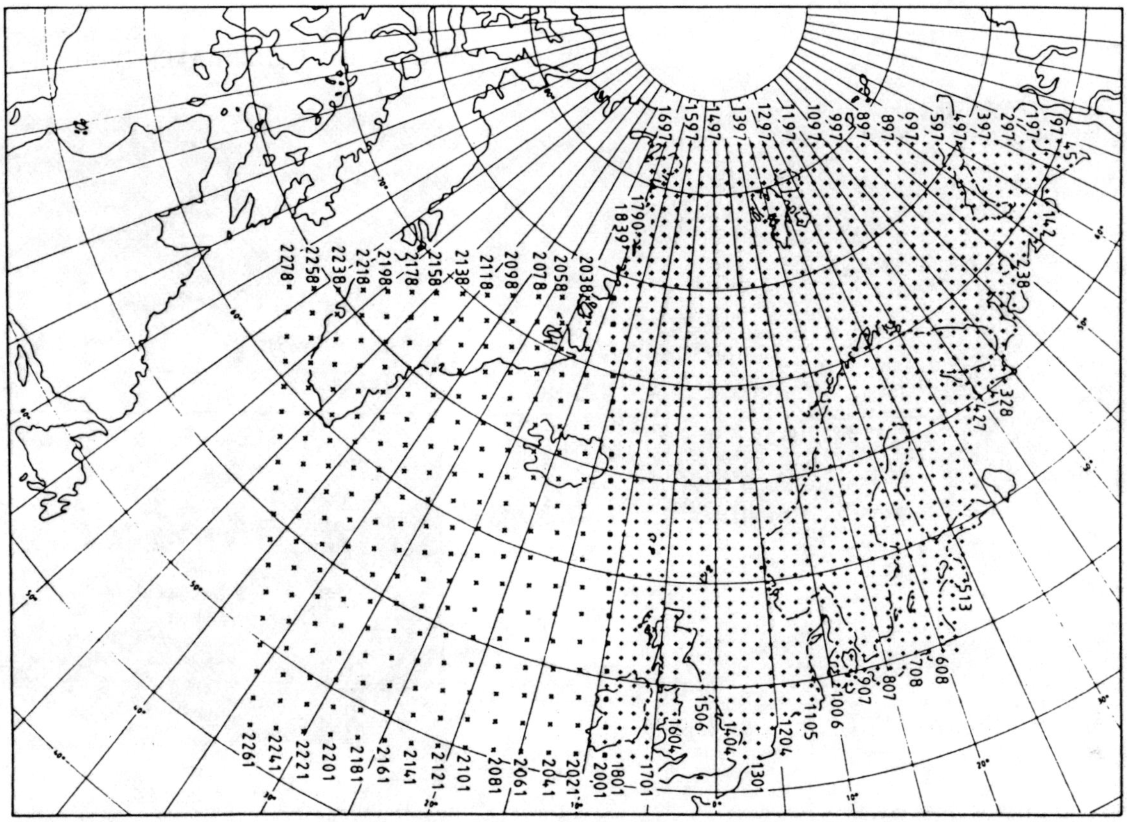

Fig. 2 DNMI model integration area

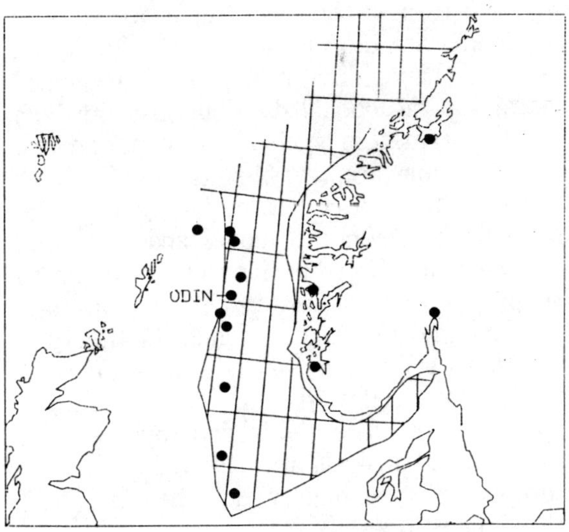

Fig. 3 The location of the *Odin* platform

surface current simultaneously. Both raw data (time-series) and processed data were logged; therefore one data cassette has sufficient capacity for approximately 75 days of operation. Esso-*Odin* personnel were instructed to change tape cassettes.

In the first part of the measurement period, the distance to the radar footprint was selected to be 487.5 m, giving a grazing angle close to 5°. Although the optimum for directional resolution (wide footprint), this low angle turned out to give shadowing problems under high wave conditions. The distance to the footprint was therefore reduced to 337.5 m, giving a grazing angle of about 8°, which is now considered sufficient for most conditions. Improved directional resolution must therefore be obtained by mounting the sensor head higher above sea level.

After completion of the measurement period, the MIROS equipment was dismantled and

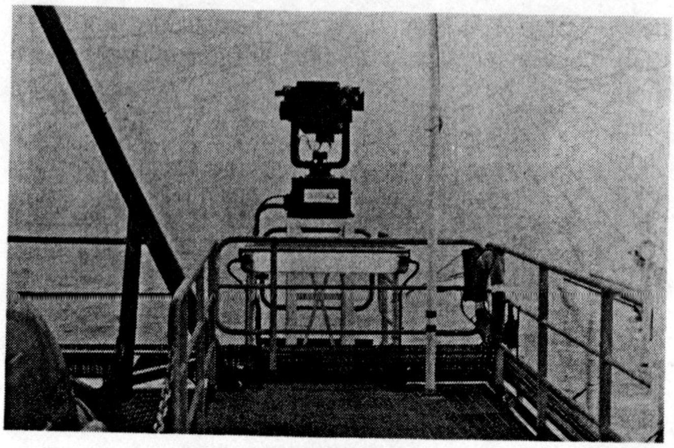

Fig. 4 MIROS at the *Odin* platform

shipped back to MIROS A/S. Having spent more than 6 months in the North Sea, the equipment was found to be in excellent condition. No significant traces of corrosion could be observed.

COMPARISON OF MEASURED AND MODEL DATA FOR SELECTED TIME PERIODS

Unfortunately, measured data from other sensors were not available at the time of writing. For this reason only model data have been compared with the MIROS data. We have selected three different periods, 6–23 February 1985, 26–31 March 1985, and 15–28 April 1985, thus covering three major storms during the measurement period. For the first period, detailed model spectra are available, while for the two last periods model output of wind speed and direction, significant wave height, peak period and peak direction are available. We have data from MIROS measurements every three hours for all periods.

The Esso *Odin* platform is located at 60.127° N and 2.166° E. The closest DNMI model gridpoint is 59.72° N and 2.67° E, which is approximately 53 km away (see Figs. 2 and 3).

Data have been compared with respect to significant wave height, spectral peak period and peak direction. In general, with respect to long-term trends, the model data are in good agreement with the measured data. For given spectrum shapes, i.e. multipeaked frequency spectra and directional distributions with high spread, parameters such as peak period and peak direction are very sensitive to small changes in the energy distribution. Some of the deviations between the model and measured data may be attributed to this fact. However, to give proper explanations in all situations one must look more into the details of the two-dimensional spectra and include more information, i.e. measured wind data.

In the present case, model data together with a simple trend analysis are used to resolve the inherent 180° ambiguity in the individual MIROS measurements.

Only a very limited set of data can, for obvious reasons, be shown here. Figures 5, 6 and 7 show the temporal progression for the significant wave height, peak period and peak direction respectively for the period 15–22 April 1985.

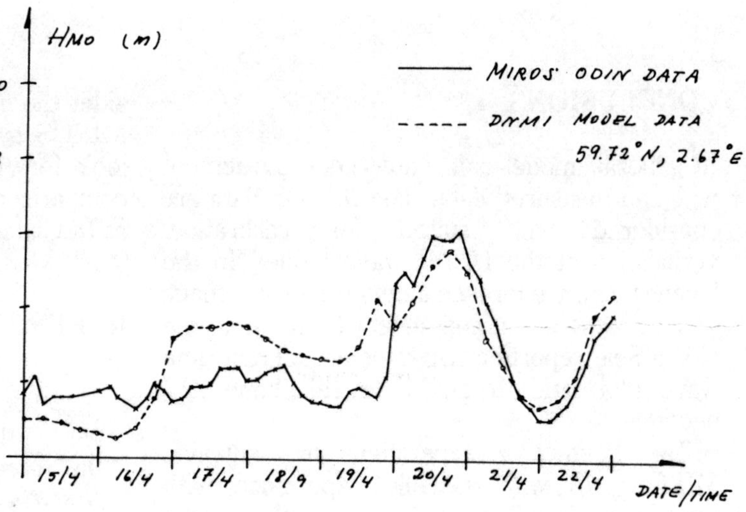

Fig. 5 Significant wave height comparison

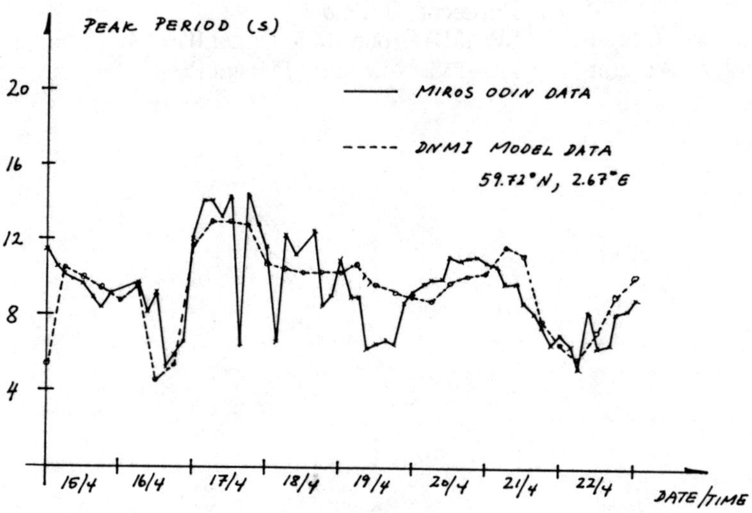

Fig. 6 Peak period comparison

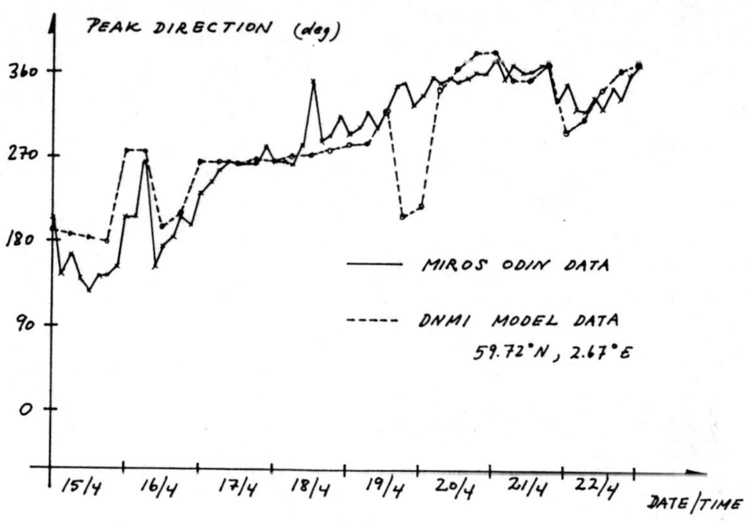

Fig. 7 Peak direction comparison

CONCLUSIONS

In general, model data show good agreement with the measured data. The MIROS data are considered well suited for calibration/ verification of the DNMI wave model. In the longer term, we foresee a network of automatic MIROS systems at convenient locations in the North Sea, reporting two-dimensional real-time wave information to DNMI via telephone connections.

The Esso-*Odin* experiment has supplied MIROS A/S with valuable experience with respect to systems offshore operation. The MIROS type of system undoubtedly represents the future in routine measurements of high-quality two-dimensional wave spectra. We consider the measurement principles applied in the MIROS system to be verified. However, we look forward to being able to perform data comparison with the reference sensors, as soon as the data becomes available.

REFERENCES

Grønlie, Ø., Brodtkorb, D. C. and Wøien, J. S. 1984. MIROS—a microwave remote sensor for the ocean surface. In *Norwegian Maritime Research No. 3,* volume 12. See also Toba, Y. and Mitsuyasu, H., eds., 1985. *The Ocean Surface.* Dordecht: D. Reidel.

SWAMP Group 1985. *Ocean Wave Modelling,* pp. 221–233. New York: Plenum Press.

ESTIMATING EXTREME STILL WATER LEVELS AND CURRENTS OFFSHORE

G Alcock, R Flather, Institute of Oceanographic Sciences, UK

Presented on **Wednesday 5 March 1986 at 11.00** in the Clarence Conference Room

EMBARGOED AGAINST DELIVERY

17

Estimating Extreme Still-water Levels and Currents Offshore

Graham Alcock and *Roger Flather*, Institute of Oceanographic Sciences, Bidston, Birkenhead

INTRODUCTION

Still-water level and currents are important parameters for the design and operation of offshore structures (Marty and Eggar, 1983; Prior-Jones, 1983). Still-water level (swl), due to tide and surge, is required for water depth and deck clearance ('air gap') studies, and it also determines the level of attack of waves on the structure. Currents contribute to the forces on the structure and may be as important as waves in coastal regions where wave heights tend to decrease (due to depth-limiting) and tidal currents are strong.

In this paper we consider the methods that are available for estimating extremes of these parameters offshore. Ideally, methods should be applied to long observational series, but there are very few North Sea sites for which good-quality long-period data are available. Therefore methods are described which are based on model simulations of tide and surge, combined with statistical analyses of onshore observations.

We have compared estimates obtained from the model with values derived directly from available observations. In particular, we have used 16 months of sea-level data from the northern North Sea, nine years of sea level and 346 days of current data from the southern North Sea (all collected by IOS), and 714 days of current data from the central North Sea, collected by MAFF. The northern, central and southern sites are referred to as A, B and C respectively. We had intended to use five years of current data from a site in the northern North Sea but, disappointingly, they proved unsuitable for analysis because of the prevalence of timing problems. This is because, unlike the other data sets, several different organizations were responsible for the data collection and analysis and consequently quality control has been poor (IOS Marine Information and Advisory Service, personal communication).

The model used here is the IOS continental shelf model (CSM) and it has been developed at IOS over a number of years. It was used for operational storm-surge prediction at the Meteorological Office from 1978 until replaced by an upgraded version in 1982. It is a two-dimensional model covering the whole of the European continental shelf and is based on non-linear depth-averaged hydrodynamical equations, which are solved numerically to give surface elevation and north and east components of the depth mean current (Davies and Flather, 1978; Flather, 1981; Proctor and Flather, 1983).

The paper is structured so that methods of estimating tidal, surge and total motion are presented in turn.

TIDAL MOTION

Basic Theory and Definitions

Tides are generated by the gravitational attraction of the moon and the sun. The tidal component of the observed water level is that part coherent with the astronomical forcing. Extraction of the tide from observed levels requires a technique of analysis. The harmonic method of analysis models the astronomical tidal level as a finite number, N, of sinusoidal motions (harmonic constituents) with amplitude H_n and angular speed σ_n,

$$x(t) = \sum_{n=1}^{N} f_n(t) H_n \cos(\sigma_n t + V_n + u_n - g_n) \quad (1)$$

where V_n is the initial phase at an arbitrary time origin $t = 0$ and g_n is the constituent's phase lag with respect to the equilibrium tide and the Greenwich meridian. Here, f_n and u_n represent the influence of changes in the Moon's orbit, varying over a period of 18.61 years. Values of H_n and g_n are derived from a least squares fitting procedure of (1) to equivalent hourly observations. The Hs and gs are known as the harmonic constants for the location. Once known, they may be used in (1) to predict the tides for any required period.

For tidal currents, the analysis is carried out for two orthogonal components usually north-directed and east-directed. The resulting harmonic constants may be combined to give the following parameters describing — the tidal current ellipse — the locus of the tip of the current vector — for each constituent:

a the semi-major axis of the ellipse
b the semi-minor axis (a positive value indicates anticlockwise rotation of the current vector, a negative value clockwise rotation)
$g\alpha$ the phase lag, relative to high water of the equilibrium tide on the Greenwich meridian, of the first of the two current maxima corresponding to the two major axes of the ellipse ($0 \leqslant g\alpha < 180$)
α the orientation of the major axis corresponding to this current maximum, measured clockwise from the north.

The accuracy of the estimated values of the harmonic constants, and hence of the tides predicted from them, depends on several factors. One year of sea-level measurements is normally required to resolve the 60 or more harmonic constituents needed to model the tides adequately. In shallow water, 120 constituents are necessary because of the large number of shallow-water tidal waves which are generated from interactions of the main astronomical tides.

With only one month of data, the influence of seasonal variations on the estimated harmonic constants may be substantial. Variations in the amplitude of M_2, S_2 and K_1 are up to 5, 8 and 50% respectively on the European continental shelf (Baker and Alcock, 1981). The effect may be reduced by making use of a long period of data, including the shorter period, from a nearby site (Alcock and Pugh, 1980).

The extreme tidal parameters are the Highest and Lowest Astronomical Tide levels (HAT, LAT) and associated currents. These are

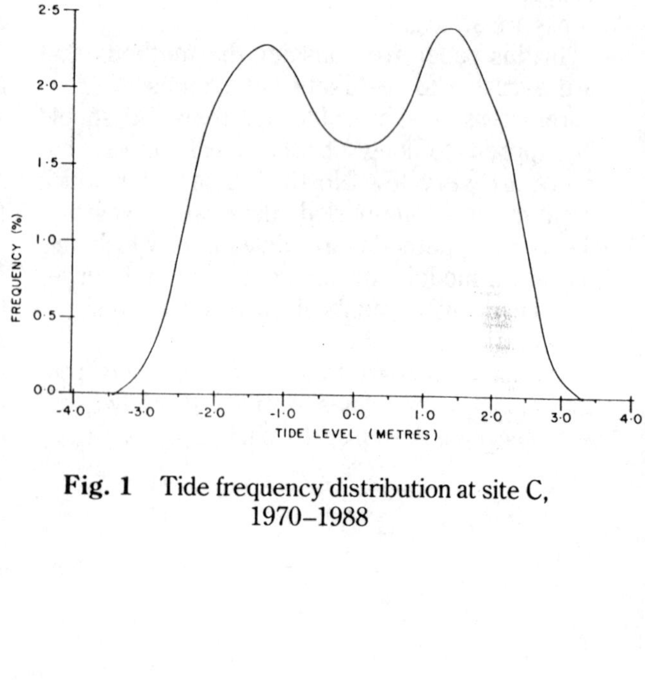

Fig. 1 Tide frequency distribution at site C, 1970–1988

defined as the highest and lowest tide respectively which can be predicted to occur under average meteorological conditions and under any combination of astronomical conditions.

Extreme Tidal Levels

HAT and LAT cannot be estimated by simple summation of the amplitudes H_n because of fixed relationships between values of α_n and g_n for some constituents in (1). They should be extracted from 19 years of predicted turning points of the tidal curve (high and low waters) to take into account the variability of the tidal constituents due to long period variations in the moon's orbit. The tails of the tidal frequency distribution (probability density function pdf) do not extend to infinity but terminate at HAT and LAT. The pdf is usually generated numerically from an hourly time series over the full 19 years (Fig. 1); the terminal values are usually within a few centimetres of HAT and LAT obtained from predicted turning points. An analysis based on one year of measurements is desirable in order to minimize the seasonal variations in the constituents, and this was tested for the offshore sites A and C using analyses based on only one month's data. Estimates of HAT and LAT extracted from 19 years' predictions differed by up to 13 and 5% at the sites A and C respectively from those obtained using predictions based on the one year's analysis, assuming knowledge of the main annual and semi-annual constituents Sa and Ssa from a nearby onshore reference site.

HAT and LAT can be obtained from predictions covering a period of a year when the astronomical positions of the moon and sun are known to be favourable in producing a high tidal range. In theory this is at the time of the equinoxes in those years when the nodal factors of the major constituents are a maximum, but Amin (1979) has shown that the times of occurrence of extreme tides may be shifted from the theoretical times of HAT and LAT by months because of a non-linear response of the water to tidal forcing. If only limited data (i.e. $<$ 1 year) at a site is available, good approximations can be obtained by scaling the HAT or LAT value obtained at a reference port by a parameter such as the spring tidal amplitude (STA $= HM_2 + HS_2$), and applying that factor to the same parameter at the required offshore site. An adjustment to HM_2 and HS_2 obtained from the offshore analysis should be made to minimize seasonal variations. This method gave HAT and LAT to within about 8% at sites A and C.

In the absence of any offshore observations, STA can be estimated from the spatial distribution contoured in Fig. 2. The distribution is based on data from 550 sites derived from sea-level measurements at coastal and island sites together with offshore seabed pressure observations. The analysis results were compared with data from tidal simulations with two-dimensional hydrodynamic models of the area, which were used to assist in drawing the contours where the observational data coverage was sparse (M. J. Howarth, personal communication). The main hydrodynamic model used was the IOS CSM, which gives good agreement with the observations — typical errors are of order 10% (Flather, 1986). The numerical mode output can be used to extract values of STA for model points corresponding to sites where interpolation of the contours of Fig. 2 is difficult, e.g. the northern North Sea east of the Shetland Islands. Values of HAT and LAT within 2% and 10% respectively were obtained for Site A and within 3% and 7% respectively for Site C using the model results.

Extreme Tidal Currents

If long-period (i.e. $>$ 1 year) current observations are available, then an estimate of the HAT current can be obtained using methods similar to those for tidal levels, i.e. from the tidal current pdf or from predictions for the relevant periods of the extreme year.

Similar methods to those for tidal levels can also be used if only limited data (i.e. $<$ 1 year) is available, with the following additional comments:

(a) The relationships for current constituents necessary for analysis of short periods of data should be derived using level relationships from a nearby port, because of the greater stability of the latter (Pugh and Vassie, 1976).

(b) There are very few estimates of extreme tidal currents based directly on observations, so that scaling factors equivalent to those for levels, but giving the ratio of extreme to spring tidal current speeds, can only be obtained at a few locations.

An alternative is to make the assumption that tidal current is roughly proportional to tidal level, so that scaling based on level may be applied. The ratio ½(HAT–LAT)/STA at the reference site is applied to the sum of the semimajor axes, a^T, of the M_2 and S_2 tidal

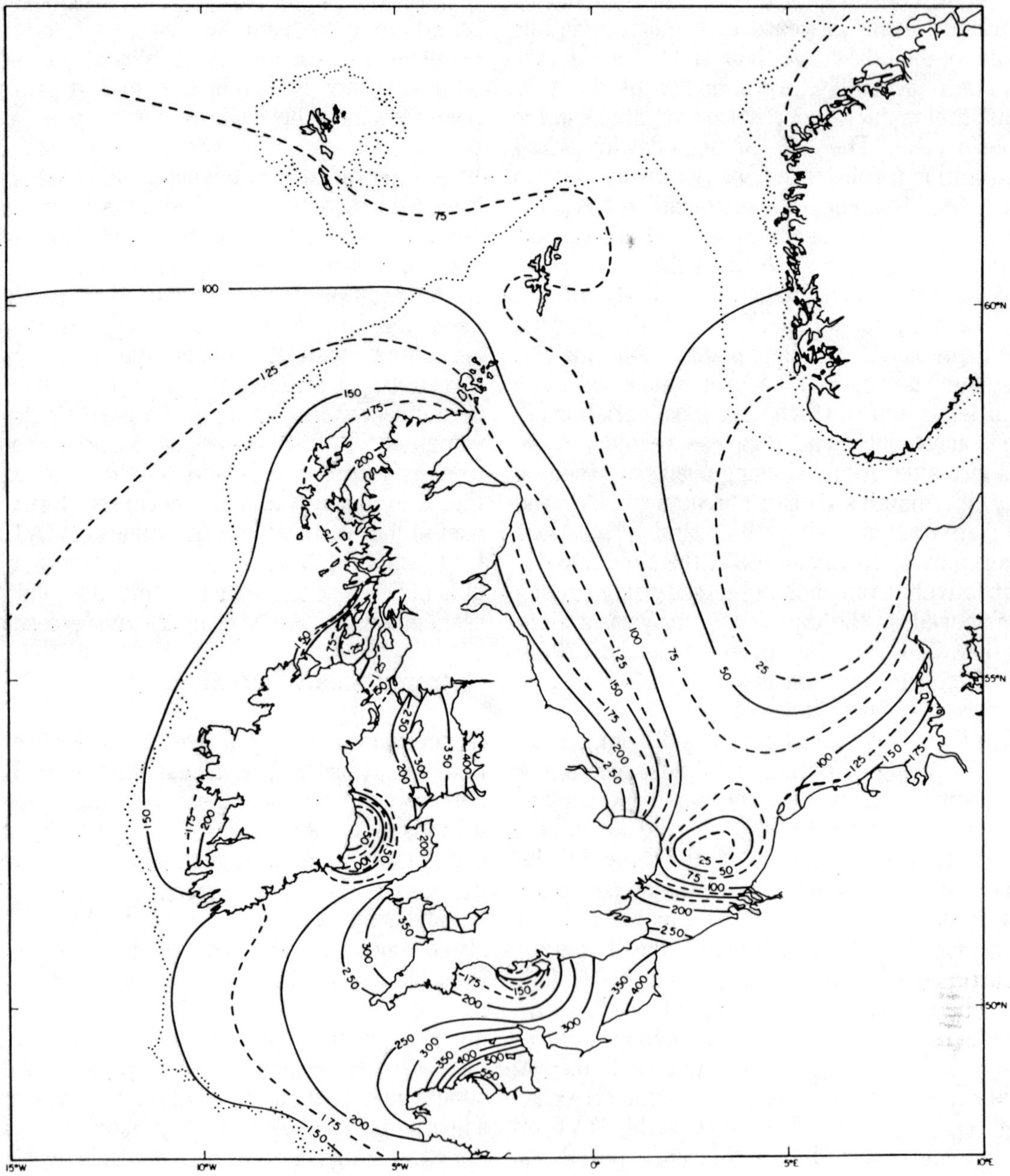

Fig. 2 Spring tidal level amplitude (m)

current ellipses at the offshore site to give the major axis of the HAT current ellipse. The same factor applied to the sum of the semiminor axes, b^T, gives the minor axes.

Directions, defined by the orientation, α^T of the tidal current ellipse can be estimated by combining the orientations of the M_2 *and* S_2 ellipses, allowing for their relative magnitudes.

(c) In the absence of any offshore observations, the spring tidal current amplitude can be estimated from the IOS CSM. The spatial

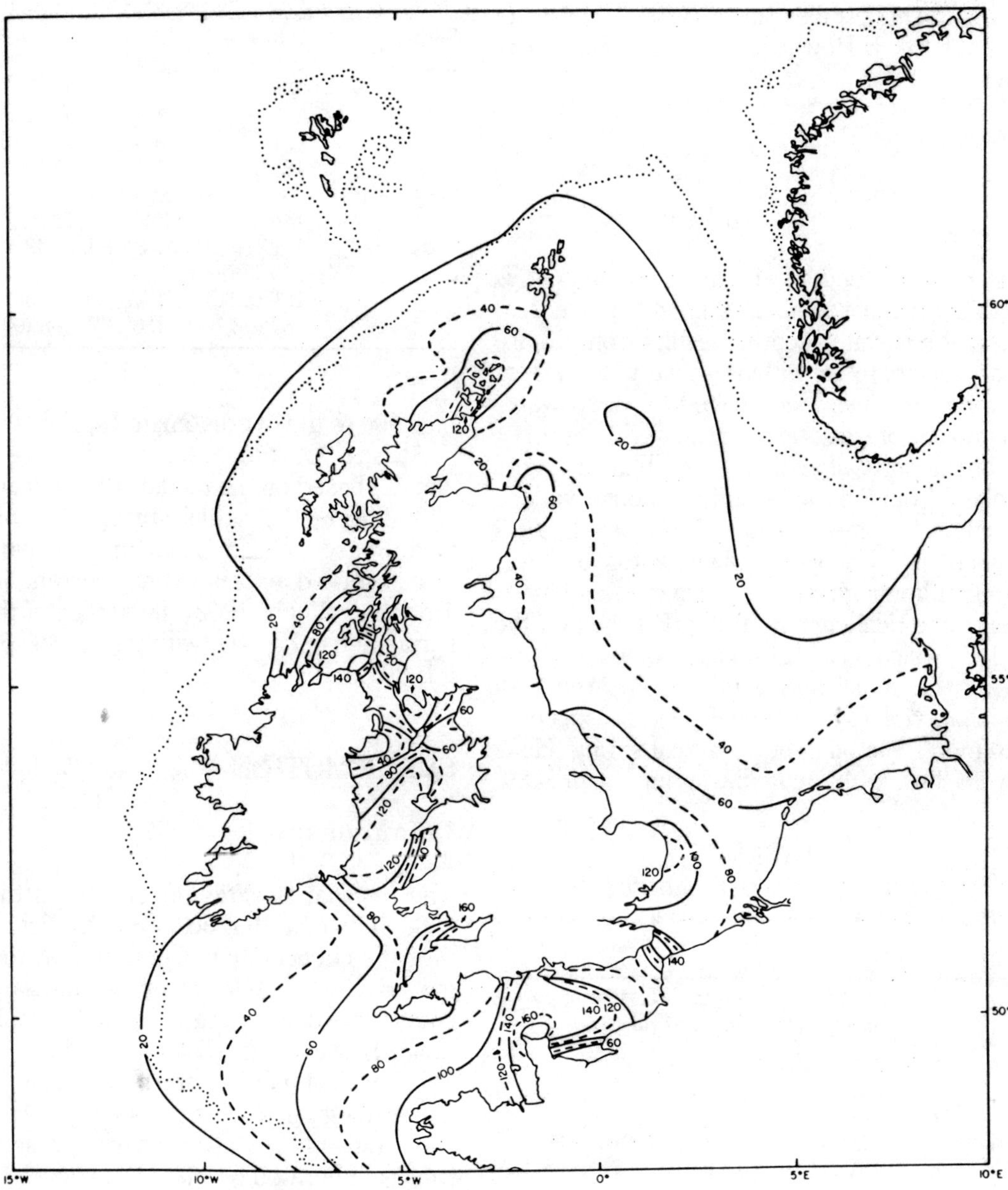

Fig. 3 Depth-averaged spring tidal current amplitude (m)

distribution is contoured in Fig. 3 and is based on both observed and modelled tidal currents.

The model results are for depth-mean currents and have been adjusted to give results applicable to the level of observations. The following formulae give a good fit to measured tidal current profiles in a wide variety of locations (M. J. Howarth, personal communication):

$$u(z) = \begin{cases} \left(\dfrac{z}{0.32h}\right)^{1/7} \bar{u} & 0 \leqslant z \leqslant 0.5h \\ 1.07\,\bar{u} & 0.5\,h \leqslant z \leqslant h \end{cases} \qquad (2)$$

where u is speed of the tidal current, $\bar{u}$ is depth-averaged speed, z is height above seabed, and h is water depth. Results from (2) are usually accurate to within ~15% but are less accurate very near the seabed, in deep water and in areas of weak tidal currents.

Comparisons of estimates of tidal currents obtained from the model and from observations are given for sites B and C in Tables 1 and 2 respectively. Using a scaling factor of 1.29, based on Immingham levels, leads to estimates of extreme tidal current at site B based on the models which are 12% high compared with scaling the actual spring tide. Compared with the observed extreme tidal current speed of 1.20 m/s, the scaling of the actual spring tide is only 4% low; with the model spring tide it is 8% high due to the overestimated spring tide.

TABLE 1

Estimates of currents at site B

(54°13.7′N 0°1.7′E, water depth = 53 m, meter height = 40 m)

		Model (depth-mean)	*Model (at 40 m)*	*Observations (at 40 m)*
Spring	a	0.93 m/s	1.00 m/s	0.89 m/s
tide	b	0.15	0.16	0.06
	α	154°	154°	148°
Extreme	a	1.20	1.29	1.15 observed
tide	b	0.19	0.21	0.08 max. = 1.20
	α	154°	154°	148°
Surge	q_s^{50}	0.76	0.81	0.75
		151°	151°	180°
Total	q_{max}	1.69(0.91)	1.81(0.97)	1.33
current		154°(64°)	154°(64°)	–
	q_{100}	1.54(0.82)	1.65(0.88)	1.80(0.44)
		154°(65°)	154°(64°)	150°(60°)

TABLE 2

Estimates of currents at site C

(53°9′N 00°35′, water depth = 20 m, meter height = 14 m)

		Model (depth-mean)	*Model (at 14 m)*	*Observations (at 14 m)*
Spring	a	0.87 m/s	0.93 m/s	0.94 m/s
tide	b	0.42	0.45	0.24
	α	188°	188°	186°
Extreme	a	1.12	1.20	1.21 observed
tide	b	0.54	0.58	0.31 max. = 1.30
α		188°	188°	186°
Surge	q_S^{50}	0.63	0.67	0.50
		145°	145°	180°
Total	q_{max}	1.50(1.05)	1.61(1.12)	1.42
current		188°(98°)	188°(98°)	–
	q_{100}	1.37(0.96)	1.47(1.05)	1.63(0.72)
		188°(98°)	188°98°)	188°(98°

At site C, the estimate of extreme tidal current based on the model is 1% low compared with scaling the actual spring tide, using the factor based on Immingham. Compared with the observed extreme tidal current speed of 1.30 m/s, (Pugh, 1982) the scaling of the actual spring tide is 7% low; with the model spring tide it is 8% low.

SURGE MOTION

Definitions

The residual elevation or current is the difference between the observed motion and the predicted tidal motion. Contributions to residual motion can be induced by waves, large-scale circulation, density gradients and inertial oscillations. However, in shallow seas like the North Sea, residual motions are predominantly due to meteorological forcing by surface wind stress and gradients of atmospheric pressure. Residuals generated by meteorological storms are known as storm-surge elevations or currents.

Extreme Surge Elevations

Surge elevations can be obtained from observations or from numerical model simulations

(Flather, 1981). The surge frequency distribution, the pdf, generated from hourly values has an approximately normal or Gaussian appearance in UK waters, but there is usually a slight positive skewness indicating that large positive surges occur more frequently than negative surges of the same magnitude. This is due to asymmetry of the meteorological forcing (Pugh and Faull, 1983). The surge distribution also has longer tails than a Gaussian distribution but, unlike the tide distribution, does not terminate at clearly definable bounds. However, the main parameters of the surge pdf, standard deviation, skewness and kurtosis, show a remarkable stability over subsets of a year for North Sea sites.

Extreme surge elevations can be estimated by extrapolating the cumulative distribution of observed surges, or by using a 'peaks over threshold' (POT) technique, or by using a method based on model simulations of surge events.

The first method is applied by linearizing the cumulative distribution of the hourly surge elevations by fitting a logarithmic curve by least squares and extrapolating to long return periods. A logarithmic fit assumes a Gaussian distribution and is probably adequate for the lower, autocorrelated, hourly surge elevations but would not hold for the higher, uncorrelated, surges. Pugh and Vassie (1979) found that estimates of long return period surges obtained from this method were about 0.3 m higher than the observed surges. For the second method, a simple POT model (NERC, 1975) has been used by IOS. This treats the number of exceedances per year of surge elevations greater than a threshold level as a Poisson variate and the magnitudes of the exceedances as an exponential distribution. A threshold level of three standard deviations (3) is used to define each positive surge event. This is the UK Storm Tide Warning Service threshold level for North Sea storm surge events (Lt Cdr J. Townsend, personal communication). The maximum surge elevation in each event is taken as the exceedance level so that relatively independent samples are considered. In fact, at Site A, lowering the threshold level to 2 altered the estimates of return period surge elevation by only a few centimetres, e.g. the 50-year return period elevation by -0.01 m. At the site C the difference was –0.23 m. Standard errors of the estimates using the 3 threshold were 0.10 and 0.16 m for the 50-year surge elevation at sites A and C respectively. The method can be improved by using a POT model which includes the seasonal variation of surges, or by using all observations exceeding a threshold rather than the peak value (Smith, 1984; Davison, 1984).

The third method is based on numerical model simulations of surge events, carried out using the IOS CSM, combined with the results of statistical analyses of observed surges at coastal sites (Flather, 1986). The model simulations included calculations to determine the sea-level (and current) response due to the main tidal constituents and to meteorological conditions associated with 16 of the most severe storms which have occurred in recent years over the North Sea, including 9 from the NORSWAM input data set (Harding and Binding, 1978). Basically, the maximum and minimum computed surges at each model point were used to interpolate the 50-year surge values derived from the coastal observations. Then the ratio $S^{+}_{N}/S^{+}_{\max}$ at the reference site is applied to $S_{\max}$ at the offshore site to give the surge with return period N years. Similarly, $S^{-}_{N}/S^{-}_{\max}$ applied to S^{-} max offshore gives the negative surge with return period N years. Figure 4 gives the spatial distribution of S^{+} 50 using the computed values of S max from the model.

The above scaling, based on distributions of maximum computed surge elevation, will generally be sensitive to the storms included in the sample. An alternative method is to use a more stable scaling parameter, S^{+}_{N}/M^{S}_{33}, where M^{S}_{33} is the r.m.s. surge elevation at the reference site. This is applied to M^{S}_{33} at the offshore site to give the extreme surge offshore. However, comparisons of estimated surge elevations, using the two scaling parameters, with those computed statistically at coastal reference sites suggest that the maximum distribution gives more accurate results (Flather, 1986). Values of the ratio $S^{+}_{50}/S^{+}_{\max}$ for coastal sites are approximately 1 in the southern and central North Sea, and 1.2–1.4 in

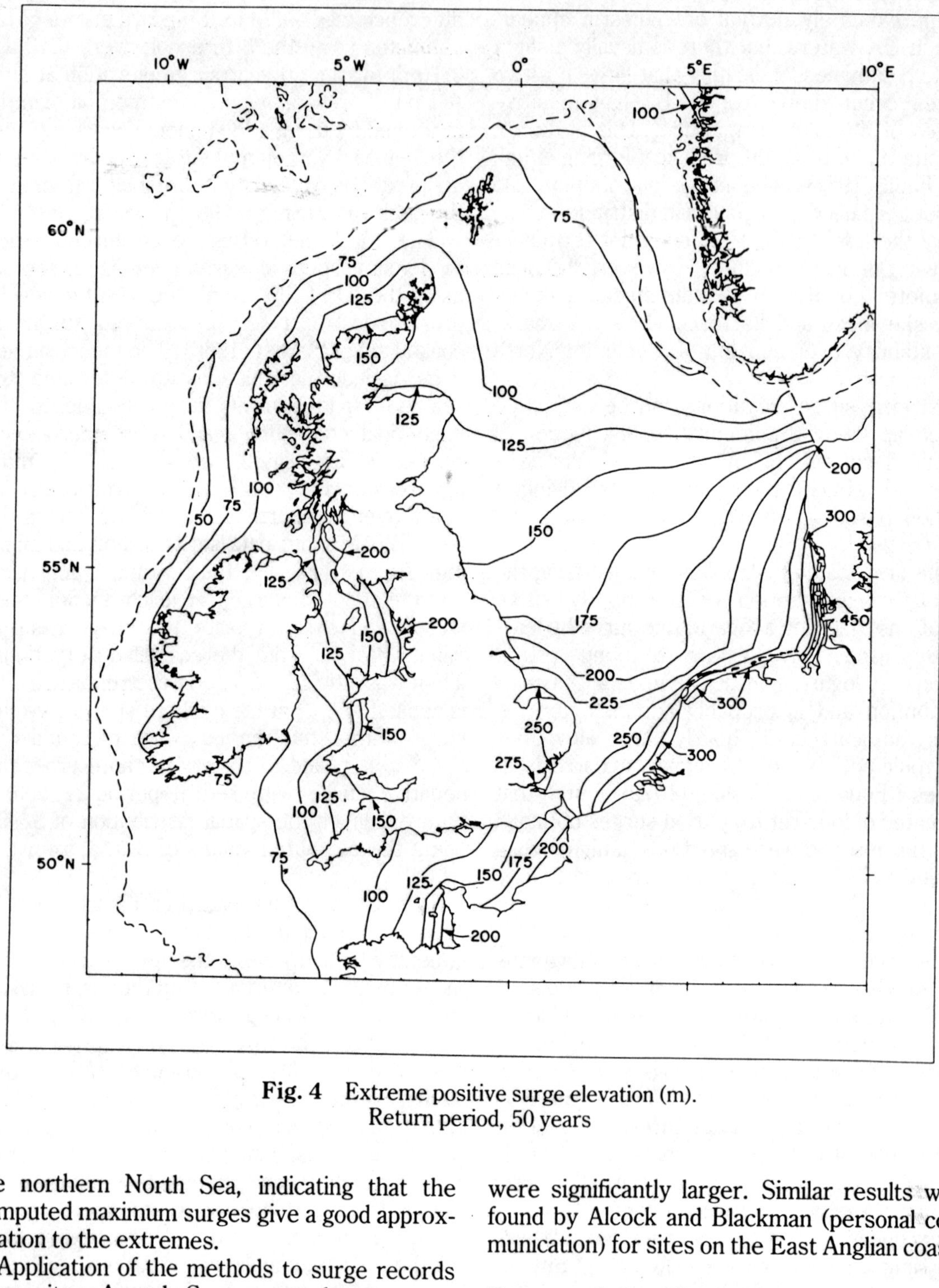

Fig. 4 Extreme positive surge elevation (m).
Return period, 50 years

the northern North Sea, indicating that the computed maximum surges give a good approximation to the extremes.

Application of the methods to surge records from sites A and C gave good agreement between the POT and numerical model results (within 3% for S^{+}_{50}) but extrapolation of the cumulative distribution gave estimates which were significantly larger. Similar results were found by Alcock and Blackman (personal communication) for sites on the East Anglian coast.

Extreme Surge Currents

In theory, extreme surge currents may be estimated by extrapolation of the observations using an extreme probability distribution, e.g. a least squares-fitted Weilbull distribution. However, estimates so derived are considered unreliable because of the difficulty in fitting and extrapolating the available data sets which are invariably small. At least one year of surge residuals is required because large currents are more probable in winter than in summer due to the strong seasonal variability of the meteorological forcing.

If short (i.e. < 1 year) data sets are available, then estimates of extreme depth-mean surge current speeds and directions can be computed using results from the IOS CSM and analyses of coastal observations. The model gives values of the maximum computed surge current speed $q^S{}_{max}$ with its associated direction, ^{S} at each grid point. The extreme surge current at some offshore site, with return period N years, is then computed by applying the factor $S^+{}_N/S^+{}_{max}$ at a nearby coastal site to $q^S{}_{max}$. The same scaling applied to maximum computed currents in 15° sectors can provide directional estimates of N-year surge currents, but with large uncertainty associated with the small sample of model currents.

An alternative method is based on the mean square surge parameters $M^S{}_{11} = <U^2>^{1/2}$, $M^S{}_{22} = <V^2>^{1/2}$, $M^S{}_{12} = <UV>^{1/2}$, where $<>$ indicates an average over all hourly values of the east (U) and north (V) directed components of the computed surge current. This then defines an elliptical distribution of surge current with major axis, minor axis and orientation given by

$$a^2 = M_{11}\sin^2\alpha + M_{22}\cos^2\alpha + M_{12}\sin 2\alpha$$
$$b^2 = M_{11}\cos^2\alpha + M_{22}\cos^2\alpha + M_{12}\sin 2\alpha$$
$$\tan 2\alpha = 2M_{12}/(M_{22}-M_{11})$$

Then the required estimate of q^S is obtained by applying the ratio S_N/M_{33} to each of a and b to give $a^S{}_N$ and $b^S{}_N$, the major and minor axes of the elliptical distribution of surge current with return period N years.

Since the observational evidence suggests that estimates of extreme elevations using the maximum distribution is the more accurate, it is considered appropriate to adopt equivalent distribution for extreme surge current. Figure 5 gives $q^S{}_{50}$ based on the modelled maximum current distribution, including arrows indicating preferred directions derived from the computed directional distributions. In the North Sea, there appear to be preferred directions only in a coastal band, probably because of the constraining influence of the coastlines. In many offshore areas of the North Sea, the strongest surge currents can occur in any direction.

An important point to be considered is that storm-surge currents vary with depth in the water column. The actual current magnitudes may thus differ significantly from the depth-mean model values. Unfortunately, the vertical structure of storm-surge currents cannot be estimated with confidence — it is determined by physical processes that are not yet fully understood. Although a logorithmic profile may be more appropriate nearer the surface and seabed, application of the 1/7 power law profile (equation (2)) is probably adequate to convert the depth-mean currents to a value appropriate to the depth of observations. The results are given in Tables 1 and 2 for sites B and C respectively, and are in reasonable agreement considering the possible significant effect of land boundaries on the model results (in particular the effects of the Wash on current propagation at site C are not modelled). Of course, the current observations are only of short duration (714 and 346 days at sites B and C) so that some uncertainty attaches to the associated extremes. The importance of obtaining long data sets of good quality with which to test the models is clear.

TOTAL MOTION

Definitions

The still-water level (swl) can be defined as the observed water level at a site when waves have been averaged out. It contains contributions due to astronomical tides $x(t)$, non-tidal residuals $y(t)$ and mean sea level $z(t)$:

$$\eta(t) = x(t) + y(t) + z(t) + xy(t) \qquad (3)$$

In the North Sea, the term $y(t)$ is dominated by meteorologically induced surge residuals, and

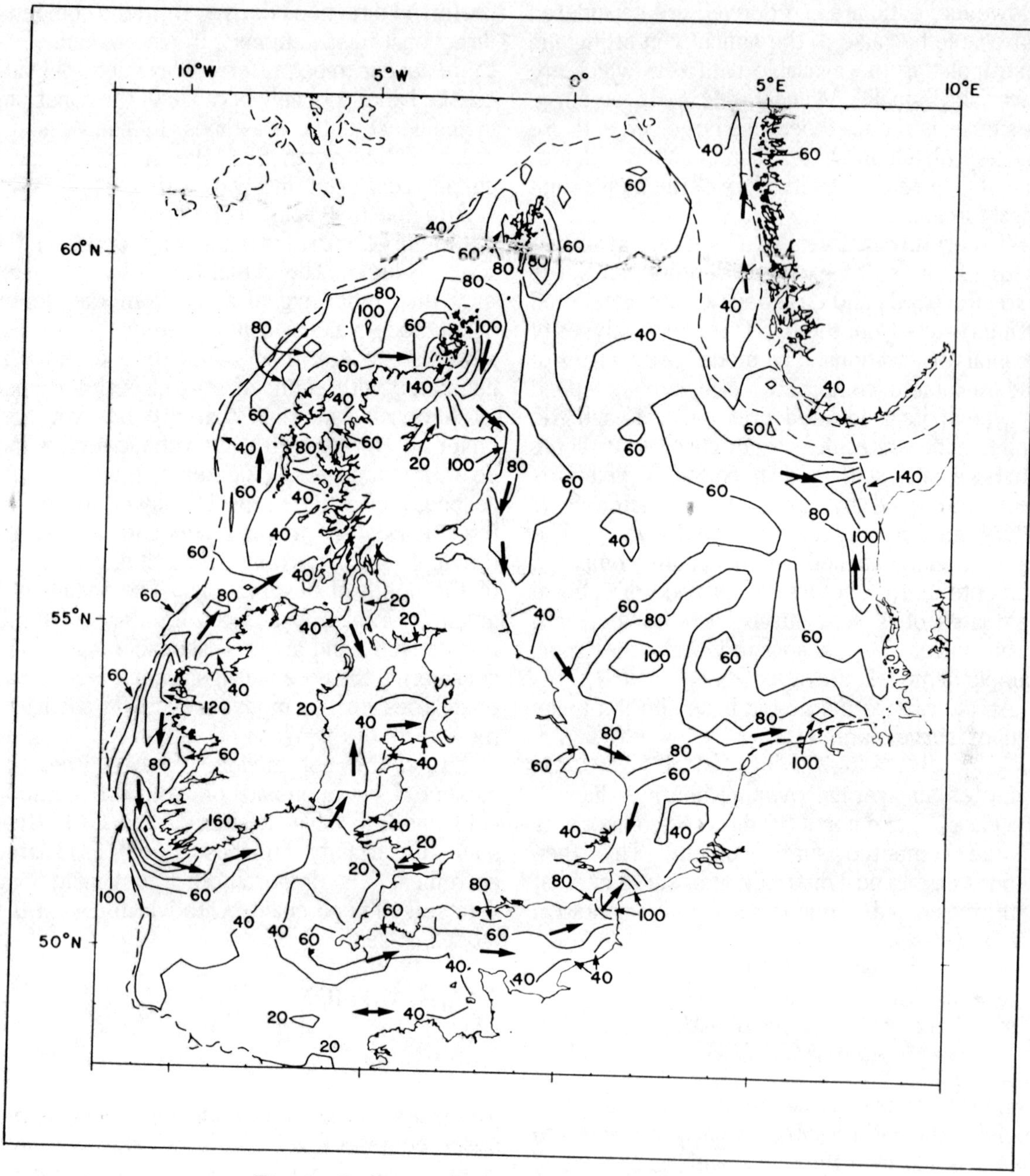

Fig. 5 Depth-averaged extreme surge current (m). Return period, 50 years

the additional term *xy(t)* represents a contribution due to tide–surge interaction which may be important in regions of extensive shallow water.

A similar expression can be considered for currents, with y(t) again being dominated by surge residuals. The equivalent of *z(t)* can be considered to be the mean current or 'circulation', usually computed as the mean of the observations (typically taken at 10-minute intervals).

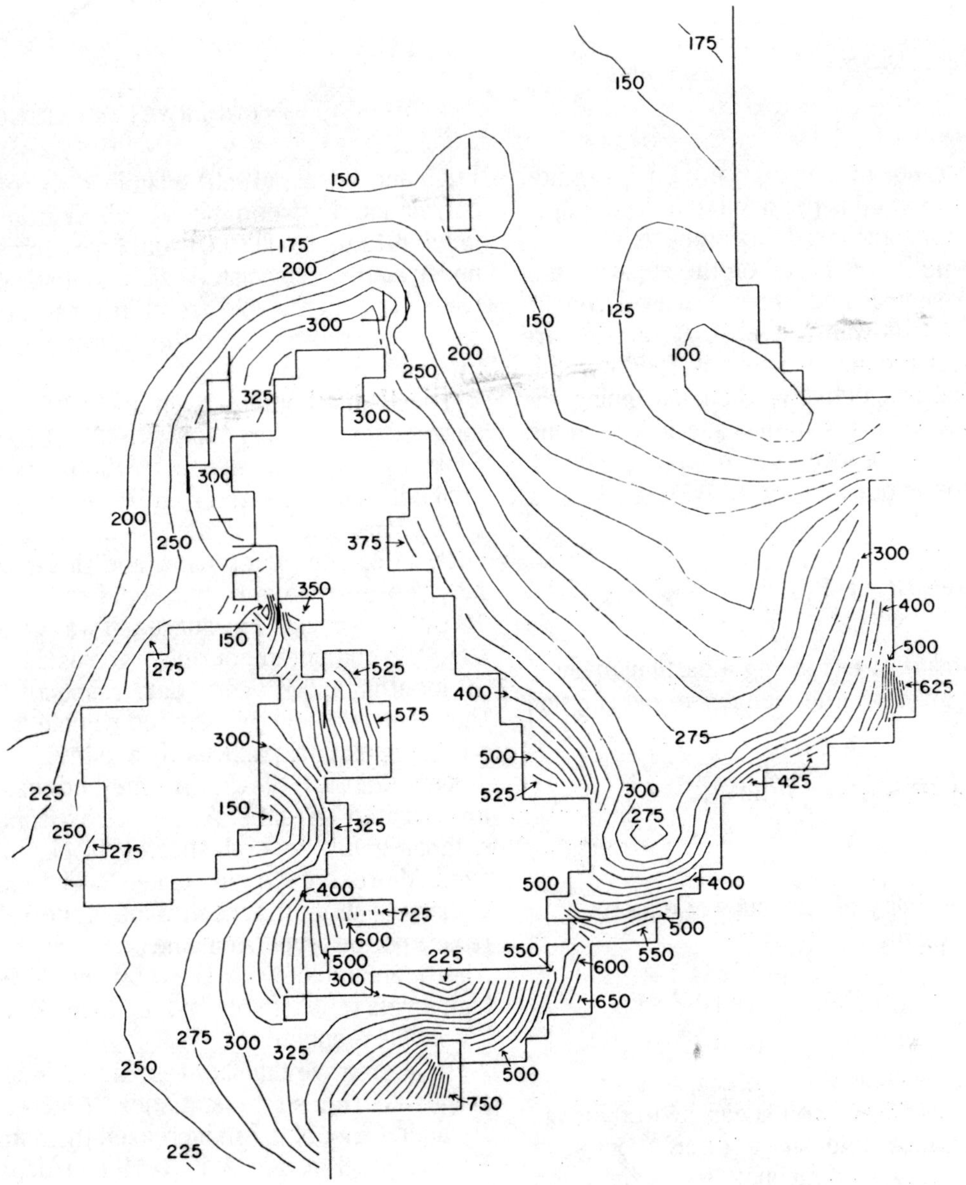

Fig. 6 Spring tidal level amplitude plus 50-year surge (m)

Extreme Total Levels

Two methods can be used to estimate extreme levels from observations: the Extreme Value (EV) method and the Joint Probability (JP) method. The EV method involves fitting the cumulative frequency distribution of the annual observed maxima (or minima) of the total level by a distribution and extrapolating to low probabilities and hence long return period values. Various distributions have been used to fit the data with various methods of fitting such as graphical, by moments, and maximum likelihood (NERC, 1975). For confidence in the results, observations extending over a period of order 100 years are desirable. Data of this duration are not available at any offshore site, so that the method is not applicable to the open sea and is not considered further here.

The JP method is based on the separation of the tide, residual, and mean sea-level components of the still-water level. If P_T and P_S are the probability density functions (pdf) for tide and residual respectively, then, assuming the tide and surge to be statistically independent, the probability of occurrence of a particular level due to the tide and surge, W, is given by

$$P(W) = \int_{-\infty}^{+\infty} P_T\,(W-y)\,P_s(y)\,dy \qquad (4)$$

The probability of exceeding a particular level, H, is then given by the cumulative distribution function

$$Q(H) = \text{Prob}\,(H > W) = \int_{H}^{\infty} P(W)\,dW \qquad (5)$$

and the probability of exposure of undercutting a level is given by

$$R(H) = \text{Prob}\,(H < W) = \int_{\infty}^{H} P(W)\,dW \qquad (6)$$

In extreme shallow water areas, such as the southern North Sea, non-linear hydrodynamic processes cause tide–surge interaction such that large surges tend not to occur with extreme tidal levels (Prandle and Wolf, 1978). Therefore the probability of and extreme total level due to a combination of extreme tide an extreme surge is lower than in the case of independence. Pugh and Vassie (1980) indicate how to allow for any tide-surge interaction by modelling $P_S(y)$ by computing separate surge pdf's for different parts of the tidal range.

The JP method produces extreme statistics in terms of the probabilities, $Q(H)$ and $R(H)$, of exceeding high levels or of falling below low levels given in (5) and (6). For sampling intervals of one hour, the return periods (rp in years), for exceeding H or for falling below H are given by

$$\text{rp} = [Q(H) \times 8766]^{-1} \text{ and } \text{rp} = [R(H) \times 8766]^{-1} \qquad (7)$$

Pugh and Vassie (1980) examined the effect of correlation between hourly observations and found it to be small (0.01–0.02 m) compared to uncertainties associated with statistical sampling. Any identifiable trend in mean sea-level can be incorporated into the extreme estimates by simple addition.

The JP method has been applied to tide and surge data measured at site A (Alcock and Stokes, 1985) and site C. In both cases, the tidal pdf was generated from 19 years of hourly predictions, using tidal constants from the analysis of one year's data, and therefore was fully representative of the tidal levels over the 18.6y nodal cycle. The surge pdf was generated from the computed hourly surge residuals from 16 months' and 9 years' data respectively and therefore is based on a limited sample from the surge population, particularly at site A.

The stability of the JP method has been investigated at sites A and C by computing estimates for several shorter blocks of data from Aberdeen and site C itself. The standard deviations about the mean value of the 100-year return period level from nine one-year blocks at Aberdeen and site C were 0.09 and 0.19 m for high levels respectively and 0.04 and 0.17 m for low levels respectively.

An alternative stability test is to add artificial surges to the surge statistics. Addition of a six-hour surge of 0.7 m increased the estimated 50-year level at site A by 0.15 m (Alcock and Stokes, 1985). However, at the nearby coastal site of Lerwick, the surge level exceeded 0.7 m on only three hourly occasions during 14 years and the main parameters of the surge pdf showed a remarkable stability over subsets of a year (J. Vassie, personal communication). Addition of a six-hour surge of 1.5 m increased the estimated 100-year level at site C by 0.02 m; such a surge has been exceeded on eight hourly occasions over 10 years at site C. Addition of a 2.0 m surge increased the level by 0.21 m, but no such observed surge has occurred at site C.

The stability of the JP method derives from the independence of the physical forcing factors of tide, surge and mean sea-level, namely tidal, weather and geological/climatic factors. In fact the method assumes the independence of tide and surge and this was investigated at the sites by studying the variance of the surges as a function of tidal level. There was only slight dependence at site A and so the tide and surge could be treated as statistically independent variables. For site C, the surge distributions are again very little affected by the corresponding tidal value (Pugh and Vassie, 1976) although the standard deviation of the surges is greater at mid-tide level than at the extreme levels, so that extreme surges are less probable at the time of high or low water level. Tide and surge were again treated as independent.

In the absence of at least one year's observations, there are two basic approaches available for estimating extremes of total level offshore. The first is to use distributions of total (tide + surge) level generated from the CSM; the second is to take suitable combinations of tide and surge values derived separately.

In the first approach, the ratio L_N/L_{max} at a reference site is applied to L_{max} offshore to give L_N, where L_N is the total still-water level with return period N years and L_{max} is the maximum level in the sample of observations or model simulations. A possible criticism of this method is that, since the computed tide is based on only M_2 and S_2, combinations of extreme tides with moderate surges (which may lead to extreme total levels) are not represented in the model distributions.

The second, preferred, approach addresses this criticism by allowing any combination of tide and surge which can approximate the extreme condition required. Thus $L_N/(T + S)$ at the reference site applied to the model value of $(T + S)$ offshore gives an estimate of L_N, where $(T + S)$ is a combination of tide and surge model values within say 10% of L_N at the reference site. A number of suitable combinations, which may include HAT + S_1, HAT + S_{10}, HAT + S_{max}, $M_2 + S_2 + S_{50}$, $M_2 + S_2 + S_{100}$, may be used for $(T + S)$, each giving an alternative estimate of L_N offshore. These may then be averaged to give a 'best' value.

At site A, application of the second approach using Lerwick as the reference site gives estimates of L^+_{100}, L^+_{50}, L^-_{100} and L^-_{50} some 2, 1, 4 and 4% lower than the values computed using the observations. At site C, using Immingham for reference, corresponding figures are 2 and 2% lower for high levels and 5 and 4% higher for low levels.

This approach is fairly lengthy and requires the computation of $(T + S)$ at the offshore site for various values of tide and surge which have themselves been estimated. Flather (1986) has shown that the combination of a spring tide and a 50-year surge gives an estimate of the 100-year level which should be accurate to within about 10%, with a slight tendency to overestimation. A more simplified method of scaling is therefore to compute the ratio of (STA + S^+_{50}) at the offshore to reference site, and apply it to L_N at the reference site.
For site A, this gives estimates of L^+_{100}, L^+_{50}, L^-_{100} and L^-_{50} which are 1, 1, 3 and 3% low respectively. Corresponding figures at site C are 3, 3, 9 and 10% high respectively. Figure 6 shows the spatial distribution of the model-based (STA + S^+_{50}), which approximates that of L^+_{100}.

Extreme Total Currents

An estimate of the total extreme current can be obtained by combining the extreme tidal and residual components. The simplest method is to add the Highest Astronomical Tidal Current (HATC) and the extreme surge current, but this will give the most conservative estimate. The next method is analagous to the JP method for elevations: combine the frequency distributions (pdf's) of the tidal and residual current speeds, assuming a joint probability of occurrence, but ignoring the direction of flow. The JP method can be modified to consider the directionality of the tidal and residual current speed, to produce directional estimates of extremes. This is done by computing and combining the two-dimensional pdf's of tidal and residual current speed and direction. The two-dimensional pdf corresponding to equation (4) is then

$$P(U, V) = \int_{-\infty}^{+\infty}\int_{-\infty}^{+\infty} P_T(U-x, V-y)\, P_S(x, y)\, dx, dy$$

where U and V *or* x and y are the east-west and north-south components of the total or residual current respectively.

The JP method has been applied to measured

currents at the sites B and C (D. Blackman, personal communication; Pugh, 1982). The basic computational methods are the same for the two sites, but the conversion from cartesian coordinates to radial components at site B is different and produces extreme statistics, which are independent of the sector size. The tidal and residual currents were considered to be independent at the two sites because neither the mean nor the standard deviation of the residuals varied significantly with the corresponding tidal current.

Any mean current, or circulation, may be included in the computed extremes, either by direct vector addition or as an offset in the U and V distributions. One problem in the JP method is the adequacy of the sample of currents. There is not sufficient data to estimate extremes based on subsets, as for the elevations, but at the very least one year should be obtained.

In the absence of at least one year's observations, then similar methods to those used in estimating total elevation can be used. The first method ignores the directional properties of the tide and surge currents, and the simplest approach would be to combine the spring tidal current (Fig. 3) and the 50-year surge current (Fig. 4) as though they were scalars, to obtain the maximum computed total current, q_{max}. Then the scalar factor $L^+_N/(\text{STA} + S^+_{50})$ at the reference site applied to q_{max} gives the total current with return period N years.

The resulting current might be assumed to occur in any direction giving a circular distribution of extreme total current. This would give fairly conservative values in cases where the orientation of the major axis of the tidal ellipse was significantly different from the preferred direction of surge currents. If the directional distribution of q_{max} can be inferred from either the tidal or residual current distribution, or both, to give a directional distribution of q_{max}, then applying the same scaling gives a directional distribution of q_N.

The alternative method used for elevations, where different suitable combinations of tide and surge parameters are used in the scaling factor, presents difficulties unless both tide and surge currents are nearly rectilinear and aligned in the same direction as the offshore site. In that case, scalar combinations may be justified, but in general it may be more appropriate to use the first approach for currents.

The results from the model are compared with those from observations in Tables 1 and 2 for sites B and C respectively. The model estimate of q_{max} has been computed as the sum of the spring tide semimajor axis, a, and q_S^{50}, with an associated direction given by the tidal current. It is 36% high and 13% high at sites B and C respectively, compared with the values from the 714 and 346 days of observations. q_{100} has been obtained by multiplying q max by the scaling factor $L^+_{100}/(\text{STA} + S^+_{50}) = 0.91$ at Immingham, for both sites. It is 8% low and 10% low at sites B and C respectively, compared with the estimated values from observations. The bracketed figures are estimates of the 100-year current perpendicular to the direction of the strongest current. They are significantly higher than the estimated values from observations. In all cases, estimates of the direction of extreme current are in excellent agreement.

CONCLUSIONS

HAT, LAT, and maximum tidal current should be extracted from 19 years' predictions, based on an analysis of at least one year's data. If estimates are extracted from predictions for a limited period, care is necessaary in selecting that period, as extreme tides may not occur at the times when astronomically favourable conditions are predicted.

If only one month of offshore data is available, the method of scaling parameters at a reference site by the observed STA gives more accurate results than using predictions based on analysing the data itself. In the absence of any offshore data, STA can be estimated, to an accuracy of about 10%, from a spatial distribution based on measured and modelled sea-level or current.

The method of estimating extreme surge motions by extrapolating the cumulative distribution of observed surge elevations or currents is not recommended. There are problems in choosing and fitting the right distribution, and an adequate data set is usually not available, especially for currents. The method using 'peaks over a threshold' is more promising, especially if the basic approach is modified to take into account the seasonal variability of the surge motion. There is a problem in choosing the appropriate threshold, in order to preserve independence of the data used but also to provide an adequate data set; further work is needed on this by investigating the 'exceedances over a threshold' technique.

In the absence of good-quality long data sets, a method of estimating extremes is available which combines model simulations of storm-surge motion with the results of statistical analyses of observed surge elevations at coastal sites. A scaling factor based on the maximum surge elevations at the offshore site gives good agreement with estimates from observations.

The Extreme Value method is not applicable for estimating extreme total motion offshore because of the lack of the long data sets necessary. The Joint Probability method gives good estimates for elevation when only data sets of a few years are available, but further investigation of the stability of the method is necessary. Application of the JP method to tide and surge current will give conservative estimates of total current, but the method can be modified, using two-dimensional pdf's, to consider the directionality of either, or both, components.

In the absence of observed data at the offshore site, a scaling method using modelled combinations of tide and surge approximating the extreme level gives estimates within 5% of those based on observations. A shortened version of this method, using $L_N/(\text{STA} + S^{+}{}_{50})$ as the scaling factor gives estimates within 10%. For currents, the IOS CSM can be used to compute the maximum total current offshore, $q_{\max}$, with directionality inferred from either the tidal or surge current or both. Applying a scalar factor to $q_{\max}$, based on the elevation parameter, gives a total extreme current which is in reasonable agreement with estimates based on observations.

The concept of a 'return period' event is used in this paper, being defined as the event which occurs, on average, once during the return period; but we must stress that such a concept should be treated with caution. There is a finite risk, or 'encounter probability', that the event will be exceeded at least once during the lifetime of the structure. This is given by

$$\text{ri} = 1 - [1 - 1/\text{rp}]^L, \tag{9}$$

where rp is the return period and L the lifetime of the structure. The appropriate design value for rp therefore depends upon the acceptable risk. If L = rp, then the encounter probability is approximately 0.635. To design to an acceptable risk or encounter probability may therefore mean adopting a return period longer than the design lifetime of the structure. For example, if the acceptable risk of encountering the 50-year return period event is only 10% for a structure with a design lifetime of 50 years, then the appropriate design return period should be 475 years.

Finally, this work has highlighted the importance of collecting good-quality long-period observations, especially of currents. This is a crucial necessity if analysis and prediction techniques are to be further improved, because there really is no substitute for collecting data.

ACKNOWLEDGEMENTS

We are grateful to colleagues at IOS for their advice and comments; in particular David Blackman, John Howarth and John Vassie. We thank Jackie Huxley and Linda Parry for typing the paper. Some of the results are from work supported by the Department of Energy and the Ministry of Agriculture, Fisheries and Food, and we thank them for their financial assistance.

REFERENCES

Alcock, G. and Pugh, D. T. 1980. Observations of tides in the Bristol Channel and Severn Estuary. *IOS Report No. 112.*

Alcock, G. and Stokes, E. G. 1985. Application of the joint-probability method of estimating extreme sea levels to observations in the northern North Sea. *E and P Forum workshop on the application of joint probability of metocean phenomena in the oil industry's structural design work,* November 1985, London.

Amin, M. 1979. A note on extreme tidal levels. *Int. Hyd. Rev. 56* (2), 133–141.

Baker, T. F. and Alcock, G. 1983. Time variation of ocean tides. In *Procs. 9th Int. Symp. on Earth tides, New York 1981.* E. Schweizerbart'sche Verlag, Stuttgart. pp. 341–350.

Davies, A. M. and Flather, R. A. 1978. Application of numerical models of the north-west European continental shelf and the North Sea to the computation of the storm surges of November to December 1973. Dt. *hydrogr. Z. A,* No. 14, 72pp.

Davison, A. C. 1984. Modelling excesses over high thresholds, with an application. In *Statistical extremes and Applications.* Reidel, Dordrecht, pp. 461–482.

Flather, R. A. 1981. Practical surge prediction using numerical models. In *Floods due to High Winds and Tides* (Ed. D. H. Peregrine). Academic Press, London, pp. 21-43.

Flather, R. A. 1986. Estimates of extreme conditions of tide and surge using a numerical model of the north-west European continental shelf. Submitted to *Estr. Coast. and Shelf Sci.*

Harding, J. and Binding, A. A. 1978. The specification of wind and pressure fields over the North Sea and some areas of the North Atlantic during 42 gales from the period 1966 to 1976. *IOS Report No. 55.*

Marty, L. and Eggar, D. 1983. An offshore operator's requirements for oceanographic data and research. *J. Soc. Underwater Tech. 9* (2), 13–21.

NERC 1975. *Flood studies report.* Vol. I. *Hydrological Studies,* The Council, London.

Prandle, D. and Wolf, J. 1978. The interaction of surge and tide in the North Sea and River Thames. *Geophys. J. R. astr. Soc.* **55**, 203–216.

Prior-Jones, R. L. 1983. Extending environmental data into design criteria. *The Oilman,* June 1983, 61–65.

Proctor, R. and Flather, R. A. 1983. Routine storm surge forecasting using numerical models: procedures and computer programs for use on the CDC CYBER 205E at the British Meteorological Office. *IOS Report No. 167.*

Pugh, D. T. 1982. Estimating extreme currents by combining tidal and surge probabilities. *Ocean Eng.* **9**, 361–372.

Pugh, D. T. and Faull, H. 1983. Tides, surges and mean sea level trends. In *Shoreline protection.* Thomas Telford, London, pp. 59-69.

Pugh, D. T. and Vassie, J. M. 1976. Tide and surge propagation offshore in the Dowsing region of the North Sea. *Dt. hydrogr. Z.* **29**, 163–213.

Pugh, D. T. and Vassie, J. M. 1979. Extreme sea levels from tide and surge probability. In *Proc. 16th Coastal Eng. conf., Hamburg, 1978.* American Society of Civil Engineers, New York, pp. 911–930.

Pugh, D. T. and Vassie, J. M. 1980. Applications of the joint probability method for extreme sea level computations. *Procs. Inst. Civ. Engs,* Part 2, **69**, 959–975.

Smith, R. L. 1984. Threshold methods for sample extremes. In *Statistical extremes and Applications.* Reidel, Dordrecht, pp. 621-638.

A FORECASTING METHOD OF TYPHOON SWELL

Nai Kuang Liang, Institute of Harbour
& Marine Technology, Taiwan

Presented on **Wednesday 5 March 1986 at
12.00** in the Clarence Conference Room

EMBARGOED AGAINST DELIVERY

18

A Forecasting Method for Typhoon Swell

Nai Kuang Liang, Director, Institute of Harbor & Marine Technology, WuChi, Taiwan,

INTRODUCTION

Taiwan is located in the subequatorial region of the Western Pacific Ocean. From June until October typhoons originate in the vicinity of Guam. These either pass by or they invade the island and cause catastrophes. If the typhoon directly attacks Taiwan, of course, it generates high waves. However, even if the typhoon does not come to the island but passes nearby, there are always swells of long wave period. The typhoon swell has a bad effect on the coastal and offshore engineering operations, such as the installation of breakwater caissons, armour blocks and underwater pipelines, as well as the dredging works. The typhoon swell may stop these maritime activities or even damage the unfinished maritime structures, especially for those works which are very weak during their construction stage, such as the underwater pipeline. There are cases when the maritime works in construction are damaged by the typhoon swell.

The usual typhoon wave prediction methods are the numerical method of Tang (1970) and the parametric method of Bretschneider (1976). The former can also predict typhoon swell. However, the calculation is quite complicated and not suitable for the climate forecasting operation. Hence, Tang's method has not been used to predict typhoon swell. Also, Bretschneider's method can only predict wind waves. The author has studied the typhoon swell prediction (Liang *et al.*, 1977).

The typhoon swell height is $H_{1/3} = 0.11H^*_R R/\sqrt{DD}$ (metres), in which H^*_R is the wave height (in feet) in the typhoon centre calculated by Bretschneider's method, R is the radius (in nautical miles) of the maximum wind speed and DD (in nautical miles) is the distance between the typhoon centre and the station. The swell period is

$$T_{1/3} = \left(0.05\,\frac{DD \times T'_{1/3}}{R \times U^*_{RD}} + 1\right) T'_{1/3},$$

in which $T'_{1/3}$ is the wave period in the typhoon centre estimated by Bretschneider's method and U^*_{RS} (in knots) is the sea surface wind speed at radius R.

WAVE HEIGHT ACCUMULATION AND EXPANSION CORRECTION FACTOR

When the distance between the typhoon centre and the station changes, the propagating duration of the swell also changes. If the duration becomes shorter, the wave height will increase, i.e. the accumulation effect. If the duration increases, the wave height decreases, i.e. the expansion effect. If more than two typhoons appear or if the propagating duration changes abruptly, for example, the typhoon suddenly becomes strong and its speed of travel increases violently, the swells may overlap. The latter can be solved simply by taking the square root of the wave height square sum. For the former, it is considered by the principle of wave energy transmission. The wave power passing through the cross section of a unit crest width is equal to $\frac{1}{8}\rho g H^2 C_g$, in which C_g is the group velocity and H is the wave height. If the typhoon is stationary, the duration of the swell is D. Owing to the moving typhoon, this duration becomes D'. However, the wave power passing that cross section is conserved. They should be equal:

$$\int_0^D \rho g H^2 Cg \, dt = \int_0^{D'} \rho g H'^2 C'g \, dt. \quad (1)$$

The wavelength is much shorter than the diameter of the typhoon. Therefore the movement of the typhoon has no influence on the wave period. Hence, $C_g = C'_g$ and

$$\int_0^D H^2 \, dt = \int_0^{D'} H'^2 \, dt.$$

Suppose H'/H is constant and equal to λ:

$$\int_0^D H^2 \, dt = \lambda^2 \int_0^D H^2 \, dt.$$

If H becomes small during this period, then $D = \lambda^2 D'$, or

$$\lambda = \left(\frac{D}{D'}\right)^{1/2}, \quad (2)$$

where λ is the wave height correction coefficient.

EXAMPLES

Two cases are studied as an example to verify the method. The wave stations were Hualien and Suau. The typhoon data are shown in Table 1.

TABLE 1
Typhoon data of Sally and Elsie

Typhoon	*Local time, date*	P_0 (mbar)	V_F *(knots)*	*Forward Direction*	*DD (nautical miles)*
Sally	2000, 26 June 1975	992	8.6	295	886
	0200, 27 June 1975	975	8.6	325	854
	0800, 27 June 1975	965	5.4	325	778
	1400, 27 June 1975	960	10.8	325	735
	0200, 28 June 1975	940	5.4	330	649
	0800, 28 June 1975	930	7.6	300	643
Elsie	1400, 11 October 1975	970	14.1	300	427
	2000, 11 October 1975	925	10.8	300	368
	0800, 12 October 1975	900	8.6	310	281
	2000, 12 October 1975	920	10.8	295	222
	0200, 13 October 1975	930	10.8	285	222
	0800, 13 October 1975	940	11.9	285	238

TABLE 2
Hindcasted swell for typhoon Sally and station Hualien

No.	Local time, date	$H_{1/3}$ (m)	$T_{1/3}$ (s)	Lag time (h)	Arrival time, date	Modified $H_{1/3}$
1	2000, 26 June 1975	0.49	8.9	64	1200, 29 June	
2	0200, 27 June 1975	0.9	11.4	48	0200, 29 June	1.6
3	0800, 27 June 1975	0.91	11.1	45	0500, 29 June	1.6
4	1400, 27 June 1975	1.18	11.6	40	0600, 29 June	2.1
5	0200, 28 June 1975	1.45	13.1	32	1000, 29 June	2.6
6	0800, 28 June 1975	1.65	13.5	30	1400, 29 June	1.7

TABLE 4
Hindcasted swell for typhoon Elsie and station Suau

No.	Local time, date	$H_{1/3}$ (m)	$T_{1/3}$ (s)	Lag time (h)	Arrival time, date	Modified $H_{1/3}$
1	1400, 11 October 1975	1.63	10.5	26	1600, 12 October	
2	2000, 11 October 1975	2.49	12.7	19	1500, 12 October	3.0
3	0800, 12 October 1975	2.93	12.5	14	2200, 12 October	2.93
4	2000, 12 October 1975	3.28	12.1	12	0800, 12 October	3.28
5	0200, 13 October 1975	3.14	11.9	12	1400, 13 October	3.14
6	0800, 13 October 1975	3.11	11.7	12	2000, 13 October	3.11

Case 1 (Hualien; Typhoon Sally)

The hindcasted swell data are shown in Table 2.

From No. 2 to No. 5, there is an accumulation effect, in which D is 24 h and $D' = 8$ h; hence the wave height correction coefficient $\lambda = 1.73$. No. 1 and No. 6 can be considered as overlapping. The recorded wave data at Hualien are shown in Table 3.

TABLE 3
Wave data at Hualien

Local time, date	$H_{1/3}$ (m)	$T_{1/3}$ (s)
0100, 29 June 1975	2.27	11.0
0500, 29 June 1975	2.25	11.0
0900, 29 June 1975	2.81	11.1
1300, 29 June 1975	2.03	11.4

One can see the modification of accumulation effect has improved the hindcasting very much.

Case 2 (Suau; Typhoon Elsie)

The hindcasted swell data are shown in Table 4.

No. 1 and No. 2 can be considered as overlapping. The others need not be modified. The recorded wave data at Suau are shown in Table 5.

TABLE 5
Wave data at Suau

Local time, date	$H_{1/3}$ (m)	$T_{1/3}$ (s)
0900, 12 October 1975	3.44	11.6
1400, 12 October 1975	3.48	11.6
2000, 12 October 1975	2.85	10.6
0000, 13 October 1975	3.34	10.8
0800, 13 October 1975	3.07	10.2
1200, 13 October 1975	2.72	9.9

The hindcasted data are very close to the recorded data.

LAND CORRECTION FACTOR

The land may affect the typhoon swell in the following two ways. First, the wave components may be blocked by the land during its propagation. Secondly, the typhoon wind may blow onto the land (no wave can be generated on the land). According to the author's elementary wave theory, the land correction factor C_e is

$$C_e = \sqrt{E_e / E} \quad (3)$$

where

$$E_e = \iint_{A^*} U^2(r, \alpha) \cos^2 [\theta (r, \alpha)]\, dA^*$$

and

$$E_e = \iint_A U^2(r, \alpha) \cos^2 [\theta (r, \alpha)]\, dA,$$

A and A^* being the wind field without, and considering, the land effect, respectively, $U(r, \alpha)$ the typhoon wind speed at (r, α) and $\theta(r, \alpha)$ the angle between the wind direction and the direction pointing to the station.

The correction factor C_e multiplying the swell wave height results to the modified swell wave height.

CONCLUSIONS

The typhoon swell prediction method is accurate and simple. This method can be used by marine engineers and meteorologists.

REFERENCES

Bretschneider, C. 1976.
Liang *et al.* 1977.
Tang, L. W. 1970.

NUMERICAL WAVE MODELS AS A SOURCE OF DATA FOR MARINE CLIMATOLOGY

Dr P E Francis, Meteorological Office, UK

Presented on **Wednesday 5 March at 12.30** in the Clarence Conference Room

EMBARGOED AGAINST DELIVERY

19

Numerical Wave Models as a Source of Data for Marine Climatology

P. E. Francis, Principal Scientific Officer, Meteorological Office, Bracknell, UK

INTRODUCTION

The forecasting of surface wave conditions has been a major feature of the services offered by the Meteorological Office since the development of the North Sea oil industry in the 1970s. Originally, forecasters made use of surface wind speed, direction and fetch as the forcing functions. The increase in demand for sea-state forecasts, linked with a growing confidence in numerically modelled surface wind fields, provided the impetus for the development of a numerical wave model. Such a model, for use in deep water areas was brought into operational use in 1976, followed by a shallow-water version in 1977. Various improvements have since been made to those models, leading to the present status of the models which are briefly described later in this paper.

In the decade since the introduction of the models the range of services based on their products has grown rapidly. Some idea of the scope of present services is presented later, illustrating that the great majority of work is done in the "operational" mode, i.e. the forecast products are used in real time. A growing amount of interest is, however, now being shown in applications for the archive of "hindcast" data, accumulated since January 1978. This archive contains the best simulation of the actual sea state that is possible with the present models, being generated by numerical wind fields that are mainly influenced by recent observations (see below). At the time of writing, nearly 8 years of this archive have been accumulated, forming a mass of data which almost rivals in length the data from a single-wave measurement location in European waters. The archive is now of such a length that climatological statistics can be inferred, albeit with reasonably wide confidence limits. As more data are routinely added the usefulness of this data source can only improve.

This presentation is completed by a brief discussion about the accuracy of the hindcast data (both the resultant wave fields and the generating wind fields), and a speculative look at some climatological applications which will be statistically of use in a few years time.

THE WAVE MODEL

The wave model at present run on an operational basis in the Meteorological Office is the result

TABLE 1

Physical specifications of operational wave models in the Meteorological Office

Model areas	*Resolution at 60°N (km)*	*Source and frequency of wind field data*	*Forecast period*	*Depth-dependent processes*
Northern Hemisphere	150	Global NWP model every 1½ h	3 days	No
Mediterranean Sea	50	Fine-mesh NWP model every 1½ h	36 h	No
European continental shelf	25	Fine-mesh NWP model every 1½ h	36 h	Yes

of an evolving series of such models, in use since 1976. The model (in three versions covering different geographical areas) is driven by wind fields extracted from the operational numerical weather prediction (NWP) models. All models are run twice daily throughout the year. The three versions of the wave model in operational use cover different areas on different scales, and are run for different forecast periods (see Table 1). The only difference in formulation between these three models is the inclusion of extra shallow-water processes in the continental shelf model.

The wind input for the wave models is taken either from a global NWP model (resolution 1½° × 1⅞°) or a limited area fine mesh NWP model (resolution ¾° × 15⁄16°), the frequency of input being governed by the requirement of numerical stability in the model calculations (Table 1). Figures 1–3 depict the geographical extent of the wave models, also showing the grid point locations at which wave information is calculated.

The theoretical basis of the wave model is the representation of the wave field as a spectrum, essentially the variance of the sea surface elevations that are present in a statistically random fashion. The behaviour of individual frequency and direction components of the spectrum can be calculated from known

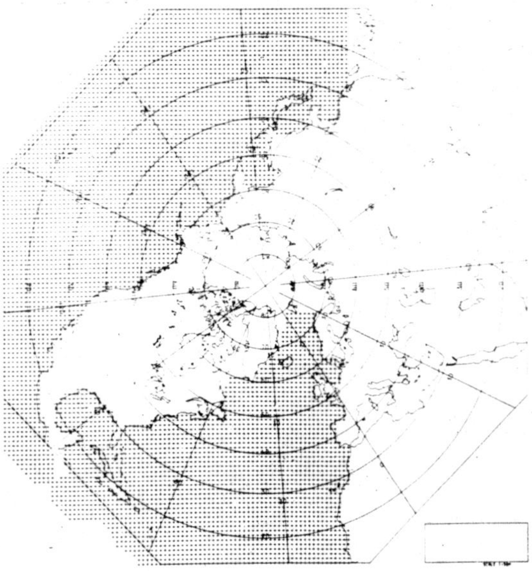

Fig. 1. Meteorological Office operational wave model, Northern Hemisphere coverage and grid.

Fig. 2. Meteorological Office operational wave model, Mediterranean Sea grid.

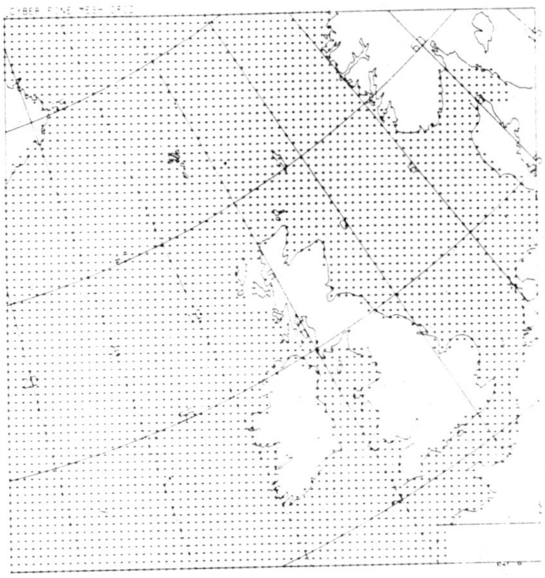

Fig. 3. Meteorological Office operational wave model. Continental shelf coverage and grid.

theories, and the more-recognized quantities such as wave height and period can be calculated from moments of the spectrum. The model at present has 16 equally spaced directional components and 14 irregularly distributed frequency values, ranging from 0.04 Hz (25 s period) to 0.324 Hz (approx. 3.1 s period). Thus at each model grid-point the sea state is represented by a total of 224 discrete spectral energy density values. The variation of these values in time and space is controlled by means of integrating the energy balance equation, which expresses the rate of change of energy density (at a point) as a balance between advection, refraction, generation and dissipation. Full details of the mathematical and physical theories used in the model may be found in the comprehensive paper by Golding (1983). It is sufficient here to outline the processes in a more general way.

The advection, or propagation, term is calculated by means of standard finite-difference techniques, with appropriate treatment at coastal and open boundaries in order to prevent reflection. The refraction term is solved by adopting a continuous form of Snell's law, more familiar in optics, in order to calculate the effects of irregular bottom topography on the component waves. The generation and dissipation terms contain a number of processes which attempt to represent the currently held theories concerning wave growth and decay. In simple terms, the growth of sea waves is due to the input of atmospheric energy through boundary-layer transfer processes which are still not completely understood. What is clearly apparent is that growth is inhibited by a limit on wave size and shape, beyond which breaking takes place with an accompanying loss of energy. The growth and decay processes in the model (for deep water states) are balanced so that the limiting form is attained but not exceeded, and also so that observed growth rates (in given wind speeds) are reproduced. In shallow water an extra dissipation term is introduced to represent the energy loss through frictional effects on the sea bed.

The other very important process that is represented (in an empirical fashion) in the model is that of non-linear interaction. It is known from an extensive series of measurements that the spectrum of a wind-sea (i.e. that portion of the full spectral range within the influence of the local generating wind) is maintained in a recognized shape by means of internal energy transfers from high to low frequency. The calculation of the exact transfer terms is, however, computationally very expensive, hence we have adopted an empirical approach which assumes a known standard spectral shape as the desired product of energy redistribution and which conserves the wind-sea energy during the calculation of the new spectrum. The main difficulty with this empirical treatment of non-linear interaction processes stems from the need to distinguish between wind sea and swell (i.e. that portion of the spectrum not under the influence of the wind). The present empirical scheme does, however, appear to work successfully, as is demonstrated in the verification results presented in a later section.

It is important in what follows that the hindcast mode of running the wave model is fully understood. Unlike an NWP model, where there are sufficient observed data to allow an analysis (or starting field) to be calculated for all of the different variables, the wave model has to

TABLE 2
Contents of the Meteorological Office wave model archive

Model	*Period*	*Data type*	*Resolution*
Northern hemisphere	1 Jan. 1978–23 Oct. 1979	Full spectrum	300 km at 60°N 12 directions × 6 frequencies
Northern hemisphere	23 Oct. 1979–26 Sept. 1982	Full spectrum	300 km at 60°N 12 directions × 11 frequencies
Northern hemisphere	1 Jan. 1983 to date	Integrated data	150 km at 60°N 16 directions × 14 frequencies
Mediterranean	16 Feb. 1983 to date	Integrated data	50 km at 60°N 16 directions × 14 frequencies
Continental shelf	26 Sept. 1978–23 Oct. 1979	Full spectrum	50 km at 60°N 12 directions × 6 frequencies
Continental shelf	23 Oct. 1979–26 Sept. 1982	Full spectrum	50 km at 60°N 12 directions × 11 frequencies
Continental shelf	1 Jan. 1983 to date	Full spectrum	25 km at 60°N 16 directions × 14 frequencies

be initialized by means of a hindcast. This is essentially the repeat of the first 12 h of the previous forecast (run 12 h earlier) but this time using improved wind fields, improved by making use of the intervening analysis runs of the NWP models. (These models run on a 6-h cycle of assimilating new data, producing high-quality surface wind fields in the process.) There are insufficient measurements of the directional wave spectrum to allow a wave analysis and, additionally, the theoretical mathematical work on which to base such an analysis has yet to be carried out. The hindcast technique yields a series of fields, 12 h apart, which are the best description of the actual wave fields that can be derived at present.

Table 2 contains a summary of the contents of the wave model archive to date, listing the resolution of the data over various periods during the evolution of the models. Surface wind speed and direction are also stored, along with the wave data. With the introduction of the higher resolution northern hemisphere model, and the Mediterranean model, in 1983, some data reduction became necessary. For these models the archive since 1983 consists of integrated data only, i.e. significant wave height, period and direction information for wind sea, dominant swell and total sea, rather than the full spectrum.

ASSESSMENT OF MODEL PERFORMANCE

The wave model can be assessed in two ways: first by comparison of values of significant wave height (the mean height derived from the zero-order moment of the spectrum — approximately the mean height of the one-third highest waves as usually estimated) as calculated by the model and as processed from buoy data or other measured information and, secondly, by means of comparison with more-detailed spectral data, both in terms of frequency and directional distributions. The first verification technique described above is performed operationally in the Meteorological Office, where incoming measured wave data from synoptic sources are routinely compared with model results. However, spectral comparisons are most usefully carried out during detailed investigations into model behaviour, such as the SWIM project (The SWIM Group, 1985), where very carefully analysed wind fields are used to minimize errors from that source.

Table 3 contains verification statistics over a period of 1 year (October 1984–September 1985) for model values of significant wave height generated in the operational hindcast mode. Data are presented here from five

TABLE 3
Error statistics of significant wave height values, generated by Meteorological Office operational wave models in hindcast mode, October 1984–September 1985

Area	No. of stations	Model	0–3 m			3–6 m			>6 m		
			N	$\bar{X}$	σ	N	$\bar{X}$	σ	N	$\bar{X}$	σ
USA, East coast	6	NH	1873	−0.1	0.5	304	+0.2	0.9	15	+0.7	1.4
USA, West coast	5	NH	1559	−0.1	0.5	982	−0.3	0.8	120	+0.4	1.2
Gulf of Mexico	3	NH	1733	−0.2	0.4	46	+0.4	1.3	0		
Northern and Central North Sea	6	CS	1512	−0.4	0.6	570	−0.4	0.9	60	0.0	1.2
Southern North Sea	5	CS	2711	−0.1	0.4	105	−0.1	0.6	0		

Wave height ranges are for measurements; N = number of measurements; $\bar{X}$ = mean error in metres, measured minus modelled; σ = standard deviation of error in metres; NH = northern hemisphere; CS = continental shelf.

aggregated areas and two wave models (i.e. one with depth-dependent terms in addition to the basic formulation). Model performance can be assessed over three ranges of measured wave height. A few general points can be inferred from the results in Table 3. Mean errors (the bias of the models) are generally small, with a tendency for slight overprediction of lower wave heights and slight underprediction of the higher ranges. The scatter of errors increases with wave height. The measured data are from buoys, which give time-integrated mean values at one location, whereas the model values are in effect areal averages as well as time means. On that basis the agreement between measured and model values is acceptable, especially since the resolution of the oceans model is of the order of 150 km, and the complex bed topography of the southern North Sea is barely adequately resolved at 25 km.

A more detailed comparison of operational archive data with buoy measurements is to be found in Houghton (1984), where data from further back in the archives is carefully examined. Houghton found that even with the coarser resolution of the Continental shelf model (50 km before 1983) at that time, the results indicated a good agreement between model and measurements, except in cases of strong tidal currents.

A recent careful hindcast study commissioned by the Department of Energy (WHIST, Wave HIndcast STudy) confirmed the general impression of good qualitative agreement between model simulation and measured values. That study also made available some more detailed spectral data against which model values could be assessed. Figures 4 and 5 illustrate model values of one-dimensional energy spectra (i.e. integrated over direction) and mean directional spectra plotted against similar data from a buoy located in the Norwegian Sea. These comparisons show that the correct spectral shape and distribution with direction can be obtained from the model, especially if extra care is taken (as in these cases) with the specification of the wind field. The agreement between model and measured mean directions (Fig. 4) is excellent, even though the validation

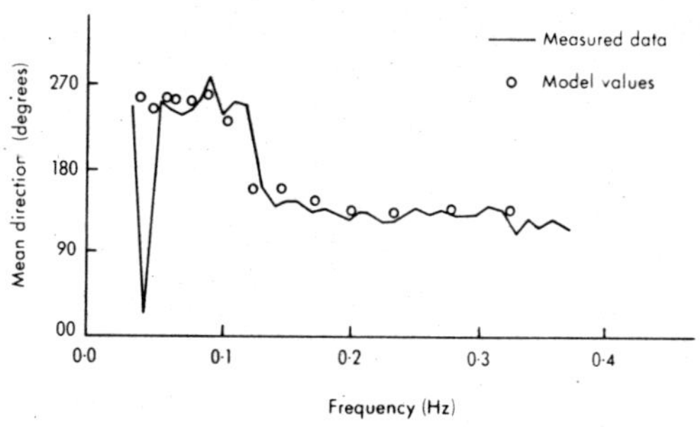

Fig. 4. Comparison of model hindcast mean direction spectrum (1800 h GMT 3 January 1983) and measured spectrum (1500 h GMT 3 January 1983) at the location Haltenbanken (65° 2′N, 7° 33′E). Measured data by courtesy of IKU, from ODAP.

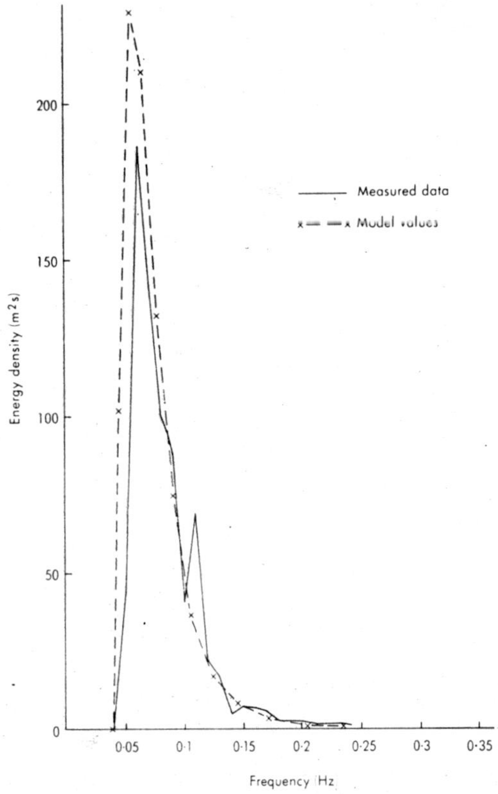

Fig. 5. Comparison of model hindcast one-dimensional energy spectrum (0600 h GMT 7 January 1983) and measured spectrum (0300 h GMT 7 January 1983) at the location Haltenbanken (65° 2′N, 7° 33′E). Measured data by courtesy of IKU, from ODAP.

times of the two sets of data are 3 h apart. The one-dimensional spectra (Fig. 5) agree as to peak frequency and general shape, the discrepancy in maximum energy being the result of comparing a site-specific value (measured) with an areal and temporal average (the model). Data from another nearby buoy (Trænebanken) indicate a very strong gradient in wave height at this time (early on 7 January 1983). Model values are only obtainable at 6-h intervals in spectral form, hence the slight gap (3 h) in comparison times.

Other comparisons of the model with spectral wave data, and integrated moment functions, are to be found in the SWIM paper (The SWIM Group, 1985), where three wave models are compared in both idealized and realistic wind and shallow bathymetry situations. The conclusions of that paper were that the 3-s generation (i.e. containing non-linear wave interaction terms) wave models give broadly the same spread of errors when assessed in a realistic hindcast, the overall results being quite acceptable.

Table 4 contains a summary of surface wind errors, evaluated from the wind fields used to drive the wave models when in hindcast mode. The modelled wind speed values are taken directly from the lowest level of the NWP models, and are thus probably relevant to a height of about 20 m above sea level. The wind measurements are from buoys in the American areas, and from a mix of buoys and platforms in the North Sea areas. The bias in wind speeds (i.e. the mean error) is low everywhere except in the Gulf of Mexico, where the model boundary layer is not correctly representing air–sea interactions in an area of warmer water. Values of bias tend to increase with the observed wind speed range, i.e. stronger winds tending to be more underestimated than lighter winds.

RECENT WAVE MODEL APPLICATIONS

A detailed description of recent applications of the wave models developed at the Meteorological Office is to be found in Ephraums (1985). Broadly speaking, the wide range of applications include forecasts for the offshore industry, shipping forecasts (espcially dedicated ship routeing services) and forecasts of coastal flooding events. A wide range of output media has been developed to facilitate these services, including both chart formats and digital transmission options.

Applications of data from the archives span an even wider range. This includes wave energy research, design studies, environmental studies, first-order verification of remotely sensed wave data, and some insurance enquiries. These applications have necessitated a range of accession software, capable of extracting data from the archive and presenting it in a useful form. However, use of archived data has been limited to what might be called single-time or short-period applications. It is only recently that

TABLE 4

Error statistics of surface wind speed values, generated by the operational NWP models of the Meteorological Office in assimilation mode, October 1984–September 1985

			0–10 ms			*10–15 ms*			*15 ms*		
Area	*No. of stations*	*Model*	*N*	$\bar{X}$	σ	*N*	$\bar{X}$	σ	*N*	$\bar{X}$	σ
USA, East Coast	6	Global	1977	0.0	1.9	261	0.9	2.6	47	1.3	2.3
USA, West Coast	6	Global	2909	0.1	1.9	560	−0.3	2.5	54	−0.1	3.5
Gulf of Mexico	3	Global	1570	0.8	1.7	183	2.2	2.4	8	3.2	2.2
Northern and central North Sea	6	LFM	1595	−0.4	1.9	593	0.0	2.4	137	0.6	2.8
Southern North Sea	5	LFM	2373	0.1	1.8	647	0.4	2.3	56	0.9	2.8

Wind speed ranges are for measurements; N = number of measurements; $\bar{X}$ = mean error in metres per second, measured minus modelled; σ = standard deviation of wind speed error in metres per second; LFM = limited area fine mesh model.

the accumulated quantity of data in the archive (every 12 h for 8 years in the longest case) has come to be recognized as a potential source for climatological applications. In the next section the possible climatological uses for the archive data are explored.

POSSIBLE FUTURE WAVE MODEL APPLICATIONS

The growth of the wave hindcast archive gives an assured source of useful wind and wave data for the foreseeable future. The exploitation of such a database is a viable proposition, since the accuracy of the component data is demonstrably good enough for most climatological applications. At present the longest archive record is 8 years, but an extensive range of analyses have been carried out, for all kinds of purposes, on wave measurement series of shorter duration than that. The strength of the archive at present lies not necessarily in its length, but in its breadth, the fact that consistent values are obtainable over a very wide area indeed.

To exploit the areal coverage of the archive it is a straightforward matter to produce time average charts for specific regions. Thus Figures 6 and 7 are mean value charts (wave height and wind speed) for the North Atlantic for the month of January 1984. Longer-period mean values can also be produced. Those

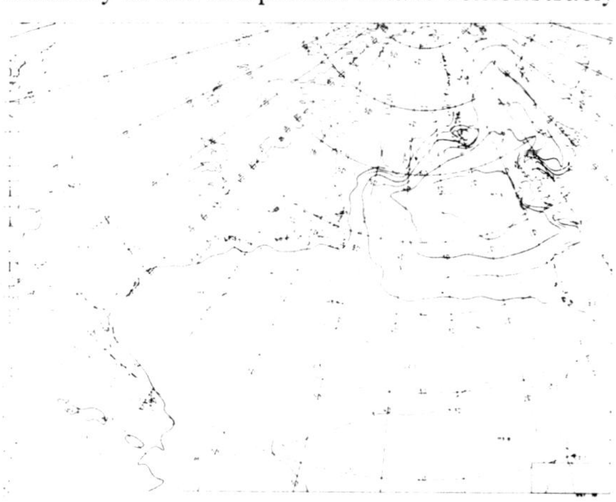

Fig. 6. Monthly mean significant wave height for the North Atlantic, January 1984, taken from the operational archive of the Meteorological Office wave model.

Fig. 7. Monthly mean wind speed for the North Atlantic, January 1984, taken from the operational archive of the Meteorological Office wave model.

charts show a surprisingly coherent picture, with maximum wind and wave activity in the track of Atlantic low pressure areas, between 50° and 60°N. The minimum activity areas are associated with the Azores high-pressure region. Areal differences are strongly highlighted in such a presentation, a feature that it is difficult to emulate with buoy data. Similar charts, based on wave power, can also be produced, which over a long averaging period are ideal for locating areas of the Continental shelf that are best suited for the exploitation of wave energy.

Many applications of climatological data make use of frequency of occurrence, or of exceedence of certain critical limits. Figures 8–10 illustrate such an approach, albeit for a short run of data (1–31 January 1984). The graphs show percentage exceedence against total wave height, for three locations, the American east coast, north-central Atlantic Ocean and northern North Sea. The wave climate during the month for these locations was obviously very different, a fact confirmed, of course, by reference to Figure 6. The American east coast location was relatively quiet, less than 5% of occurrences exceeding 5 m. The North Sea location (near the Brent and Statford platform) was more energetic, 50% of occasions exceeding 5 m and 12% being greater than 7 m. The Atlantic location is near to the Ocean Weather ship "L", and reference to Figure 6 puts the site in the area of maximum wave activity for that month. The percentage exceedence wave is not surprisingly less steep than for the other locations with about 15% of occurrences greater than 10 m.

Areal differences are again highlighted by this kind of presentation, which gains in statistical significance with longer runs of data included. At present the relevant software is only in a development stage, hence the somewhat limited applications that can be shown. Other parameters can, of course, be depicted in this way for any location, e.g. wind speed, wave period or wave direction. In principle it would also be possible to break wave height or period data into separate distributions for wind sea and swell.

The most interesting applications are probably in the area of joint distributions, e.g. of wave height against period, height against direction or wind speed against direction. Table 5 illustrates a simple example of such a

Fig. 8. Percentage exceedence plot of significant wave height taken from the Meteorological Office wave model archives for the location 38.0°N, 70.4°W, data from the month of January 1984.

Fig. 9. Percentage exceedence plot of significant wave height taken from the Meteorological Office wave model archives for the location 57.4°N, 19.7°W, data from the month of January 1984.

presentation, containing pairs of wave height and period data for a 50-day period. Even with the limited number of data pairs presented it is apparent that a clearly defined distribution exists, with longer periods associated (usually) with higher waves. There are, however, one or two interesting exceptions. When the appropriate software is available it should be possible, for instance, to extract such data for every January over 8 years, or every winter period, and to more significantly establish such bivariate frequency distributions. Such information is used extensively from available buoy records and other measurements. The wide areal coverage of the wave model archives should make them a very attractive source of such information presented in ways which logically should be based on user requirements.

FUTURE DEVELOPMENTS

Two interesting development areas in this field are worth mentioning in the conclusion of this paper. First, the Meteorological Office is shortly (Spring 1986) to introduce global coverage for its wave-modelling capabilities, taking the opportunity to reorganize the model archives for more flexible future exploitation.

Secondly, there is now a good chance that an international wave hindcasting programme will be funded in order to establish an archive back to 1956. A well-established second generation wave model will be used for the purpose, driven by carefully analysed wind fields, reconstructed for the programme. The hindcast archive will initially, of course, be confidential to the programme sponsors and the modelling groups, but eventually the resources of such an archive will be available to a wider group of potential users. In the meantime the somewhat more modest length of data in the growing Meteorological Office wave model archive is available for those who wish to use it.

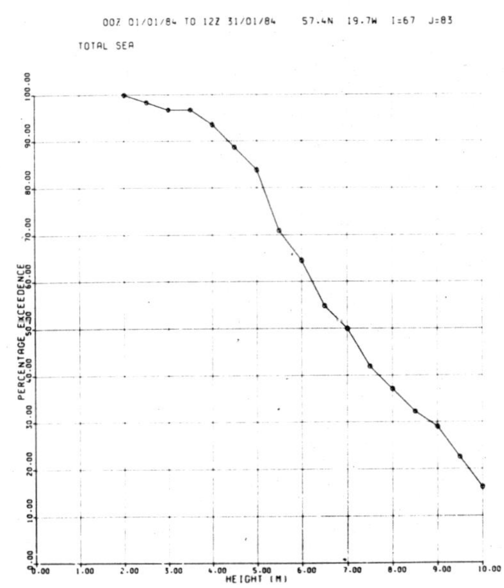

Fig. 10. Percentage exceedence plot of significant wave height taken from the Meteorological Office wave model archives for the location 61.2°N, 1.9°E, data from the month of January 1984.

TABLE 5

Scatter diagrams of the frequency of joint occurrence of wave height and period values, extracted from the fine-mesh wave model archive; location 61.2°N, 1.9°E, period 1 January 1984–19 February 1984

	Period (s)										
Height (m)	4	5	6	7	8	9	10	11	12	13	All periods
0											0
1											0
	2		2								4
2		2	2	1			1				6
		1	4	1	2						8
3		1	2	3		2					8
			2	4	2	1				1	10
4			1	5	4						10
				1							1
5				3	7						10
					9	3					12
6					6	2	1				9
					2	3					5
7						5					5
						2					2
8											0
						1	1				2
9							1				1
							1				1
10											0
All heights	2	4	13	18	32	19	5	0	0	1	94

REFERENCES

Ephraums, J. J. 1985. Applications of wave and surge models. Met. Mag. 114, 282–291.

Golding, B. 1983. A wave prediction system for real-time sea state forecasting. *Q. J. R. Met. Soc.* **109**, 393–416.

Houghton, I. 1984. A long time-series verification of hindcasts from the Meteorological Office wave model archive. *Met. Mag.* **13**, 317–328.

The SWIM Group (Bouws, E., Ephraums, J. J., Ewing, J. A., Francis, P. E., Gunther, H., Janssen, P. A. E. M., Komen, G. J., Rosenthal, W. and de Voogt, W. J. P.) 1985. A shallow water intercomparison of three numerical wave prediction models. *Q. J. R. Met. Soc.* **111**.

Part V

Geology/Geographics/Geotechnics

20

A New Opto-electronic Technique for the Measurement of Seabed Sand Transport

J. Hardisty, Marine Morphodynamics Unit, Department of Geography, Royal Holloway & Bedford New College, University of London, Egham, Surrey TW20 0EX

The movement of sand under the action of waves and tidal currents is of considerable importance to hydrographers, engineers and geologists because it may lead to sediment accumulation, bank and channel migration and structural or pipeline scour.

Although a range of predictive sediment transport equations exist, direct determination of sediment transport rates in the marine environment is presently limited to low-frequency measurements with traps, tracers or transmissometers. This chapter describes a non-intrusive probe which utilizes optical fibres to direct a modulated infra-red light beam over a short seawater path length. Interruption of the beam by moving sand grains is recorded and the transport rate is estimated. The instrument has the advantage of being able to resolve the movement of individual grains very close to the sea bed. The results from preliminary trials in Carmarthen Bay are described.

INTRODUCTION

Formulae Which Need Not Exist

Sediment transport — and we shall use the phrase here to refer to the movement of loose sand and gravel under the action of waves and tidal currents — has been studied for practical reasons for at least 20 centuries: documents show that the Greeks were concerned with a problem of harbour siltation in the Mediterranean (Graf, 1971). More recently, others have addressed the so-called two-phase problem; ranging from geographers working on landform evolution, geologists interpreting bedforms in ancient strata, oceanographers interested in bed roughness and drag, as well as a wide variety of engineers concerned still with siltation, but now also scour, channel maintenance, coastal erosion, winning aggregates and related problems. However, very few of these groups

A NEW OPTO-ELECTRONIC TECHNIQUE FOR THE MEASUREMENT OF SEABED SAND TRANSPORT

Dr J Hardisty, University of London, UK

Presented on **Wednesday 5 March 1986 at 14.30** in the Norfolk Conference Room

EMBARGOED AGAINST DELIVERY

would wish to be concerned with the process of sediment transport itself. They are generally concerned with the result of sediment transport, for example the infilling of a dredged channel by tidal flows, and need an understanding of the process only to link the known hydraulics with the expected change in the hydrography of the channel. In an ideal world, the knowledge of the flow would lead directly to a prediction of the hydrography. Such a solution is, of course, presently not available, although recent developments in the application of non-linear systems theory to morphodynamic models may improve the position (e.g. Hardisty, 1986a).

Existing Formulae

Presently, the problem is tackled by substituting some characteristic of the flow and of the bed sediment into one of the sediment transport formulae, and inte grating the resulting rate over a chosen time interval to determine the hydrographic change. A large number of these formulae, and integrating the resulting rate 26 which have been applied, with varying degrees of success, to the marine environment.

Research Objectives

The Marine Morphodynamics Unit at RHBNC has been examining these formulae with a view to improving their predictive capability by achieving a more fundamental grasp of the physics of the processes involved. It appears that, largely because sediment transport does reflect the inherent turbulence of the forcing flow, the group of formulae which originated from the work of Einstein (1950) includes a probabilistic element. An alternative group of formulae appears to presume a purely deterministic relationship, and these are exemplified by the energetics approach of Bagnold (1966). It seemed that the two approaches might be ratified by comparing the spectrum of the bedload rate with that of the nearbed turbulence. A sufficiently close cross correlation would suggest that a deterministic relationship would predict the apparently probabilistic phenomena, provided that a suitably high frequency of turbulence was employed in the calculations.

Existing Technology

It is apparent, however, that although electromagnetic, ultrasonic and hot-wire devices have vastly improved our understanding of turbulent flow, bedload theory is restricted by the lack of measurements at comparable frequencies.

The existing technology for the measurement of bedload has been reviewed by Hubble (1964) and Salkield (1978) and is summarized in Table 1, which attempts to show the maximum operating frequency of each technique, based largely on the author's experience, though the values for surveying, tracers and traps will depend upon the operating procedure. The quoted values assume tidal surveys and sample collections for these techniques, and 10-minute intervals for the pumping techniques. Clearly both the theory and the measurement of sediment transport lags considerably behind that of hydrodynamic phenomena where frequencies of tens or even hundreds of Hertz are commonly used. The present sediment transport technology might be analogous to investigating turbulence with a floating orange.

TABLE 1
Estimated response frequencies of existing sediment transport devices

Technique	*Example reference*	*Typical frequency (Hz)*
Hydrographic Surveying	Langhorne (1982)	0.00002
Tracers	Heathershaw (1981)	0.00002
Fixed traps	Hardisty and Hamilton (1984)	0.00002
Automatic traps	Reid and Frostick (1984)	0.002
Transmissometers	McCave (1979)	0.1
Electromagnetic	Reid and Frostick (1984)	0.01
Acoustic noise	Heathershaw (pers. comm.)	1

The Marine Morphodynamics Unit therefore proposed a new device, called the Optical Transport Sensor (OTS) which utilizes optical fibres to detect the passage of individual grains, and therefore to provide the required measurement speeds for comparing the bedload and flow spectra. The design of the OTS and the results of preliminary trials are described here.

EXPERIMENTAL TECHNIQUE

Analogue Optical Methods

Optical methods have been employed for some time to measure the volume of sediment in a water body. Such instruments are generally used for the measurement of suspended load and are known as transmissometers. McCave (1979) describes the principles and the theoretical basis for these techniques. In general, a light beam is directed between a source and a detector and is arranged so that sediment-laden water passes through the light path. The solid particles absorb and deflect the light, so that the transmitted energy decreases with increasing sediment concentration.

A number of important practical problems are, however, associated with the use of transmissometers in the marine environment. First, the light extinction depends not only upon the sediment concentration, but also upon the mineralogy, shape and grain-size distribution of the sediment grains. These problems are usually addressed by careful calibration of the instrument with the particular sediment type under investigation. More importantly for the present work, the maximum operating frequency is inherently restricted for two reasons. First, the light detector is necessarily sensitive to small changes in the incident radiation. Therefore daylight can, in all but the deepest deployments, induce erroneous readings. Commercial instruments overcome this problem by shrouding the light path, but this results in the sediment concentration within the instrument lagging behind the ambient conditions by an appreciable and variable flushing interval. Secondly, the cross-sectional area of the light path and sensor is typically three or four orders of magnitude larger than the particle size, resulting in a mean beam concentration measurement. Both of these problems result in a smoothed analogue signal, with concentrations averaged over intervals of many seconds. The result is suitable for general suspension measurements but the technique is clearly inappropriate for the higher-frequency measurements required here.

Digital Opto-electronic Methods

An improved optical technique was proposed in an attempt to increase the frequency of sediment transport measurements and, ultimately, to allow the comparison of the bedload and turbulence spectra. Clearly, the spatial resolution of transmissometers could be improved by reducing the physical size of the light source and detector, and by reducing the length of the light path. The recent development of optical fibres and of the associated signalling technology appeared to provide the opportunity to considerably reduce the size of these components and to greatly increase the sampling rate. Two types of optical cable are presently available: the larger polymer type has a typical diameter of 1 mm while the glass type has a typical diameter of 100 μm. Thus a light path could be achieved which is similar in size to the actual sand grains under investigation. Furthermore, these cables can be attached to very high speed switching devices which may permit the detection of individual sand grains.

The OTS System

Fig. 1 Detail of the Optical Transport Sensor head showing the two aligned optical fibres and the light path

A prototype Optical Transport Sensor was assembled, largely to examine the possibilities for further development. The sensor consisted of an Omron Fibre Optic Amplifier Unit (Farnells catalogue number 170–210) connected to a separate — as opposed to a diffused — polymer cable and optic head (Farnells catalogue number 170–526). The amplifier unit consists of a modulated infra-red source and photocell with a variable threshold direct d.c. output load switching option; the modulated beam allows the system to be used in daylight without the necessity of shrouding the beam, and therefore obviates some of the problems which were discussed above.

The terminations of the optical fibre were aligned and arranged to direct the light path about 0.5 mm above and parallel to the surface

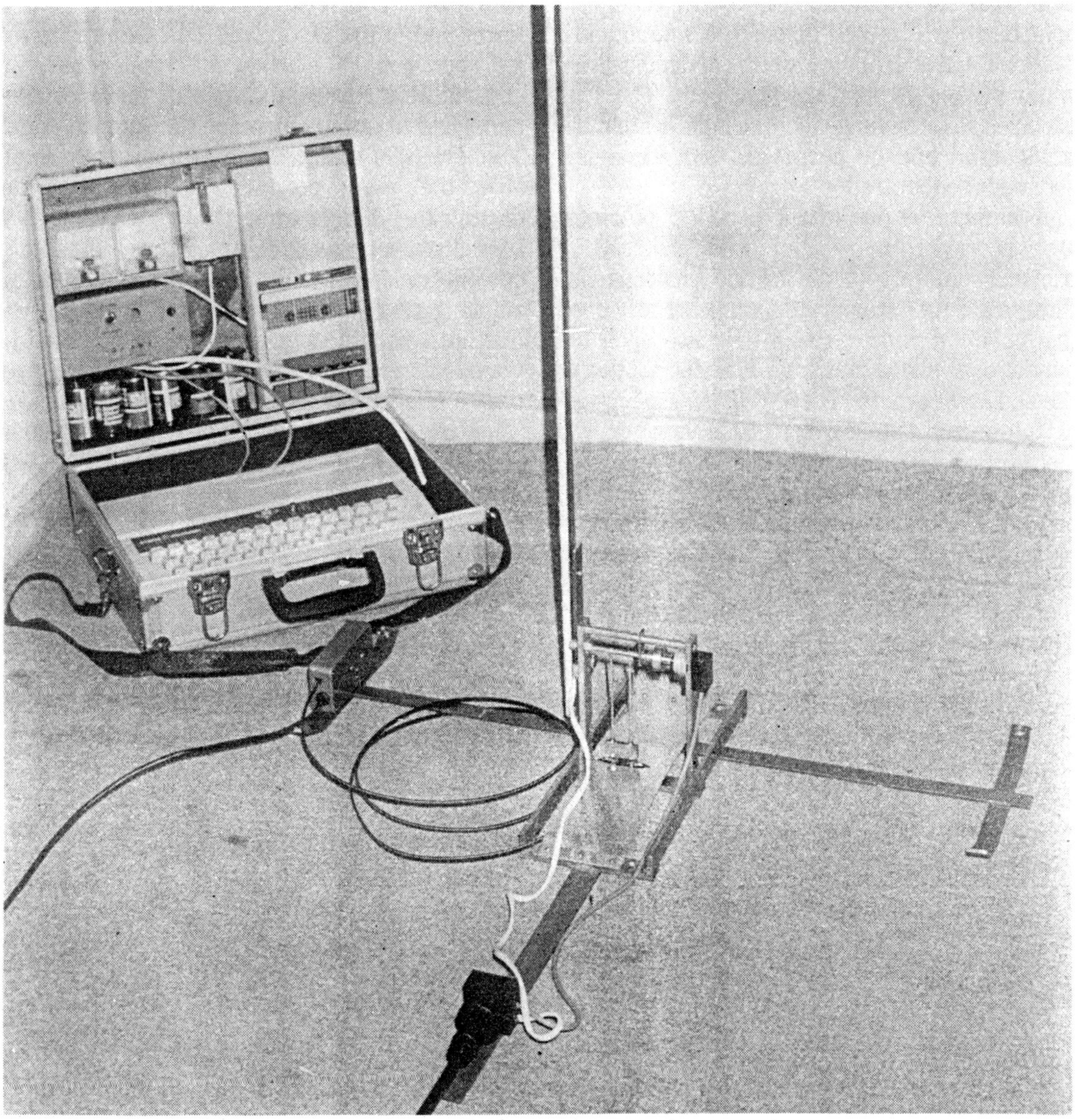

Fig. 2 General view of the OTS mounted on a vane flow sensor and shallow-water waves staff. The amplifier and signalling system is housed within the enclosure at the left-hand side of the apparatus. Data is cabled to the battery-powered microcomputer for initial analysis and storage on cassette tape.

of a slightly domed aluminium sheet (Fig. 1) which was mounted below the vane type of flow sensor shown in Fig 2. The flow sensor and wave staff consisted of a shaft-mounted potentiometer which sensed the angle of deflection of a small vane. The system was designed to measure shoaling wave transformations and nearbed flows within the breaker and swash zones, and is described in detail elsewhere (Hardisty, 1986b). The optical terminations were mounted 10 mm apart within specially shaped housings which were intended to direct bedload grains through the light beam. Extinctions of the beam due to the passage of sediment particles were initially counted using a high-speed TTL LCD counter, though the system was designed for direct input into a computerized logging system at a later stage.

Field Measurements

A trial deployment with the equipment was undertaken at Pendine Sands on Carmarthen Bay, south-west Wales. The sedimentology of the area is described in Jago and Hardisty (1984). The OTS was placed in the mid-swash zone and the counter reading was recorded at 5 s intervals, for periods ranging from 1 to 5 minutes. Typical results are shown in Fig. 3. Observations of the equipment during these experiments confirmed that peak count rates did coincide with greater numbers of sand grains passing between the sensors, which occurred slightly before the passage of a wave trough, and to a lesser extent before the passage of a wave crest. The results also appear to exhibit a slower overall variation in

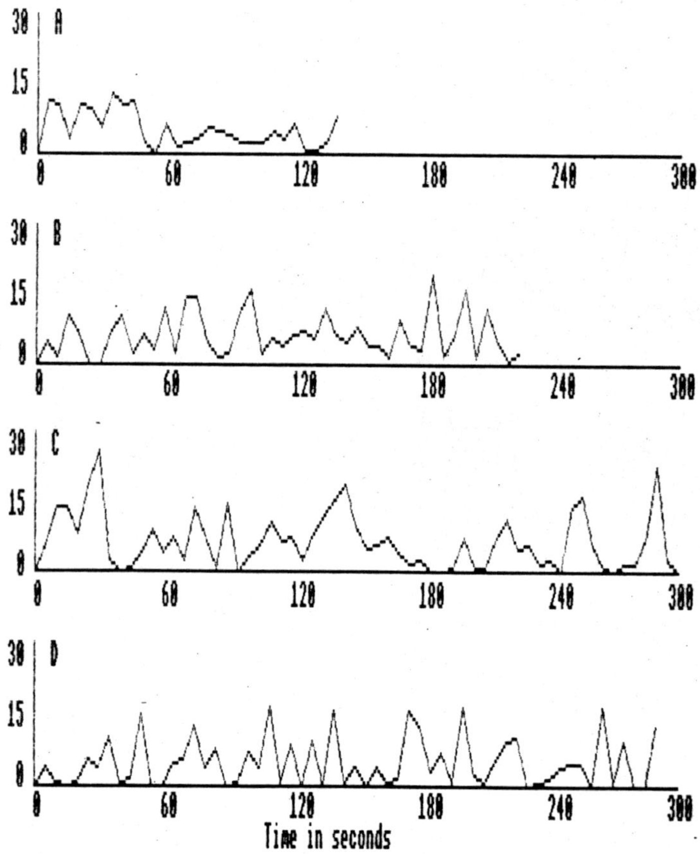

Fig. 3 Four typical counter readings from the first field development of the OTS: the vertical axes are all counts per second

the bedload rate, which may be attributable to the longer infra-gravity waves present in a resonant nearshore wave field.

CONCLUSIONS

Initial trials with the Optical Transport Sensor suggest that an optical device incorporating glass or polymer fibres, a modulated light source, and digital switching techniques are capable of measuring bedload sand transport with finer spatial resolution, and at higher frequencies than has hitherto been possible.

REFERENCES

Bagnold, R. A. 1966. An approach to the sediment transport problem from general physics. *U.S. Geological Survey Professional Paper 422-I.*

Einstein, R. A. 1950. The bedload function for sediment transportation in open channels. *Tech. Rep. 1206.* Washington, DC: U.S. Department of Agriculture,

Graf, W. H. 1971. *Hydraulics of Sediment Transport.* London: McGraw-Hill.

Hardistry, J. 1986a. A morphodynamic model for beach gradients. *Earth Surface Process & Landforms,* in press.

Hardisty, J. 1986b. A deductive determination of the effect of slope on the bedload rate from field measurements in the Western Mediterranean. *Sedimentology,* in press.

Hardisty, J. and Hamilton, D. 1984. Measurements of sediment transport on the seabed southwest of England. *Geo-Marine Letts* **4**, 19–23.

Heathershaw, A. D. 1981. Comparison of measured and predicted sediment transport rates in tidal currents. *Mar. Geol.* **42**, 75-104.

Hubble, D. W. 1964. Apparatus and techniques for measuring bedload. *U.S. Geological Survey Water Supply Paper 1748.*

Jago, C. F. and Hardisty, J. 1984. Sedimentology and morphodynamics of a macrotidal beach, Pendine Sands, SW Wales. *Mar. Geol.* **60**, 123–154.

Langhorne, D. N. 1982. A study of the dynamics of a marine sand wave. *Sedimentology* **29**, 571–594.

McCave, I. N. 1979. Suspended sediment. In: K. R. Dyer (Ed.) *Estuarine Hydrography and Sedimentation.* Cambridge: CUP, pp. 131–185.

Reid, I. and Frostick, L. E. 1984. Particle interaction and its effect on the thresholds of initial and final bedload motion in coarse alluvial channels. In: Koster, E. H. and Steel, R. J. (Eds) *Sedimentology of Gravels and Conglomerates.* Canadian Society of Petroleum Geologists, Memoir 10. pp. 61–68.

Salksfield, A. P. 1978. Sensors for the measurement of the transport of sand as bedload. *Institute of Oceanographic Sciences Report no. 73,* 37 pp.

Sleath, J. A. 1984. *Seabed Mechanics.* New York: Wiley, 335 pp.

21

RALPHNET: An Underwater Local Area Network

Dave Heffler, Geological Survey of Canada

INTRODUCTION

Many instruments that operate in the marine environment use microcomputers for control and data-logging. RALPH, designed to study seabed sediment transport processes, is such an instrument. Although the design was started in 1978 and useful data have been collected since 1980 (DeIure, 1983; Boyd et al., 1983), RALPH, like many research tools, has been in a constant state of development. This development has mainly involved the addition of sensors and the provision for more data-logging. These two thrusts have led to RALPHNET.

To reduce the number of cables and connectors, the initial criteria of RALPH's design called for housing all electronics and most sensors in one large pressure vessel. This also facilitates interfacing the various components to the computer which controls all sensors and the data-logging. However, some of the sensors (e.g. flow meters) are required to be outside the pressure vessel in contact with the environment. As RALPH grew, more cables and more computer interfaces were required. In 1984 the concept of RALPH as an underwater local area network of special-purpose computers was adopted. This concept solved the problems cited, gave other unexpected advantages and has also found application in unrelated instruments.

This paper first gives a brief description of RALPH and the components which have been added to it. The details of RALPHNET are then described, its advantages listed and some other applications cited. An Appendix contains a more detailed description of RALPH and several of its components.

RALPH AND DEVICES

RALPH (Heffler, 1984) may be used from the outer limits of the surf zone (5 m depth) to the edge of the continental shelf (200 m depth). It is operated for periods of up to a month, and may be independent of cables to the surface or to the shore. Power is supplied by internal batteries and data are logged internally. This enables RALPH to continue sampling during a storm, the time when sediment transport is a maximum and when most instruments cannot be used.

RALPH is an instrument system which

RALPHNET: AN UNDERWATER LOCAL AREA NETWORK

D E Heffler, Geological Survey of Canada, Canada

Presented on **Wednesday 5 March 1986 at 15.00**
in the Norfolk Conference Room

EMBARGOED AGAINST DELIVERY

includes a main pressure vessel housing a computer, battery power and some sensors. Additionally, various external devices are connected to this main pressure vessel. All components are mounted on a tripod structure. The external devices contain microprocessors which are interfaced to the computer in the main pressure vessel. RALPHNET is a network of the RALPH computer, the external devices and an operator's terminal. Figure 1 shows a block diagram of RALPH.

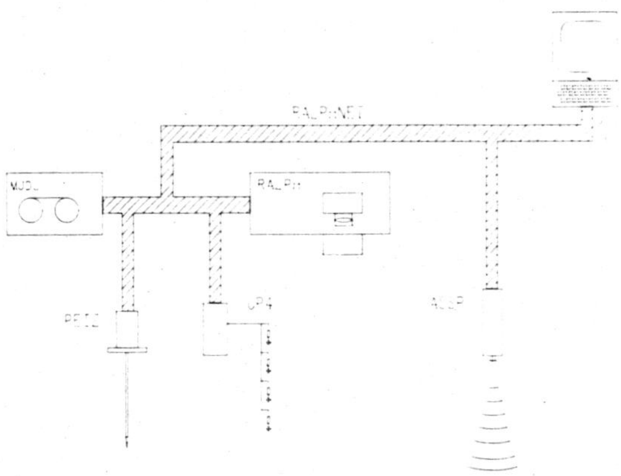

Fig. 1. A block diagram of RALPH components showing RALPHNET.

RALPHNET

A Local Area Network (LAN) is a method of connecting a variety of computers so they can transfer information. Often one device is considered to be the master, but other devices — slaves — are able to transfer data without intervention of the master. Three simple topologies are obvious when connecting several slaves to a master. Figure 2 shows a star, a ring and a bidirectional line. The star uses many connections and interfaces to the master. The ring is sensitive to the failure of one device and, without complex protocol, its operation is dependent on the position of the slaves around the ring. The bidirectional line was adopted for RALPHNET. It has reduced the underwater connections needed and is versatile.

RALPH, as the master, and all the other devices are connected to a pair of wires (signal and ground). All communication uses standard

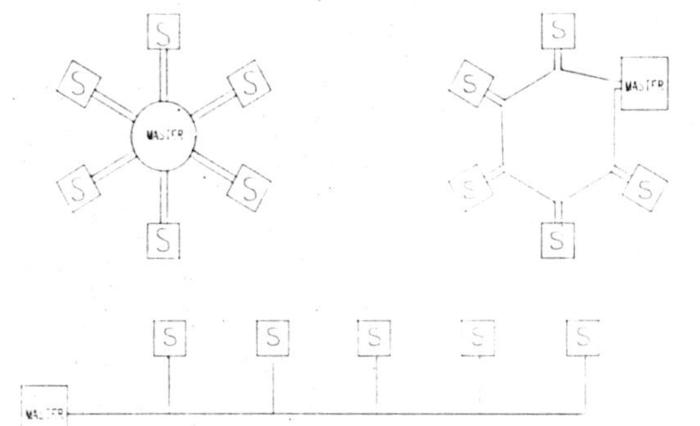

Fig. 2. Three possible Local Area Network topologies.

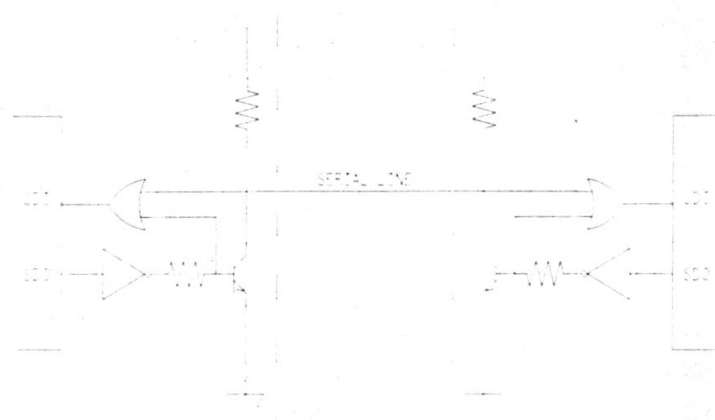

Fig. 3. The circuit for the RALPHNET bidirectional serial line.

Fig. 4. Hardware reset, the device on the left is protected from negative voltages on the line, the one on the right generates a reset signal (active low) when the line is forced below ground.

ASCII characters at 9600 baud. A hardware reset of the components on the network is also possible using these two wires. This reset is particularly useful for developmental devices which have not been fully debugged.

All computers which use serial communication have two signals, serial data out (SDO) and serial data in (SDI). Inside the computer, SDO is normally at a logic high (5 V) and goes low (0 V) for signal transmission. The circuit of Fig. 3 shows a simple method of connecting two such computers on a single bidirectional line. This circuit works if both computers do not try to transmit at the same time. The gates are used to prevent the SDO of a computer from appearing at its own SDI. Figure 3 shows a serial line which is normally high. It is also possible to change a few components and use a line which is normally low.

Both the normally high and normally low lines are always in the range 0–5 V. An external hardware reset may be implemented by externally forcing the line to a negative voltage. A diode must be added in each RALPHNET device to protect the transistor and, if the hardware reset is to be used, that device must have a few extra components, as shown in Fig. 4.

THE USE OF RALPHNET

There is a potential problem with data collisions on RALPHNET. If two devices send data simultaneously, random signals will appear on the line. It is possible that such signals be interpreted as an unintended command for another device and cause it to send unwanted data. For this reason, the master must be careful not to allow two devices to send simultaneously. The external devices should only send data (numbers) and text (letters) and never send the punctuation marks (@,#,!,%,etc.) used as control codes for the external devices.

It is convenient to include a data-logger (MUDL or SUDL, see the Appendix) as one of the devices on RALPHNET. The logger is plugged into the serial line and records everything that is sent. The main RALPH computer can send time, data from internal sensors and commands to the other devices. MUDL or SUDL record all of this and any data returned by the external devices. RALPH's computer can ignore the data, knowing that it is being recorded elsewhere.

An operator, watching a terminal interfaced to RALPHNET, can see the commands and the data in real time. For data playback, the operator powers-down the devices and then commands MUDL to retransmit the data. RALPH and the other devices are put on standby so none of the recorded commands causes a response on the network. For test and program development purposes, an operator can connect a single device to the terminal, give it commands and observe the response. The operator functions as an extra master and must take care not to cause data collisions.

An example of a RALPH experiment may serve as a useful illustration of RALPHNET. Let us assume that RALPH with MUDL, CP4 and PIEZ are in water of depth 10 m, 1 km from the shore (see the Appendix for a description of MUDL, CP4 and PIEZ). A RALPHNET cable is secured to a surface buoy, but does not go ashore. RALPH is programmed to take a photograph every 15 min and a time series of data every 2 h.

Every 15 min, when the RALPH computer operates the camera, it also sends a message on RALPHNET, to be recorded by MUDL and ignored by the other devices. This message contains the time and the fact that a picture was taken. It also includes other data, such as battery voltages and light levels. The message contains alphanumeric characters and carriage returns, but no punctuation marks (which might be interpreted as commands to the other devices).

Every 2 h, RALPH powers-up CP4 and PIEZ and waits for a 30 s warm-up period. RALPH sends a header with time information to MUDL. At quarter-second intervals, RALPH sends "*2" to tell CP4 and PIEZ to sample their sensors but not to send any data. RALPH samples the transmissiometers and pressure gauge and sends these data on RALPHNET (to be recorded by MUDL and ignored by the others). RALPH then sends "#4" to tell CP4 to send its data for all eight channels in hexadecimal, and waits while the data are sent. RALPH sends "%4" to tell PIEZ to send its data. After a short delay, RALPH sends a carriage return so the data will be readable on playback. After 1024 such samples RALPH will power-down CP4 and PIEZ and send a "!S" to tell MUDL to write its entire data buffer on tape.

During good weather, it is possible to go out in a small boat and connect a personal computer as a terminal to the RALPHNET line secured to the surface buoy. Data-logging will be observed as it happens or MUDL will be commanded to replay selected blocks of data and save them on floppy disks. The data-logging parameters (time series length, rate, etc.) may be changed. The performance of the sensors may also be checked.

On completion of the experiment, RALPH will be recovered and the data recorded in MUDL be fully replayed and analysed.

ADVANTAGES

The advantages of using RALPHNET are the following.

(1) The number of underwater connectors required is almost half that required if each device were separtely connected to the master.

(2) Each device, including the master, needs only one serial interface.

(3) A device can be tested in a stand-alone mode or in a limited configuration of devices.

(4) It is easy to arrange a data-logger to be on the network and record all the data.

(5) For shallow-water or near-shore experiments, a simple cable may go to a boat or to the shore. An operator can check the operational status of each device, obtain real-time data or request data replay. Testing that all components are working after deployment has been useful.

(6) Only one device transmits at a time, so long cables do not introduce cross-talk.

(7) The system lends itself to an external hardware reset, especially useful for prototype instruments which are not fully tested.

(8) The system can easily be expanded by the addition of other devices on RALPHNET and trivial software modifications.

OTHER APPLICATIONS

A system of Ocean Bottom Seismometers (OBS), each with a computer for time-keeping and control, has been developed at AGC. Each OBS is configured as a RALPHNET device, and many may be connected to a single cable (signal and ground) and controlled by a single terminal. A new timed electrolytic release unit under development will use only one single-pin connector. The serial line (which is normally low) will also supply the voltage to electrolytically corrode a small nickel alloy wire and cause the OBS to release a ballast weight and float to the surface.

An attitude-sensing and logging package to operate on a towed camera sled is also being designed. The components of the system will be compatible with RALPHNET, so various sensors and loggers can be interchanged.

CONCLUSIONS

RALPHNET has proved to be a good solution to the problem of connecting small computer-based instruments in the marine environment. The unexpected advantages of ease of testing individual devices and simple data-logging have also been important.

A similar concept has also been successfully implemented in several unrelated instruments.

REFERENCES

Boyd, R., Forbes D. L. and Heffler, D. E. 1983. Fairweather sediment dispersal on the shoreface of the Nova Scotia Atlantic Coast (abstract). *EOS, Trans. Am. Geophys. Un.* **64**, 1102.

Delure, A. 1983. *Storm effects in Halifax harbour.* M.Sc. Thesis, Dalhousie University, Halifax, Canada.

Heffler, D. E. 1984. RALPH — an instrument to monitor seabed sediments. In *Current Research, Part B, Geol. Survey of Canada, Paper 84-1b,* pp. 47–52.

APPENDIX

This Appendix contains a short description of RALPH, highlighting the devices of particular interest. For a description of RALPH before RALPHNET, see Heffler (1984).

The sensors included in the main pressure vessel are the following.

(1) A Super-8 movie camera used in the single-frame mode, which can capture 3600 images of the sea bed.

(2) A pressure gauge, used to sense both tides and waves.

(3) Two external optical transmissiometers which are mounted on the tripod are used to estimate suspended sediment concentrations. These are directly interfaced to the main computer.

The external devices connected to RALPH are the following.

MUDL (Massive Underwater Data Logger). This is a self-contained data-logger (including power) in a pressure vessel 20 cm in diameter and 80 cm long. It will operate for 1 month and store several megabytes of data. The capacity may soon be expanded to several hundred megabytes. It uses a streaming tape drive and a low-power computer with a large data buffer. It has been developed by Seimac Ltd (Bedford, Nova Scotia, Canada).

SUDL (Small Underwater Data Logger). This is a logger using two 0.5 Mbyte DRAM (Dynamic Random Access Memory) boards. SUDL stores 1 Mbyte of data, is small, and draws very little power.

CP4. This is a small package (15 cm in diameter and 25 cm long) which contains the electronics for four Marsh McBirney (model 512) electromagnetic current meters, an analogue-to-digital converter and a computer interface to RALPHNET. It will transmit data in decimal, hexadecimal or a more compact code. Separating the current meter electronics from the rest of RALPH's electronics appears to have reduced problems of noise and drift commonly associated with electromagnetic current meters.

ASSP (Acoustic Suspended Sediment Profiler). This is a 1-5 MHz ultrasonic sonar with an analogue-to-digital converter and computer used to measure quantitatively the sediment concentration in the bottom 1 m of the water column with a resolution of 1 cm. It is currently being tested and calibrated.

PIEZ. A pore pressure meter in a separate pressure vessel is currently under development.

Each of these external devices (MUDL, SUDL, CP4, ASSP and PIEZ) responds to two character codes. The first character, a punctuation mark (!,?,@,&,#,etc.), identifies the device and is followed by a single digit command character (e.g. 2—sample data or 4—send data in decimal format). In this way the master can control each device separately.

THE BENIGRAPH SYSTEM - FIELD TRIAL RESULTS

E Hammerstat, Bentech A/S, Norway

Presented on **Wednesday 5 March 1986 at 15.30**
in the Norfolk Conference Room

EMBARGOED AGAINST DELIVERY

22

The Benigraph Surveying System: Results of Field Trials

Erik Hammerstad, Arne Lovik, Leif Krane, Magne Steinseth, Svein Halvorsen and *Svein Minde*, Bentech A/S, Tromsö, Norway

INTRODUCTION

In the field of seabed surveying for nautical charting and offshore engineering applications, information on the topography is traditionally acquired by line surveys. The instrumentation is normally an echosounder with the required positioning, storage and presentation systems. Subject to geological aspects and survey specifications, the echosounder may be supplemented by a sidescan sonar, perhaps with fish positioning to obtain bottom information in between lines. To overcome the many shortcomings and uncertainties of these systems, the Benigraph seafloor surveying system has been specified and developed.

The Benigraph system was conceived as a pipeline inspection system, designed to be able to assess coverage and free spans. The system contains a multibeam sonar with very high resolution both angularly and in range. It is fish mounted so as to allow an economic operating speed. This required the development of a new high-accuracy fish-positioning system. To allow quality control, the bottom is shown in real-time on a graphical display in 3D form. All sonar data are given with position in a global reference frame and stored on computer tape both in processed and raw form as is all the positioning data.

The Benigraph project started four years ago, and the first prototype went through an intensive field trial period outside Tromsö in northern Norway this summer. In this paper, an overview to the system is given, followed by description of the subsystem, with performance figures, and concluding with the results of field trials.

THE BENIGRAPH SYSTEM

The objective in the original pipeline inspection idea was found to be eqivalent to producing a high-resolution real-time topographic seafloor mapping system, suitable for a wider field of application. The high accuracy in *X*, *Y*, and *Z* determination and the complete bottom coverage that the Benigraph achieves is essential in hydrographic survey work. For preroute pipeline surveys the system gives the information needed for safe determination of the pipeline route. The system was intended for work in the North Sea and is therefore designed to map the seafloor at water depths down to at least 300 m without loss of accuracy.

Figure 1 shows an overall block diagram of the Benigraph. The system is completely modular and is divided into the following subsystems:

- surface positioning;
- towfish positioning;
- scanning sonar;
- data storage and presentation;
- towfish and handling equipment.

The system used for surface positioning may be any type of high-quality radio positioning system, for example Syledis or Hyperfix. The fish-positioning system, FIPOS, consists of two sub-units:

- the acoustical positioning reference system (APR);
- the inertial positioning reference system (IPR).

The measuring parts of the Benigraph are all mounted on the towfish while the towing vessel carries the data-processing equipment, displays, handling equipment, and a beacon for ship-to-fish range measurement. The measured data are transmitted on digital links up to the surface, except for the hydrophone signals in the APR which are only amplified before being transmitted up the tow cable.

The APR processor determines the relative position of the towing vessel with respect to the fish by correlating signals transmitted through the water to hydrophones on the fish. In the IPR processor the APR measurements, fish depth measured by the DVU, and the ship's position are used together with measurements on the inertial platform in the fish to calculate its absolute position. The output data from the IPR processor also contain the fish attitude. The IPR processor runs a Kalman filter which updates the inertial platform in the fish. The sonar processor performs bottom detection while beamforming is done in the fish. The range and beam position of the detected data are geometrically converted to rectangular form, and fish movements are taken into

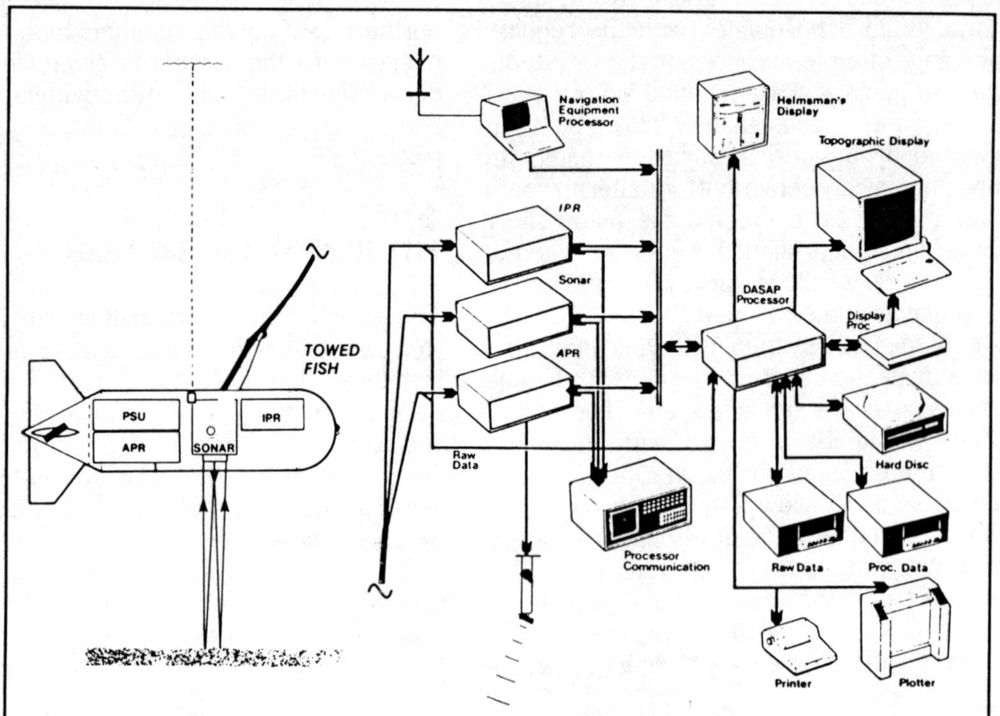

Fig. 1 Block diagram of the Benigraph system

account using IPR data.

The three processors for SONAR, APR, and IPR communicate with the data storage, acquisition and presentation system, DASAP, on an IEEE 488 bus (GPIB). The DASAP stores both raw and processed data and presents the seafloor on the topographic display in a 3D format with echo intensity added as a fourth dimension in colour. DASAP also drives the helmsman's display, which provides information on desired fish track and the position of both the towfish and the ship.

THE BENIGRAPH SONAR

The Benigraph sonar was designed to measure the seafloor with the necessary spatial resolution to allow quantitative assessment of the condition of the pipeline. The design goal set to achieve this purpose was a multibeam sonar with angular resolution of 0.5 × 0.5 degrees, coverage of 90–100 degrees, slant range of 40–50 m, and a range resolution better than 0.10 m. The fish tow speed was nominally set at 6 kn and the pulse repetition rate at 10 Hz, giving a horizontal resolution of 30 cm.

The operating frequency has been chosen as a compromise between transducer size, beamwidth, and slant range. The former led to a choice of 1 MHz, but in order to operate at greater ranges the sonar is designed also to work at 500 kHz and 740 kHz. In the field trials, damage to the transmitting transducer made the source level of 1 MHz too low to be usable, and therefore all performance figures reported are for the 740 kHz frequency. At this frequency the beamwidth is 0.75 degrees in the central beams. A block diagram of the sonar is shown in Fig. 2.

The receiver-transducer array consists of 256 elements on a 200 mm square piezoelectric plate. The elements are cut with thickness resonance at 1 MHz and are matched to an absorbing epoxy backing by quarter-wave transformers. All transducer element signals are fed into separate channel receivers, each functionally equivalent to an ordinary sonar receiver. They are constructed in thick film to achieve the neccessary size, power consumption, and cost reduction. The total gain of the receivers is 110 dB, of which 80 dB is digitally controlled for time variable gain (TVG). The outputs of the channel receivers are quadrature-detected and lowpass filtered before being sampled, time multiplexed, and fed to the beamformer.

The beamformer is realized as a chirp Fourier transformer with two SAW (surface acoustic waves) dispersive delay lines, as shown in Fig. 3. This realization, which is analog with continuous output, allows much more rapid beamforming than the digital one more commonly used. Especially important, since the sonar is fish mounted, is that it also gives a very small and power-efficient beamformer. The beamformer has a maximum capacity of 300 beams, which may be formed every 20

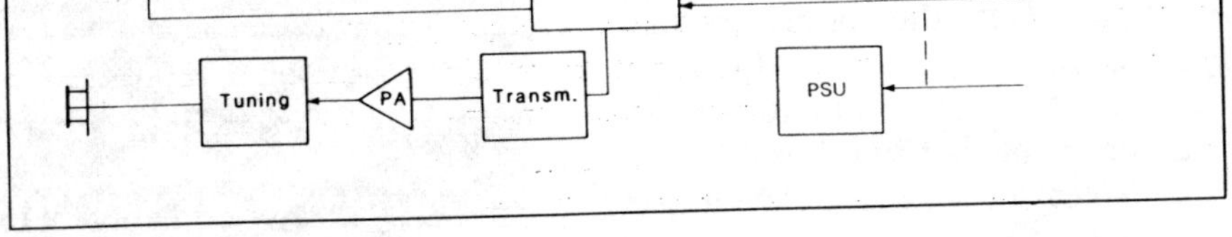

Fig. 2 Block diagram of the sonar

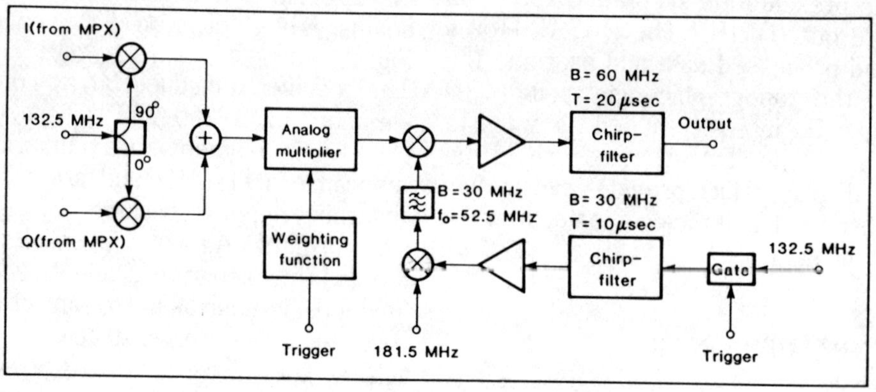

Fig. 3 Beamformer block diagram

ms. This allows sampling in a range of down to 1.5 cm, and there is no limit on the sampling range. This is in contrast to digital beamforming, where a compromise has to be found between the along-track resolution given by transmitter pulse repetition frequency, depth resolution, and depth range for which data can be collected. The Benigraph beamformer thus has a processing capacity of the order of 100 times that of typical digital beamformers in use today.

The output of the beamformer is envelope detected before being A/D converted at a sampling frequency of 20 MHz. The data rate at this stage is too high to allow direct transmission to the surface. Instead samples whose intensity levels are below an operator-settable threshold are discarded. The remaining data are transmitted up on a PCM link which has a capacity of approximately 4 Mbits/s.

Bottom detection is performed in a specially built bitslice processor. A detection is accepted when the echo samples have intensity levels which stay above a certain value for a number of consecutive samples. Both the sample number and threshold value are operator settable, as are the bottom detection end parameters which are used in determining bottom echo length. The processor also calculates the mean intensity of the bottom echo.

The bitslice processor is integrated with the sonar processor which is realized with a MC 68000 microprocessor on a VME bus. This processor performs geometrical corrections and transformations to determine the absolute position of every detected bottom echo. The sonar is thus electronically stabilized both in roll and in pitch. The fish attitude and position are received from the FIPOS system on the GPIB, and the processed data are transmitted on this common bus to DASAP. Raw data, which have been defined as data transmitted from the fish, are time sampled and sent via the GPIB to DASAP for storage.

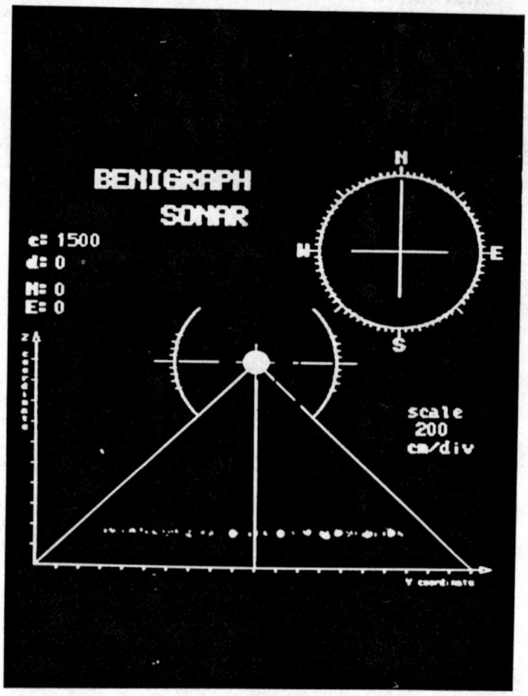

Fig. 4 Sonar display of sea bottom with fish attitude and position

The sonar processor also supports operator interaction to allow control of the hardware parameters in the fish and has a local graphical display for sonar quality control. A picture of the sonar display is shown in Fig. 4. The position of the fish in relation to the sea bottom, fish attitude as an artificial horizon, and fish heading, together with other IPR and APR data, are shown. This display proved invaluable when optimizing coverage by controlling the flying height of the fish above the bottom via the winch remote control.

As far as could be determined in the field trials, the sonar has the expected angular and range resolution. The damaged transmitter transducer had less source level than planned, and this might explain why the angular coverage was seldom more than 90 degrees, although the element polar response measurements indicated that 100 degrees is achievable. Another observation, which might be explained by the reduced source level, was that, typically, bottom echoes were only detected in half the beams, although this could also be due to a large variability in bottom reflectivity which would be much more pronounced than usual because of the very small area seen by each beam.

THE FISH-POSITIONING SYSTEM

The fish-positioning system is composed of two main units: an inertial system, IPR, which measures the fish movements and gives its position and attitude; and an acoustic positioning system, APR, which determines the ship's position relative to the fish by measuring the range and direction in relation to the ship. In addition there is a sensor package, DVU, in the fish for measuring fish depth and sound velocity in the sea.

The DVU contains two sensors: a very accurate digiquartz pressure sensor and a temperature sensor. Depth and sound velocity at the fish are calculated from the data given by these sensors. A velocity profile is also recorded as the fish is lowered or otherwise moved up and down in the water column. These data are used by the APR in transforming time of arrival measurements to range and angle.

The APR uses two hydrophones on the fish fins to measure the times of arrival of an acoustic signal transmitted from a transducer mounted on the towing vessel. The mean time of arrival gives the range from the fish to the ship, while the time difference gives the angle to the ship as seen from the fish. The measurement of time delay in the APR is not carried out through a phase measurement, as in more conventional acoustic positioning systems, but through the use of a correlation technique. This allows the use of a much larger bandwidth, which again represents a very large increase in signal-to-noise ratio with a corresponding increase in accuracy obtainable under practical operating conditions. Because the system uses a large bandwidth signal, it is not possible to characterize it according to baseline length as is usual. Rather the system has a varying baseline – at low frequencies it is super short, at high frequencies it is super long.

The APR system has another unique feature in that it may operate on any kind of signal format, even on the noise emitted from the ship's machinery and propellor. Thus when operating at a distance where sound ray bending is not too large, the addition of a third channel in the APR would allow the position measurement to be done without requiring a separate measurement of ship-to-fish range. This makes it possible to avoid any underwater installation on the towing vessel.

The time delay difference, and thus the angle, is found by cross-correlating the hydrophone signals, as shown in Fig. 5. The correlation is implemented as a smoothed coherence transform (SCOT). The correlation is calculated stepwise:

(1) A blockwise time-to-frequency transform.
(2) Cross and auto-spectrum estimation.
(3) Averaging.
(4) Whitening.
(5) Frequency-to-time transform.

The correlation demands a large number of calculations, and, to get the neccessary computer power, a MC 68000 microprocessor together with a special purpose array processor on a VME bus is used.

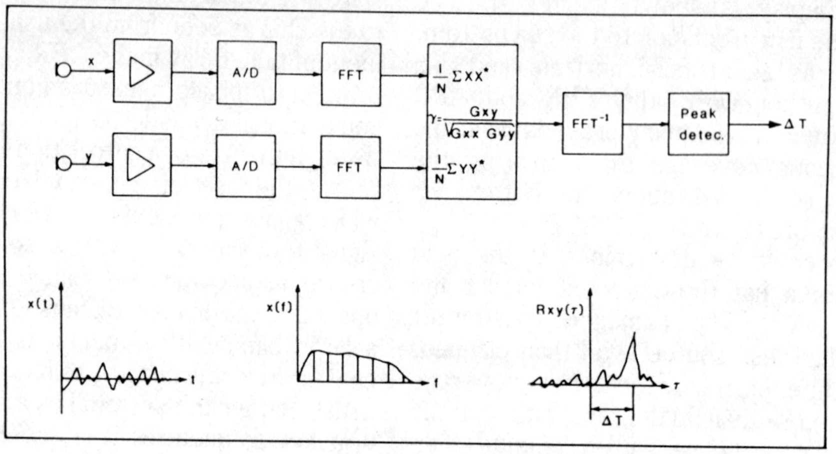

Fig. 5 APR correlation scheme

The resolution of the APR is expected to be 0.1 degrees, which has been confirmed in the laboratory using real data, including the underwater noise of a ship. However, in the field trials an angular resolution of only 0.5 degrees was achieved. The cause of this is at the moment under investigation, but it seems to be due to an unhappy interaction between two different faults in the electronics. The range resolution accuracy seems to be as expected, in the order of 0.1% of slant range. The update rate from the APR is once per second, which may be increased by a factor of two.

In order to estimate positions a reference frame has to be established in the fish. This is done by the use of the inertial system, IPR, which consists of a sensor unit in the fish and a processor on-board the ship. The fish unit is a full three-axes inertial platform made up of a gyrocompass, RSG-82, and a vertical reference unit, MIREF, both supplied by Robertson Tritech of Egersund, Norway. The gyrocompass measures the heading with respect to true North, and the vertical reference measures the attitude and accelerations of the fish. Within MIREF is a navigational computer which transforms the data to the UTM reference frame and also runs a simple Kalman filter for platform alignment. The data from the platform are used in a twenty-state Kalman filter, which runs on the topside computer. Figure 6 shows the functions of the IPR system.

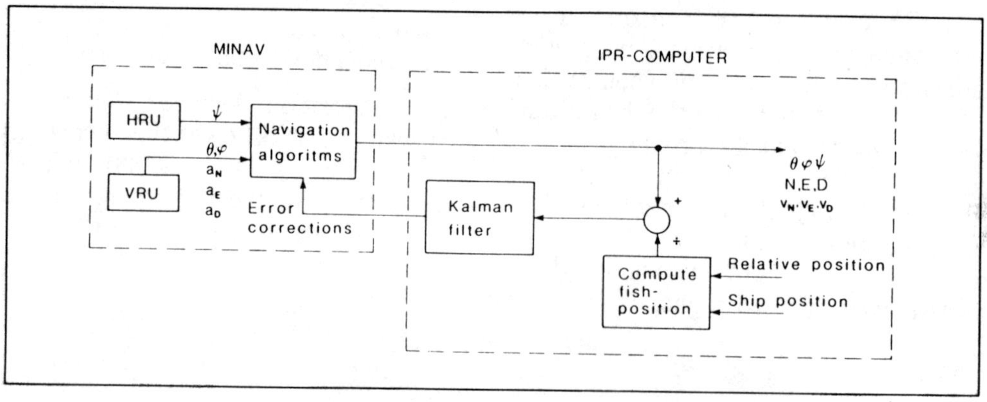

Fig. 6 IPR functional description

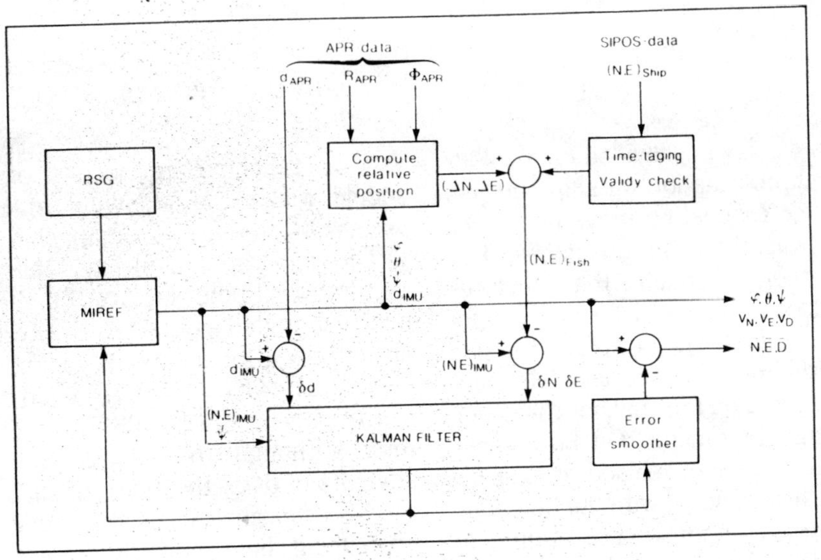

Fig. 7 IPR main computational tasks

The main Kalman filter measurements are generated in the following way. Data from the APR are combined with heading and attitude data of the fish to compute the relative position between the fish and the towing vessel. The relative position is then added to the absolute position of the ship given by the surface-positioning system to give absolute position fixpoints for the fish. The Kalman filter is used to measure the difference between these fix-points and the position given by the inertial platform. Figure 7 shows the main computational tasks of the IPR system.

The estimated position errors are smoothed before they are added to the data output from the IPR. This is done in order to avoid discontinuities in the position data from the IPR.

The operation of the IPR was done with surface positioning both from Syledis and Polarfix systems during the field trials. Software bugs and the reduced APR resolution did not allow a complete verification in accuracy then, but the results achieved, and later measurements in the laboratory, seem to confirm the expected accuracy. This is at 60 degree latitude ±7 m (1 σ) long term when the surface-positioning error is 5 m. The relative accuracy over the short term is of the order of 40 cm. The heading accuracy is better than 0.5 degrees and the roll and pitch accuracy is better than 0.05 degrees.

The fish-positioning system represents a considerable advance in accuracy due to using the two subsystems both separately and in combination. For example, the fish-mounted acoustic positioning with widely separated transducers makes it quite easy to align the acoustic axis with the heading reference. The heading reference is also very accurate and the acoustic system operates in the fish at a much lower noise level than does a ship-mounted system. Although very extensive calibrations could eliminate most of the error sources of a ship-mounted super-short system, the Benigraph fish-positioning system should still have a significant advantage and would typically have 3–10 times better accuracy.

DATA STORAGE AND PRESENTATION

All data from the subsystems are directed to a HP1000 A900 series minicomputer on the GPIB. On this computer the data are stored in standard format on two 1600 bpi ½ in tape stations, one for raw and the other for processed data. The capacity of the tape reels is 40 Mbytes, which is sufficient for at least one hour of surveying between tape changes.

A helmsman's display is provided on a colour monitor where all information on ship and fish positions and their respective movements are presented along with the desired runlines. The runlines are given through the surface-positioning interface system which was a Geofix Scope 26 in the field trials. Two modes are available on the helmsman's display. The one shown in Fig. 8 is used on the runline: the other is for approaching the runline. On both, data from the ship's echosounder is used to indicate any danger of the fish colliding with the bottom. The bottom topography is shown in real-time on a graphical display manufactured by Ican or Horten, Norway. The bottom is shown in pseudo 3D as a waterfall display, as shown in Fig. 9. The scan lines are derived from 5–10 different pings in the sonar, which allows using a constant horizontal distance between the lines and a correct vertical scale, regardless of fish movements or varying tow speed. The echo intensity is colour coded into the display as a fourth dimension, and, in addition to indicating varying reflective properties of the bottom, should make the following of a pipeline or visualization of the bottom much easier.

The topographical display, while giving a very exciting real-time picture of the bottom, is mainly intended as a quality-control display. It is not intended for on-line assessment of pipeline conditions, which is more a job for the post-processing on the client's own computing system. Although the Benigraph collects vast amounts of data, these are of such good quality and so easily accessible that the post-processing should go fairly simply.

FIELD TRIALS

The Benigraph field trials took about two months and ended in September. They consisted of subsystems debugging, integration, and a system test. Unfortunately these activities took so much time that a planned pipeline survey had to be postponed until the spring of

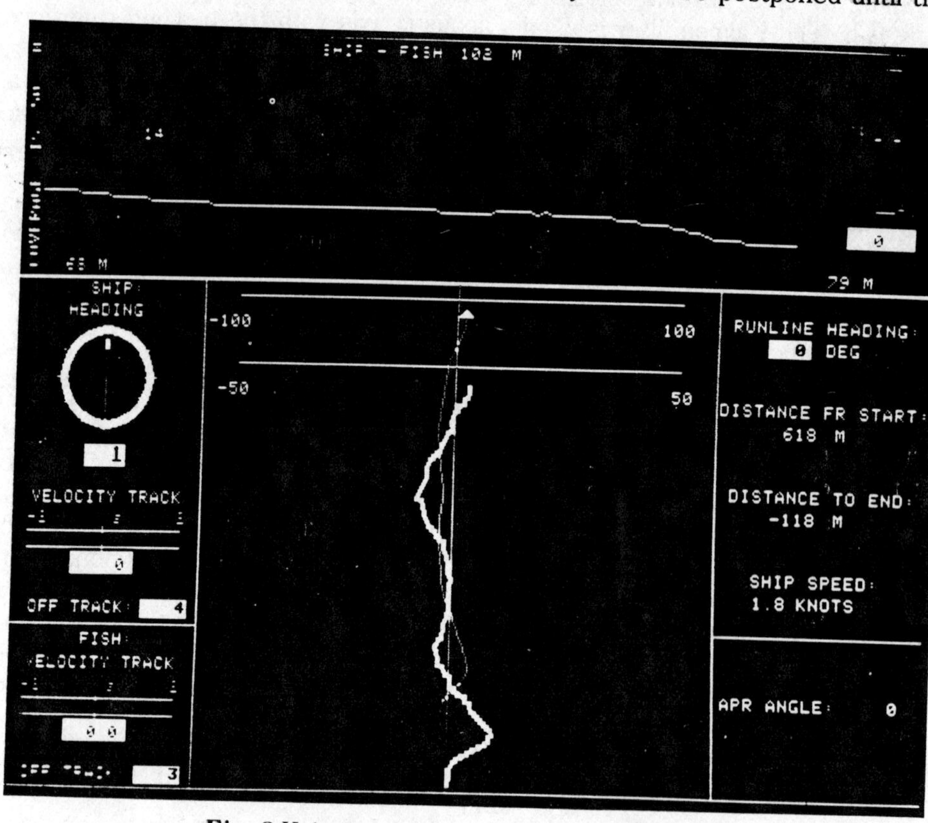

Fig. 8 Helmsman's display in runline-following mode

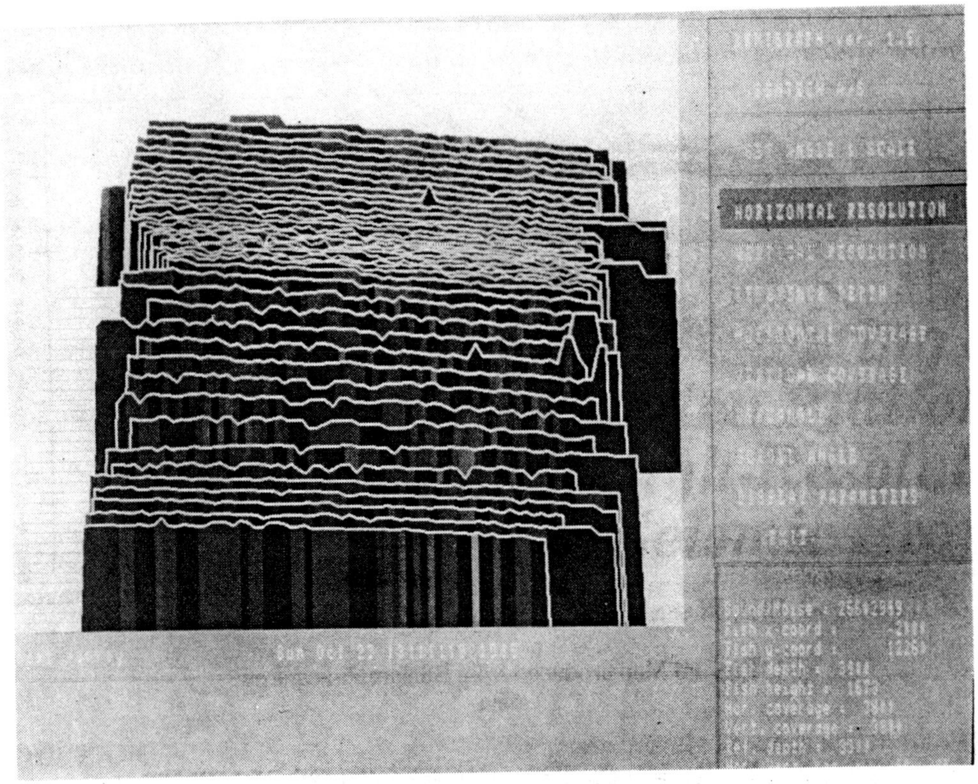

Fig. 9 Benigraph topographic display

1986. Instead, a complete survey of an area near Troms was carried out. Maps from this survey were afterwards processed with Syss-can cartographic software at Norsk Hydro in Oslo. Despite the enormous amounts of data, approximately half a million samples of the bottom in a 500 m^2 area, the processing procedure went very smoothly. After a first attempt which failed due to wrong tape format, less than one day of data preparation and one day of processing was sufficient for producing the map shown in Fig. 10.

The Benigraph was operated in depths down to 180 m. With remote control on the winch and the very comprehensive data presentation, it was easy to control the fish to obtain good and constant coverage from the sonar. This was the case even in areas where the bottom was extremely silty with little reflectivity, making it neccessary to fly the fish 10 m above the bottom at 180 m depth. The fish was 380 cm long and 80 cm in diameter and had a weight during the field trials of 700 kg.

CONCLUSIONS

The field trials of the Benigraph showed that despite minor problems the system worked according to the most optimistic expectations and produced data which enabled high-resolution charting. In contrast to conventional systems, the retrieved information contains absolute positioning data allowing a complete quantitive mapping of the sea bottom, including areas in between runlines. With regard to pipeline inspection, this will allow assessment of the pipeline conditions such as measurements of free spans. It is thus believed that the system will have a major impact on the ease and quality of hydrographic surveying and pipeline inspection. The system is currently undergoing accuracy verifications and minor modifications. It is intended that it will be ready for commercial pipeline surveying in the North Sea early next spring.

In a separate and linked project a scanning, shallow seismic source is being developed for

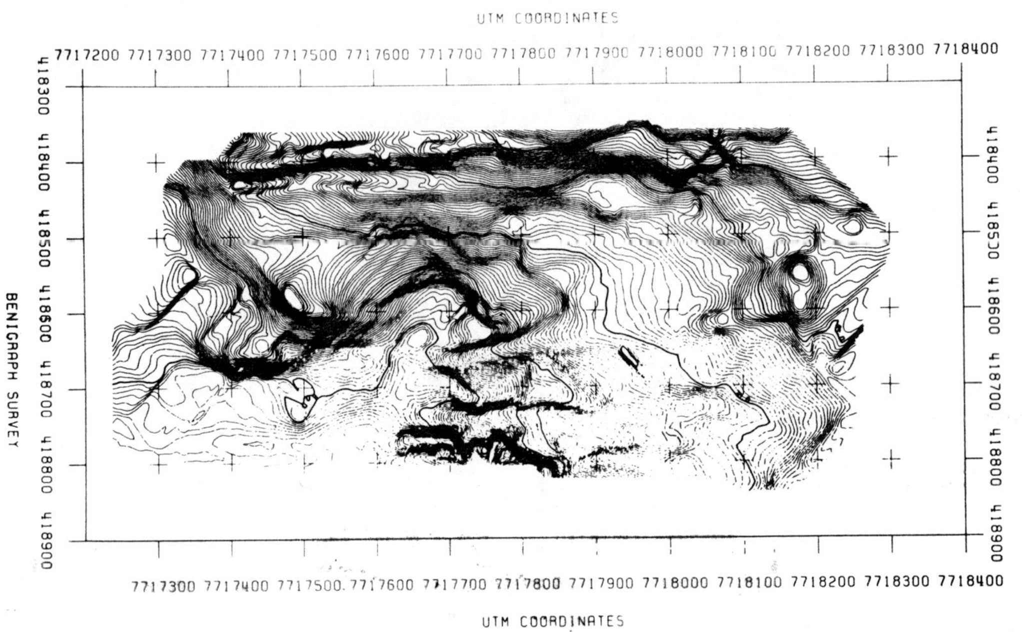

Fig. 10 Map produced from Benigraph-acquired data

use in conjunction with the Benigraph. This source will be capable of high-resolution data acquisition to a penetration depth of the order of 100 m, and with a pipeline following mode it will ultimately provide the final link to a fully integrated remote sensing survey package.

ACKNOWLEDGEMENTS

This work has been carried out under a contract from A/S Norske Shell. The advice and guidance of Dr C.D. Green and Ir. J.G. Riemersma of Shell Internationale Petroleum Maatchasppij, the Netherlands, has been of invaluable importance in the project.

TWO SHIP UNDERSHOOTING OF A PRODUCTION PLATFORM

A W Hill, A/S Geoteam, UK

Presented on **Wednesday 5 March 1986 at 16.30**
in the Norfolk Conference Room

EMBARGOED AGAINST DELIVERY

23

Two-ship Undershooting of a Production Platform: The Magnus High-Resolution Seismic Project

Andrew W. Hill, Geoteam (UK) Limited

INTRODUCTION

BP Petroleum Development (UK) Limited's Magnus Field is the most northerly field in the North Sea to have been developed so far. Magnus lies mainly in block 211/12A of the British sector of the North Sea, with a slight northerly extension into block 211/7A. The field is approximately 16 km in length with an almost north-south trend (Fig. 1). The development plan called for one large, fixed, central production platform. A total of 16 producing wells are to be drilled from the platform; in addition, seven subsea completions, about the field, are to be tied back to the platform by buried flowline and control bundles. Lying in 186 m of

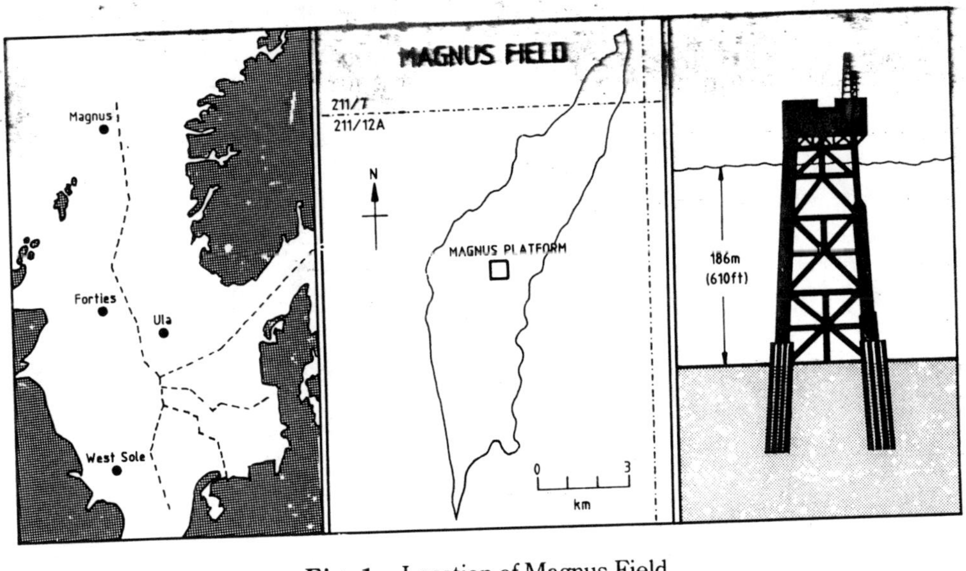

Fig. 1 Location of Magnus Field

water, the Magnus platform is a piled steel jacket and is the largest single-piece steel platform in the world. At deck level the platform is approximately 100 m square. The four corners of the platform point approximately north, south, east and west.

In 1983, before the commencement of production drilling from the platform, a review of the shallow geology across the field was performed by BP Dyce. The study entailed a study of all well and seismic data that was available for the field to a depth of 1000 m. (All depths in this paper refer to metres below sea-level.) This depth is the approximate depth of the boundary between the Nordland group, in the shallow section (Middle Miocene to Recent) and the Hordaland group (Eocene to Early Miocene), the two groups are separated by a Middle Miocene unconformity. The Nordland group generally occupies the geological column to a depth of at least 900 m (metres subsea) and occasionally exceeds 1000 m. The group consists mainly of mudstones and clays with occasional silts. More important, however, is the presence of localized, poorly consolidated sand bodies in the sequence. It was the presence of the latter that the study was particularly interested in. In the early 1970s, during the development of BP's Forties Field, similar sand bodies were found to be gas bearing and to present a threat to drilling in the shallow section. Indeed, the pioneering work into high resolution seismics, in the North Sea, was carried out by BP in the Forties area to map these sand units (Lucas, 1974). The techniques put forward at this time have become the standard techniques, for rig and platform site surveys and have changed little. With the commencement of drilling at Magnus imminent, BP considered final drilling plans for which the presence of shallow gas sands was important.

From the 15 well logs available for the area, it was apparent that there were two particular sand bodies that were correlatable across the field. The upper sand unit lay at approximately 450–540 m and varied from 10 to 60 m in thickness. This bed was also seen to be locally lignitic. The lower of the two sand units was typically 20–30 m in thickness and occurred in the 750–820 m interval. Other sand bodies were present in the logs; however, they did not seem to be correlatable and were thought to be fairly localized.

During exploration drilling across the Magnus area most of the wells encountered gas shows in the shallow section. In all cases no problems were experienced from the presence of gas, which was thought to be locally generated and normally pressured. The sands that were encountered were unconsolidated and highly porous, leading, in a few cases, to problems with loss of circulation. Only in one well was the gas level anomalously high; however, as already mentioned, the gas was normally pressured and did not cause any problem to drilling. From a study of 'D'-exponent data for the shallow section it was concluded that the whole of the Nordland group column should be normally pressured, though the possibility of a localized sand body being overpressured could not be ruled out.

These findings were compared to all the digital seismic data available for indications of anomalous high-amplitude events and events tying to the two main sand bodies observed from the well logs. No anomalous high-amplitude events were identified from the sections. However, two events were observed which were thought to require closer observation. The first lay at approximately 500 ms (Two-Way-Time, TWT). This it was thought corresponded to the sand-lignite body observed in well logs around the 450–540 m level. In places this event coincides with the seabed multiple, and from the data available it was uncertain as to whether lateral changes in amplitude were due to multiple interference or a real change in reflection properties. A second event seemed to be associated with the second sand body seen across the area. Again, though, the event did not show as an apparent hazard on the sections. Overall the study pointed to the presence of sands below the platform. However, the quality of the seismic data in the platform's immediate environs, though not revealing any hazards, was not thought to give enough coverage under the platform nor to resolve entirely the nature of the potential of the upper sand body delineated by the reflector at the 500 ms level. It was therefore proposed that an attempt should be made to gather high-resolution seismic data from beneath the platform itself.

Thus in early April of 1984 BP Petroleum Development (UK) Ltd approached Geoteam (UK) Ltd regarding the feasibility of undertaking a high-resolution undershoot project off the Magnus platform.

OBJECTIVES, PLANNING AND SURVEY PERFORMANCE

When the contract for the project was awarded there was not much lead time available for planning technical details and no time for system trials.

The survey's objectives were to gain data over a 600 m square centred on the platform. The size of the area was defined as the lateral extent that wells would deviate from the platform at the 800 m level. The survey was to collect data over a dense line spacing of 25 m for north-south lines with five lines running east-west, giving ties at 150 m spacing. Some of these lines would be recorded conventionally, by a single ship, while the central eleven north-south lines, which ran close or under the platform, would be recorded by a two-ship undershooting method. The central East-West line would also require undershooting. Though the single-ship lines presented no technical difficulties, they shared the same environmental problems. When recording any seismic data the levels of background noise are highly important if reflections are not to be blanked out. Collecting high-resolution seismic data within 100 m of a production platform immediately presents difficulties. Noise levels throughout the project, be it from the platform or standby ships, were to be a constant problem. However, when it came to processing, only on one line was noise seen to effect data quality. Similarly, the potential problem of side-swipe off the platform's legs was not encountered. The fear of difficulties from current swirl about the platform did result in problems for streamer control. It was found that steaming north, current swirl around the platform invariably took the streamer feather past acceptable limits. As a result all north-south lines had to be shot in a southerly direction.

The short lead time did present problems for the development of a system for recording two-ship undershoot data. These problems can be split into two areas.
Firstly, there is the problem of coordinating the two ships so that the shot firing and recording are synchronized, and the ships keep the same relative position along their respective tracks. The second problem is of obtaining accurate shot and streamer positional data for use in the processing of the seismic data to final sections.

SEISMIC DATA ACQUISITION SYSTEM

The basic recording parameters used were the same for conventional and undershoot lines. The Minisleeve exploder system was selected as source for the project and was programmed to fire all six guns simultaneously, giving an output of approximately 5 bar metres. The recording system was a Texas Instruments DFS V configured to record 56 channels at a 1 ms sampling rate, the maximum number of channels, at this sample rate, that can be recorded on a two-analogue box system. As noise was considered a serious problem beforehand, recording as large a number of channels as possible was considered important, so as to achieve a greater improvement in the signal-to-noise ratio as the data went through stack. The recording of 56 channels rather than the more conventional 48 channels would finally result in 28-fold CDP coverage, giving a S/N enhancement factor of 5.2, compared to a factor of 4.9 with a 48-channel system. Though only of marginal apparent benefit, the extra channels were considered important. The DFS V recorded the input from a GECO US streamer of 56 groups, with a group interval of 12.5 m giving an active streamer length of 700 m. The added offset length from the extra channels would also aid in determining better velocity information to a greater depth in data processing.

The problem of coordinating shot firing and

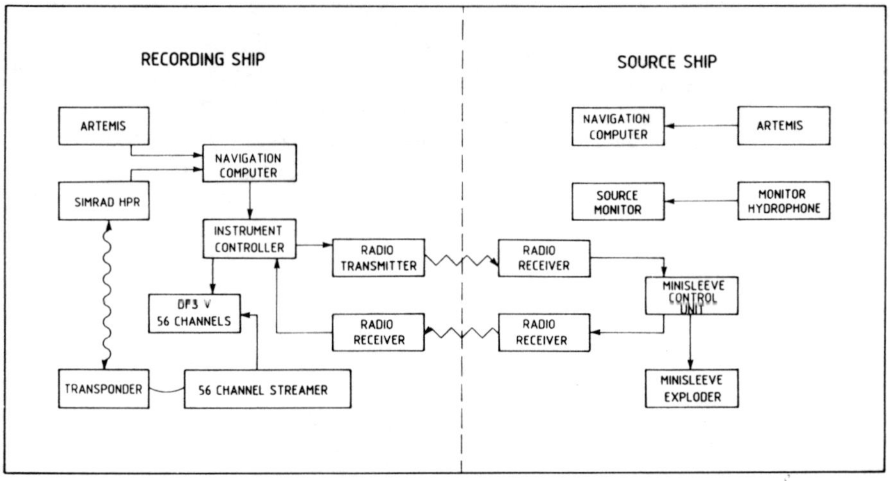

Fig. 2 Two-ship undershooting system schematic

data recording between the two ships was solved by way of a two-way telemetry link capable of sending commands between the two ships with minimal delay. It was not possible in the time available to have the two ships' navigation systems linked on-line. Thus the decision was taken to run one of the ship's navigation systems in overall control of operations.

The M/V Geo Scanner as recording ship for the undershoot lines was chosen as the control vessel for navigation purposes. The M/V Geoboy's navigation system was run on an independant basis, on a time mode, recording a position every 5 s. With the Geoboy's navigation clock synchronized by Magnavox satellite receiver clock to that of the Geo Scanner's navigation system, the Geoboy's position at each shot point could be recovered accurately. The Geo Scanner's navigation system was run in a distance mode outputting a fix every 12.5 m along line. Figure 2 shows a simple block diagram depicting the system configuration for two-ship operations. When the Geo Scanner navigation system calculated 12.5 m had been travelled along line, a command was sent to the Geo Scanner's instrument room control box. This command was then transferred by radio link to the Geoboy via a telemetry link where, on arrival, it triggered the minisleeve control unit to load the minisleeve exploder. After the loading period the minisleeve controller sent a fire command to the minisleeve, which was simultaneously sent back to the Geo Scanner's instrument room by telemetry link. This was used to give a command to the DFS V for defining time zero, the start of recordings. The measured delay between the minisleeve fire command and the DFS V receiving a command was 4 ms. This constant delay was recovered in post-processing by application of a static correction to raw shot records.

Coordination of the two ships' relative positions along line was important to maintain a near-constant offset between source and streamer. With two navigation systems running independently, a long run onto each line was taken to achieve alignment in along-line position and constant speed between the two ships. The two navigators compared along-line positions via VHF. If any change in speed was required, to maintain alignment, this was performed by the Geoboy. In this way the Geo Scanner was able to keep a constant speed so as to maintain a steady streamer depth.

In addition to a system for coordinating firing and recording and maintaining the relative position of the two vessels, a system was required to give data on streamer position for each shot. This was needed for computer processing of the recorded data.

When the geometry of a conventional digital seismic line is considered, the shot receiver ray paths can be displayed in a simple sketch (Fig.

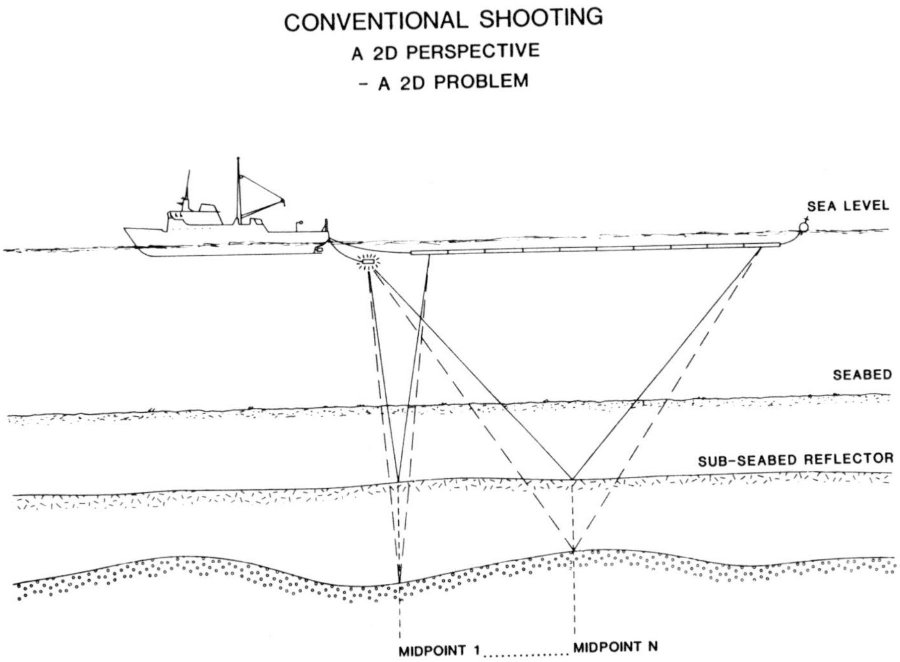

Fig. 3 Conventional seismic acquisition geometry and ray paths

3). This sketch shows a single plane on which the ship's path and streamer path lie. If streamer feather and side echoes from structures are ignored, it can be seen that all shot-receiver ray paths lie on the same plane. This plane represents a single slice down through the Earth along the ship's track. As a result all common depth points, into which data is sorted, also lie on the same plane. Again, if streamer feather is assumed to be within reasonable limits, it can be seen that receiver positions will always lie on the same plane and will move along a distance equal to the distance moved by the source between each shot. Due to this, in conventional surveys the only measurement that needs to be made to ensure correct processing of data is the source-to-near-trace offset; subsequent channel positions can be calculated by the addition of the group interval distance.

With two-ship undershoot surveys this is not true. If a 2D sketch similar to that in Fig. 3 is drawn to try and illustrate the problem (Fig. 4), it can only illustrate a single shot-channel configuration with associated ray paths. The

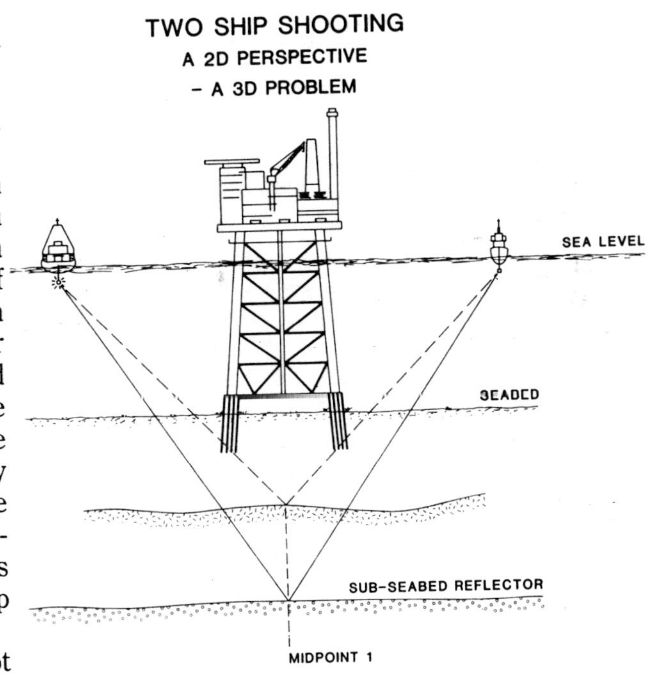

Fig. 4 Two-dimensional representation of undershoot data ray paths

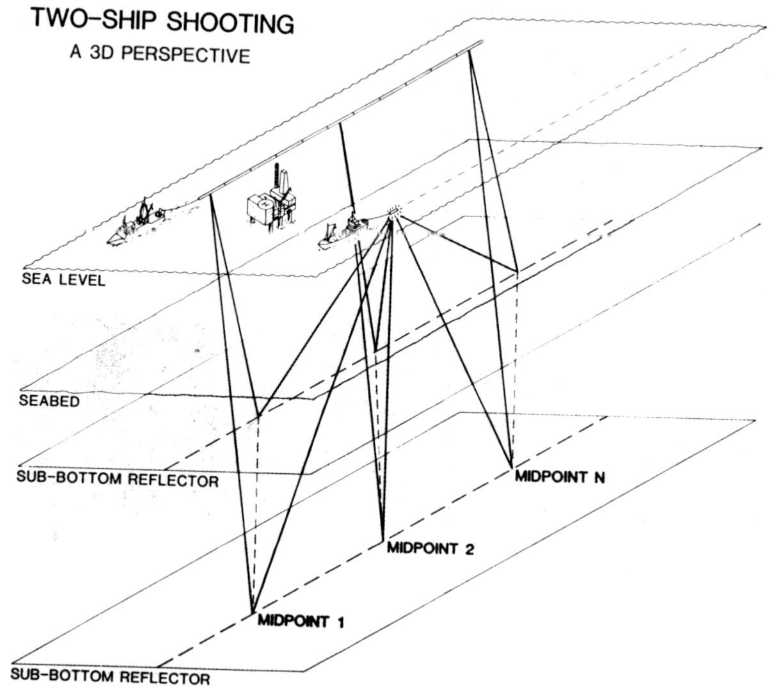

Fig. 5 Three-dimensional representation of geometry and ray paths for undershoot data

problem is now a three-dimensional one; in the two-ship case shot-receiver ray paths that lie on 56 different planes must be considered. In addition to this problem, common depth points lie on a line that is halfway between the two ships, Figure 5 shows a simple diagram depicting shot-receiver ray paths for this situation. To be able to carry out a correct processing of the recorded high-resolution data would require the calculation of the correct position of each shot-channel common-depth point in x and y coordinates. To be able to calculate these figures the positions of each shot and each channel for each shot must be known. This points to a highly accurate primary navigation system being needed, together with a system for streamer positioning at each shot.

For primary navigation purposes the Artemis system was selected as being the most suitable and accurate. The range and bearing system's accuracy of 1–2 m gave excellent positional data throughout the survey period. Each ship had it's own master beacon positioned on the side of the platform off which it was working.

Onboard the Geo Scanner the navigation computer was also linked to a Simrad HPR system. This system had been selected to provide positional data for the streamer. An HPR transponder attached to the tail of the streamer was regularly interrogated by the HPR, and range and bearing data were logged by the navigation computer. In general the HPR gave reasonably good data. This was especially pleasing due to the shallow depth of the transponder (2–3 m), forcing the command and response pulses to pass through the ships wake, and the high background noise levels. As a check against the system, the tail buoy feather was regularly checked by radar for correlation.

Using the systems described, the fieldwork ran quite smoothly with no problems occuring during production with fix sending, shot firing, or on-line navigation. The problems experienced, such as ship noise from other ships, rig noise and weather, were outside survey control.

Throughout two-ship operations tight quality control was maintained on both seismic and navigational data. To check midpoint line posi-

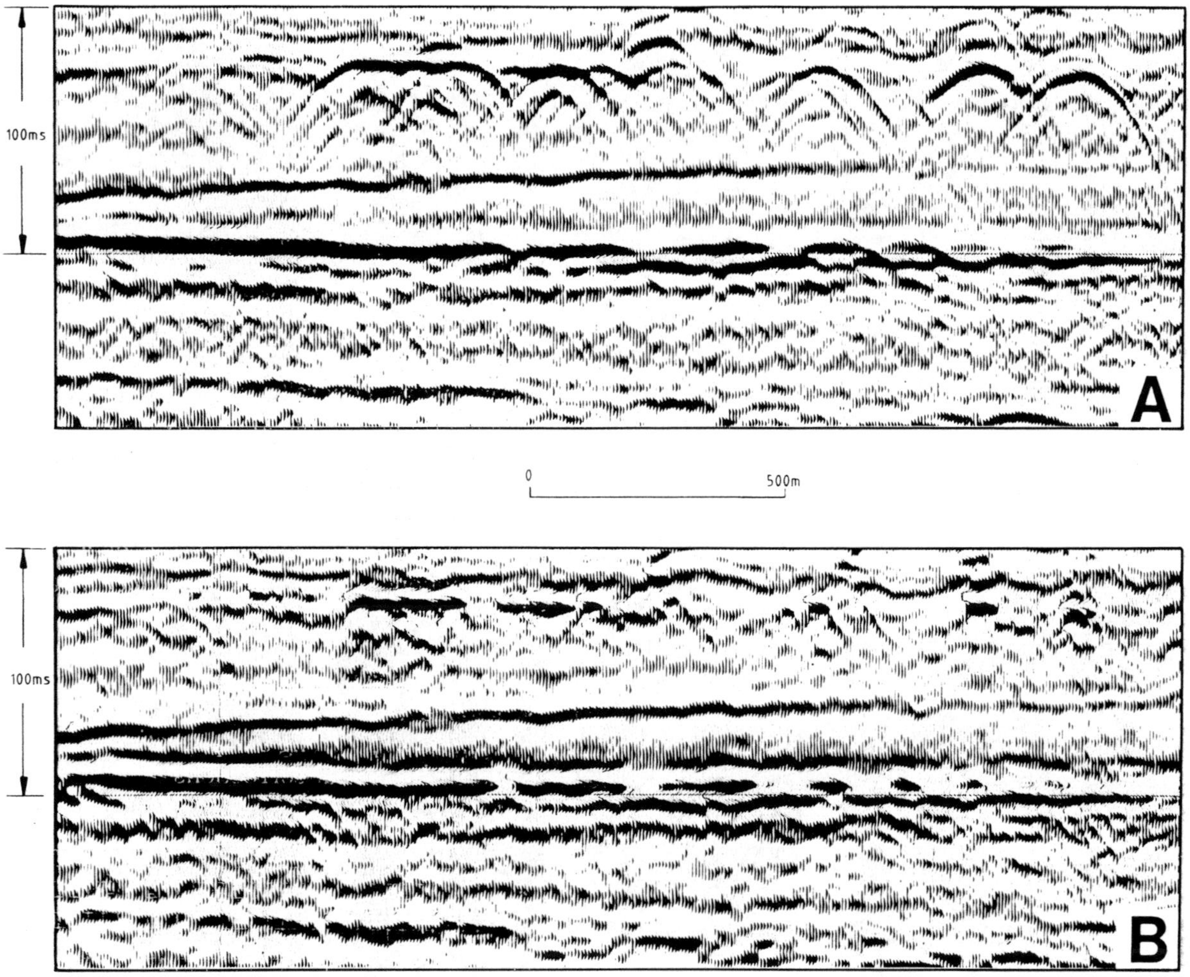

Fig. 6 Geometry scatter plot showing shot, receiver and midpoint positions

tions of completed lines, navigation data were sent via a data-link from the Geoboy to the Geo Scanner after line completion. Once on board the Geo Scanner the two navigation data sets were combined to determine common midpoint positions for the shot to first channel. The calculated positions were automatically plotted to show line coverage in the undershoot area. In this way the two clients, party chief, and senior surveyor could check line spacing and evaluate whether any reshoots or extra lines were required. Overall, however, line control was good and kept well to the planned line spacing.

DATA PROCESSING

Processing of navigation and seismic data took place at A/S Geoteam's processing centre in Oslo. Processing of single-ship data followed a conventional route. Throughout data processing there was close cooperation on both quality control and ideas between Geoteam's seismic processors and geophysicists at BP's Exploration Department in Dyce. With high resolution of great importance, a heavy processing sequence was applied to data to extract maximum vertical and lateral resolution.

The main problems faced in processing were

the initial definition of shot and channel positions and thence common midpoint positions. Special software was written which took in, as raw data, minisleeve sledge position, channel one position of the streamer and HPR streamer azimuth to the streamer tail and output data in a format useable by the seismic computer; a scatter plot was output showing shot, receiver and midpoint positions defined in UTM coordinates. Figure 6 shows a detail from ai portion of one line. To the right the line of squares represents individual shot positions, to the left the dark swath represents individual channel positions at each shot. Down the centre runs another line, the midpoint line, marked by small triangles each representing a shot-channel midpoint position. Superimposed on the midpoint line (though difficult to make out) are the bounds of defined common midpoint bins which define into which bin a particular shot-channel midpoint will finally be sorted. To cross-check the HPR calculated offset positions, another program computed streamer feather, by means of checking the seabed arrival time, of the first, middle and last channel of the streamer. Results were generally good, though in a few cases redefinition of geometries had to be made.

After the data had been sorted into common midpoint gathers, the output gathers were studied to check the accuracy of the data sort. There were generally some inaccuracies remaining requiring adjustment by a static correction. On completion of this static application the data were now in a position to undergo normal seismic data processing similar to that for the conventional data. The following data processing sequence was used to produce final sections:

- velocity analysis
- gapped deconvolution
- spherical spreading compensation
- normal moveout corrections
- mute

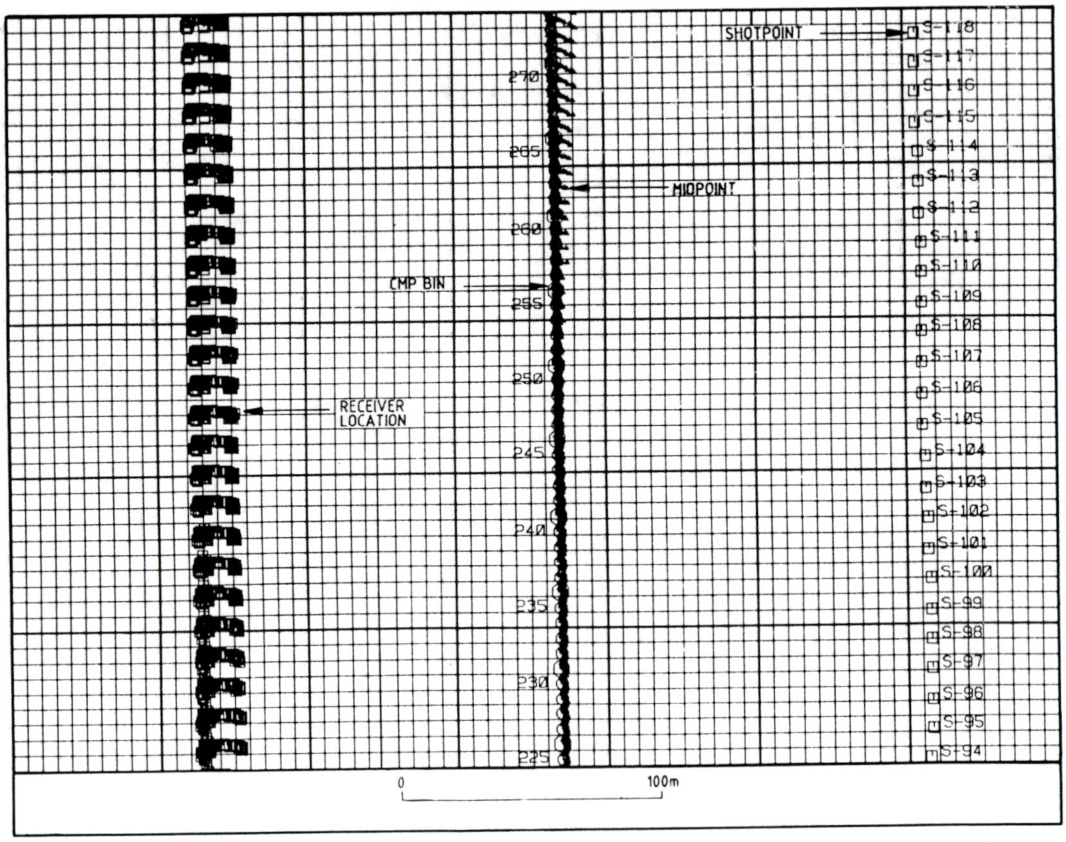

Fig. 7 The use of migration to improve lateral resolution of anomalies: A, stacked section; B, migrated section

- 28-fold stack
- wavelet processing to produce zero phase section
- time migration (finite difference approach)
- time variant filter

As mentioned earlier, great attention was paid to maximizing vertical and lateral resolution. With the single-ship lines this was easier than with the two-ship lines. To maximize vertical resolution wavelet shaping was used to move energy back into the pulse. On the single-ship lines this was performed using far field signatures for each shot. A filter designed from these recordings was applied pre-stack; a second filter was then applied post-stack to convert the section to zero phase. On the two-ship lines the lack of a far field recording for the lines precluded the use of similar techniques. Instead a wavelet was extracted from the seabed and used in filter design. Though a more inaccurate method, results were nonetheless good, providing a zero phase section. With the application of wavelets the expected maximum resolution in the upper section is approximately 1 m.

Enhancement of lateral resolution was achieved by use of migration to move energy back into it's correct lateral origin. In this way edge effect diffrators are restored to their point of origin. The effect of this in properly defining the lateral extent of some bright spots can be seen in Fig. 7. The figure shows the same section pre (A) and post (B) migration. The effect on the lateral definition of the bright spots is clearly seen. This permitted highly accurate spatial delineation of bright spots in interpretation.

Overall the standard of sections was very good. In comparing one- and two-ship lines a slight difference in seismic character can be seen. Figure 8 shows matching portions from

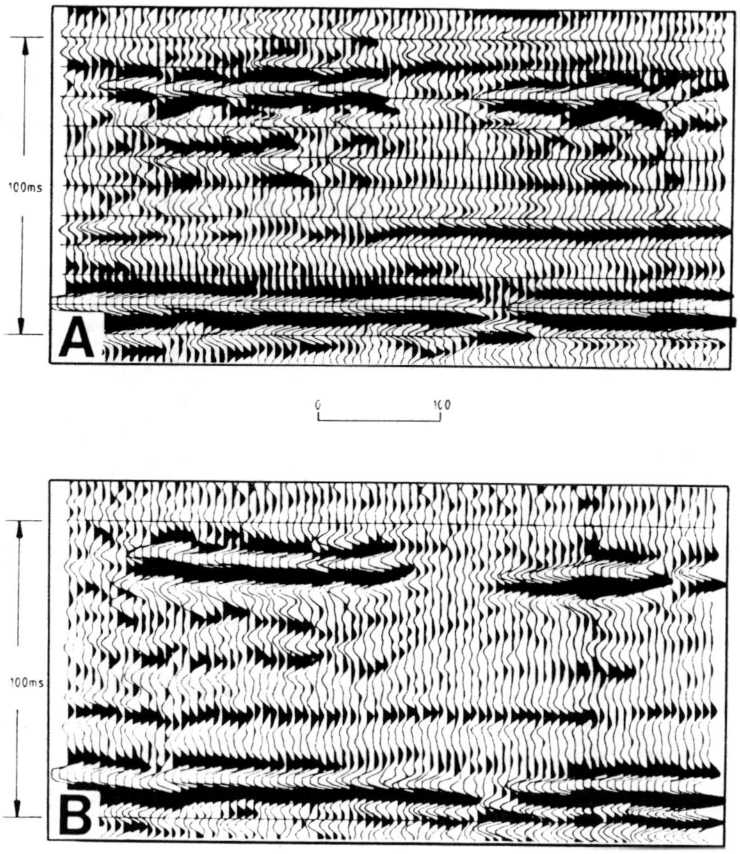

Fig. 8 Data comparison of conventional data (A) to undershoot data (B)

one- and two-ship lines lying along the same track. It can be seen that the two-ship line has a slightly lower frequency content than the conventional line. This is to be expected due to the large offset from source to near trace on the two-ship line (approximately 425 m) compared to the very short offset of the conventional line (30 m). In addition, more accurate wavelet processing techniques were applied to the conventional lines. Though there is a difference in the quality of the vertical resolution, this does not detract from the data to too great an extent. The lateral resolution is comparable between the two sections.

INTERPRETATION

The lithological log for the first well to be drilled from the platform was made available for interpretation of the sections. By use of this, the velocity log, and check shot survey for the same well a detailed data evaluation could be made. Towards the end of interpretation the lithological results from the M2 well were also made available to aid interpretation.

Figure 9 shows a detail of a north–south line across the area, and Fig. 10 shows an interpretated profile with interpreted lithologies for the same line. The section shown shows the section from seabed down to approximate base Quaternary, top Pliocene. The sequence is mainly made up of clays and sands of glaciomarine origin. Two layers exist where strong, anomalous, reflectors can be seen.

From the seabed to a depth of 280 m a series of silty clays with sand layers interspersed exist, belonging to the Tampen, Sperus, Cape Shore and Ferder formations as outlined by Rise et al (1984). These sediments were laid down in glaciomarine to marine conditions. No anomalous reflectors were seen in this interval.

At a depth of 280 m a strong reflector is seen lying flat across the whole area. In many parts of the survey area this reflector can be considered anomalous. This reflector marks a thin sand covering an erosion surface. It marks the top of the Mariner Formation. The erosion

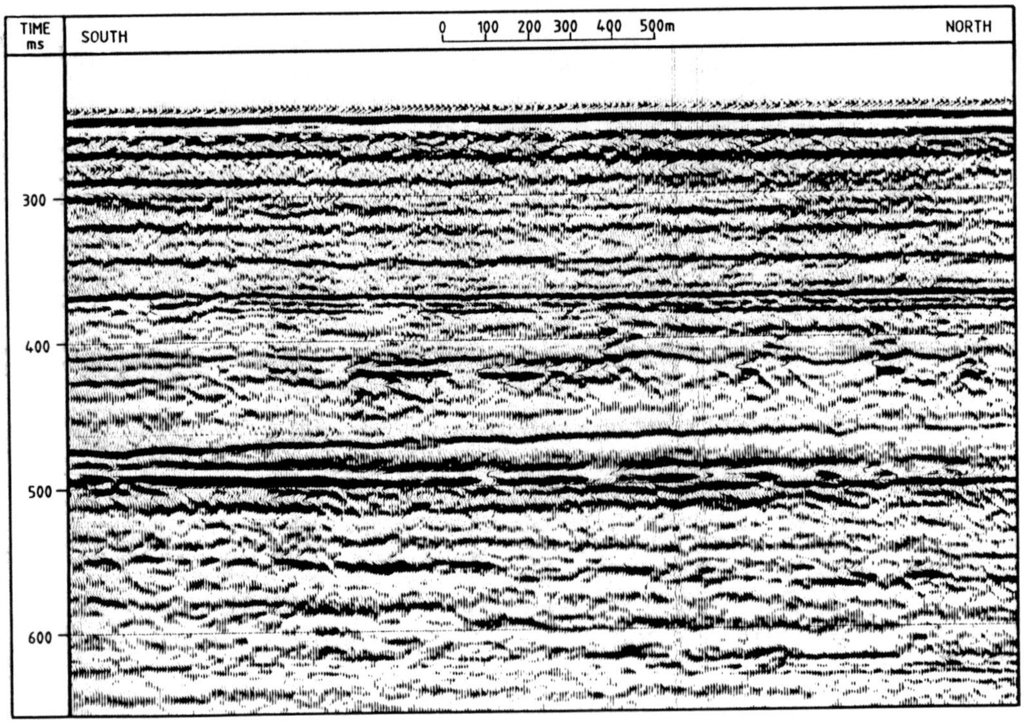

Fig. 9 North-south line across Magnus survey area

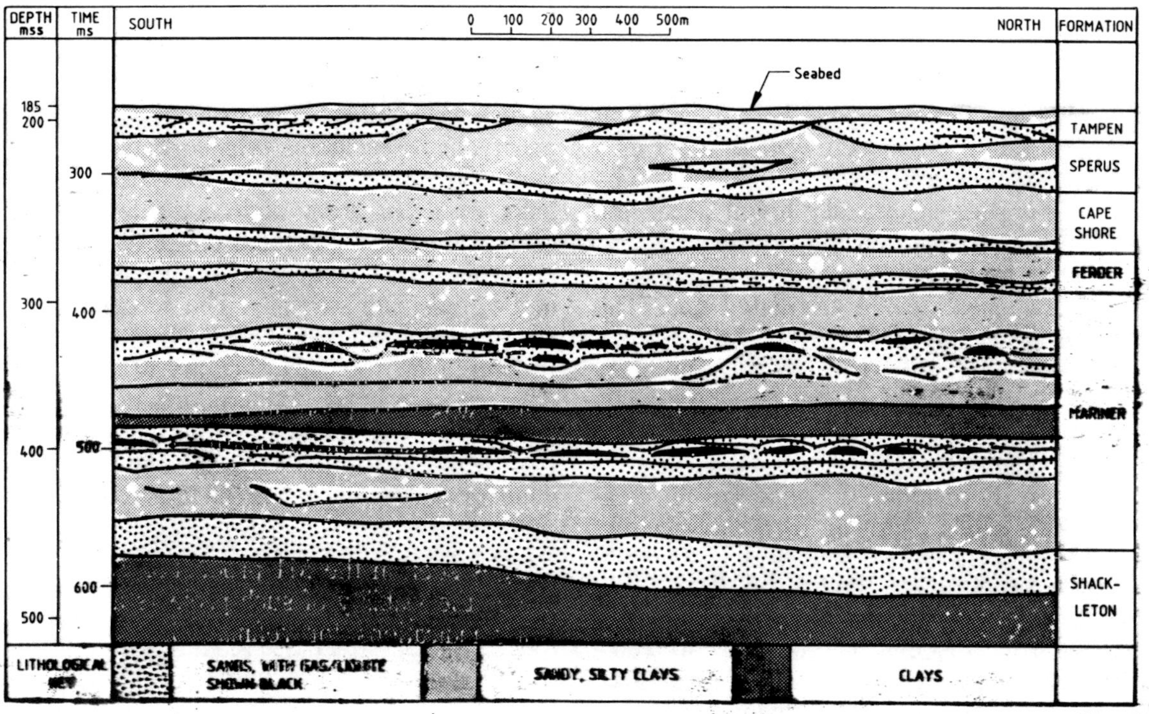

Fig. 10 Interpreted lithological profile for the section in Fig. 9

surface is thought by Rise *et al.* to mark a glacial erosion phase during the Saalian glacial maximum (128 000 years).

The Mariner Formation is a unit of glacial and glaciomarine silty, sandy clays with sand layers interbedded. It is in this series from 280 to 450 m that the two most significant intervals are seen on the sections.

The first interval of interest lies at approximately 320 m; it is marked by an irregular reflector covering very strong localized anomalous reflectors below. This reflector correlates to a sand unit seen in the M1 production well at this depth.

It can be seen that the anomalies closely correlate to the highs in the reflector above. These anomalies were interpreted as localized gas sands, where the gas has migrated to the structural high within the sand units and has been trapped by an overlying clay unit. There are no anomalies at this level below the platform, though anomalies are seen within 80 m of the west side of the platform. These sands are thought to be well sorted, highly permeable, unconsolidated and marine in origin. It is possible that they were laid down in tidal conditions and a regime similar to the initial stages of a sand wave development. Such sands would be well sorted and highly permeable. This sand sequence, with interlying silts and clays, is possibly up to 20 m thick in places.

The second interval of interest lies at a depth of 385 m and correlates to the strong event described in the previous study of the area. At this depth in the M1 well, however, no sand interval was observed, though the sonic log shows a complex series of kicks across the interval. On the sections a strong anomalous reflector is seen at this depth almost across the whole area and is anomalous below the platform. Despite the lack of a definite indication from the lithological log, this was interpreted as a sand and lignite body. It is possible that this layer has a gas presence. However, on the basis of subsequent seismic attribute processing, after completion of the project, this seems improbable due to the relative strength compared to the layer seen at 320 m.

At approximately 450 m, a weak, though occasionally anomalously strong event is seen across the area. This is thought to mark the top of a sand unit and correlates with a sand unit seen in the first production well. In this well a slight gas peak was logged at this point. The reflector is seen acoustically bright under the eastern corner of the platform. This unit is thought to have lignites present in it, together with small quantities of associated gas. The base of this sand unit is thought to lie at approximately 480–500 m. This unit is probably the sand layer seen in almost all previous exploration wells at the 450–540 m level. The top of this unit is thought to mark the top of the Shackleton Formation, a sequence of well-sorted fine sands overlying hard clays below. The sands and clays are marine in origin and are the base Quaternary unit in the area.

Below this level no other reflectors of distinct interest were seen to exist across the area.

The four main sand bodies described above were mapped in a series of Isopach charts, showing depth to the horizon. The acoustic anomalies were also mapped spatially at the different levels described.

SUMMARY

The Magnus two-ship high-resolution seismic survey proved that it is possible to record good quality high-resolution seismic data from beneath a production platform. Despite the problems expected from high noise levels, data quality was very good and permitted accurate mapping of sand units beneath the platform and in it's immediate environs. The accuracy of the methods can be guaged by the close correlation of events on the final sections to lithological changes seen in the M1 production well. This proves the viability of such projects as well as giving indications for refining the techniques used for similar projects in the future.

ACKNOWLEDGEMENTS

Geoteam would like to thank the Exploration, Development and Survey Departments at BP Dyce for permission to present data from this survey and to publish this paper.

REFERENCES

Lucas, A. L. 1974. A high resolution marine seismic survey. *Geophysical Prospecting* **22**, 667–682.

Rise, L., Rokoengen, K., Skinner, A. C and Long, D. 1984. Northern North Sea. Quaternary Geology map between 60°30′ and 62°N, and East of 1°E. IKU, Norway.

A THREE COMPONENT OCEAN BOTTOM MICROPROCESSOR-BASED SEISMOMETER

G Pascal, B R Person, Institut Francais de Recherche pour l'Exploitation de le Mer, France

Presented on **Wednesday 5 March 1986 at 17.00** in the Norfolk Conference Room

EMBARGOED AGAINST DELIVERY

24

A Three-component Ocean Bottom Microprocessor-based Seismometer

Georges Pascal, Département des Sciences de la Terre, Faculté des Sciences et Techniques, Brest, France, *and Roland Person*, IFREMER, Brest, France

INTRODUCTION

Ocean bottom seismometers (OBS) have been widely used during the last thirty years to collect seismic data for the determination of the structure of oceanic lithosphere and earthquake activity. Understanding the process of the creation of crust and lithosphere at mid-ocean ridges is a major goal of marine geophysics. Seismic refraction and reflection studies provide information on crustal structure. The distribution of earthquake activity in time and space contains information on the forces active in the spreading process. To achieve these important scientific goals, a number of research groups have developed ocean bottom seismometers (Ambuter and Solomon, 1974; Prothero, 1974,1976,1979,1981; Johnson *et al.*, 1977; Mattaboni and Solomon, 1977; Sutton *et al.*, 1977; Avedik *et al.*, 1978; Latham *et al.*, 1978; Heffler and Varret, 1979; Kasahara *et al.*, 1979; Koelsch and Purdy, 1979; Moore *et al.*, 1981; Kirk *et al.*, 1982; Whitmarsh and Lilwall, 1983). Recent developments in electronic devices and improvements in battery design have allowed the ocean bottom seismometer to become an operational tool. Considerable flexibility is provided by using low-power microprocessors. So the instrument complexity has shifted from hardware that is difficult to modify to software algorithms. This paper describes a new instrument which uses current electronics technology to achieve significant improvements in performance. This instrument is the first one of a planned network which will be used by the French scientific community.

SENSORS: SEISMOMETER LEVELLING

A seismometer measures the motion of the ground on which it is emplaced. A well-designed instrument will detect this motion with high fidelity and low noise over a broad range of amplitudes and frequencies. Data quality from previous experiments has often been disappointing because of poor signal quality and high noise levels. The coupling problem is much more serious in the ocean than on land and is quite complicated. Many authors have dealt with it and proposed some solutions to eliminate both the noise and the cross-coupling problem which occurs between the vertical motion of the seafloor and the horizontal motion of the sensors (Duschenes *et al.*, 1981; Sutton *et al.*, 1981; Trehu and Solomon, 1981; Prothero and Eickemeyer, 1982; Trehu, 1985). The best way to reduce the mechanical noise is to separate the geophones from the main pressure vessel and to place them at a distance on the seafloor. This small pressure package must be designed to reduce the noise from all potential sources to acceptable level. The main instrument must also be designed not to generate too much noise. We have chosen a compact instrument with a large baseplate on the seafloor to avoid oscillations of the OBS induced by ocean currents near the seafloor (Kasahara *et al.*, 1980; Duennebier *et al.*, 1981; Kasahara *et al.*, 1981).

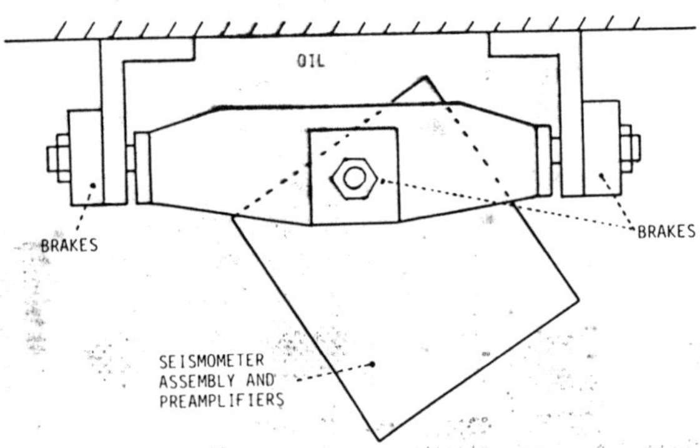

Fig. 1 Schematic drawing of the geophones mounting/levelling arrangement

Horizontal seismometers must be levelled to less than 2°, once the OBS is on the sea bottom. Figure 1 is a cutaway drawing of the deployed geophone package. The ground motion sensors are three orthogonal 2 Hz geophones. Sensors and preamplifiers are housed inside a titanium case. Gimbals and brakes are balanced pressure mounted. The brakes are always clamped except when the sensors are to be levelled. Then a current is supplied to the brakes for about 5 s which levels the seismometers from any starting position. The range of the levelling is more than ±25° in any direction. A mercury contant closure is tested by the program to detect anything wrong in the levelling process. This is achieved through acoustic transmission. When sensors are not levelled, the OBS must be recovered and relaunched.

A fourth vertical geophone is provided for *in situ* calibration (Zelikovitz and Prothero, 1981; Sauter and Dorman, 1983).

INSTRUMENT DESIGN

The OBS is a free-fall pop-up instrument which is designed for long-time and deep-sea immersion. Its external design is shown in Fig. 2. A modular structure has been chosen to use preexisting devices, such as transponders, release, etc.

The recording and timing electronics and power supply are housed in two cylindrical pressure vessels which sit vertically on an aluminium structure. Attached to this structure are also a standard acoustic release, another one modified for data transmission, an indepen-

Fig. 2 The ocean bottom seismometer, ready for launch

dent timer release, a flashing light, a radio beacon and a high-sensitive hydrophone (OAR type E4S-D).

That structure is wrapped in with syntactic foam cylindrical floats providing buoyancy and lies on a circular baseplate which serves as ballast and remains on the seafloor after recovery of the OBS. The separate sensor package is deployed on the seafloor about 1 m from the main unit. After the OBS reaches the seafloor, a rigid arm swings down with the sensor package when an explosive bolt is shot; then the pin holding the geophone package to the arm is pulled out by another explosive bolt when the package is about 30 cm above the seafloor. After dropping the package, the arm is automatically retracted firmly against the main unit by the pull of a stainless steel spring.

RECORDING SYSTEM

The block diagram of the microprocessor-based ocean bottom seismometer is shown in Fig. 3. The signals from the four sensors (hydrophone, three geophones) are preamplified and notch filtered with a high cut-off frequency of 30 Hz. The signal from each channel is then passed through a microprocessor gain control circuit, a sample and hold circuit and a 12 bit analog to digital converter. The digital signal from each channel is loaded into a 2944-word memory (16 bits) which is equivalent to 23 s of data at a sampling rate of 128 samples per second. The 12-bit word length results in a resolution of over 66 dB. Combined with the 8 bits of gain control this leads to a total dynamic range of 108 dB. The four gains are in the ratio of 4, so 4 bits are sufficient to give the range. For each sample two convertions are triggered. The first one with the minimum gain gives the 4 bits of range. The second one with the chosen gain gives the 12 low-weight bits. The signal from the vertical component is passed through an event detection algorithm which every second compares the long-term average to a short-term average. This algorithm will be turned into a Walsh transform algorithm in the future

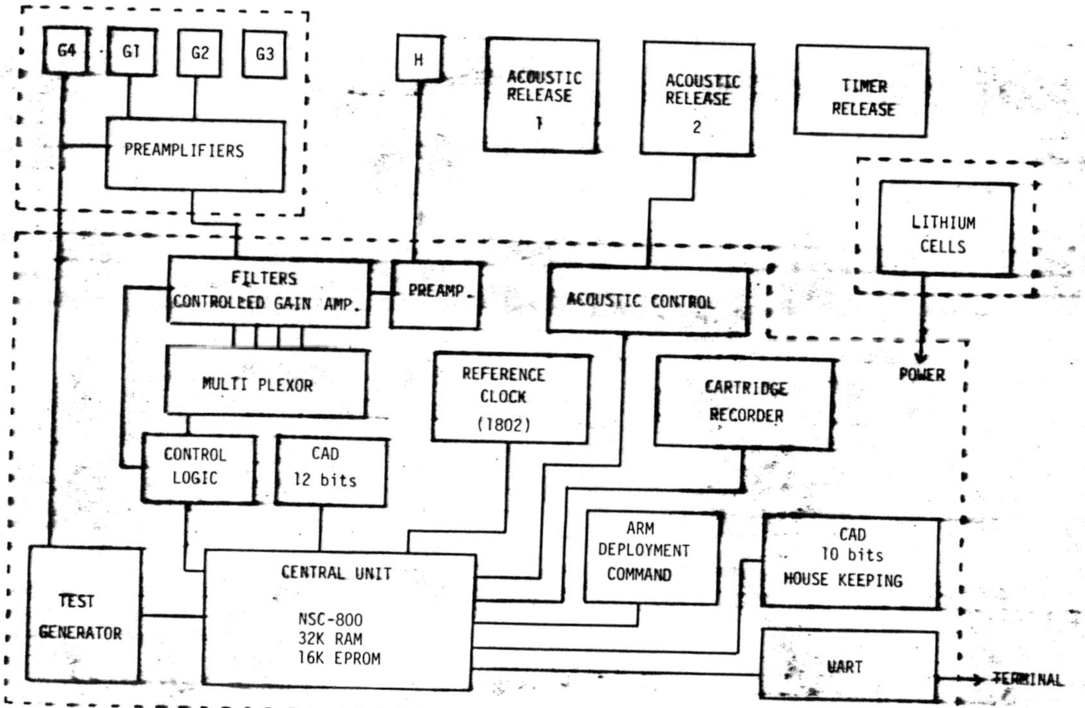

Fig. 3 Block diagram of the ocean bottom seismometer

(Goforth and Herrin, 1981). When the short-term average exceeds the long-term average according to the programmed set fraction, the time of the detected event is stored. Data acquisition is then done for about 18 s. Then the 23 s of the signals stored in the memory are written on the DC-300 cartridge tape along with the detection time, the LTA and STA values, the experiment code and all useful informations.

The central processor unit consists of a NSC-800 with a 6 MHz quartz crystal oscillator. The program is stored in a 14K EPROM, including a monitor program. The RAM capacity is 34 Kbyte.

Reference time for the instrument is provided by a low-power digitally temperature-compensated crystal oscillator with an accuracy of one part in 10^8 per day. This clock, implemented on two boards, is power saved for one year by lithium cells. Digital compensation is controlled by a RCA-1802 microprocessor. Every 20 seconds, it gives a reference pulse which interrupts the NSC-800 processor; then the day, hour, minute, second and millisecond can be read on the clock. The timer is used to refer accurately any event between two clock pulses. Synchronization and stability control are achieved under NSC-800 microprocessor control. After recovery of the OBS, the absolute accuracy of the clock can be checked using radio time signals or a shipboard standard clock if the main power supply needs replacing.

A commercial digital DC-300 cartridge tape recorder which stores 17 Mbytes is used. Data are formatted in 8 Kbytes blocks, so a record consists of three blocks. Data are transferred between recorder and central units at 192,000 bits per second. The same tape recorder can be used to play back the data and a directory of the recorded events is then provided. Three hundred and thirty-six events of 23 s and *in situ* calibration sequences are recorded in the same format.

A standard acoustic release is used for recovery of the intrument. A second one has been modified to provide data link capabilities along with the releasing function. From the sea surface it is possible to command different functions of the OBS: arm swinging out, geophone package pulling out, detection parameters modifying (ratio LTA/STA). The OBS responds in a specified code. It is also possible to get some additional data from the OBS: geophone levelling, voltage and current information, STA and LTA values, last recorded block number. An 8-bit word is transmitted every 4 s; only 4 bits are data information. This function can be switched off at the end of the deployment phase to avoid misfunction during long-term immersions.

Seventy-two lithium cells of 10 A h under 3,5 V (2520 W h) are used for a three-month deployment. For short-time immersions (refraction studies for example) CdNi batteries can replace the lithium cells. Acoustic release and the reference clock are meant to work for one year.

OPERATION AT SEA

Operations at sea are a very important factor in the success of an OBS experiment. The instrument is closed up in the laboratory and then quickly checked out prior to deployment without opening it. To do so, a seven-wire cable connects the instrument to a test cabinet via opto-electronical couplers. Three wires supply a serial data link. So from a terminal it is possible to test all of the seismometer: memories, voltage current, recorder, analog digital converties, filters. An internal digital analog card generates the analog signal necessary to these functions. Reset and clock synchronization can also be externally achieved. All the tests and the deployment preparation are performed under microcomputer control.

PLAYBACK

Data can be played back at sea for instrument checkout and the planning of further deployments. You can either use the OBS recorder for playback without opening the instrument by using the seven-wires cable or playback the cartridge on another reader recorder connected to the test microcomputer. In the land-based laboratory, data are played back on a reader recorder connected to an HP-1000 computer.

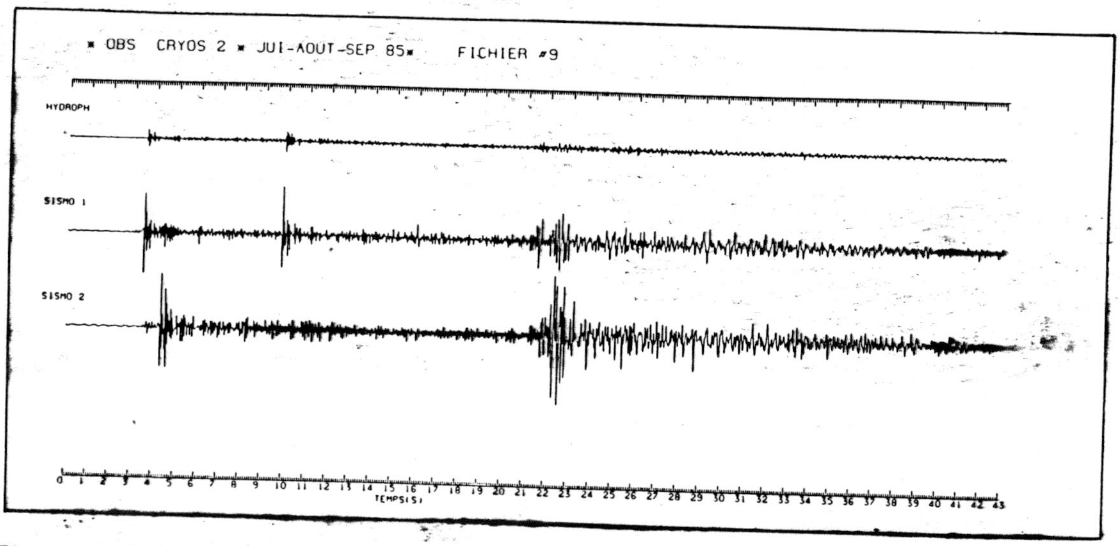

Fig. 4 Example of earthquakes recorded during the OBS test deployment along the Azores-Gibraltar plate boundary. Sismo 1 is the vertical geophone record, and Sismo 2 is one of the horizontal geophones.

All the events are then stored on a nine-track magnetic tape. In that form the data can be plotted, filtered and otherwise analysed. The whole process takes only one day.

TEST CRUISES AND SCIENTIFIC RESULTS

In November 1984 we went on a cruise near Brest in shallow waters to test the mechanical functioning of the instrument and we made the first recording tests. The purpose of another test cruise in the Mediterranean Sea in December 1984 was to test the acoustical system and the event detection software. New tests were conducted in the Mediterranean Sea, in May 1985, in still deeper waters (2500 m) and shots up to 15 miles were recorded. The first purpose of the cruise in the Azores-Gibraltar area (depth 5000 m) was to record another set of shots with a 1000 in 3 air gun. The instrument was then

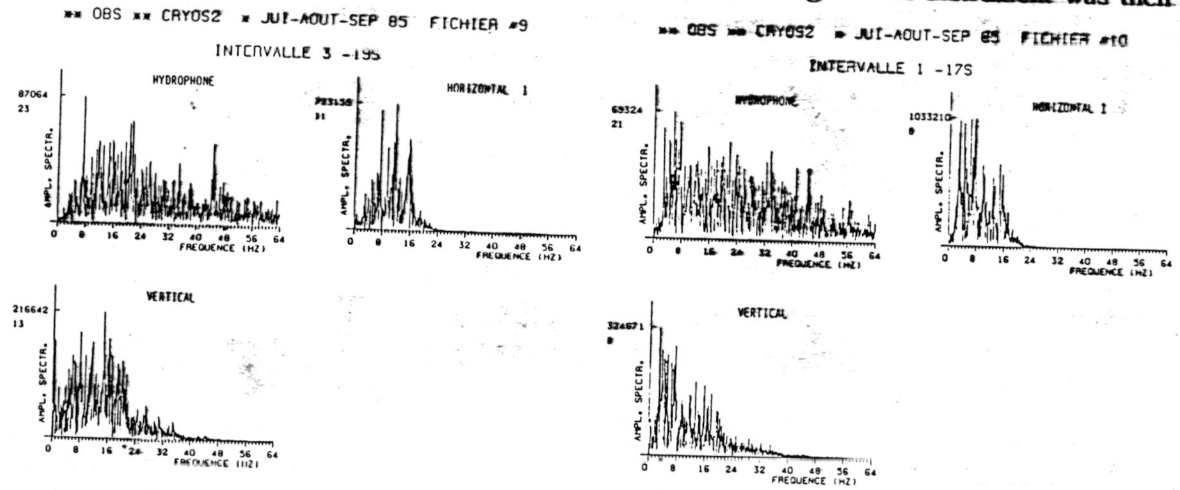

Fig. 5 Corresponding amplitude spectra for the seismic event shown in Fig. 4. Files 9 and 10 are the respective spectra for the 23 first and 20 last seconds

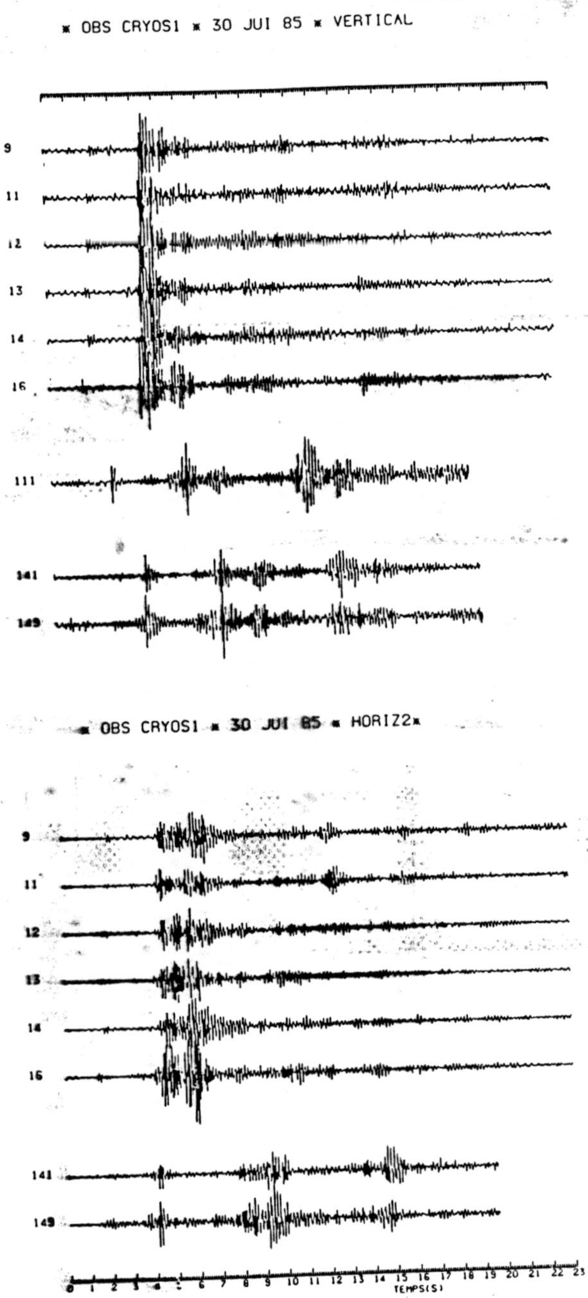

Fig. 6 Examples of records of 1000 in^3 air gun shots. Distances from the OBS decrease from 15 to 10 nautical miles (shot 9 to shot 149). Plots have not been connected for navigation and therefore do not show a true distance scale

deployed on the sea bottom for several weeks. It was picked up after 46 days and it successfully recorded some local earthquakes of the Azores–Gibraltar plate boundary. Figure 4 shows examples of micro-earthquakes recorded by the instrument during the six weeks' experiment. The earthquakes were located at 25 to 160 km distance from the instrument. The corresponding spectra on the P and S phases are shown in Fig. 5. Although the data collected during the last test are under evaluation, the preliminary results do not exhibit strong resonances.

The examples of shots (Fig. 6) from 15 to 10 miles were recorded during the last cruise in the Atlantic Ocean. The spectra of the vertical (Sismo 1) and the horizontal (Sismo 3) components are quite satisfactory (Fig. 7).

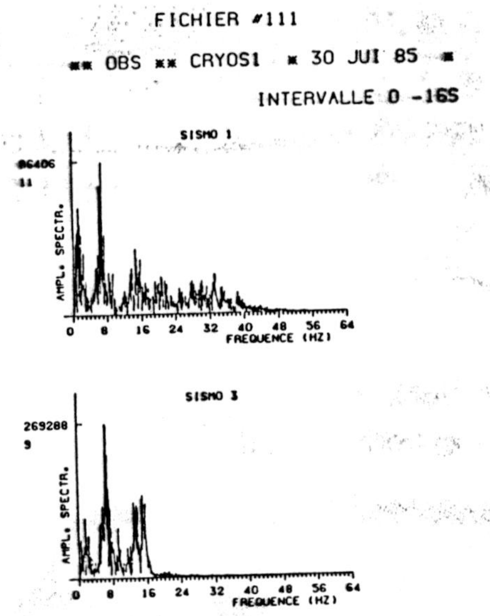

Fig. 7 Amplitude spectra relative to shot 111. Sismo 1 is the vertical; component. Sismo 3 is one of the horizontal components (a 20 Hz low-pass hardware filter has been tested on this channel)

CONCLUSION

At present one OBS of this design exists. During the test cruises some improvements were successively made to the instrument and we will improve the software in the next few months. Two other instruments will be built in 1986 and used with the first one in scientific programs. We plan to use them during cruises on the continental margin of the Eastern Atlantic, in connection with the French national program of exploration of the deep crust.

ACKNOWLEDGEMENTS

During the development and the test cruises of the seismometer described here, the work of a number of people has been crucial. We thank F. Le Foll, Y. Raer and A. Echardour of IFREMER for the mechanical design, P. Leon, B. Ollivier, J. Savary, J. F. Karczewski, H. Floch of UBO, INSU and IFREMER for their work on the electronic systems. J. Goslin, F. Marillier, M. Dietrich and C. Guennou of UBO have also made important contributions to this work. We would also like to acknowledge with gratitude the assistance in sometimes difficult circumstances of the captains and crews of the research vessels Thalia, Catherine Laurence, Cryos and Noroit.

This work was supported by the PIRO, INSU and IFREMER.

REFERENCES

Ambuter, B. P. and Solomon, S. C. 1974. An event-recording system for monitoring small earthquakes. *Bull. Seis. Soc. Am.* **64**, 1181-1188.

Avedik, F., Renard, V., Buisine, D. and Cornic, J. Y. 1978. Ocean bottom refraction seismograph (OBSR). *Mar. Geophys. Res.* **3**, 357.

Duennebier, F. K., Blackinton, G. and Sutton, G. H. 1981. Current- generated noise recorder on ocean bottom seismometers. *Mar. Geophys. Res.* **5**, 109-115.

Duschenes, J. D., Barash, T. W., Mattaboni, P. J. and Solomon, S. C. 1981. On the use of an externally deployed geophone package on an ocean bottom seismometer. *Mar. Geophys. Res.* **4**, 436-450.

Goforth, T. and Herrin, E. 1981. An automatic seismic signal detection algorithm based on the Walsh transform. *Bull. Seis. Soc. Am.* **71**, 1351-1365.

Heffler, D. E. and Barret, D. L. 1979. OBS development at Bedfort Institute of Oceanography. *Mar. Geophys. Res.* **4**, 227-245.

Johnson, R. V., Lister, C. R. B. and Lewis, B. T. R. 1977. A direct recording ocean bottom seismometer. *Mar. Geophys. Res.* **3**, 65-85.

Kasahara, J., Koresawa, S., Nagumo, S., Daikuhara, T., Hara, T. and Ando, S. 1979. Pop-up OBS (ERI Type P-79). *Bull. Earthquake Res. Inst.* **54**, 515-530.

Kasahara, J., Nagumo, S., Koresawa, S., Daikuhara, T. and Miyata, H. 1980. Experimental results of vortex generation around ocean bottom seismograph due to bottom current. *Bull. Earthg. Res. Inst.* **55**, 169-182.

Kasahara, J., Koresawa, S. and Nishi, Y. 1981. Preliminary report for large bottom current at the deep ocean floor inferred from ocean bottom seismic observations. *Bull. Earthg. Res. Inst.* **56**, 207–221.

Kirk, R. E., Langford, J. J. and Whitmarsh, R. B. 1982. A three-component ocean bottom seismograph for controlled source and earthquake seismology. *Mar. Geophys. Res.* **3**, 215–226.XI

Koelsch, D. E. and Purdy, G. M. 1979. An ocean bottom hydrophone instrument for seismic refraction experiments in the deep ocean. *Mar. Geophys. Res.* **4**, 115–125.

Latham, G., Donoho, P., Griffiths, K., Roberts, A. and Ibrahim, A. K. 1978. The Texas ocean-bottom seismograph. *1978 Offshore Tech. Conf.*, Vol. 3, p. 1467.

Mattaboni, P. J. and Solomon, S. C. 1977. MITOBS: a seismometer system for ocean-bottom earthquake studies. *Mar. Geophys. Res.* **3**, 87–102.

Moore, R. D., Dorman, L., Chin-Yen, H. and Berliner, D. L. 1981. An ocean bottom, microprocessor based seismometer. *Mar. Geophys. Res.* **4**, 451–477.
Prothero, W. 1974. An ocean bottom seismometer capsule. **Bull. Seis. Soc. Am. 65**, 1251–1262.
Prothero, W. 1976. A digital event-recording ocean bottom seismometer capsule. *Mar. Geophys. Res.* **3**, 119–141.
Prothero, W. 1979. An operationally optimized ocean-bottom seismometer capsule. *Phys. Earth Planet. Int.* **18**, 71–77.
Prothero, W. 1981. Earthquake signal processing and logging with a battery powered microcomputer. *Bull. Seis. Soc. Am.* **70**, 2275–2290.
Prothero, W. and Eickemeyer, J. M. 1982. Ocean bottom seismometer response studies. *Trans. AGU*, Fall 1982 meeting.
Sauter, A. and Dorman, L., 1983. Ocean bottom seismometer calibration. *Trans. AGU*, **63**, 1026.
Sutton, G. H., Kasahara, J., Ichinose, W. N. and Byrne, D. A. 1977. Ocean bottom seismograph development at Hawaii Institute of Geophysics. *Mar. Geophys. Res.* **3**, 153–177.
Sutton, G. H., Duennebier, F. K. and Iwatake, B. 1981. Coupling of ocean bottom seismometers to soft bottom. *Mar. Geophys. Res.* **5**, 35–51.
Trehu, A. M. 1985. Coupling of ocean bottom seismometers to sediment. Results of tests with the U.S. Geological Survey OBS. *Bull. Seis. Soc. Am.*, in press.
Trehu, A. M. and Solomon, S. C. 1981. Coupling parameters of the MIT OBS at two nearshore sites. *Mar. Geophys. Res.* **5**, 69–78.
Whitmarsh, R. B and Lilwall, R. C. 1983. Ocean-bottom seismographs. In *Proceedings of Nato Advanced Research Institute on Structure Development of the Greenland-Scotland Ridge – new methods and concepts, NATO Conference, series IV: Marines Sciences.*
Zelikovitz, S. and Prothero, W. 1981. The vertical response of an ocean bottom seismometer – analysis of the Lopez Island vertical transient tests. *Mar. Geophys. Res.* **5**, 53–67.

Part VI

Wave Directionality

25

The Wavescan Second Generation Directional Wave Buoy: from Design Concepts to Field Testing

Stephen F. Barstow, Oceanographic Center, Trondheim, Norway
Gard Ueland, Continental Shelf Institute (IKU), Trondheim, Norway
Bjørn A. Fossum, Seatex Ltd., Trondheim, Norway

ABSTRACT

In March 1980 the Continental Shelf Institute (IKU) deployed its first buoy with wave direction measurement capability. This buoy was a result of a joint development by the Christian Michelsens Institute in Bergen and IKU. Since then IKU, latterly the Oceanographic Center, has carried out oceanographic and meteorological measurements with several buoys of this type totalling over 10 years of operations at a number of locations from the tropics to the high Arctic. Operating these buoys in remote areas can be very expensive and IKU saw the need for bringing the data collection costs down. One way of reducing the cost is to increase the maximum operational period for the data buoy and thereby reduce the number of expensive field cruises to a minimum. In order to achieve this without reduction in data recovery it is a requirement that the measurement process be carefully monitored.

The first generation wave direction buoys utilized the Argos satellite system for monitoring the meteorological sensors and some housekeeping functions. However, the wave measurement and recording process could not be monitored well enough to justify a long operational deployment. As a first step to overcome this problem, IKU developed an *in situ* wave processor in 1981-82. This processor was capable of computing the *l*-dimensional wave spectrum in addition to some zero-crossing parameters based on the wave elevation time series. A prototype version was installed in a buoy and was tested on Haltenbanken offshore mid-Norway and provided real time wave data for six months. However, the system still lacked wave direction data in real time. This first prototype was based on a first generation CMOS microprocessor and was quite bulky, the software being written in assembly language. To implement wave direction processing using this system would have required too large an effort. IKU therefore decided to make use of the latest processor technology and high level real time language for the next generation *in situ* processor. This together with the need for a more flexible data acquisition system and a more lightweight buoy with better dynamic response, lead to the development of a completely new data buoy, WAVESCAN.

In the paper we describe how the final

THE WAVESCAN SECOND GENERATION DIRECTIONAL WAVE BUOY; FROM DESIGN CONCEPTS TO FIELD TESTING

Dr S F Bastow, Oceanographic Centre; G Ueland, Continental Shelf and Petroleum Technology Research Institute Ltd; B A Fossum, Seatex A/S, Norway

Presented on **Wednesday 5 March 1986 at 14.30** in the Clarence Conference Room

EMBARGOED AGAINST DELIVERY

Fig. 1 The WAVESCAN prototype (left) and the NORWAVE ODAS 492 buoy at Haltenbanken during the October 1984 field tests.

system design evolved from design concepts through a prototype field tests. A historical review is first presented discussing the various design solutions which have been employed for the collection of directional wave data from buoys over the past 25 years. With this in mind the WAVESCAN design aims and the methods used to fulfil these aims are discussed. This includes a description of the series of model wave tank tests carried out in order to try and quantify the response and stability characteristics of a number of possible design solutions. Next, the "brain" of the WAVESCAN system – the data acquisition and processing unit – is described. This system includes real-time directional wave processing by standard FFT methods. The results may then be transmitted to land by way of the ARGOS satellite data collection system. Meteorological and wave sensors employed in the prototype are also described.

The first WAVESCAN prototype was completed during autumn 1984. After a series of function tests in the fjord off Trondheim, the buoy (ODAS 499) was deployed on 18 October approximately 150 km off the Norwegian coast on Haltenbanken. The location is one of the long-term stations for acquisition of environmental data served since spring 1980 by IKU's original directional metocean data buoys. One of these buoys (ODAS 492) was located ca 1 km from the WAVESCAN mooring collecting data as part of ODAP (Oceanographic Data Acquisition Project). Both systems are shown in Fig. 1 below during the first field test. The location is situated in one of the most exposed areas of the Norwegian Shelf.

Data recorded by the metocean buoys during the first five years in this area has produced so far significant wave height up to 14 m and single waves up to 24 m. The area is also exposed to energetic long period swells propagating in from the Atlantic. WAVESCAN was positioned on Haltenbanken for two month-long deployments during the 1984-85 winter, the second deployment lasting from 16 February to 26 March. Operational experiences have been good with near 100% wave data return for the two periods. We present results from the intercomparison between the ODAS 492 and WAVESCAN from these tests. Both buoys were equipped with the real-time processors allowing us to follow the intercomparison in some detail in real time.

The first production version of WAVESCAN was delivered by the manufacturer Seatex Ltd in Trondheim on 1 September 1985 to the Oceanographic Center. During September/October the buoy has been deployed off the Norwegian coast at Stadt as part of a measurement programme to map wave conditions in an area prone to an unusually high number of unexplained shipwrecks thought to be due to the occurrence of extreme breaking waves. In November 1985 Wavescan joined a large directional field test near the EDDA Platform (Ekofisk field) in the North Sea alongside a number of other directional wave buoys (Wavetrack, Marex, Wadibuoy, Wavec and Norwave). This is the WADIC (Wave Direction Calibration Project) experiment, which aims to evaluate the above commercially available wave data buoys for measuring ocean wave heights and directions under severe open-ocean conditions. The WAVESCAN development project has been supported by Norske Shell, Norsk Hydro, Statoil and the Royal Norwegian Council of Scientific and Industrial Research.

FIELD RESULTS COMPARING TWO DIRECTIONAL WAVE BUOYS

B-A Juszko, B de Lange Boom, D Green, R Brown,
Seakem Oceanography Ltd, Canada

Presented on **Wednesday 5 March 1986 at 15.00**
in the Clarence Conference Room

EMBARGOED AGAINST DELIVERY

26

Field Results Comparing Two Directional Wave Buoys

Barbara-Ann Juszko, Bodo de Lange Boom, David R. Green and *Robin Brown*,
Seakem Oceanography Ltd, Sidney, British Columbia, Canada
Currently at the Institute of Ocean Sciences, Patricia Bay, British Columbia, Canada

INTRODUCTION

In the past few years, increasing demand for directional wave information has led to the development of new instrumentation and analysis procedures. Two basically different designs of directional wave buoys are currently available on the market. The first is the traditional surface slope-following buoy including the Norwave (Bergen Ocean Data), Wadibuoy (CNEXO and Néréides) and Wavec (Datawell) buoys. The second, exemplified by the Wave-track Type 956 (Endeco Inc.) buoy, is the wave orbital-following buoy. An understanding of the operation, reliability and applicability of these two instrument types would prove useful when designing field experiments. Two major studies have recently been conducted in Canadian waters, whose goals were in part to assess the performance of the Datawell Wavec and Endeco Type 956 Wave-track buoy. Part of this assessment included comparison of the derived information with visual observations and hindcast model results. This paper details some of their findings.

STUDY BACKGROUND

Two studies were conducted in Canadian coastal waters employing Datawell Wavec and Endeco Type 956 Wave-track buoys. The first occurred along the northern coast of British Columbia from October 1982 to May 1984. The Wavec buoy was moored in 189 m of water at 52°06.8′N, 128°57.5′W, while the Wave-track was located in 156 m of water at 53°21.2′N, 130°46.7′W. Although neither buoy operated continuously over the 2 years, at least 10 months of data were obtained from both.

The second study was of shorter duration, February to March 1984, but was more intensive as it was designed for specific intercomparison between these two buoys as well as against a standard Waverider (Datawell), visual observations and hindcast model results. Two Wavec buoys and one Wave-track buoy were moored in approximately 85 m of water on Newfoundland's Great Banks, off Canada's East Coast. Their positions were, respectively, 46°45.8′N, 48°48.8′W, 46°44.83′N, 48°49.75′W and 46°45.67′N, 48°45.75′W with less than 1 nautical mile separating them from each other. The visual observations were obtained from a nearby manned mobile drilling unit (the West Venture, operated by Mobil Oil Canada Ltd) and the model results supplied by

F. G. Bercha Ltd of St John's, Newfoundland.

The Wavec is a slope-following buoy, approximately 2.5 m in diameter, consisting of a central instrument canister to which are strapped peripheral flotation segments. The Wavetrack has an inverted-pendulum design composed of a surface sphere and subsurface shaft to which is attached the sensor package. The wave direction components are obtained not from east-west and north-south surface slopes as for the Wavec, but from the angular tilt of the shaft induced by the wave orbital velocities.

Standard time-series analyses methods were used to process the data to spectral components and the approach of Longuet-Higgins *et al.* (1963) was taken to calculate directional spectra. A second processing scheme, a frequency bandpass analysis, as described in Brainard (1982) can also be used for Wavetrack data. However, this method proved inefficient and the results were of limited usefulness.

BUOY OPERATION

Over the two studies, the Wavecs were generally reliable in terms of hardware, telemetry and data quality. The major concern was its ruggedness, as on two occasions during the first study the peripheral flotation segments were dislodged or lost during storms (significant wave height of >7.0 m). Communication with other users has indicated similar problems.

The Wave-track is a small buoy (less than 1.0 m in diameter), which permits ease in handling. However, the small size results in reduced battery capacity and it cannot transmit continuously, as does the Wavec, which is a desirable feature during storms. Its design and sensor systems are quite rugged, as they survived beaching on at least one occasion. During the winter deployment on the Grand Banks it suffered severely from icing of the exposed surface interrupting data transmission or resulting in useless data. There were no apparent icing problems with the Wavec.

In order to assess the buoy response to the wave field, a linearity analysis on the data was performed. Using the equations set out by Longuet-Higgins *et al.* (1963) and the wave dispersion equation, it can be shown that under linear wave theory

$$K^2\, CO_{11}(f) = CO_{22}(f) + CO_{3}3(f), \qquad (1)$$

where K is the wavenumber at frequency f, and CO_{11}, CO_{22} and CO_{33} are the co-spectral estimates obtained from the heave and two slope signals (north-south and east-west slope). For an example derivation, see Audunson *et al.* (1982).

The ratio

$$R(f) \left[\frac{K^2\, CO_{11}(f)}{CO_{22}(f) + CO_{33}(f)} \right]^{1/2}$$

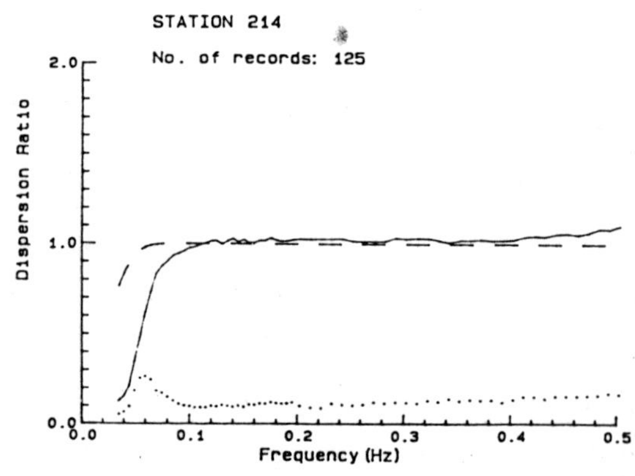

Fig. 1. Wavec buoy response under normal operating conditions (solid line, experimental; broken line, theoretical; dotted line, one standard deviation).

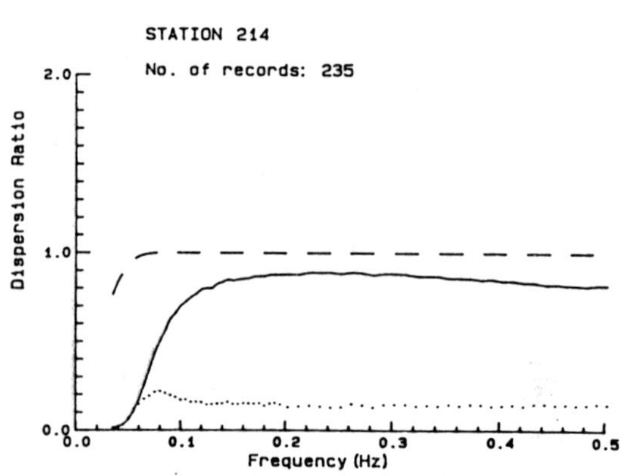

Fig. 2. Wavec buoy response after dislodgement of the peripheral flotation segments (solid line, experimental; broken line, theoretical; dotted line, one standard deviation).

can be defined and should be equal to 1.0 at all frequencies if the wave field is behaving according to linear theory and if the buoy system is measuring it properly. Hence, it can be used to assess the instrument operation. Audunson *et al.* (1982) were able to pinpoint the time of loss of a mooring chain in their system using this method. Ezraty and Cavanie (1981), for example, used this analysis to assess the influence of Doppler shifts, induced by tidal currents, on their data.

Figure 1 shows the standard Wavec response. As can be seen, the Wavec is measuring the wave field well through the entire frequency range. The drop off at low frequencies is a reflection of water depth. Figure 2 shows the linearity analysis results on data from a buoy which had dislodged two of its peripheral flotation segments, and illustrates how this problem can be diagnosed. It should be mentioned that during this period there was no obvious indication of a problem in the spectra.

The response curve of the Wave-track buoy is shown in Fig. 3. It is obvious from this figure that the amplitude response of the buoy is not linear. There is increased energy at low frequencies (*R(f)* should equal 0.0) below water depth cut-off and increased noise for frequencies greater than 0.3 Hz. Low-frequency energy contamination was common during both studies.

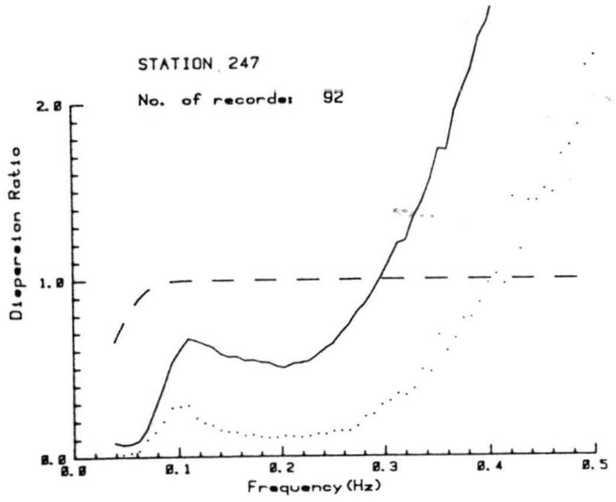

Fig. 3. Wave-track buoy response under normal operating conditions (solid line, experimental; broken line, theoretical; dotted line, one standard deviation).

To obtain *R(f)*, the wavenumber, *K*, is calculated using an interation procedure on the wave dispersion equation. Conversely, one may calculate *K* from eqn (2) assuming *R(f)* = 1.0 and then use this derived wavenumber in the equations of the Longuet-Higgins *et el.*, thereby providing a derived instrument correction. This is a standard procedure, but it is often used without consideration of instrument problems of a non isotropic nature. Such a problem was encountered during buoy icing.

Equation (2) supplies only an amplitude correction. Phase corrections are also required for Wave-track data whose tilt and heave signals are 90° out of phase compared with surface slope and heave signals.

DATA RELIABILITY

In terms of data recovery, during the long-term deployment from 1982 to 1984 there was a 95.6% recovery of Wavec data and 65.7% for Wave-track data, when the buoys were in operation. Both buoys supply good heave spectral estimates, provided, in the case of the Wave-track, that they are corrected for any

Fig. 4. Time-series of apparent direction overlayed on observed winds. (a) February 1984; (b) March 1984; (c) April 1984. Squares, wind; triangles, apparent direction; crosses, apparent direction for periods less than 7.0 s.

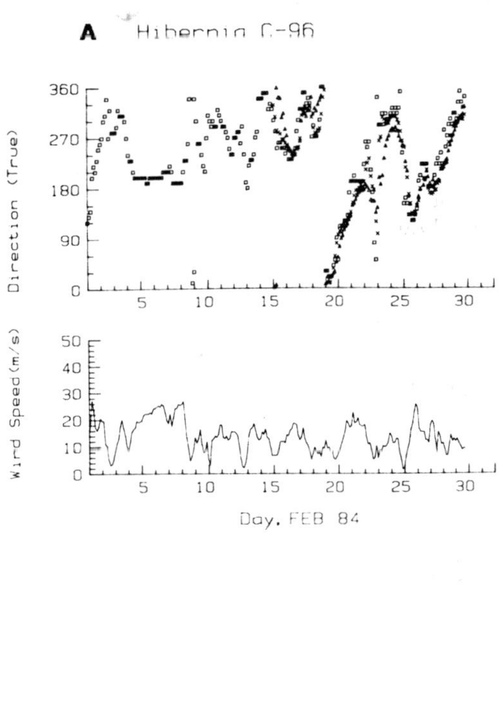

Fig. 4 (cont.)

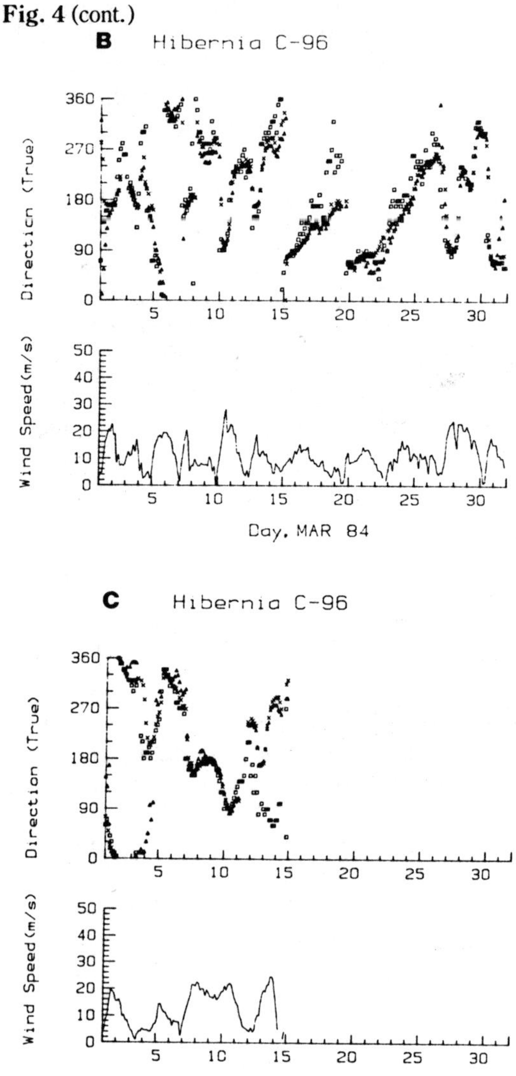

contains time-series of apparent directions, observed from one Wavec buoy, overlayed on local wind directions. The agreement is very good, particularly with "sea"-derived directions. Any discrepancies occurred during swell-dominated conditions, when local winds are not expected to represent local wave directions.

There were problems with the direction statistics obtained from the Wave-track buoy. Figure 5 contains sample heave spectra and corresponding mean direction by frequency plots. The heave spectra for the different instruments agree well, though with some site variability and a slight rise at low frequencies for the Wave-track. The mean directions also agree over the mid-frequencies (spectral peak to approximately 0.3 Hz). There is the expected increased scatter at higher frequencies. Correlation with nearby Wavec data of peak directions always shows greater scatter than spurious low-frequency energy. The direction statistics obtained from the Wavec data were also good, showing agreement between buoys and with visual observations, expected directions from wind measurements and hindcast model results. Various direction statistics can be examined, including the mean wave direction with frequency, as defined by Longuet-Higgins *et al.* (1983), the peak direction (i.e. mean direction at the spectral peak) and the apparent direction. The apparent direction, as defined in Juszko (1985a) consists of a weighted averaging across frequency of the mean direction with the weighting supplied by the energy density or the directional spread at each frequency. Figure 4

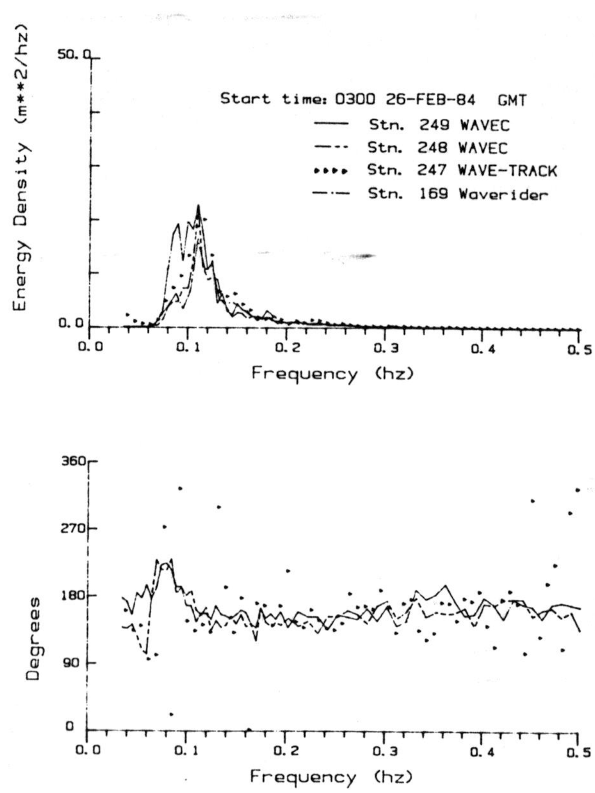

Fig. 5. Sample heave spectra (upper) and mean direction by frequency (lower) plots of Wavec, Wave-track and nearby Waverider.

the apparent directions ($R = 0.29$ versus $R = 0.73$). In comparison, similar regressions between the two Wavec buoys result in $R = 0.86$ and $R = 0.99$ for peak and apparent direction, respectively. Use of the bandpass analysis showed similar problems, including long-period swell directions coming from shore and disagreement with directions calculated using the method of Longuet-Higgins *et al.* Some of these problems may have been made more severe by the physical conditions.

An examination of the directional spread about the mean at each frequency also shows that this is generally larger for Wave-track data than for Wavec (example in Fig. 6). Both buoys do suffer from the inherent large spread associated with the analysis using the method of Longuet-Higgins *et al.*, also improved by applying the linearity correction. The use of other fitting techniques can reduce the spread such as the cosine-spread model of Hasselmann *et al.* (1980), or the use of a different weighting scheme as suggested by LeBlanc and Middleton (1982). However one artificially reduces the spread, its applicability to the real world can be questioned. In fact, the choice of models to describe the spread can vary and no testing method currently exists, as the approach set out by Long (1980) has recently been rejected by Hodgins *et al.* (1985).

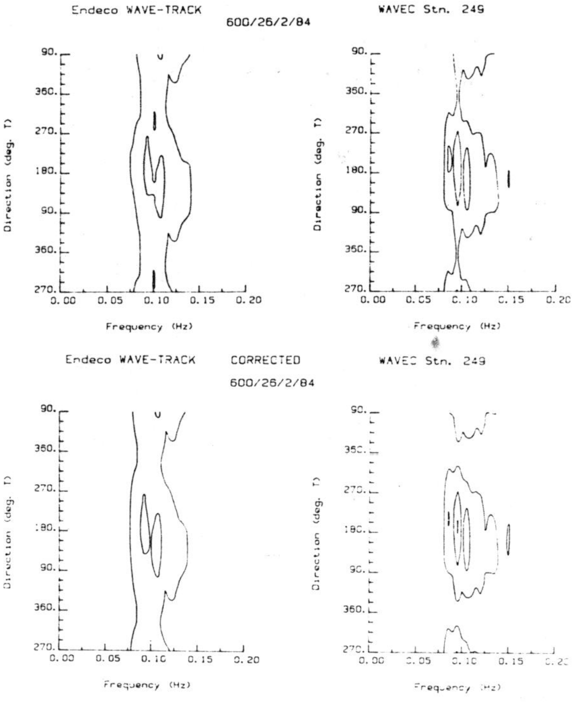

Fig. 6. Sample directional energy plots for Wave-track and Wavec data. Upper plot: expansion using method of standard Longuet-Higgins *et al.* Lower Plot: corrected using linearity analysis. (Contours represent: 1, 5 and 10 m 2/Hz energy levels.)

MODEL RESULTS

Because of the great interest in hindcast modelling, data from the Wavec buoy were compared with model results. The model used was a discrete spectral model developed by D. T. Resio at the US Army Waterways Experiment Station in Vicksburg, Mississippi and described in Resio (1981). Mixed wind values, consisting of a weighted average of kinematic winds from local drilling units and pressure-based winds, were used as input. Basically, the model consists of a discrete spectral portion (up to 0.19 Hz) through which energy is propagated and a non-propagating, high-frequency parametric model (frequency > 0.19 Hz) which handles all of the energy dissipation by forcing the tail of the spectrum to an f^{-5} equilibrium dependence. For details in the specific application of the model in this study see Penicka *et al.* (1985).

Figures 7 and 8 contain time-series plots, for two hindcast storm periods, of significant wave height, peak period and peak direction. The agreement between the observed and model results are generally quite good. Model spectra extended to longer periods, as shallow-water effects were not taken into account. By examination of individual spectra, it was observed that the model resulted in a more rapid development of the spectrum than was observed. A series of modelled and observed spectra through a storm which occurred on 27 March to 1 April 1984 are included in the Appendix. Winds during this period can be seen in Figure 4(b). The model was capable of separating wave components below 0.19 Hz, but the lack of information for higher frequencies severely limited the analysis. This was particularly true when assessing spectral de-

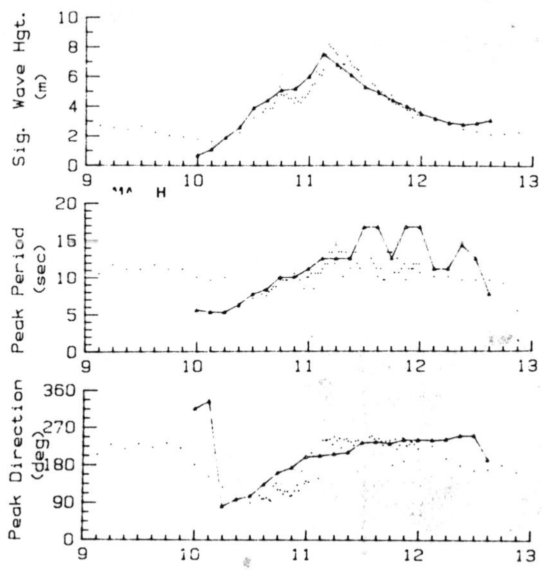

Fig. 7. Storm I time-series of Wavec (dotted) and hindcast model (solid) derived significant wave height, peak period and peak direction statistics.

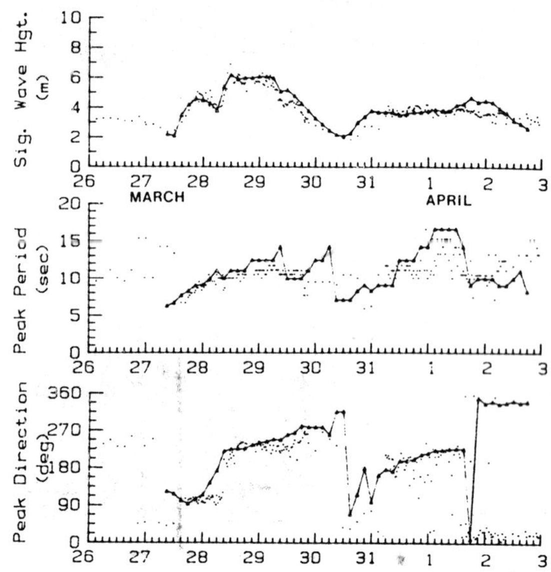

Fig. 8. Storm II time-series of Wavec (dotted) and hindcast model (solid) derived significant wave height, peak period and peak direction statistics.

velopment, as observations showed energy entering at high frequencies (>0.2 Hz) which moved to lower frequencies with time while all model results showed the initial peak developing at approximately 0.15 Hz. A too rapid development of the spectrum was shown by Komen *et al.* (1984) to be consistent with a model lacking any energy dissipation at low frequencies, such as the Resio model. For further details see Juszko (1985).

SUMMARY

In summary, both the Wavec and Wave-track buoys can be used to provide wave heave and directional statistics. However, the Wavec was generally more reliable in terms of operation and data quality. Corrections for the amplitude and phase transfer function of the Wave-track are necessary, and they greatly improve the data quality. The large directional spreads are inherent to the analysis of Longuet-Higgins *et al.*, although they are more severe in Wave-track data. The Wave-track did suffer some operational problems, perhaps reflected in the poorer quality data.

When compared with visual observations, wind directions and hindcast model results, Wavec-measured wave properties showed consistent agreement. The discrete spectral model used for hindcasting, reproduced wave statistics (significant wave height, peak period and peak direction) well, but the spectra developed too quickly, energy appeared at mid-frequencies as opposed to the higher frequencies observed in the buoy data and contained energy at much lower frequencies than was expected, given water depth. The model may be improved by including shallow-water effects, low-frequency energy dissipation and, possibly, reduced levels of non-linear energy transfer across frequency bands.

ACKNOWLEDGEMENTS

Funding for the two studies was supplied by the Marine Environmental Data Service (MEDS), Fisheries and Oceans and the Environmental Studies Revolving Fund, Department of Energy, Mines and Resources of Canada. Equipment was supplied by MEDS and Petro-Canada Inc. Numerous individuals from MEDS, Atmospheric Environmental Service, Canadian Coast Guard, Datawell, Endeco Inc., Petro-Canada Ltd and Mobil Oil Canada Ltd provided information and assistance in the performance of the two studies. We would particularly like to thank Dr R. Wilson, B. Kelly, J. Murphy and J. Minaker (MEDS), Dr J. R. Buckley (Petro-Canada Ltd) and J. Ransom (Mobil Oil Canada Ltd).

REFERENCES

Audunson, T., Barstow, S. F. and Krogstad, H. E. 1982. Analysis of wave directionality from a heave, pitch and roll buoy operated offshore Norway. *Ocean Sci. Engng* **7**(8), 291-319.

Brainard, E. C. 1982. High resolution directional wave spectra using the F.F.T. digital band pass filter method. Endeco Inc. (unpubl.).

Ezraty, R. and Cavanie, A. 1981. Evaluation of wave direction measurements using a pitch and roll buoy. *Oceanol. Acta* **4**(2), 139–149.

Hasselmann, D. E., Dunckel, M. and Ewing J. A. 1980. Directional wave spectra observed during JONSWAP 1973. *J. Phys. Oceanogr.* **10**, 1264–1280.

Hodgins, D. O., LeBlond, P. H., Dunbar, D. S. and Niwinski, C. T. 1985. A wave climate study of the northern British Columbia Coast. Volume II. Wave properties and wave predictions. *Contractor Report,* prepared for the Marine Environmental Data Service Brank, Fisheries and Oceans, Canada, 235 pp.

Juszko, B.-A. 1985a. Wave field properties obtained from directional wave buoy data (in preparation).

Juszko, B.-A. 1985b. Directional wave spectrum intercomparison study. Volume III. Intercomparison of results. *Contractor Report,* prepared for the Environmental Studies Revolving Fund, Department of Energy, Mines and Resources, Canada, 340 pp.

Komen, G. J., Hasselmann, S. and Hasselmann, K. 1984. On the existence of a fully developed wind-sea spectrum. *J. Phys. Oceanogr.* **14**, 1271–1285.

LeBlanc, L. R. and Middleton, F. H. 1982. Storm direction wave spectra measured with a single buoy. Paper presented during *Oceans 82,* 20–22 September 1982. *Conference Record,* pp. 885–892.

Long, R. B. 1980. The statistical evaluation of directional spectrum estimates derived from pitch/roll buoy data. *J. Phys. Oceanogr.* **9**, 373–381.

Longuet-Higgins, M. S., Cartwright, D. E. and Smith, N. D. 1963. Observations of the directional spectrum of sea waves using the motions of a floating buoy. In *Ocean Wave Spectra,* Prentice-Hall, Englewood Cliffs, New Jersey, pp. 111–136.

Penicka, F. X., Resio, D. T. and Worsfold, R. D. 1985. Wave directional spectrum intercomparison study. Volume II. Wave directional spectrum hindcasts. *Contractor Report,* prepared for the Environmental Studies Revolving Fund, Department of Energy, Mines and Resources, Canada, 255 pp.

Resio, D. T. 1981. The estimation of wind-wave generation in a discrete spectral model. *J. Phys. Oceanogr.* **11**(4), 510-525.

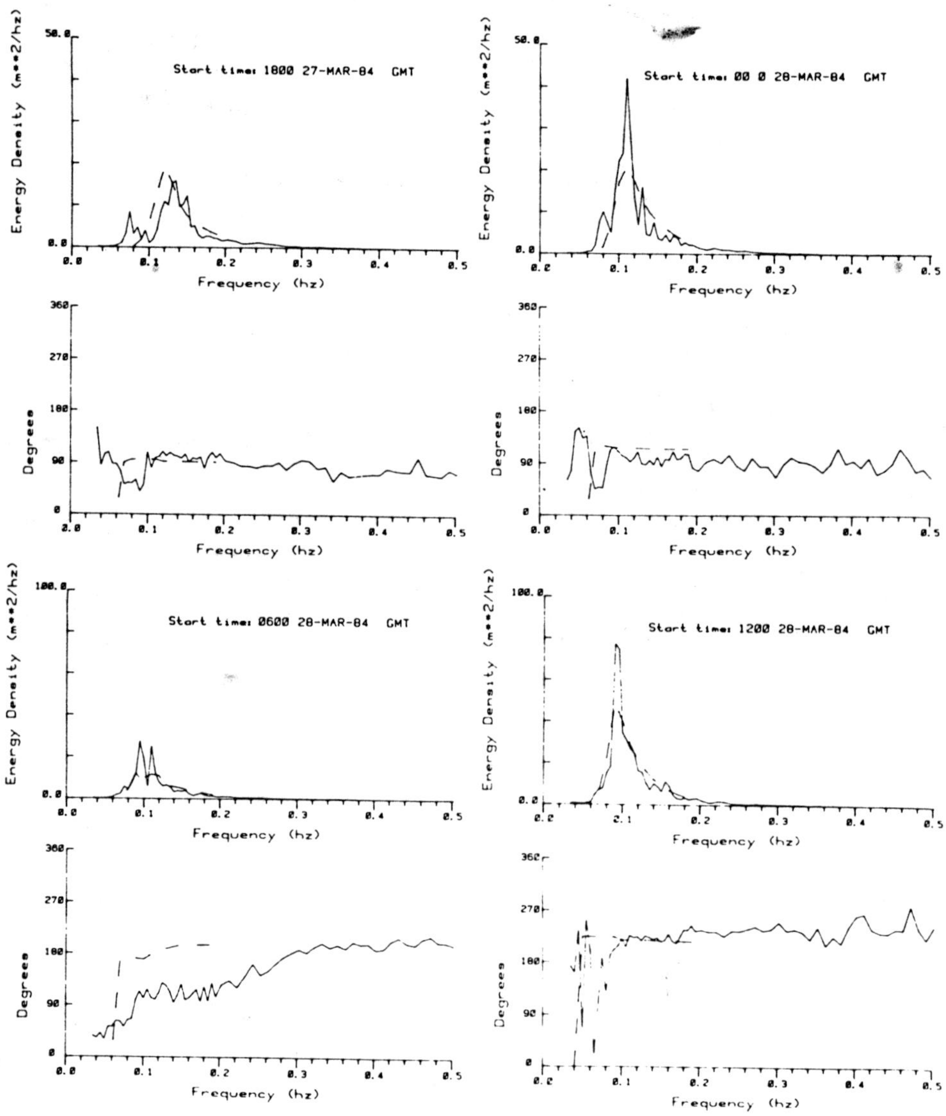
Start time: 1800 27-MAR-84 GMT
Start time: 00 0 28-MAR-84 GMT
Start time: 0600 28-MAR-84 GMT
Start time: 1200 28-MAR-84 GMT
Energy Density (m**2/hz)
Degrees
Frequency (hz)

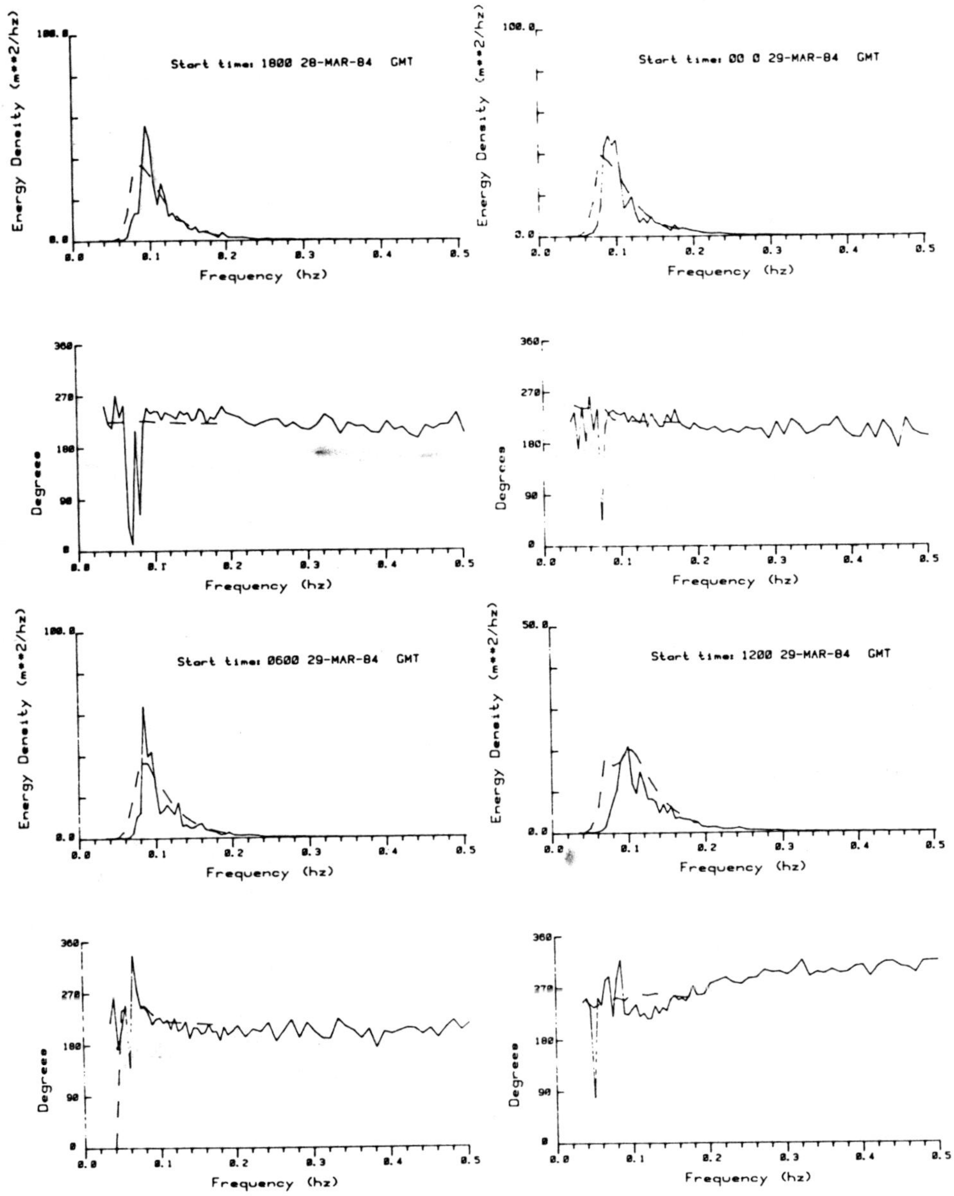
Start time: 1800 28-MAR-84 GMT
Start time: 00 0 29-MAR-84 GMT
Start time: 0600 29-MAR-84 GMT
Start time: 1200 29-MAR-84 GMT
Energy Density (m**2/hz)
Degrees
Frequency (hz)

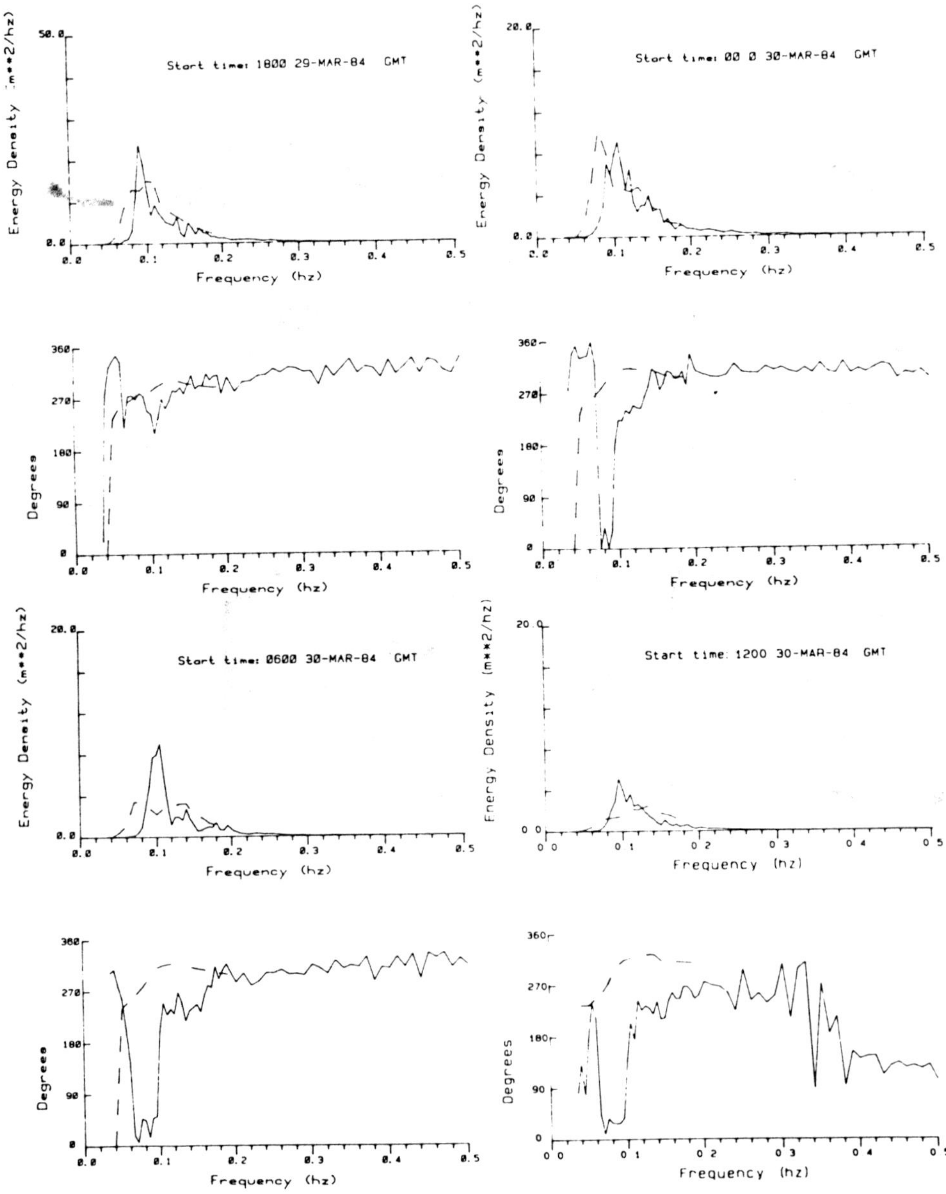
Start time: 1800 29-MAR-84 GMT
Start time: 00 0 30-MAR-84 GMT
Start time: 0600 30-MAR-84 GMT
Start time: 1200 30-MAR-84 GMT
Energy Density (m**2/hz)
Frequency (hz)
Degrees

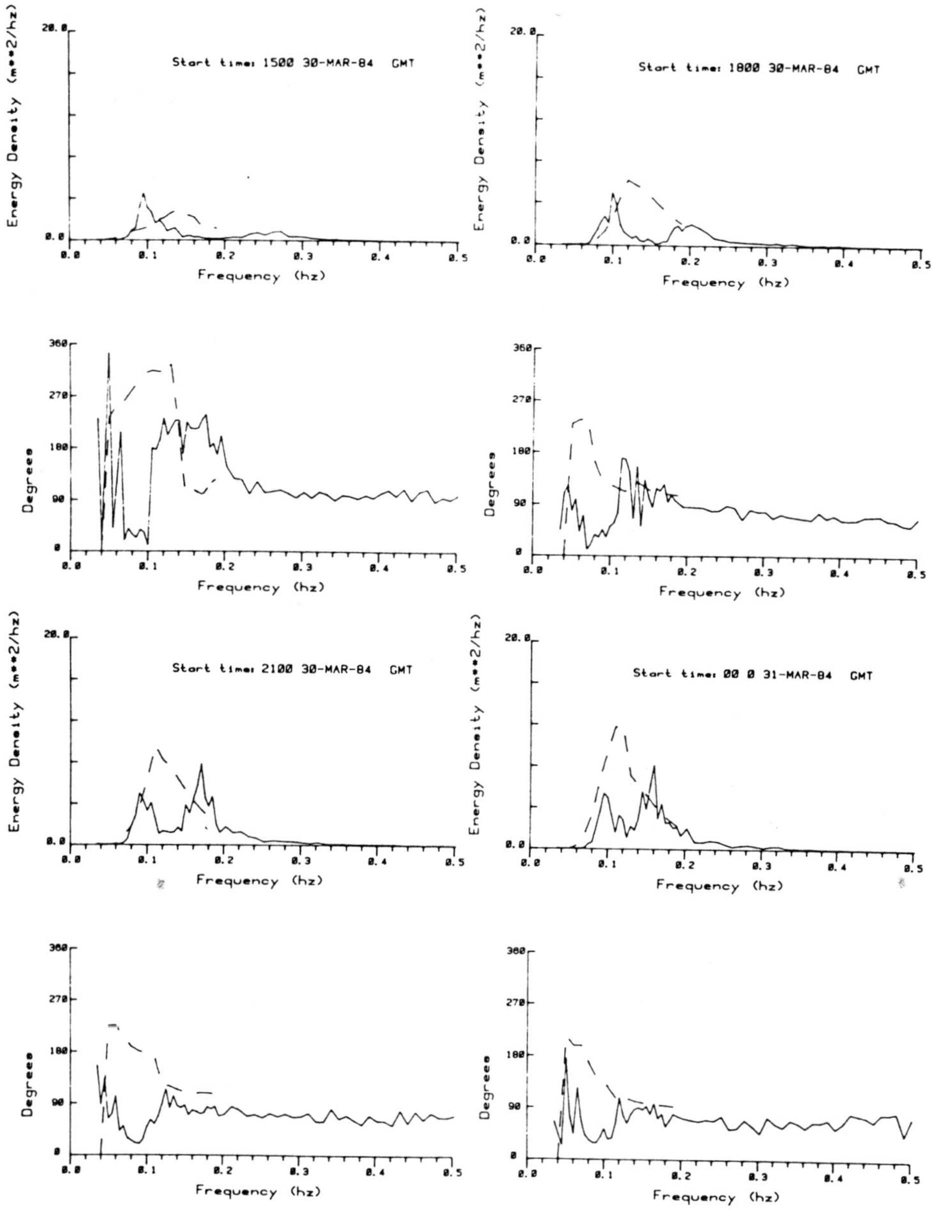
Start time: 1500 30-MAR-84 GMT
Start time: 1800 30-MAR-84 GMT
Start time: 2100 30-MAR-84 GMT
Start time: 00 0 31-MAR-84 GMT
Energy Density (m**2/hz)
Degrees
Frequency (hz)
20.0
0.0
360
270
180
90
0
0.0
0.1
0.2
0.3
0.4
0.5

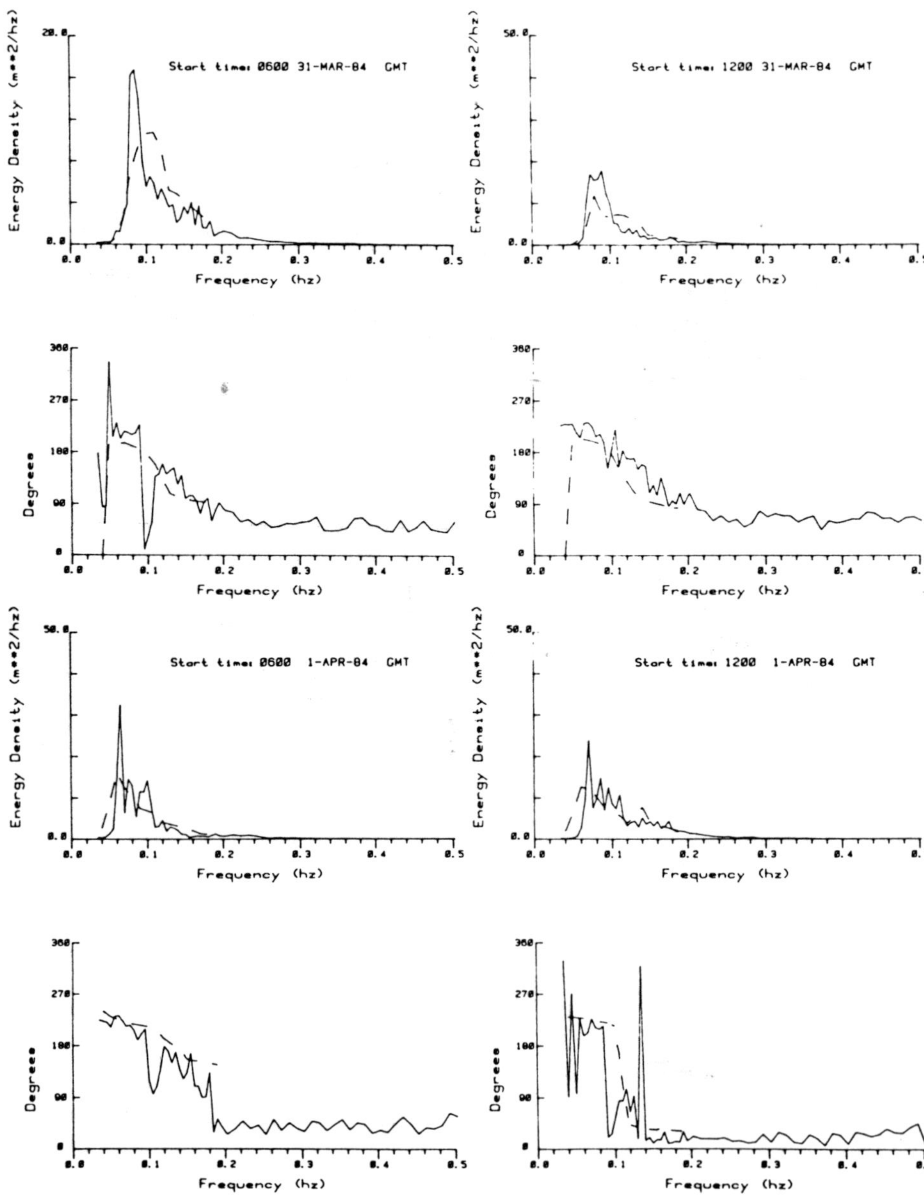
Start time: 0600 31-MAR-84 GMT
Start time: 1200 31-MAR-84 GMT
Start time: 0600 1-APR-84 GMT
Start time: 1200 1-APR-84 GMT
Energy Density (m**2/hz)
Degrees
Frequency (hz)

SPEAR-F (A WAVE HEIGHT SPECTRUM BUOY VIA ARGOS) AND THE DIRECTIONAL WAVE HEIGHT MEASURING BUOYS AT IFREMER

G Ayela, R Ezraty, J P Hue, J M Coudeville,
Institut Francais de Recherche pour
l'Exploitation de le Mer, France

Presented on **Wednesday 5 March 1986 at 15.30** in the Clarence Conference Room

EMBARGOED AGAINST DELIVERY

27

SPEAR-F (a Wave Height Spectrum Buoy via ARGOS) and the Directional Wave Height Measuring Buoys at IFREMER

G. Ayela, R. Ezraty, J. P. Hue and *J. M. Coudeville,* Institut Français de Recherche pour l'Exploitation de la Mer, BP 337, 29273 Brest Cédex, France

INTRODUCTION

The need to measure wave height and wave direction with no radio range limitation led the French Ocean Agency, l'Institut Francais de Recherche pour l'Exploitation de la Mer, to design and test new products.

The first SPEAR buoy, a wave height spectrum buoy transmitting via ARGOS, was used to evaluate the wave energy resource along the French Atlantic coast (Ollitrault and Coudeville, 1982).

The Wadibuoy (Néreides, France) uses the heave, pitch and roll Hippy sensor (Datawell, the Netherlands) and a Thomson flux gate compass, processes the data in-board and transmits energy and mean direction in 19 selected period bands via the ARGOS system. This wave directional spectrum buoy has been used in several experiments and the efficiency of the system has been proved in high sea-states. (Ezraty and Cavanie, 1981; Cavanie *et al.*, 1982; Ezraty *et al.*, 00).

In both cases, the use of the ARGOS system to transmit the processed data and to locate the buoys has been very satisfactory. On the other hand, the hard-wire technology to build the SPEAR buoy in the late 1970s led to complex equipment which has to be set and maintained by skilled people.

This paper describes these buoys; emphasis is put on the new SPEAR-F buoy.

The SPEAR-F buoy can be easily used and maintained; its acquisition and processing sub-system is Waverider-compatible. Intercomparisons between the SPEAR-F spectrum and the standard 27 MHz Waverider buoy data have been made. At-sea test results are presented. The data collected from the two wave height buoys agree very well in different sea-state conditions.

Besides this work, a new static wave directional sensor has been designed by IFREMER to eliminate the drawbacks of non-static sensors which are more fragile. It is based on three accelerometers and three flux-gate sensors. The data collected with the static sensor are very promising. An intercomparison between this sensor and the Hippy sensor (Datawell) was made during the winter of 1985/6.

THE WADIBUOY

This 2.5 m diameter, flat-bottom fibreglass toroid buoy is equipped with a Datawell Hippy 40 sensor and a Thomson flux-gate compass. The total weight, including the 1-year life battery is about 800 kg (Fig. 1). Furthermore, to prevent the buoy from capsizing a cone-shaped structure was mounted above the main body.

Fig. 1. The Wadibuoy — mooring operation.

The buoy was first fitted with a radio transmitter to send the raw data to shore. This configuration was successfully tested during two 3-month campaigns off Brittany. The received data were processed in real time, using fast Fourier transforms according to the method proposed by Longuet-Higgins *et al.* (1963). The heave and slope information which are redundant were used to compute slope transfer function of the anchored buoy: results showed that it was flat and close to unity for periods from 2.5 to 16 s (Fig. 2). (Ezraty *et al.*, 00, 1983.)

To extend the potential use of the buoy to the open ocean, the ARGOS satellite data transmission and positioning system was used, this having already proved its ability for collecting non-directional wave spectra. Since an ARGOS message is 256 bits long, the 100 000 bits of raw data needed for the analysis had to be processed on-board to obtain estimates of energy and mean directions, in 19 selected period bands. This leads to a data chain which is compatible with two successive ARGOS messages (Becq *et al.*, 1983).

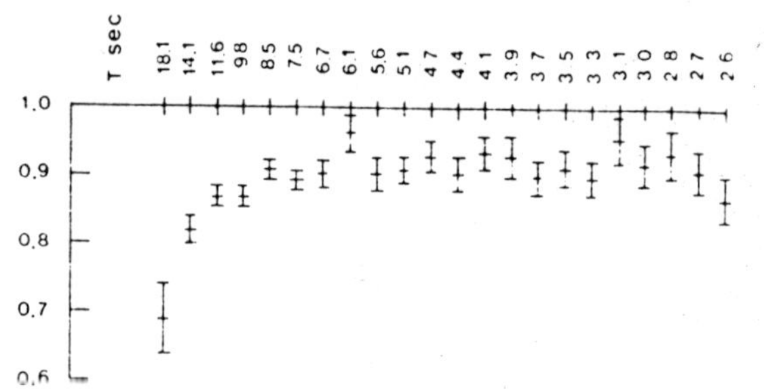

Fig. 2. The ratio of the heave energy obtained from vertical displacement to the heave energy computed from slopes.

System Overview

The system controller is a low-power RCA 1802 microprocessor; a Z80 subsystem is used for computing. The sensor outputs are sampled twice per second during 21 min 20 s with a 10-bit resolution. Raw data are converted to physical units using calibration relationships. The following sensor control values are computed: mean tilt angles, mean heave signal, mean radius of the trigonometric circle computed from the sine and cosine information of the compass. The sines of tilt angles, from the Hippy sensor, relative to the buoy axes, are converted to slopes and rotated to geographic fixed frame using the compass data. The significant wave height H_s is then computed.

The spectrum is divided into 19 period bands from 3 to 18 s; the 3 dB bandwidth is $\triangle f = 5/64 f_0$. For each period, the variance of the heave filtered signal is computed and converted into a percentage of the total variance. The "mean" component direction is then computed.

The total information (38 values plus H_s plus the four sensors control values and housekeeping parameters) is then sent to the ARGOS transmission pack.

Sea-state Measurements

A 3-month experiment took place, from August to October 1982, near the island of Ushant, at the entrance of the English Channel. The buoy was anchored using a loose mooring, in waters of nominal depth 50 m.

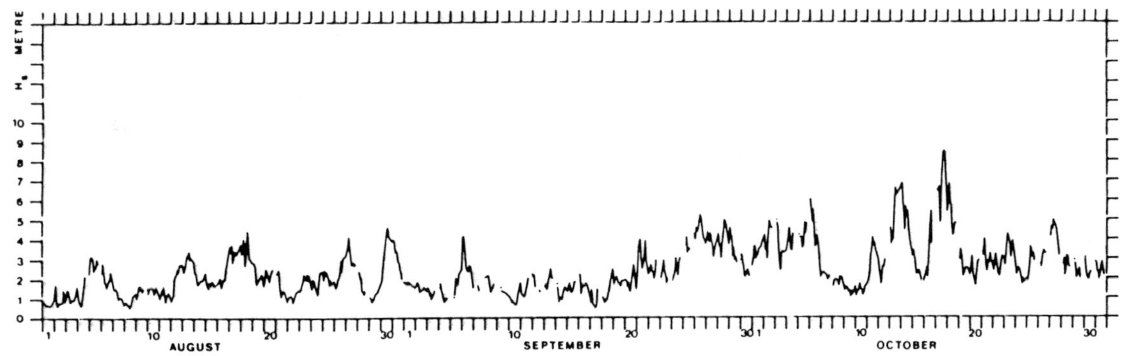

Fig. 3. Significant wave height from 1 August to 31 October 1982 at 48°26.7′N, 5°8.7′W.

Figure 3 shows the evolution of the significant wave height, for the 92 days of data. Strong sea states occurred (up to 9 m H_s) on 17 October and Fig. 4 shows a typical example of wave directional spectra transmitted via ARGOS.

Since then, the Wadibuoy has been used by IFREMER in several instances and it appeared to be very reliable. The buoy location has always been monitored thanks to the ARGOS location capability.

The Wadibuoy was the first operational directional wave spectrum measuring buoy and its "R-ratio" is one of the best of the known buoys on the market. It has been fully qualified and its ARGOS link enables the user to moor the buoy anywhere around the world.

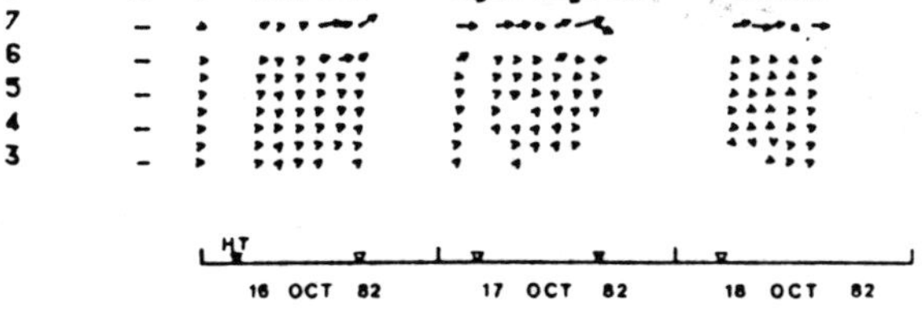

Fig. 4. Wave component directions and energies obtained during strong sea-states, showing tidal modulations in direction.

THE SPEAR-F BUOY

The acronym SPEAR-F means Spectre de l'Etat de mer via ARgos/Fast Fourier Transform (Sea State Spectrum via ARGOS/FFT).

SPEAR-F Components

(Figure 5 is a picture of the SPEAR-F package and Fig. 6 shows a schematic of the buoy.)

The *buoy hull* is the well-known 0.70 m diameter Waverider sphere of Datawell (the Netherlands) without its 27 MHz transmitter and antenna.

The *sensor* is the Datawell accelerometer fitted inside the Waverider, which delivers a

Fig. 5. The standard 27 MHz Waverider hatch-cover and the SPEAR-F package.

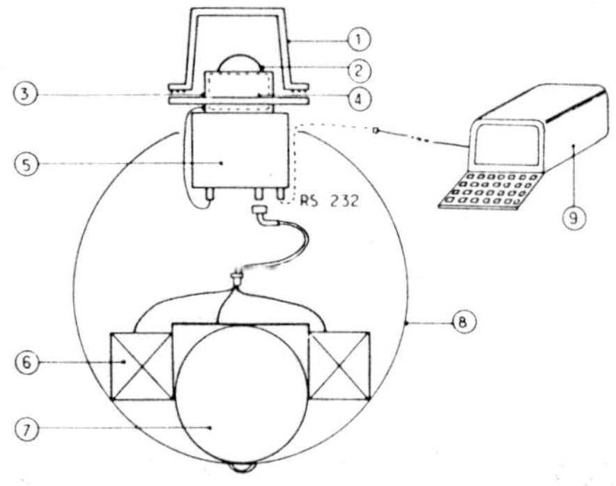

Fig. 6. SPEAR-F buoy schematic. 1, Protection hat; 2, ARGOS antenna; 3, transmitter container; 4, ARGOS transmitter; 5, acquisition and processing system; 6, batteries; 7, heave sensors; 8, Waverider hull; 9, CRT display (Minitel).

frequency output proportional to the wave height (central frequency 259 Hz, sensitivity 1.86 Hz/m).

The *transmitter* is the UHF-83 ARGOS transmitter of CEIS–ESPACE (Toulouse, France); it is made of two small circular boards (125 mm dia; height 90 mm). It can be located by the ARGOS system and can transmit up to 256 bits of data.

The *antenna* is a low cost, low wind-drag ARGOS antenna (CIT–ALCATEL, Lannion, France). A fibreglass hat has been built to protect it (TECHNICAP, Cap d'Ail, France). Its ground plane is made of stainless steel and is used to keep the buoy watertight.

The *power supply* can be made of any type of primary or secondary batteries. Duracell primary batteries (6 V, 40 Ah, type 1D 9180) have been chosen. Twenty of them can be installed in the buoy. Ten batteries power the ARGOS transmitter, the other 10 power the acquisition and processing system. They give an autonomy of more than 1 year.

The *acquisition and processing system* is the heart of the SPEAR-F. It has been designed to be compatible with the Waverider frequency output, but any type of sensor which delivers an analogue output proportional to the wave height can be used. It has only to be compatible with the analogue-to-digital converter (±2.5 V).

This is one of the main advantages of this equipment: any Waverider owner can transform his buoy into a SPEAR-F by taking out the 27 MHz transmitter and plugging-in the SPEAR-F package with the same plug. This transformation is reversible and does not affect the sensor.

Many of these components are supplied by well-known manufacturers, and will not be described in more detail.

The Acquisition and Processing System

To convert the analogue data given by the double integrated accelerometer output into a 32 × 8-bit word ARGOS message, several operations are needed. Every time the clock gives a pulse to the processor, the following sequence begins.

Acquisition

A phase look loop (PLL) whose centre frequency is 259 Hz transforms the Waverider frequency-modulated output (sensitivity 1.86 Hz/m) into an analogue signal proportional to the wave height. This output has been chosen to avoid any technical intervention on the Datawell electronic printed-circuit board and to keep a full Waverider compatibility.

An analogue-to-digital converter (ADC) digitizes this analogue output into 12-bit words at a 2 Hz sampling frequency. These words are stored in C-MOS memories after they have been converted into the 32-bit floating Zilog format to keep the overall accuracy as good as possible.

During the acquisition phase, which lasts 34 mm and 8 s, 4096 samples are stored.

FFT Algorithm

A radix 2 Sande-Tukey fast Fourier transform algorithm is used over the 4096 points, leading to 2048 energy values from 1/2048 Hz to 1 Hz, with an interval of 1/2048 Hz.

Correction and "Data Compression"

These 2048 values are corrected for the frequency-dependent transfer function of the double integrator of the Datawell sensor between 0.0426 and 0.1094 Hz (e.g. 25.27 to 9.14 s) according to the manufacturer's specifications (see the Waverider instruction manual).

To be compatible with the 32 × 8-bit word

ARGOS message, a "data compression" has to be made. The above corrected energy rays are added in frequency bands of predetermined variable width.

Given the frequency distribution of the sea surface energy and the equally spaced frequency information obtained by an FFT algorithm, the groups listed in Table 1 have been chosen (ray no. 1 is at 1/2048 Hz).

TABLE 1

Frequency bands chosen for the data compression

Band number	*Centre period of the band (s)*	*Noo of added rays*	*Added rays nos.*
1	23.14	16	81–96
:	:	:	:
21	5.01	16	400–416
22	4.74	32	417–448
:	:	:	:
28	3.28	32	609–640
29	3.08	48	641–688
30	2.62	144	689–832

A special band called the "low-frequency band" (LF band) is computed by adding the first 80 rays with a centre period of 51.2 s.

Since the energy in the bands can vary by a factor of 10^5 depending on the sea-state, and because the maximum value to be transmitted is 255 (8 bits), the range of energy has to be decreased. Instead of transmitting the computed values of E, a factor called α is computed so that the maximum value of $2^\alpha \sqrt{E_{max}}$ can be coded with the maximum significant bits.

Each "normalized" value is then transmitted as $2^\alpha\sqrt{E(i)}$, α being coded with a 4-bit resolution.

The ARGOS message

Besides the computation of the 30 normalized bands and the "LF normalized band", the power supply voltage is digitized with a 3-bit resolution and a counter is incremented every time the acquisition phase begins.

So the ARGOS message is composed of:

Words 1–30: 30 values of the normalized square roots of energy at the selected periods.

Word 31: 5 M.S. bits: "LF normalized band".
4 L.S. bits: power supply voltage.

Word 32: 4 M.S. bits: factor α
4 L.S. bits: counter value.

The full cycle (acquisition + FFT + correction and "data compression") lasts less than 1 h. The cycle recurrence is a user-selectable parameter and can be set at 1, 2, 3 or 6 h depending on the user's experiment (buoy autonomy, location of the buoy with respect to the ARGOS satellite trajectories, etc.). Figure 7 is a block diagram of this component and Fig. 8 shows the ARGOS message.

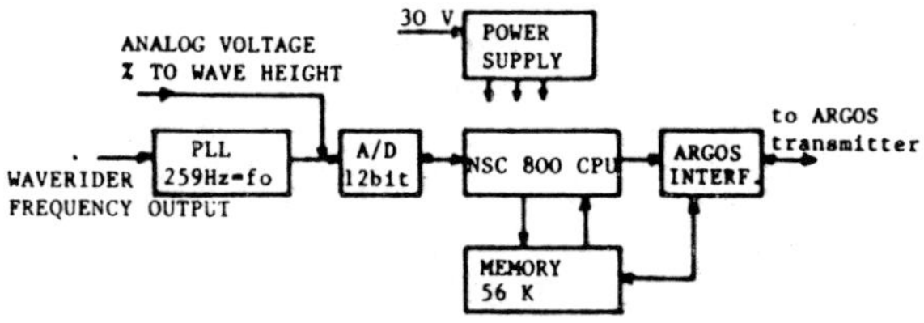

Fig. 7. SPEAR-F electronic block diagram.

23.14	19.6	17.0	15.0
13.43	12.15	11.10	10.21
9.46	8.81	8.24	7.74
7.30	6.91	6.55	6.23
5.94	5.68	5.44	5.22
5.01	4.74	4.41	4.12
3.86	3.65	3.46	3.28
3.08	2.62	LF BAND / POWER SUPPLY	α / COUNTER

Fig. 8. ARGOS message (central periods are in seconds).

Internal Tests and Controls

A set of interactive test programs has been developed to help in the calibration and the maintenance of the SPEAR-F electronic circuits. They are stored in ROM and can be activated by a standard terminal via an RS 232 serial data link. Several programs are available.

Auto-test. The RAM, ROM, input/output, ADC and ARGOS interface are checked. A specific diagnostic is given for each tested parts.

User-selectable simulation signal. Starting from basic signals such as sine, square signals with or without an offset voltage, the user can synthesize complex test signals to test the FFT module. A known internally-synthesized "test signal" can also be used to perform the same test and to control the whole buoy with an ARGOS receiving unit and the terminal.

Control and setting of:

- The frequency output of the Waverider, to check that the Waverider sensor is working properly;
- The gain of the ADC part, using an internal or external voltage reference, or the battery voltage.

Other functions are also available thanks to the "Monitor", which is stored in the ROM (step-by-step or breaking point functioning, etc.).

SPEAR-F Circuits: the IFREMER 800 Card Set

A set of cards based on the NSC 800 CMOS microprocessor has been designed by IFREMER and called the IFREMER 800 card set. These can be installed in a piece of 15 cm diameter tubing and have low power consumption. Many of them are operational (acoustic data transmission, CCD camera acquisition, image processing, heat flux measurement, etc.) and their reliability has been proven.

Five of these have been used to build the SPEAR-F electronic circuits:

- CPU (NSC 800) with 6k memory;
- acquisition (eight channels, 12 bits per channel);
- 56k memory board;
- power supply board and
- ARGOS and Waverider interface.

The mean power consumption is 40 mA/5 V. The PASCAL language was used to develop the specific software. A CRT display (ref. 254 CIT Alcatel, Paris, France) with an RS 232 serial data link has been chosen to be used with the SPEAR-F buoy.

Laboratory and At-sea Tests

The testing laboratory of IFREMER operates a device which simulates sine movements within the wave range (height, 0–4 m; period, d.c. to 1 s at reduced amplitude). This apparatus has been used to test the entire buoy under different conditions. It has been proven that, because of the "normalization", very low motions can be analysed.

To validate the SPEAR-F data, a standard 0.90 m diameter, 27 MHz Waverider has been moored near the SPEAR-F buoy (48.231°N, 4.730°W), close to the Centre Océanologique de Bretagne (Brest, France).

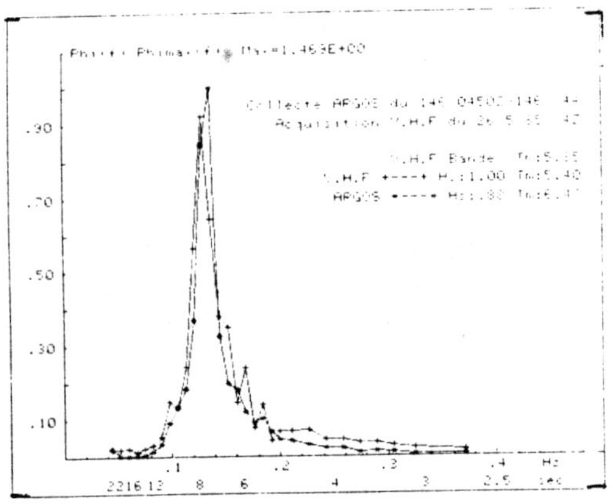

Fig. 9. SPEAR-F and 27 MHz Waverider spectrum intercomparison ($H_s \simeq 1$ m).

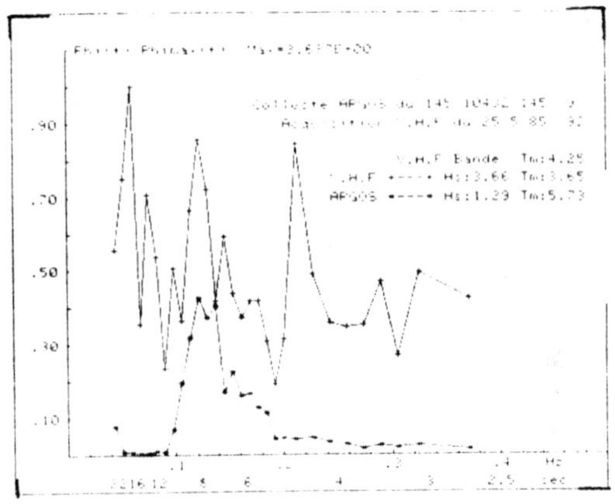

Fig. 10. 27 MHz radio interference on the Waverider data. Good SPEAR-F spectrum.

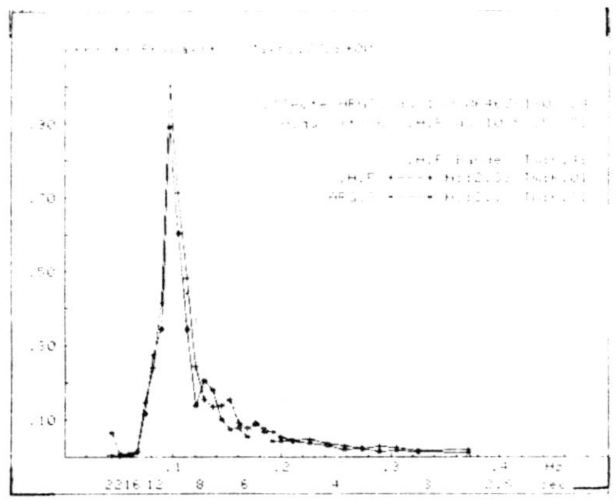

Fig. 11. SPEAR-F and 27 MHz Waverider spectrum, intercomparison ($H_s \simeq 2$ *m)*.

The SPEAR-F ARGOS messages were collected by interrogating the ARGOS processing centre in Toulouse (France); a 27 MHz Waverider receiver (WAREP) was connected to a data acquisition and processing system. Data reduction was achieved using a constant bandwidth of 16/2048 Hz from 1/2048 Hz to 1 Hz and then summing to obtain the same format as the one used in the SPEAR-F. Significant wave height H_s and mean period T_m are computed.

Figures 9 and 10 show the typical spectra which were obtained at very low and medium, sea-states. Figure 11 shows the "0.90 m Waverider spectrum" when radio interferences occur on the 27 MHz band. The SPEAR-F spectrum is not influenced by this kind of radio interference and can be received directly by means of an ARGOS receiving unit if the buoy is within radio range of the receiving antenna (400 MHz).

For this experiment, the full processing capacity was chosen and a new spectrum was computed every hour. Different sea-state conditions have been recorded (swell only, and swell and wind sea). The shapes of the spectra of the two buoys agree very well under these different conditions. The values of H_s *and* T_m are very close.

Future Uses of the SPEAR-F

This NEW SPEAR-F buoy has been designed by IFREMER to help in the calibration and the validation of satellite altimeters such as that of ERS 1, within the Toscane Project (Ezraty and Ayela, 1985).

It can be used by any Waverider owner with no complex change because of the Waverider plug-to-plug compatibility. It can solve the 27 MHz radio interference problem in coastal areas and eliminates the need of a costly, manned receiving and processing shore station.

ARGOS data collection can be optimized by the location of the buoy, which helps in the retrieval when the buoy has "left" its mooring line or when the mooring line has drifted away.

About 12 SPEAR-F buoys will be ordered by IFREMER, which has sub-contracted the manufacture of these buoys to Néréides (Orsay, France). One will be used as an operational product during the winter of 1985/6.

THE STATIC WAVE DIRECTIONAL SENSOR

Everybody who uses sensors looks forward to static ones because of their many advantages.

The Agence Nationale pour la Valorisation de la Recherche (ANVAR, France) has partially funded the design of a static wave directional sensor. A patent has been taken out for the product (Patent No. 82 175 84, 10 October 1982).

The Principle of this Sensor

The sensor is composed of three orthogonally mounted accelerometers (Schaewitz) and of a triaxial magnetometer (Thomson, France) whose axes are collinear with the axes of the accelerometers. The sensor is rigidly mounted at the rotation centre of a toroidal buoy, the z-axis being the axis of symmetry of the buoy. It is supposed that this buoy perfectly follows the sea surface, which is true for the Wadibuoy hull.

To transform the measurements made aboard the buoy into values related to a geographical referentiel (Ox, Oy, Oz), a reference has to be found. The "Earth magnetic field" vector can be taken as a reference. The following parameters are measured:

B_x, B_y, B_z: the Earth's magnetic induction components and

Γ_x, Γ_y, Γ_z: the acceleration components of the centre of rotation of the buoy.

The data processing consists of the computation of a set of representative parameters such as the unit vector of the vertical axis, the vertical speed of the buoy, the wave slope and the wave direction. It can be shown (Hue, in press) that, using the triaxial magnetometer data, it is possible to find the roll and pitch angles of the buoy. It is then possible to compute the wave slope; the acceleration of the buoy, which follows the sea surface, is measured by the accelerometer whose axis is the vertical axis of the buoy.

Validation of the Principle

In order to validate the principle and the theoretical aspects of the sensor, a prototype has been built and tested at sea. The raw data were recorded and processed ashore. The good results obtained led to the specification of the preprocessing system. A prototype has been built, which includes:

the sensor itself (three accelerometers + triaxial magnetometer);

the processing system to end with the heave, pitch, roll and heading data, the same type of data as is given by the Hippy sensor;

the post-processing system to compute the wave directional spectrum using a digital band-pass recursive filter technique to obtain estimates of energy and mean direction in 19 selected period bands (Ezraty *et al.*, 1983);

an ARGOS transmitter to send the computed data to shore.

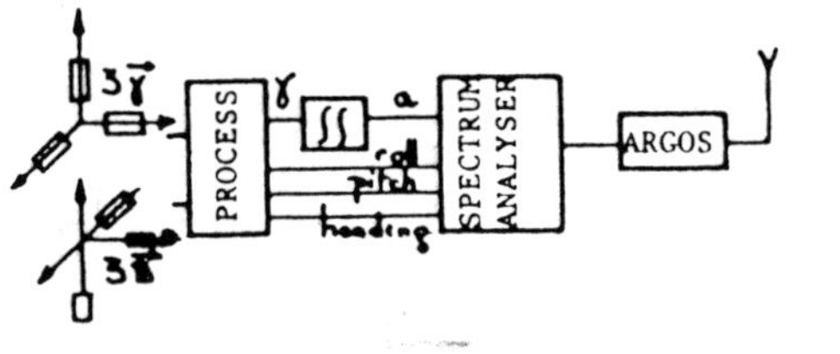

Fig. 12. Block diagram of the static sensor system.

Figure 12 shows a block diagram of the prototype. A Wadibuoy hull (Néréides, France), the 2.5 m diameter, flat-bottomed, fibreglass toroidal buoy has been equipped with two parallel data acquisition and transmission systems.

The first system includes the Hippy sensor (Datawell), the post-processing system described in Ezraty et al. (1983) and an ARGOS transmitter.

The second system is the new prototype of the static wave directional sensor, and its processing, post-processing and ARGOS system. The only common component was the power suppy.

Figure 13 shows the buoy just after the deployment.

Fig. 13. The Wadibuoy with the static wave directional sensor after deployment.

At-sea Test

This buoy has been moored near Brest. The two ARGOS transmitters were supposed to transmit the data of the two parallel systems for directional spectrum intercomparison.

Unfortunately, the ARGOS transmitter of the "Hippy Channel" which was going to be used as the reference channel, failed just after the vessel departure. It has since been impossible to compare the two channels. This kind of failure was the first one experienced in the use of this Wadibuoy.

The data which have been sent by the static sensor system are very promising, but cannot be compared with the reference, because of the failure. The buoy will be moored during the winter of 1985/6 to get spectra from the two systems; intercomparisons will then be made to validate the static sensor.

CONCLUSIONS

The Wadibuoy is fully operational and can be used to make scientific studies or to help in industrial works. it is one of the best buoys available on the market, with a very good "*R*-ratio".

This paper insists on the new SPEAR-F product which is presently operational and which will be intensively used by IFREMER. It has been proven by at-sea intercomparisons that the computed and transmitted data are valid. The plug-to-plug Waverider compatibility enables any Datawell buoy owner to transform a limited range buoy easily into a no range-limitation buoy with the biggest advantage of obtaining the wave height spectrum without maintaining any shore station.

The validation of the new static wave directional sensor and of its processing system has begun. The effort will go on, and an intercomparison campaign will take place during the winter of 1985/6. The industrialization of this sensor will end with a product which will have many advantages over the presently available sensors.

REFERENCES

Becq, D., Ezraty, R. and Hainsselin, J. P. 1983. Wave directional and amplitude measuring system using ARGOS positioning and transmitting capabilities. In *ARGOS Users Conference,* London, 27–28 September.

Cavanie, A., Ezraty, R. and Gouillou, J. P. 1982. Tidal current modulations of wave directional spectra parameters measured with a pitch and roll buoy west of Ushant in winter. In *First International Conference on Meteorology and Air/Sea Interaction of the Coastal Zone,* The Hague, 10–14 May.

Ezraty, R. and Ayela, G. 1985. On board spectral sea-state analysis — the SPEAR-F buoy inputs to ERS 1 — calibration and validation phase. In *ARGOS Users Conference,* Kiel, May.

Ezraty, R. and Cavanie, A. 1981. Evaluation de la mesure de la direction des vagues à partir des données d'une bouée instrumentée. *Oceanol. Acta,* **4**(2), 139–149.

Ezraty, R., Ollitrault, M. and Ayela, G. 1983. A wave directional spectrum buoy using ARGOS. In *Proceedings of Oceans,* 23 August – 1 September.

Ezraty, R., Cavanie, A. and Gouillou, J. P. Système de mesure de la houle directionnelle. Resultats de campagnes hivernales aux abords d'Ouessant. In *Rapports Scientifiques et Techniques du CNEXO.*

Hue, J. P. (in press) Un nouveau capteur statique de houle directionnelle. *Rapport Technique de la Direction à l'Ingenierie et à la Technologie.*

Longuet-Higgins, M. S., Cartwright, D. E. and Smith, N. D. 1963. Observations of the directional spectrum of sea waves using the motions of a floating buoy. In *Ocean Wave Spectra,* Prentice-Hall, Englewood Cliffs, New Jersey.

Ollitrault, M. and Coudeville, J. M. 1982. Evaluation de la ressource énergétique des vagues sur la facade atlantique francaise. In *Conference Utilisateurs ARGOs,* Paris, 20–22 April.

Part VII

Geology/Geographics/Geotechnics

SEALION: A SNELLIUS-II EXPEDITION AUTOMATIC SYSTEM FOR IN-SITU GEOTECHNICAL TESTING IN WATER DEPTHS OF 6000M

J van der Wal, Free University of Amsterdam;
C N van Bergen Henegouw, University of Amsterdam; H G Hoogendoorn, A F Richards, Fugro BV, The Netherlands

Presented on **Thursday 6 March 1986 at 10.00** in the Norfolk Conference Room

EMBARGOED AGAINST DELIVERY

28

Sealion: A Snellius-II Expedition Automatic System for In Situ *Geotechnical Testing in Water Depths of 6000 m*

*Jan van der Wal**, Free University, Amsterdam, The Netherlands, *Cok N. van Bergen Henegouw,* University of Amsterdam, The Netherlands, *Hage G. Hoogendoorn* and *Adrian F. Richards,* Fugro BV, Leidschendam, The Netherlands

INTRODUCTION

Deliberations between the Dutch and the Indonesian governments led to the organization of the Snellius-II Expedition, which took place in 1984 and 1985 in Indonesian territories. The Netherlands Council of Oceanic Research (NRZ) is the Dutch coordinating agency. Cooperation between the Free University of Amsterdam and Fugro BV personnel led to the development of a Geotechnical Project, as part of the NRZ's scientific programme, to be included on one of the geological cruises in the eastern Banda Sea. This region is the location of three active plate boundaries and offered a unique opportunity for geotechnical investigations. Previous studies in other active margin areas had showed the influence of tectonic forces on the consolidation (dewatering) of the sediments (Carson, 1977). Excess pore-water pressures were thought to play an important role in these processes (von Huene and Lee, 1982). To derive in situ information on pore pressures, and to provide in situ shear strength data for comparison with laboratory data obtained from piston cores, an automated piezocone probe system, called the Sealion, was conceived. This was part of the spin-off from the feasibility study for the Deep Ocean Research Apparatus (DORA) (Richards, 1985) and partly from other corporate research and development. The development and testing of the Sealion (Table 1) in late 1984 and early 1985 for the NRZ is the subject of this chapter.

DEVELOPMENT OF THE SEALION

The initial design of the Sealion was influenced by three principal factors: (1) the environmental conditions, (2) the ship's handling capabilities, and (3) the engineering requirements. The maximum water depth in which the Sealion would be deployed was fixed at 6000 m. At this depth, the experience of the Netherlands Institute for Sea Research was that it was impractical to use an electromechanical Kevlar (aramid fibre) cable because the wires broke prematurely when the cable was tensioned. Only a purely mechanical Kevlar cable would be available on the NRZ ship, the *Tyro.* This restraint required the design of a fully automated

* Present address: NAM, Assen, The Netherlands.

TABLE 1
Sealion specifications

Dimensions	base	2.4 × 2.4 m
	height	6.2 m
	effective bearing area	3.4 m
	extra bearing plates	2.7 × 0.3 m
	optional skirt	2.4 × 0.5 m
Weight	in air	1400 kg
	submerged	900 kg
Conditions	maximum water depth	6000 m
	temperature range	0–40°C
Drive unit	type of drive	drive wheels
	penetration depth	5 m
	penetration force	5.7 kN
	penetration speed	50 mm/s
	capacity	6 penetrations
	motor and pump	2 kW
	batteries (2)	24 V, 230 Ah each
Piezocone	cone diameter	44 mm
	cross-sectional area	15 cm 2wb
	cone-resistance range	0–1 MPa
	pore-pressure range	0–600 kPa
Automatic control	Programmable Logic Controller (PLC)	
Data storage	eight-channel solid-state recorder with 64 k × 12 bits of RAM memory	
Sensors for	inclination full penetration full retraction bottom contact	
Acoustic telemetry	12 kHz pinger with adjustable pulse intervals for different functions	

machine without direct control from the surface. The Sealion would be designed having the capability (1) of establishing its degree of inclination on the seabed, (2) of automatically penetrating and retracting, and (3) of storing data in a solid-state recorder attached to the frame. The Sealion would also be capable of acoustically notifying the ship in which mode it was operating. Using these design criteria, the following list of basic elements was formulated.

Mechanical elements: frame, drive unit and powerpack (power supply and pump unit).
Electronic elements: control and data storage unit, acoustic transmitter, piezocone, inclinometers and penetration sensor.

Those elements having less common usage in geotechnical ocean engineering are emphasized in this chapter.

Frame

The ship's hoisting capabilities and deck space confined the base dimensions of the Sealion to 2.4 × 2.4 m and the maximum weight to about 1000 kg when submerged in seawater. A desire for a penetration depth of 5 m necessitated a frame height of about 6 m for the intended configuration. Stability calculations indicated that for very soft soils extra bearing capacity would be desirable. Since a large and permanently installed baseplate would reduce hoisting speed, extra bearing plates would be made an option for use in very soft sediments. An optional skirt having a penetration of 0.5 m was designed to avoid horizontal movement in very soft sediments on slopes.

Special attention was paid to the mounting place of the hoisting lugs because of the rather high length-to-width ratio of the steel frame. The best position for these was found to be about 2.5 m above the baseplate, which is well above the centre of gravity.

Drive Unit

Early in the Project it was decided to abandon the concept of a double-acting hydraulic jack for the drive system in favour of a continuous wheel-drive system. Richards (1985) discusses the background of wheel-drive units and gives the design parameters for the DORA, which, as previously stated, was one of the conceptual predecessors of the Sealion. The Sealion wheel-drive system provided friction between the wheels and the 36 mm diameter test rods using a large adjustable spring. The wheels are driven by a gearbox and a hydraulic motor. These were filled with oil and externally connected to a flexible oil reservoir located on the frame. A drive speed of 50 mm/s was selected over the customary 20 mm/s for reasons of convenience.

Powerpack

An electrohydraulic drive system was chosen for the powerpack because it was an easy and highly adaptable system to develop in the very short time period available. While the pressure-compensated use of DC motors is not common in very deep ocean technology, they have previously been used in similar applications (e.g. Inoue et al., 1978). It was decided to use a 24 V DC motor as a simpler and more expedient alternative to having a more customary DC to AC converter coupled to a synchronous AC motor. Tests were made of the DC motor operating immersed in oil in a test chamber pressurized to 60 MPa over an extended time. There was insignificant brush fouling and the 50% reduction of efficiency could easily be countered by using a motor having an appropriately higher power output. The DC motor was directly connected to the gear pump and a normal hydraulic control system was incorporated for controlling the machine.

Ordinary lead-acid car batteries were used in a pressure-compensated mode as the power source. Tests of battery performance conducted at 2°C and subjected to a hydrostatic pressure of 60 MPa indicated that 80% of the expected output was lost under these conditions for the batteries that were chosen. (Subsequent testing in late 1985 suggested that this initial test may not have been valid because of oil fouling the plates of the battery when it was tested.) As a result, with two 12 V 230 Ah batteries, power was believed to be available for only six full penetrations at 6000 m. The two batteries and their individual cell pressure-compensation equipment, the DC motor, the pump, and the valves were enclosed in a pressure-compensated, lightweight steel container filled with light hydraulic oil. This container also served as an oil reservoir for the hydraulic motor in the drive unit.

Control and Data Storage Unit

A programmable Logic Controller (PLC), contained in a pressure-protected steel cylinder, was designed to control the Sealion's automatic system, including all sensors, directional valves for oil flow, acoustic transmission and operation of the data recorder. A built-in voltage monitor was capable of stopping penetration when a preselected point of minimum drive-battery power was reached, at which time automatic retraction would immediately occur.

An eight-channel solid-state recorder (Schaap and Liefting, 1984) was selected to collect the data. Four channels were assigned to piezocone data, two to record the inclination in two directions, one to monitor the penetration and one to record time. The data recorder was programmed to record all channels every 1 cm of penetration during the penetration phase. In all other operating phases, the channels registered information every 30 s. In this manner, the behaviour of the entire system could be monitored from the time the Sealion left the ship until the time it was brought back on board.

The Sealion could automatically move from one mode of operation to an other based on sensor information: vertical orientation, penetration, pore-pressure dissipation test (optional), and retraction. A deep-sea pinger signalled, in preselected pulse intervals, the most important operation modes to the ship. In this way, the operators would be informed of the machine's status and would be able to react as required.

Piezocone, Inclinometers and Penetration Sensor

A newly developed deep-sea piezocone was built for this Project to measure total pore pressure and cone resistance. The design included having the transducers automatically zeroed before the Sealion started penetration. The piezocone was pressure tested at 60 MPa. The transducers performed well at this pressure for short-term geotechnical testing; however, the pore-pressure transducer proved to be insuffiently stable to be used for dissipation tests of extended duration.

A bottom-touch sensor under the base plate was designed to trigger the penetration cycle. Before penetration, a check on the inclination in two horizontal directions would ensure that the vertical inclination of the Sealion was less than

10°. Time was programmed to allow the machine to stabilize itself on the seabottom before starting penetration. The inclinometers checked the verticality of the rig during penetration and provided a signal to stop penetration and retract the test rods and piezocone if the inclination exceeded 15°. Automatic retraction was also initiated under any of the following circumstances: when a predetermined force on the cone of 5.7 kN was reached, when the bottom-contact lever lost contact with the seabed and after a certain time had elapsed without full penetration being attained.

Vertical movement of the test rods during penetration and retraction was sensed from a wheel that generated magnetic pulses. Magnetic switches switched the drive system when full penetration or full retraction of the test rods was reached.

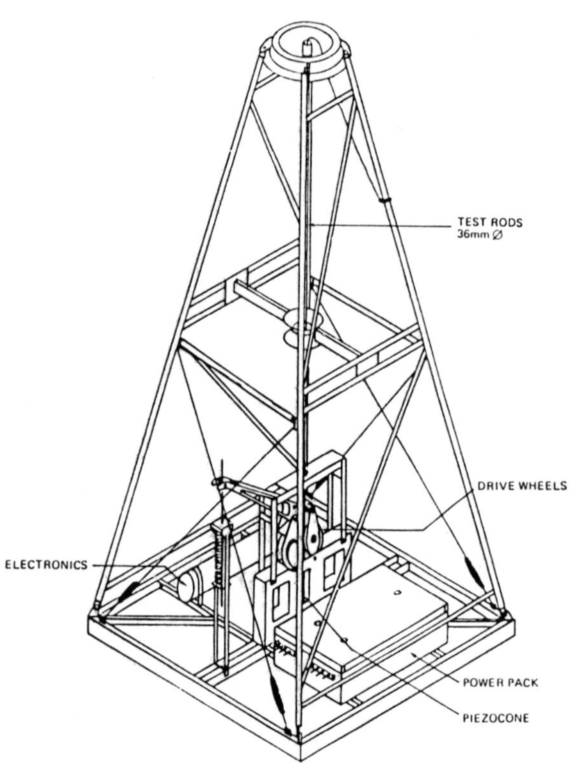

Fig. 1 Sealion major component arrangement

TESTING AND OPERATIONS

Tests of the completed Sealion (Fig. 1) on land and in Rotterdam harbour indicated that the basic elements functioned as designed. Minor problems with the seabed detector and with test-rod slippage in the drive-wheel assembly and subsequent deformation of test-rod joints caused by the slippage were corrected by enlarging the contact area of the detector and adjusting the drive-wheel tensioner, respectively. Following the harbour tests, the Sealion was dismantled and the frame cut for air shipment to Jakarta and Ujung Pandang, Indonesia, where the rig was reassembled on the quay and aboard the NRZ's research vessel, the *Tyro* (Fig. 2).

Fig. 2 Photograph of the Sealion with test rods and piezocone partly extended on the NRZ ship, *Tyro*, in the Banda Sea, Indonesia

The first test was made in May 1985 under ideal weather and sea conditions in a small sedimentary basin, having a water depth of 4200 m, in the central Banda Sea. The *Tyro*, a converted cargo and animal-transport vessel, was very noisy. As a result, the pinger's 91 dB 12 kHz acoustic signal could not be detected at depths greater than about 3000 m. However, since the wire length to the bottom had been previously determined when box and piston cores were raised, the test proceeded without further feedback from the Sealion. The Sealion was gently set on the flat seafloor and the sensitive electronic dynamometer registering cable tension indicated a high probability that the rig remained vertical after touchdown. Two penetration-retraction sequences were made at two slightly different seabed locations before the Sealion was raised to the surface.

The Sealion was deployed using a Kevlar cable 15 mm in diameter, which was also used for piston coring. Because of concern about cable chafing, about 100 m of steel wire attached to a frame of three glass floats was inserted between the compliant braided-rope sling, which was attached directly to the lugs on the frame, and the ship's Kevlar cable. It was thought by the ship's experienced winch and coring operator that this rigging method would minimize the possibility of cable chafe. Unfortunately, the Kevlar cable wrapped around the float assembly and, as this assembly was being lifted out of the water, the Kevlar cable broke. It was assumed that upon the unexpected return to the seabed the Sealion made one final penetration and retraction. No performance data were retrieved as a result of this accident. There was no back-up equipment.

EPILOGUE AND CONCLUSIONS

A number of design elements developed for the Sealion were incorporated into a new machine, called the Seasprite, which was built in mid-1985. It was successfully deployed on a number of commercial pipeline operations in deep water in the North Sea during 1985. This demonstrated the operational suitability of the frame assembly, the reliable functioning of the wheel drive, and the general arrangement of various mechanical and electrical components. The Sealion tests had previously demonstrated the suitability of pressure-compensated DC motors and the performance of piezocone transducers under very great hydrostatic pressures.

ACKNOWLEDGEMENTS

This research has been carried out as a part of the Snellius-II Expedition, organized by the Netherlands Council of Oceanic Research (NRZ) and the Indonesia Institute of Science (LIPI). The writers wish to express their gratitude to the NRZ, and particularly to Dr J. Stel, for major financial support of the Free University-Fugro BV Marine Geotechnical Project. Additional financial support was made by Fugro BV. Prof. J. E. van Hinte, Free University, provided encouragement and interest. Ir. W. Schrier, Ing. L. Schaap and Ing. J. Wijsman, Fugro BV, made substantial contributions to the development of the Sealion. The first two engineers, and Ing. H. Zuidberg, are thanked for a critical review of this paper.

REFERENCES

Carson, B. 1977. Tectonically induced deformation of deep-sea sediments off Washington and northern Oregon: mechanical consolidation. *Marine Geology* **24**, 289–307.

Huene, R. von and Lee, H. 1982. The possible significance of pore fluid pressures in subduction zones. In: *Studies in Continental Margin Geology,* Memoir 34 (Eds. J. S. Watkins and C. L. Drake). American Association of Petroleum Geologists, Tulsa, Oklahoma, pp. 781-791.

Inouc, E., Maruyama, S. and Yuasa, M. 1978. Submerged rock drill, 'MD-500H'. *Fifth International Ocean Development Conference Preprints,* B-2, pp. 41–46.

Richards, A. F. 1985. *In Situ* testing and sampling offshore in water depths exceeding 300 m. In: *Offshore Site Investigation,* Graham & Trotman, London, pp. 129–163.

Schaap, L. H. J. and Liefting, J. H. 1984. Geotechnical applications of solid state recorders. In: *Field Measurements in Geomechanics,* Vol. 1 (Ed. K. Kovari). A. A. Balkema, Rotterdam, pp. 325–333.

EFFECT OF TIME AFTER SAMPLING ON LABORATORY TEST RESULTS

T Lunne, A Kleven, Norwegian Geotechnical Institute, Norway

Presented on **Thursday 6 March 1986 at 10.30** in the Norfolk Conference Room

EMBARGOED AGAINST DELIVERY

29

Effect of Time After Sampling on Laboratory Test Results

Tom Lunne and *Arne Kleven,* Norwegian Geotechnical Institute

EXTENDED ABSTRACT

Until very recently, standard procedure in North Sea oil investigations has been to perform mostly simple index tests on soil samples in the offshore field laboratory. More sophisticated laboratory testing on samples is carried out in the onshore laboratory with the purpose of determining soil design strength and deformation parameters for foundation design. These onshore laboratory tests are normally carried out one to several weeks, sometimes months, after sampling (see Andresen et al., 1979).

It has not been clear what effects the transport of the samples to the onshore laboratry and the age of samples have on the derived soil properties. Until recently this unknown effect has been considered both theoretically and in the laboratory (e.g. Kirkpatrick and Khan, 1984).

Over the last couple of years several major North Sea soil investigations have been carried out with a comprehensive programme of offshore laboratory testing including:

(1) radiography of samples in the sample tubes;
(2) consolidated and unconsolidated undrained triaxial tests;
(3) consolidated constant volume direct simple shear tests; and
(4) constant rate of strain consolidation tests.

Radiography, direct simple shear tests and constant rate of strain tests have been performed in special containers built at NGI and brought on board the drilling vessel (Lacasse *et al.*, 1984; Sandbaekken *et al.*, 1985). Offshore triaxial testing has been provided by McClelland and Fugro.

For one particular job at the Troll East Field in the Norwegian Sector of the North Sea all of the above special tests were performed both offshore and onshore on parallel samples. For this job performed in the Summer of 1984 Norsk Hydro contracted NGI to be main soil investigation consultant while McClelland Ltd was subcontracted by NGI to perform sampling, offshore triaxial testing and *in situ* testing.

Special care was taken to test samples offshore quickly after sampling and to perform onshore tests on selected parallel samples (Amundsen *et al.*, 1985).

The purpose of performing the advanced

offshore laboratory testing programme was to:

(1) check quality of samples offshore so that, if necessary, sampling procedures and equipment could be modified;
(2) facilitate presentation of preliminary soil design parameters during and shortly after the end of field work;
(3) check the effect of transport and storage time on laboratory-determined strength and deformation soil parameters;
(4) define stress history and strength characteristics so that the consolidation stresses and procedures to be used for the onshore testing are as correct as possible. Usually very little time is available from end of the field work to when the laboratory report must be submitted. In the present paper emphasis is placed on the effect of storage time.

Based on a comparsion between results of onshore and offshore tests it can be concluded that it has not been possible to find any significant differences between the offshore and the onshore test results with regard to the following parameters:

- water content;
- volume change in consolidation to p_o' in the CRSC tests;
- preconsolidation pressures determined by Casagrande and Janbu methods from the CRSC test results;
- constrained modulus in the first loading from p_o' to p_c', the dimensionless modulus number, m, and the coefficient of consolidation, c_v, in the normally consolidated range ($>p_c'$) for the CRSC tests; and
- undrained shear strength from the direct dimple shear tests and the triaxial tests.

There appear to be some small differences for the following factors:

- the soil unit weight was on the average slightly lower for the onshore tests; and
- the shear strain to failure in the DSS tests was on the average slightly lower for offshore tests relative to onshore tests.

The overall conclusion, however, must be that there are no significant differences between the offshore and the onshore laboratory test results from the Troll East Field soil investigation.

REFERENCES

Amundsen, T., Lunne, T., Christophersen, H. P., Bayne, J. M. and Barnwell, C. L. 1985. *A deep water soil investigation at the Troll East Field.* Paper presented to International Conference on Offshore Site Investigation '85.

Andresen, A., Berre, T., Kleven, A. and Lunne, T. 1979. Procedures used to obtain soil parameters for foundation engineering in the North Sea. *Marine Geotechnology*, **3**(3), 201–226. Also published in Norwegian Geotechnical Institute. Publication, 129.

Kirkpatrick, W. M. and Kahn, A. J. 1984. The reaction of clays to sampling stress relief. *Geotechnique*, **34**(1), 29–42.

Lacasse, S., Iversen, K., Sandbaekken, G. and Morstad, P. 1984. Radiography offshore to assess sample quality. Nordiska Geoteknikermötet. Linköping, Statens Geotekniska Institu. Proceedings, Vol. 2, 575–584. Also published in Norwegian Geotechnical Institute. Publication, 153.

Sandbækken, G., Berre, T. and Lacasse, S. 1985. *Oedometer testing at the Norwegian Geotechnical Institute.* Presented at ASTM Symposium on Consolidation of Soils Laboratory Testing. Fort Launderdale, Florida, USA, January, 1985.

PUPPI - A FREE-FALL SEABED PIEZOMETER FOR GEOTECHNICAL STUDIES

S D McPhail, Dr P J Schultheiss, Institute of Oceanographic Sciences, UK

Presented on **Thursday 6 March 1986 at 11.30** in the Norfolk Conference Room

EMBARGOED AGAINST DELIVERY

30

PUPPI — A free-fall Seabed Piezometer for Geotechnical Studies

S. D. McPhail and *P. J. Schultheiss,* Institute of Oceanographic Sciences, Wormley, near Godalming, Surrey GU8 5UB, UK

A free-fall piezometer (Pop Up Pore Pressure Instrument, PUPPI) has been developed to measure excess pore pressures in seabed sediments. The instrument is capable of operating in water depths of up to 6000 m for a period of one year. Excess pore pressure is measured with a resolution of 0.05 kPa at a port located near the tip of a 4 m long lance. The pressure caused by insertion, the insertion pressure decay curve and any residual (ambient) excess pressure can be used to assess some of the geotechnical properties of *in-situ* sediments. The acoustic, mechanical, and electronic systems of the PUPPI are described, and examples from deployments into soft sediments in oceanic depths are presented.

INTRODUCTION

The impetus for the development of an instrument to measure pore pressures in sediments stemmed from the wide-ranging study into the feasibility of disposing of high-level radioactive waste in deep ocean sediments. In this study, accurate estimate of *in-situ* permeability and pore pressures are required to quantify any naturally occurring advection of pore waters at sites designated as study areas. Pore-water advection is one of many factors that could potentially accelerate or retard the release of dissolved radionuclides to the ocean waters from buried canisters.

Geotechnical characterization of the *in-situ* properties of a soil or sediment cannot be complete without some knowledge of the pore pressure. Soil behaviour depends to a large extent on the effective stress $\sigma' = \sigma - u$ (where σ is the total stress and u is the pore pressure). Consequently, pore-pressure measurements also could be of significant interest to those concerned with the offshore construction industry.

Effective stresses will also influence many geological processes such as consolidation, lithification, sediment stability and liquefaction potential. Consequently there is a wide interest in measurements of pore pressure and the techniques used to determine *in situ* values.

PRACTICAL AND THEORETICAL DESIGN CONSIDERATIONS

The primary requirement of the instrument is that it should accurately measure the ambient excess pore pressure, u (expressed as a head of water) at some depth, z, beneath the sea floor. (The excess pore pressure at any point in the sediment is defined as the difference between the actual pressure and the hydrostatic pressure at the depth of the measurement.) To achieve this, a pore-pressure port has to be inserted into the sediments on a lance which inevitably causes some disturbance. A pressure pulse will be generated which decays at a rate that depends on the geometry of the lance and on the physical properties of the sediments. For clays it may take many hours, or even days, for this insertion pressure to decay to a residual ambient value. Therefore, it was considered impractical to develop a tethered instrument and the free-fall Pop-Up-Pore-Pressure-Instrument (PUPPI) was conceived. A two-part system was designed which consists of a recoverable bouyant section and a disposable section including the lance and pore-pressure port.

The initial design requirements dictated that a pore-water advection velocity as small as 1 mm/year should be resolved. The advection velocity in permeable porous media can be calculated from the simple expression

$$V = ku/nz$$

assuming Darcian flow, where V is the velocity of an element of water within the sediment of permeability k and fractional porosity n (Lambe and Whitman, 1979). For sediments typical of present study areas (where $k = 1 \times 10^{-7}$m/s and $n = 0.7$) a differential pressure resolution of ~50 Pa is required for a lance of length 4 m.

Other geotechnical parameters are potentially available from an instrument of this type. The decay of excess pore pressure, caused by insertion, is governed by the consolidation process around the probe. The horizontal coefficient of consolidation can be obtained from the decay curve (Gillespie and Campanella, 1981) which in turn can yield an estimate of horizontal permeability. It has also been suggested that the undrained shear strength of the sediments can be predicted from the maximum induced pore pressure caused by probe insertion (Esrig *et al.*, 1977). However, it is the ambient pore pressure that is necessary to calculate not only the vertical pore-water advection velocity but also to calculate the in-situ effective stress. The ultimate stability of sediments depends on the effective stress and most pore-pressure measurements may be made in the future for this reason.

Although a few seabed piezometers have been built elsewhere (Richards *et al.*, 1975; Bennett et al, 1976, 1982) none has the versatility to work autonomously in water depths down to 6000 m.

MECHANICAL DESIGN

The "free-fall" penetration and pop-up requirements of PUPPI meant that the hydrodynamic drag should be minimized and, to make the instrument simple to handle at sea, the weight of the recoverable instrumentation and buoyancy package should be as small as possible. To allow this light construction, the instrument is assembled, handled and launched in the vertical position in order that bending moments, and hence weight, can be kept to a minimum.

Figure 1 shows the assembled instrument. It consists of three sections:

- a top section, containing the glass sphere buoyancy, the command unit tube, the data logger tube, a flashing light and the radio beacon, which are mounted around a central lifting rod;
- the centre section, comprising the differential pressure transducer, pipe cutter, reed switch for penetration indication and the acoustically triggered mechanical release assembly, all of which are housed in a cage of aluminium rods;
- the bottom section, consisting of the disposable lead weights, the 4 m lance with the pressure port near the tip, and the penetration indicator.

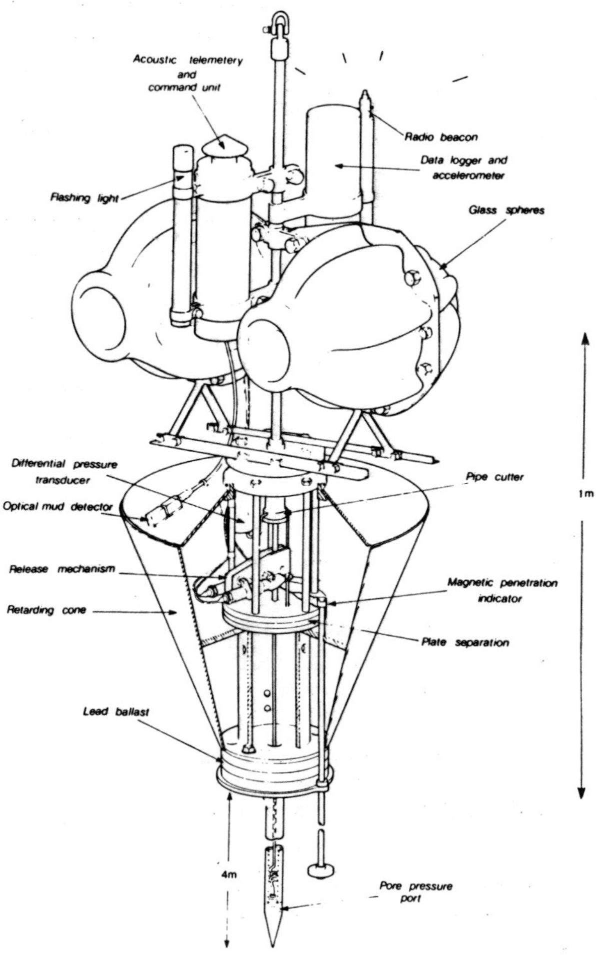

Fig. 1. Pop-Up-Pore-Pressure-Instrument (PUPPI)

A polypropylene cone covers the bottom and centre sections. This is to prevent sediment fouling the release mechanism if overpenetrations occurs. It is left behind with the ballast weights on recovery (Fig. 2).

The buoyancy of the top and centre sections, which are recovered, is provided by two 0.43 m diameter glass spheres, each providing 23.3 kg of buoyancy. The excess buoyancy available for ascent to the surface is 13.6 kg. The command pinger is positioned to allow a clear acoustic window to the surface with the data logger pressure housing mounted diametrically opposite to maintain symmetry. Fins, mounted on the buoyancy hard hats and outside the logger and acoustics, aid the hydrodynamic stability of the instrument.

The disposable weight stack consists of a variable number of lead weights, each 2.54 cm thick, 30 cm in diameter, and weighing 18.8 kg in air. This adjustable driving force is to compensate for differing sediment strengths. It is desirable that the probe should penetrate up to the lead weight stack for maximum stability in the sea floor. On the Madeira abyssal plain, where most of the deployments have been made, it has been found that three weights are sufficient for full penetration (Schultheiss *et al.*, 1985).

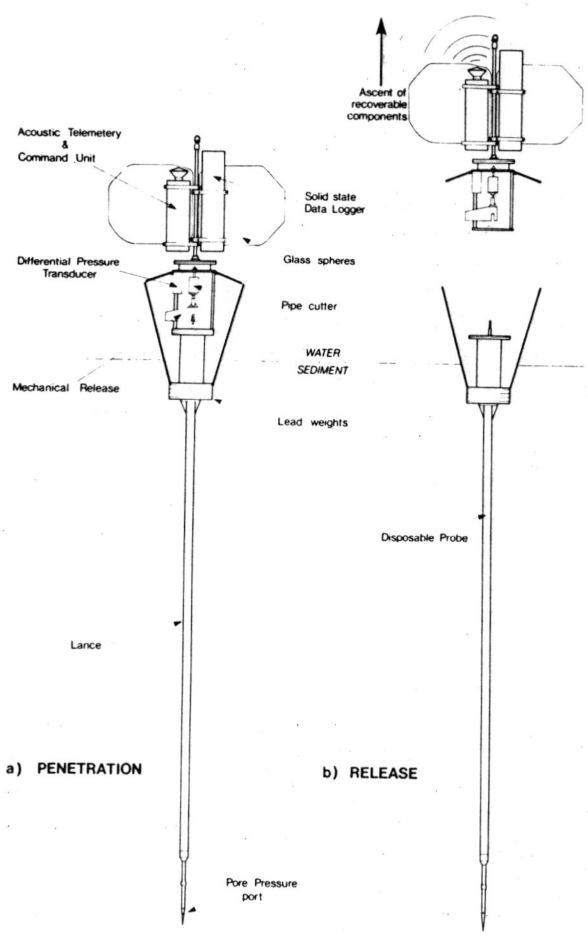

Fig. 2. Penetration and release of PUPPI from sea bed

The 4 m lance is a thick-walled steel pipe with a solid tip which contains a porous stainless steel pressure port. This pressure port is connected to a 3.2 mm diameter nylon tube which passes up through the inside of the lance, the lead weights and an acoustically triggered pipe cutter. It terminates at the differential pressure transducer. When the acoustically triggered mechanical release operates, it separates the top and centre sections from the lead weights. The release plates (Figs 1 and 2) have been designed to prevent movement in both vertical and horizontal planes during all operations prior to release, thereby giving stiffness to the system. Separation is designed to be possible even at angles greater than 45° to the vertical. Simultaneously with this release, the spring-loaded cutter severs the pipe, allowing the top and centre sections to ascend freely to the surface as shown in Fig. 2.

To ease relocation of the surfaced instrument, a radio transmitter (Novatech, 150 MHz) and flashing light (Oceania Inc.) are attached. Both are activated by a pressure switch. A direction finding receiver/antenna (Datong) is used to home in on the instrument.

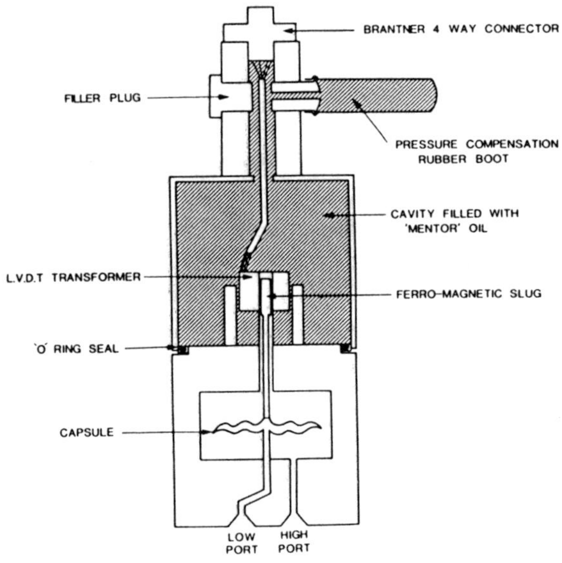

Fig. 3. Differential pressure transducer

THE DIFFERENTIAL PRESSURE TRANSDUCER

The differential pressure transducer (Appleby and Ireland, Multitron 741DD) was chosen for robustness and compatibility with the harsh operating environment. Figure 3 is a simplified drawing of the transducer with the modifications necessary for use at ambient pressures of 60 MPa.

Differential pressure is applied across the capsule with compresses or expands, moving a ferromagnetic core within a Linear Variable Differential Transformer (LVDT). The primary coil of the LVDT is excited by a 1 kHz sinusoidal current, and the output current from the secondary is proportional to the displacement of the core.

The transducer has no mechanical linkage between the moving part (the core) and the electrical part (the LVDT). This allows the cavity around the LVDT to be filled with oil at ambient pressure. A rubber boot compensates for changes of volume due either to thermal expansivity or bulk modulus of the oil, or that of any entrapped air. 'Mentor' oil was chosen for its low thermal expansivity, and its relative ease of handling. Tests of the transducer have been carried out at pressures equivalent to a depth of 6000 m.

Calibration

A transducer of range ± 200 kPa was used to accommodate the possibly large differential pressure produced when the probe penetrates the sediment. The residual excess pore pressure is expected to be only a small fraction of this range, and hence the zero offset drift and hysteresis are of most concern.

Figure 4 illustrates the zero offset temperature drift which is about 80 Pa/°C. For a 13-bit digitization the resolution is 50 Pa, so the temperature must be known to 0.6°C to allow correction for the zero drift to within the bit resolution. This is possible, but unfortunately, the zero shift caused by a pressurization/depressurization cycle (~ 1% full scale) is not fully reversible, and hence there is a small permanent shift in zero which cannot be sub-

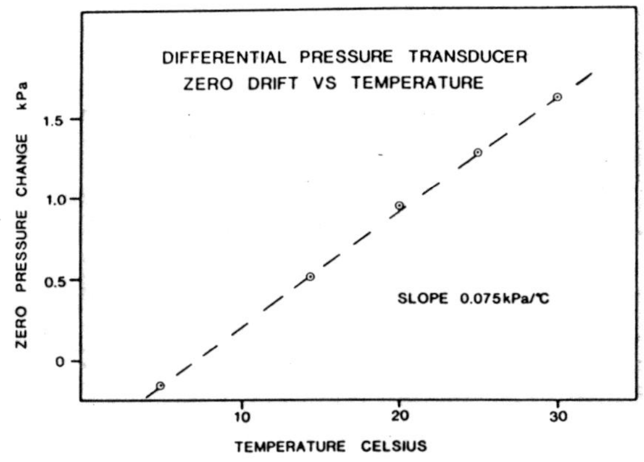

Fig. 4. Differential pressure transducer zero drift vs. temperature

tracted from the data with sufficient accuracy. An *in-situ* zero reference is therefore required.

Originally, the zero reference was obtained from the severed pipe as the instrument was released from the sea bed. This is potentially prone to error because of flow-induced pressure on the transducer ports if the instrument starts to move before a zero reading is made. A pipe cutter, based on a modified IOS retractor, and fired by acoustic command some time before release, gives a much more reliable zero reference. The temperature on the deep ocean bed in GME is very stable (to a few millidegrees (Saunders, 1983)) and hence this zero reference obtained at the end of the experiment is reliable. By cutting the pipe some time (possibly a day or so) before the instrument is released, the long-term drift of the transducer

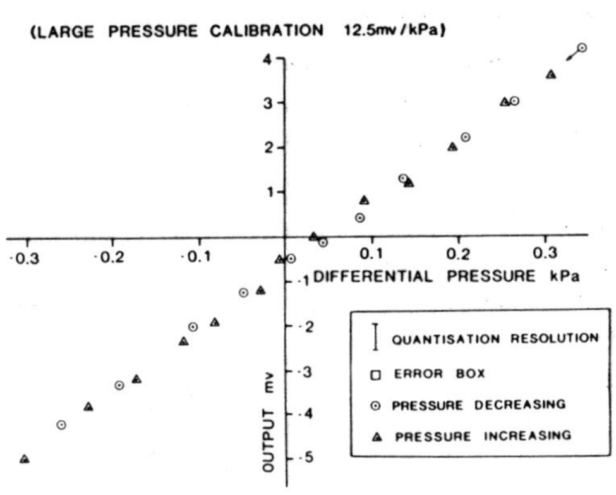

Fig. 5. Small differential pressure linearity

can be ascertained, and subtracted from the pore-pressure data.

Transducer hysteresis could seriously affect the measurement of small pore pressures, but laboratory tests (Fig. 5) show the hysteresis to be less than the digital quantization.

The range of the transducer can be altered by changing the capsule (Fig. 3). In a recent deployment, a transducer of range +60 kPa (resolution 0.015 kPa) was used. Results were encouraging and so, in future deployments, transducers of this range are to be employed.

A possible source of error (±100 Pa) identified was that the sea water in the nylon pipe must have a density similar to that of the pore water. The density of water used to fill the pipe is closely matched to the density of sea-bottom water (Saunders, 1983) which has very nearly the same composition as pore water (Wilson, 1985). Consequently, the errors caused by any density mismatch have been reduced to less than the bit resolution of the logger.

THE DATA LOGGER

The requirement of the data logger is to record the differential pore-water pressure induced in the sediment as the probe is inserted, to measure the decay as this pressure pulse dissipates, and to quantify any residual differential pressure associated with pore-water movement.

The data recorder is required to log several thousand pressure readings over a period of several days (perhaps weeks or even months), and hence it was decided to use a solid-state EPROM logger to record the data.

The advantages of an EPROM logger over the alternative tape cassette logger are:

- robustness (there are no moving parts);
- ease of replay (data can be transferred quickly with a minimum of additional hardware);
- savings in weight and volume;
- cost (the replay systems for cassette loggers are particularly expensive)

An EPROM logger has the disadvantage of having a much smaller data storage capacity (the PUPPI logger has a capacity of 256K-bit compared to 14M-bit for the "Sea Data" logger). However, for the PUPPI the capacity afforded by the EPROM logger is more than sufficient.

The Transducer Interface

The transducer interface electronics provides a.c. excitation to the transducer and detects the return signal, generating an output voltage proportional to differential pressure.

The relatively high power consumption necessitates that the interface electronics is powered on for only one second before the pressure reading is digitized and stored. A programmable clock circuit determines when a pore-pressure measurement is to be made, and also controls a power supply switching circuit providing power to the appropriate circuits only when needed.

The Analog-to-Digital Converter

An ICL7901 integrating analog-to-digital converter (ADC) giving 13-bit resolution is used. The ADC works ratiometrically with the transducer interface, as the reference for the stabilization of the LVDT excitation, and the ADC reference are both the +5 V supply. As with the transducer interface, power is applied to the ADC for only one second for each measurement.

The Clock and Programmable Divider

The clock circuit sets the sampling rate, which may be preset to vary throughout the instrument's deployment to give the most efficient use of the available memory.

A one-second pulse, derived from a crystal oscillator, is divided down according to a matrix, and preprogrammed with diode shorting pins. The deployment is split into ten "time frames". For each frame, the sampling interval can be set from 2 to 2049 s. The number of measurements within that frame can vary from 10 to 1280.

The EPROM Logger

Addresses in two EPROMs are programmed in parallel for each word which is 16 bits long. Four NMOS 2764, 64K-bit EPROMs can be programmed during each different deployment, giving a storage capacity of 16K, 16-bit words. As the stand-by power consumption of the EPROMs is high (100 mA at 5 V), power is applied to the EPROMs only for the 50 ms required for programming an address.

Zero insertion force sockets allow the EPROMs to be easily interchanged.

The Power Supply

The power supply, under control of the clock, provides a switched ±5 V for the ADC and transducer interface, and the programming voltages (5 V and 21 V for 50 ms) for the EPROM logger.

The quiescent current of the CMOS logic is very low (c. 200 μA), so a low quiescent current regulator is used for the logic supply.

A 10 ampere hour lithium sulphur dioxide battery pack (three L026S and six L032S batteries) provides 12 months' duration.

The Replay System

The most convenient way to display the data for a quick look is to play it out as an analog record on a pen recorder. The replay system plugs into the EPROM logger board and converts the binary words back into analog voltages. Data is output at a selectable rate, with a linear but compressed timescale. A time-coded trace is generated and played out alongside the pore-pressure trace.

For a more detailed look at the data, the EPROMs are plugged into a reader which interfaces with a BBC B microcomputer. Data is transferred to floppy disk, and is output to the screen, or to a digital plotter.

THE COMMAND PINGER

The acoustic command pinger and release system enables the instrument to be monitored through the water column and while in the sediment. It also provides the current required to operate the pyro-releases on command from the ship, enabling the instrument to ascend to the surface. A second channel fires the retractor, cutting the pipe to give a zero-pressure reference. All aspects of this system are described in detail by Phillips (1981).

Acoustic Telemetry of Tilt and Penetration

Each deployment of PUPPI will involve a degree of uncertainty as to what it may hit on the sea floor. Consequently, real-time information is required about its tilt angle and penetration. On the basis of this information, it can be quickly decided whether to release immediately and recover the instrument or to leave the instrument to record the pressure data.

Tilt is telemetered acoustically by gating extra pulses through the command pinger. The tilt sensors use mercury and two electrodes in a glass phial to make or break an electrical circuit (Phillips, 1981). Five discrete sensors are used to indicate angles greater than 10, 15, 20, 30 and 60 degrees from the vertical.

An infra-red emitter and detector are used to detect gross overpenetration, when sediment obscures the light path. This is telemetered as an extra pulse. At penetrations greater than 3.5 m a plate attached to the bottom of a vertical bar slides up the lance. A magnet, mounted at the top of the rod, moves away from an encapsulated reed relay. This provides another gated pulse which is telemetered by the command pinger.

Figure 6 illustrates the acoustic and data recording systems of the PUPPI. Note how two pipe cutters are used; one to cut the pipe on acoustic command, the other which cuts the pipe simultaneously with instrument release.

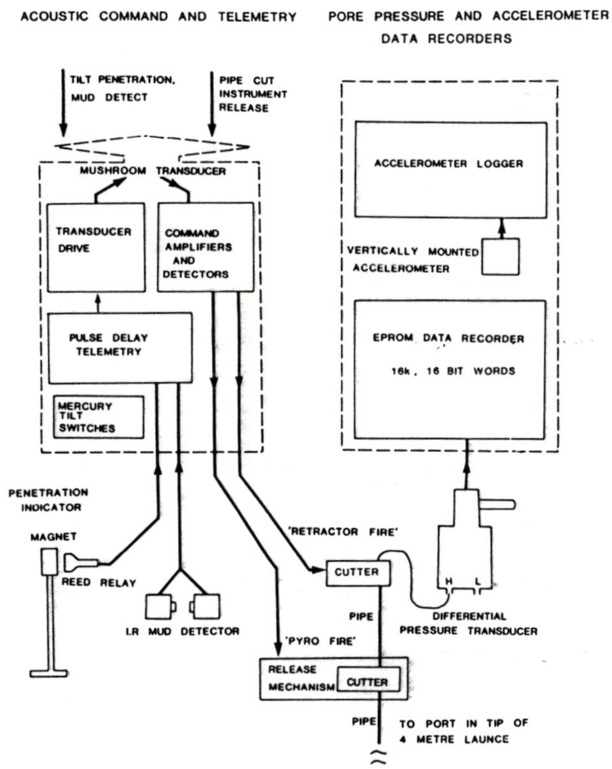

Fig. 6. Acoustic and data recording systems of PUPPI

ACCELEROMETER

To monitor the dynamic behaviour of the PUPPI accurately, it was instrumented with a vertically mounted accelerometer. It was intended that the deceleration record prior to, during and after penetration would provide valuable information on the behaviour of the instrument both in the water and the sediments.

The numerical modelling of the penetration event had indicated that a 1 g accelerometer would have an adequate range to cover the expected peak decelerations. The accelerometer trace of the penetration event would, on integration (w.r.t time) give the velocity profile and, on double integration, the depth of penetration. (Terminal velocity could also be obtained from the acoustic pinger Döppler shift.)

One immediate problem realized was one of data storage requirements. A sampling rate of greater than 50 Hz would be required and almost an hour's worth of data would have to be stored, assuming recording was started at the beginning of the descent. This works out at over 100K bytes of data, which cannot be stored in the EPROM logger, or any other practical solid-state logger. A cassette logger would have the required capacity but would be a rather expensive and bulky addition to the electronics package.

The solution was to store the data in a RAM which is continually overwritten, so that, at any one time, the last minute of the record is stored in the memory. The penetration of the probe is detected and the logger stops sampling, so that the last minute of data, which should include the deceleration of the probe, is "frozen" in the memory.

The penetration of the probe is "sensed" indirectly by a circuit which looks for the period of low vibrational noise (from the output of the accelerometer) experienced after penetration (when the instrument has come to rest) compared to the relatively high levels of vibration during the descent or when on the ship. An accelerometer (variable reluctance type Schaevitz A411-005) is vertically mounted. The output is anti-alias filtered at 30 Hz and converted to 8-bit words (at a rate of 128 Hz) which are stored on a 8K-byte RAM with its own battery back-up. The memory is continuously overwritten so that, at any one time, it contains 64 s of data.

A threshold detector resets an interval timer when the vibration exceeds a presettable level. If the vibration does not exceed the threshold for a period of 36 s, the timer times out, resetting a latching relay, switching off power to the logger. The last 64 s of data are now stored in the logger. If the noise detector has functioned correctly, the penetration event should be 36 s before the turn-off.

Upon recovery, the logger is switched to the read mode, the data is converted back to an analog signal and replayed on a pen recorder. Integration of the penetration can then be done graphically.

EXAMPLE RECORDS

Data has been collected on *Discovery* cruises 141, 149, 153 and *Darwin* 6 within an area known as the GME (Great Meteor East) study area (around 32°N 25°W).

Station D11184 is used to illustrate the pressure–time record which is shown in Fig. 7 (note the changes in timescale necessary to illustrate the behaviour). When the instrument penetrates the sea floor, a large pore-pressure transient is created (c. 40 kPa) which decays with time. After 10 minutes it has decayed to 10% of the peak value and after 3 hours it is less than 1% of the peak. No detectable decay

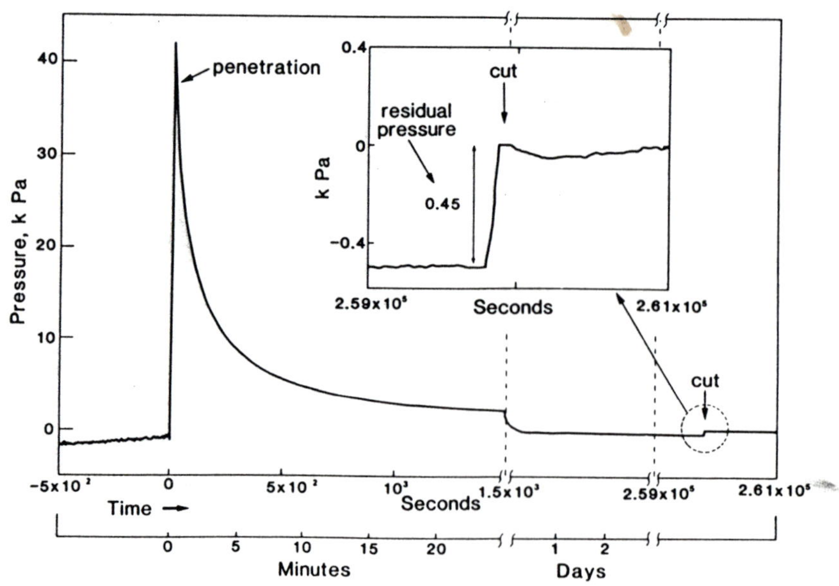

Fig. 7. Pore pressure record for deployment in GME

occurs during the final 24 hours.

A residual pressure of –0.45 kPa was recorded for this deployment, indicating that a small downward advection of pore water is occuring (Schultheiss and McPhail, in press). To date there have been eight successful deployments and recoveries of PUPPI in the GME area. The results in terms of the dynamic behaviour, the accelerometer records and the pore pressure have been consistent. Other deployments have recorded small positive pressures and zero residual pressures (to the resolution of the instrument).

The accelerometer record (Fig. 8) is typical of deployments in GME. Figures 9 and 10 illustrate how the depth of penetration is obtained from a double integration of the deceleration record.

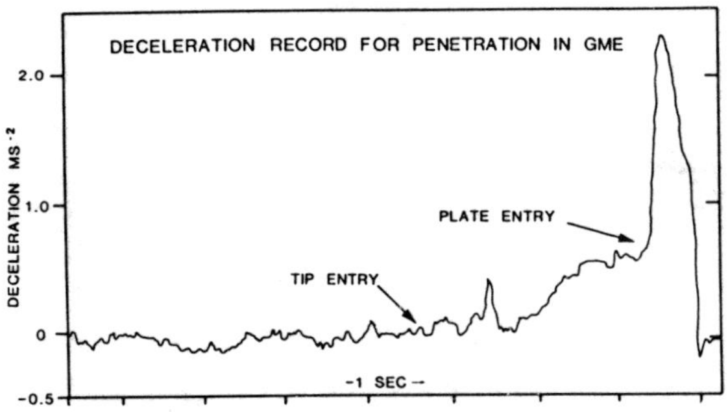

Fig. 8. Deceleration record for penetration in GME

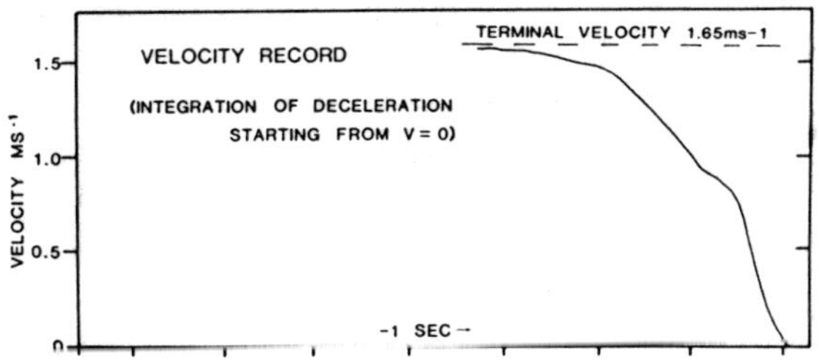

Fig. 9. Velocity record

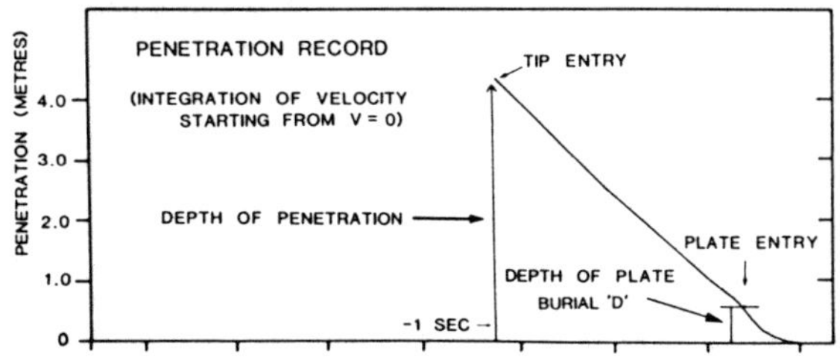

Fig. 10. Penetration record

Integration is started at the instant the instrument has come to rest, working back in time. The computed velocity at tip entry agrees closely with the terminal velocity from acoustic Doppler shift (1.65 m/s).

The depth of penetration is also obtained by integrating back in time. Integration to the point of entry of the weight stack (as seen by the increase of deceleration on Fig. 8) gives the depth of burial of the base of the stack (D in Fig. 10). Further integration allows the instant of lance tip entry to be determined, as P = 4 m (length of lance) + 0.5 m (depth of plate burial). Reference to Fig. 8 shows that this is a good estimate.

The degree of tilt is indicated by the displacement from zero of the accelerometer record, after the instrument has landed. In the example shown there is no apparent tilt, which is typical, although the resolution is poor for small tilts ($< 7°$).

The accelerometer gives a reasonably accurate penetration depth, provided the instrument does not settle down appreciably after penetration.

CONCLUSIONS

The concept of a free-fall, pop-up instrument for measuring pore pressures in the upper 4 m of deep-sea sediments has proved successful. Differential pore pressures can be measured over long periods of time, if necessary, with a resolution of better than 50 Pa. The autonomous and versatile nature of the PUPPI makes it suitable for use in different sediment types irrespective of the water depth.

ACKNOWLEDGEMENTS

We would like to thank many IOS staff for help with the concept, design and testing of the PUPPI. In particular we are indebted to B. Hart, Dr A. Packwood, Dr R. B. Whitmarsh, R. E. Kirk, D. E. Gunn and the Master, Officers and Crew of *R.R.S. Discovery*. We would also like to thank Dr R. B. Whitmarsh and G. R. Phillips for reviewing this chapter.

This research has been carried out under contract to the Department of the Environment, as part of its radioactive waste management programme. The results will be used in the formulation of government policy, but at this stage they do not necessarily represent government policy.

REFERENCES

Bennett, R. H., Burns, J. T., Clarke, T. L., Faris, J. R., Forde, E. B. and Richards, A. F. 1982. Piezometer probes for assessing effective stress and stability in submarine sediments. In: *Marine Slides and Other Mass Movements,* eds. S. Saxov and J. K. Nieuwenhuis, Nato Conf. Ser.iv, Marine Sciences. New York: Plenum Press, pp. 129-61.

Bennett, R. H., Burns, J. T., Nastav, F. L., Lipkin, J. and Percival, C. M. 1985. Deep-ocean piezometer probe technology for geotechnical investigations. *IEEE J. Oceanic Engng* **OE-10**(1), 17–22.

Esrig, M. I., Kirby, R. C. and Bea, R. G. 1977. Initial development of a general effective stress method for the prediction of axial capacity for driven piles in clay. In: *Proc. 9th Annual Offshore Technology Conference, pp.* 495–501.

Gillespie, D. and Campanella, R. G. 1981. Consolidation characteristics from pore pressure dissipation after cone penetration. Vancouver. *Soil Mech. Series No. 47,* Dept. of Civil Engineering, University of British Columbia, 17 pp.

Lambe, T. W. and Whitman, R. V. 1979. *Soil Mechanics,* SI Version. John Wiley, New York: Section 17.3.

Phillips, G. R. J. 1981. The IOS acoustic command and monitoring system. Part 3. Releases, beacons and transponders. *IOS Report No. 96/3.*

Richards, A. F., Oien, K., Keller, G. H. and Lai, J. Y. 1975. Differential piezometer probe for an *in-situ* measurement of seafloor pore pressure. *Géotechnique* **25**(2), 229.

Saunders, P. M. 1983. Benthic observations on the Madeira Abyssal Plain: currents and dispersion. *J. Phys. Oceanogr.* **13**(8) 1416–29.

Schultheiss, P. J., McPhail, S. D., Packwood, A. R. and Hart, B. 1985. An instrument to measure differential pore pressures in deep ocean sediments: Pop-up Pore Pressure Instrument (PUPPI). *IOS Report No. 202.* DoE/RW/84/165.

Schultheiss, P. J. and McPhail, S. D. (in press). Direct indication of pore water advection in the Madeira Abyssal Plain sediments from pore pressure measurements. *Nature.*

Wilson, T. R. S. 1985. Personal communication.

CONVENTIONAL GEOTECHNICAL APPROACH TO MARINE GEOTECHNICAL APPLICATIONS: A PERSPECTIVE

Dr U Dayal, Indian Institute of Technology,
Kanpur, India

Presented on **Thursday 6 March 1986 at 12.00**
in the Norfolk Conference Room

EMBARGOED AGAINST DELIVERY

31

Conventional Geotechnical Approach to Marine Geotechnical Applications: A Perspective

Umesh Dayal, Civil Engineering Department, I.I.T., Kanpur-208016, India

The major difference in classification and testing procedures of terrigenic and submarine soils is reviewed, and suggestions have been made for modifications which may be necessary for better understanding of the behaviour of submarine soils. Among the subtopics discussed in this chapter are (a) submarine soils, (b) sampling disturbances, (c) laboratory tests, and (d) *in-situ* tests.

INTRODUCTION

Some of the earlier developments in offshore technology took place in the relatively shallow and friendly waters of the Gulf of Mexico. However, in the past three decades, offshore structures have been built in deeper waters, sometimes even more than 300 m deep, and in more difficult soil profiles and very hostile environments. Since land-based geotechnical engineering is the root of marine geotechnology, many of the developments in the latter were extensions of basic conventional geotechnology to suit oceanic environments.

Over the past 25 years, several state-of-the-art reviews have been presented on various sub divisions of the marine geotechnology (Noorany and Gizienski, 1970; Noorany, 1972; Silva, 1974; Eide, 1974; Sangrey, 1977, Fotch and Kraft, 1977; Anderson *et al.* 1979; Sulivan, 1980; Marshland and Windle, 1982; De Ruiter and Richards, 1983; Lee, 1985; Richards and Zuidberg, 1985; McClelland, 1975; Dayal, 1978). In most of the reviews, sampling and *in-situ* tests of submarine sediments have been emphasized, perhaps because they involve unique and expensive equipment, and much of the thrust of marine geotechnology has been towards the development of this equipment. However, very little is said about the applicability of the conventional geotechnical approach to marine geotechnology. This is not to say that there is any fundamental difference between terrigenic soils and submarine soils. Both obey the same fundamental principles of particulate mechanics, and both consolidate, shear and allow water passage in essentially the same way. The differences are mainly: the origin of submarine soils, the performing of laboratory and field tests, and loading conditions.

The major differences between geotechnical engineering on land and in marine environments are in the following areas: (a) submarine soils,

(b) sampling disturbances, (c) laboratory testing and (d) *in-situ* testing. This chapter reviews these differences and summarizes a comparative study between submarine soils and terrigenic soils. The emphasis is on those areas where substantial differences exist between the two types of soils, and the objective is to outline the modification which may be necessary for classification and testing of submarine soils.

SUBMARINE SOILS

The classification and testing of submarine soils implies some knowledge about the properties of the "soil" material, its distribution, depositional environments and state of stress. Although much of this information may be well known, nevertheless it may be useful to summarize certain aspects prior to focusing on the main points.

The sea floor is primarily formed from depositional processes, in contrast to the land which is formed primarily in an erosional environment. As a consequence, marine sediments exhibit more uniformity than is normally found on land (Lee and Clausner, 1979). Submarine sediments can be broadly classified on the basis of whether the sediments are land derived (terrigeneous) or are the result of marine activity (pelagic). The terrigeneous sediments are extensively found on continental shelves and slopes. Since those sediments have originated from the land we can expect them to behave much like similar soils on land, and our engineering experience and empirical relations developed for terrigenic soils should be useful in understanding the sediment behaviour of continental shelves and slopes. Fortunately, most of the current offshore development problems are within the continental shelf and slope only.

The pelagic sediments can be further divided into inorganic and organic materials. Inorganic pelagic materials are typically wind-blown soil of clay size. The organic materials are primarily the skeletal remains of marine organisms. These materials are either calcium carbonate or silica.

The pelagic sediments are predominantly found in deep seas away from continents, but may also be present on the shelf. Their behaviour is dominated by chemical effects such as cementing, which often results in very different behaviour compared to terrigenic soils. Therefore, the experience and empirical relationships derived from terrigenic soils may not hold good for pelagic sediments.

SAMPLING DISTURBANCES

Sediment sampling is an integral part of the process of classification and testing of submarine soils. Many of the inaccuracies involved in predicting field behaviour are introduced through physical disturbances caused by sampling. Any sampling technique will cause both stress release and some mechanical disturbances which alter the shear strength and compressibility characteristics of the soil. This is a well known problem of terrigenic soil sampling, and there is no reason to expect that the problem is fundamentally any different offshore. However, the marine environment imposes special constraints on soil sampling that cannot be overcome, even at great expense. A few of these, which lead to a much lower soil-sampling quality than is normally possible on land, are listed below:

- core deformation caused by removal of *in-situ* stresses (Richards and Parker, 1967);
- piston movement during sampling (Anderson *et al.*, 1965);
- change of geochemical properties of the sample (Jerbo, 1967);
- organic growth caused by temperature and other environmental changes (Noorany, 1972);
- expansion of water in the pore caused by pressure reduction (Vey and Nelson, 1967);
- formation of gas bubbles in the pore water, due to changes of total stress (Sangrey, 1977; Vey and Nelson, 1967);
- overstressing of the soil sample produced by drilling mud pressure (Sangrey, 1977).

The last two factors appear to be major causes of submarine soil-sample disturbance, and therefore these are elaborated further.

The problem of total stress change on a soil sample is very critical offshore compared to on land. For example, samples obtained on land at 10 m below the ground surface and offshore at same depth below the sea bottom in 250 m deep water, would be in the same state of effective stress. However, the state of total stress for offshore samples would be approximately 15 times more than the sample from on land. Due to large stress release, the gases present within the submarine soil samples either expand or come out of the sample (Sangrey, 1977).

When gas expands within a sample of submarine soil, important changes in physical and engineering properties take place. The unit weight and degree of saturation decreases and the strength and compressibility are irreversibly altered (Sangrey, 1977; Lee, 1985; Vey and Nelson, 1967). These changes are essentially inevitable, given the present state of technology. However, research is currently under way and soon we may have pressurized samples to minimize gas expansion.

Mud is invariably used in offshore drilling to stabilize the borehole. This poses a peculiar problem in offshore environments, specially if the water depth is very high and/or the upper submarine soil layer is too weak. The effective drilling pressure acting at any soil depth in a borehole would be the total head due to the operating height above sea level plus water and soil penetration depths. When the effective pressure exceeds the ultimate bearing capacity of the soil, the bottom of the borehole will fail or yield. Samples taken under such conditions will be highly disturbed and may be contaminated by drilling fluid. The severity of overpressure increases as the water depth and the height above sea level increases. The same problems may also exist in shallow water depths in areas of soft and weak submarine soil deposits (Sangrey, 1977).

LABORATORY TESTING

As much testing as possible should be carried out on board ship in order to provide adequate data on which decisions relating to the drilling and sampling programme can be made, and to reduce the effects of stress release, transportation and storage of the samples. Until recently, onboard tests have been limited to classification tests, water content, density and determination of the undrained strength of cohesive soils. The majority of undrained strengths are determined using hand penetrometers, miniature vanes, torvanes and occasionally fall cone penetrometers. In addition to these tests, relatively small numbers of unconfined compression or quick undrained triaxial tests have been made. While such tests provide a useful measure of consistency, they only provide relevant strengths in a very narrow range of soils.

Virtually all of the investigators (McClelland, 1967; Ladd and Azzouz, 1983) recommend extrusion of sample on board ship for the following reasons:

- if there is any delay, stiff clay may "set up" in the tube and require larger extrusion forces;
- extruding allows one to test and evaluate the samples and better define the stratigraphy;
- if samples are extruded, the engineer in charge at the site can better assess the need to obtain another sample before drilling to the next sampling interval.

The onshore laboratory testing is concerned with those tests which are either too time-consuming or too complex to be made offshore. These can include (Marshland and Windle, 1982):

- more detailed sedimentological studies;
- additional undrained, unconsolidated tests, to supplement the offshore tests and provide more complete profiles;
- undrained strengths and moduli using different types of tests and stress paths;
- tests to estimate the previous stress history and the present *in-situ* stresses;
- measurement of strengths and moduli on samples reconsolidated to or above the estimated *in-situ* stresses, to check unconsolidated, undrained tests and provide

effective stress parameters over a range of stresses to be supplied by the foundation;
- cyclic tests, to study changes in deformation and strength parameters;
- centrifugal models, in certain cases.

CLASSIFICATION AND TESTING PROCEDURES

Laboratory tests used on submarine soils, in general, follow existing laboratory procedures applicable for terrigenic soils. While there have been no major developments in marine geotechnical laboratory testing over the past few years, there has been an increasing awareness of some of the limitations of methods currently used, and the need for a more critical assessment to be made of measured data and its relevance to offshore foundation design. In this section of the chapter only those techniques which deal specifically with the problems of submarine soils will be emphasized. The standard techniques which are routinely used for terrigenic soils and are also being practised for submarine soils will be given very little attention, as an abundance of literature is available on these topics.

Identification of Soils

Water content, density, porosity, relative density, grain size distortion and Atterberg limits of submarine soils are found as in the case of terrigenic soils.

Electrolyte

Many authors have recommended that the electrolyte (saline water) should be taken into account in calculating those values. A factor that is always present in submarine soils and seldom present in terrigenic soils is a high pore-water salinity (of the order of 35 ppt). Most investigators (Inderbitzen, 1974) feel that the water-content measurements should at least be corrected for salt content.

Soil Fabric

Recent studies have shown that detailed sedimentological and fabric studies can be of considerable value in submarine site evaluations (Marshland and Windle, 1982). The new approach integrates the result of standard particle size distribution curves and index tests with descriptions and measurements of macro- and microfabric features in samples. The objective is to assess the probable depositional and post-depositional processes which have occurred at particular horizons in the soil profile, and to obtain as complete a picture as possible of the soil fabric features present, or likely to be present, in the ground so that their possible effects on engineering properties of submarine soil can be evaluated.

Calcareous Sediments

Many submarine sediments contain significant amounts of calcium carbonate in the form of coral, inorganic precipitates or biogenic oozes. Although calcareous materials are common on land also they are much more frequently found offshore. A recent ASTM symposium (Demars and Chaney, 1982) provided a forum for many papers dealing with the engineering behaviour of calcareous soils. The carbonate content of sediment is being suggested as an engineering index property, and rapid techniques for determination of carbonate are discussed by Chaney *et al.* (1982).

Shear Stress Testing

The shearing resistance of sediments is the most important characteristic for engineering design. Varieties of methods are used to describe strength and shearing resistance. Because submarine soils found in continental shelves and slopes are not all that much different from terrigenic soils, few modifications in the strength testing methodology have been needed. It would be beyond the scope of this chapter to review all of those methods. It is important to appreciate the fundamental interrelationships between those methods, and to understand the advantages and disadvantages in their use. The basic methods are (a) the total stress approach, and (b) the effective stress approach. As pointed out earlier, the majority of the strength tests performed are based on the total stress approach: only recently has the importance of the effective stress approach been recognized by marine geotechnical engineers.

It is also only recently that the potential value of measuring pore-water pressure in unconsolidated, undrained triaxial tests on offshore soils has been realized (Sangrey, 1977; Marshland, 1977). Such measurements are particularly important in offshore investigations where the high costs limit the number of samples which can be obtained. These pore-pressure measurements help to distinguish different soil behaviours, provide a check on sample quality and provide effective stress parameters which supplement those obtained from tests on samples which are further consolidated in the laboratory. The procedure for carrying out this type of test in connection with North Sea investigations has been outlined by Marshland (1977). With modifications to suit the local site, his procedure can be adopted for other offshore regions.

Lee (1985) has described procedures for direct shear and triaxial tests of soft sediments. Special care and some modifications in the testing procedures are needed for low-strength sediment samples (for example, trimming and placement of sample, saturation and consolidation). In marine geotechnology, triaxial testing was certainly the favourite in the 1970s, although direct shear testing is now gaining in popularity (Andersen *et al.*, 1979; Ladd and Azzouz, 1983; Quiros *et al.*, 1983).

Depending on the magnitude and nature of consolidation stresses applied, it is possible to obtain practically any strength in direct shear or triaxial tests. Therefore, the strategy followed in selecting the consolidation stresses may be more important than any other single factor. Of the possible strategies and philosophies available, three major groups stand out:

- no consolidation, the unconfined compression (UC) and unconsolidated-undrained (UU) triaxial tests;
- consolidation to the *in-situ* overburden stresses;
- consolidation at stresses well above the *in-situ* overburden stresses, to obtain normalized soil parameters.

The first group, consisting of the UC and UU tests, falls close to such strength index tests of the vane shear. Both these tests are commonly performed offshore. The UC test typically underestimates the field strength, sometimes by a factor of two or more. Resaturating the sample in the UU test causes some consolidation and strength increase. As a result of compensating errors, the UU strength may roughly equal the field strength in many cases (Lambe and Whitman, 1969). Results of these tests are typically taken as strength indices rather than accurate measurements of *in-situ* strength.

The second group of triaxial and direct shear tests, consolidating to the *in-situ* stresses and failing under undrained conditions, should yield about the same value as would have been obtained *in-situ,* provided that samples are of high quality. Such a procedure is almost mandatory for certain types of soils which include highly structured or quick clays and naturally cemented clays (Ladd and Foott, 1974). If there is a disturbance during the sampling process, as there is in all cases, but severely with marine samplers, interparticle bonds will be broken. Reconsolidation to the *in-situ* stresses will cause a partial collapse of the sediment structure and a reduction in void ratio. The net effect is almost always a higher, possibly much higher, laboratory shearing strength than the soil had *in situ.* The only way to avoid this difficulty is to use a very high-quality push sampler.

The final group of triaxial and direct shear test requires consolidation beyond the *in-situ* stress state and derivation of normalized soil parameters. This procedure partly overcomes the effects of disturbance and provides a more rational basis for strength parameter determination (Ladd and Foott, 1974).

Another aspect of strength testing is cyclic loading, which dominates much of the marine environment. In the fairly shallow water in which most development is occurring, storm waves generate large cyclic shearing stresses in the sea floor, either directly or through loading a structure. No analogy to storm-wave loading exists on land. In deeper water, where storm waves may no longer be a significant factor, seismically induced shearing stresses can have a more important influence than they would have in most locations on land. This is because, on the sea floor, normally consolidated soils have a strength increase with depth that is proportional to the submerged weight of the overlying soil. On land, stronger overconsolidated soils typically exist because of dessication and active erosion processes. If the land soil is normally consolidated, its strength will increase with depth at a rate that depends on the submerged unit weight, total unit weight and depth of water table. Only if the water table is at the surface and has always been at the surface, will the shearing strength of a normally consolidated soil be as low on land as it is offshore. On the other hand, the seismically generated shearing stresses that need to be resisted by soil strength vary with the total unit weight, both on land and offshore. The ratio of applied stresses to resisting strength will typically be greater offshore. The analysis and testing procedures of submarine soils for those stress states is a topic for separate discussion, and readers are advised to refer to Sangrey (1977) and Chaney and Fang (1983).

Compressibility Testing

The settlement of sediment masses, either under their own weight or under the applied weight of structures, is controlled by the consolidation compression characteristics of the soil. The consolidation tests are carried out in the laboratory, either using the conventional oedometer or using the triaxial test by means of a long-duration drained test. A good resumé of graphical and mathematical methods used for the interpretation of results is contained in Winterkorn and Fang (1975). Because of sample disturbance during collection or in laboratory treatment, the laboratory consolidation falls below the theoretical curve. In such cases the point of maximum curvature is often extremely difficult to define. In order to establish a true field consolidation curve, a technique proposed by Schmertmann (1955) is generally followed by marine geotechnical engineers. Driven samples generally yield lower preconsolidation pressures due to sample disturbances. In contrast, push samples generally give good estimates of *in-situ* vertical effective stresses (Quiros *et al.*, 1983).

Marshland and Windle (1982) have reported a new system for measuring the axial strain directly, on the sides of the samples of clay obtained from thin-wall push samplers. The internal measurements were made by measuring the change in the length between pins embedded in the quarter points of the 150 mm long specimen. In this test, moduli were also estimated from the overall axial strain measurements, in which corrections were applied to allow for the compressibility of the load cells and other parts of the apparatus. Those results are in surprisingly good agreement with those obtained from internal measurements. This approach is a promising one, but extensive development and calibration is required before they can be usefully applied on a regular basis in offshore investigations.

IN-SITU TESTS

It appears that, no matter how carefully an offshore sample is collected, its quality would be lower than is normally possible on land. At present, the amount of information obtained from offshore samples and the confidence level of such information are not comparable with the results of similar work on land. In order to enhance the data quality, it is a common practice to test the sample *in situ*. In fact, *in-situ* tests have become an integral part of any comprehensive offshore soil investigation program.

A large number of devices have been developed in over two decades to measure engineering properties of submarine soils *in situ*. The tests include measuring fundamental properties such as shear strength and deformation characteristics, and indirectly related properties such as soil density and acoustic velocity. Several review articles describe those test devices (Noorany, 1972; Marshland and Windle, 1982; De Ruiter and Richards, 1983; Richards and Zuidberg,1985; McClelland, 1975; Dayal, 1978; Richards *et al.*, 1972). An exhaustive list of *in-situ* strength testing devices has been compiled by Richards and Zuidberg (1985). Out of over a dozen devices developed, the equipment most commonly used by the industry (about 95%) are the cone penetrometer, piezocone and vane tests. The use of the pressuremeter is relatively very low (about 2 to 3%): all other devices combined together are used even less in industry. In summary, only three *in-situ* tests to measure strength are in common industrial use, the vane, the cone penetrometer, and the piezocone. These devices could be operated by the wireline system, the seabed jacking system or the seabed tethered system. The literature cited above and other recent conferences (Norris and Holtz, 1981; Verruijt *et al.*, 1982) have been devoted to these devices and, consequently, further description and interpretation of test results is deemed unnecessary here.

COMPARISON OF *IN-SITU* AND LABORATORY TEST RESULTS

In marine geotechnology, the most common comparison that has been made is between the laboratory miniature vane shear test and the *in-situ* vane shear test. Based on the analysis of data from several sites in the Gulf of Mexico, Kraft *et al.* (1976) have concluded that *in-situ* vane measured strength is not a constant multiple of the vane strength measured in the laboratory on collected cores. The ratio of *in-situ* vane strength to laboratory vane strength ranged from one to more than three. McClelland (1975) and Lee (1973) have reported *in-situ* vane test results 30–50% higher than laboratory test results. The exceptions are in very plastic sediments, where laboratory vanes give slightly higher strengths compared to field tests. Based on all the available data, a rough average of field over laboratory vane results is approximately 1.4. It appears that the ratio increases as plasticity decreases or as stiffness increases.

Another comparison that has been attempted in a few cases offshore is between the *in-situ* cone penetration test and the vane shear test. Demars and Taylor (1971) have reported a ratio of cone pressure to *in-situ* vane shear strength which ranged between 8 to 16, with 13 as a representative value for clayey sediment. The ratio of cone pressure to laboratory vane shear strength is reported by Lee (1985) as ranging between 15 and 20, with 17 as the value yielding the best agreement. The higher values are attributed to sample disturbances caused to the laboratory values. Zuidberg (1975) has also recommended a ratio of cone pressure to laboratory vane strength of about 18. By taking Demars and Taylor's ratio of 13 between cone pressure and field strength, and increasing by 40% to account for the sample disturbances, one obtains 18. All the values, therefore, are roughly consistent. The ratio of cone pressure to laboratory vane strength is in the range of 15–20, while the ratio of cone pressure to field strength is around 10–15.

Another type of comparison has been between laboratory strengths of samples taken with different samplers. Lee (1985) has compared several different types of gravity-driven coring devices with standard specimens taken with a corer specifically designed to obtain high-quality samples. He reported strengths 20% higher than those obtained from conventional oceanographic corers. Sullivan *et al.* (1979) found no difference between unconsolidated undrained strengths of pushed and driven samples of overconsolidated clay. Nevertheless, the estimated overconsolidation ratio and deformation moduli were lower for driven samples. However, a ratio of about 1.4 between the miniature vane strengths of pushed and driven samples has been reported in the literature (Quiros *et al.*, 1983; Doyle *et al.*, 1971). The strengths of pushed samples agreed very well with the *in-situ* vane shear strengths. Similarly, the ratio of unconfined compression strength of pushed and driven samples varied from 1.7 to 3.3 (Quiros *et al.*, 1983). It appears that the ratio depends on soil type and depositional history. However, it is well established that the difference between unconfined compression strengths of pushed and driven samples is greater than the difference in vane strengths.

The available data suggest considerable uncertainty as to which strength values are correct when only driven samples are available. One way to place greater reliance on the laboratory test results is to use high-quality samples obtained by push samplers. However, as described earlier, no matter how carefully an offshore soil sample is collected, it is bound to be much more disturbed in comparison to samples collected on land. The answer is to test the soil *in situ*. It is usually best to employ both methods to gain optimum information return from a cost-effective offshore subsurface exploration program.

CONCLUSIONS

- The sea floor is primarily formed by depositional processes, in contrast to land which is formed primarily in an erosional environment. Submarine sediments can be broadly classified as to whether the sediments are land-derived (terrigeneous) or are the result of marine activity (pelagic).
- The offshore samples are, additionally, disturbed due to the sampling process, gas charging, overstressing from drilling mud, changes in the total stress state and the presence of electrolyte. In order to partly solve the sample disturbance problem, it is recommended to perform triaxial or direct shear tests on a consolidated soil sample. Some engineers advocate consolidating the sample to *in-situ* overburden stress, while others recommend using the normalized soil parameter approach. Both approaches require good-quality samples, but it appears that the normalized soil parameter approach can handle somewhat more disturbance.
- The average ratio of field and laboratory vane tests is roughly 1.4. The ratio of cone pressure to field vane strength is approximately 13, whereas the ratio of cone pressure to laboratory vane strength is approximately 17. Depending upon the type of laboratory test, pushed samples yield strengths 1.4 to 3.3 times greater than driven samples. These data suggest considerable uncertainty as to which strength should be used for the design; hence measurement of the soil strength properties *in situ* is suggested as a means of enhancing data quality. It is usually best to employ both methods, sampling and *in-situ* testing, to gain optimum information return from a cost-effective offshore subsurface program.

REFERENCES

Andersen, A., Berre, J., Kleren, A. and Lunne, T. 1979. Procedures used to obtain soil parameters for foundation engineering in the North Sea. *Marine Geotech.* **3**,3.

Anderson, A., Sollie, S. and Richards, A. F. 1965. N.G.I. gas operated sea-floor sampler. In: *Proceedings of Sixth International Conference on Soil Mechanics and Foundation Engineering,* vol. 1. Montreal.

Chaney, R. C. and Fang, H. Y. 1983. Static and dynamic properties of marine sediments. *First Shanghai Symposium on Marine Geotechnology, and Nearshore/Offshore Structures.* Shanghai, People's Republic of China.

Chaney, R. C., Slonim, S. M. and Slonim, S. S. 1982. *Determination of Calcium Carbonate Contents in Soils, Geotechnical Properties, Behaviour and Performance of Calcium Soils.* ASTM STP 777.

Dayal, U. 1978. Recent trends in underwater *in-situ* soil testing, *IEEE J. Oceanic Engng* **OE-3**, 4.

De Ruiter, J. and Richards, A. F. 1983. Marine geotechnical investigation, a mature technology. *Proceedings of Geotechnical Practice in Offshore Engineering.* Austin, Texas: ASCE.

Demars, K. R. and Chaney, R. C. (eds) 1982. *Geotechnical Properties, Behaviour and Performance of Calcium Soils.* ASTM STP 777.

Demars, K. R. and Taylor, R. J. 1971. Naval seafloor soil sampling and in-place test equipment: a performance evaluation. *Tech. Report 730.* U.S. Naval and Civil Engineering Laboratory, Port Hueneme, California.

Doyle, E. H., McClelland, B. and Ferguson, G. H. 1971. Wireline vane probe for deep penetration measurements of ocean sediment strength. In: *Proceedings of the 3rd Offshore Technology Conference,* Vol. 1, OTC 1327. Houston, Texas.

Eide, O. 1974. *Marine Soil Mechanics.* NGI publication 103. Oslo, Norway.

Fotch, J. A. and Kraft, L. M. 1977. Progress in marine geotechnical engineering. *J. Geotech. Engng Div., ASCE* **103**, GT10.

Inderbitzen, A. L. (ed.) 1974. *Deep-sea Sediments, Physical and Mechanical Properties*. New York: Plenum Press.

Jerbo, A. 1967. Geotechnical and strength aspects of bothnian clay sediments. In: *Marine Geotechnique*. Urbana, Ill.: University of Illinois Press.

Kraft, L. M., Ahmad, N. and Focht, J. A. 1976. Application of remote vane results of offshore geotechnical problems. In: *Proceedings of the 8th Offshore Technological Conference*. OTC 2626, Houston, Texas.

Ladd, C. C. and Azzouz, A. S. 1983. Stress history and strength of stiff offshore clays. In: *Proceedings of Geotechnical Practice in offshore Engineering*. Austin, Texas: ASCE.

Ladd, C. C. and Foott, R. 1974. New design procedure for stability of soft clay. *J. Geotech. Engng Div., ASCE* **100**, GT7.

Lambe, T. W. and Whitman, R. V. 1969. *Soil Mechanics*. New York: Wiley.

Lee, H. J. 1973. Engineering properties of some North-Pacific and Bering Sea soils. *Tech. Note N-1283*. U.S. Naval and Civil Engineering Laboratory, Port Hueneme, California.

Lee, H. J. 1985. Laboratory determination of the strength of marine soils. In: *Proceedings of Testing of Marine Sediments: Laboratory and in-situ Measurements*. ASTM STP 883.

Lee, H. J. and Clausner, J. E. 1979. *Seafloor Sampling and Geotechnical Parameter Determination Handbook*. TR 873. Port Hueneme, California: Civil Engineering Laboratory.

McClelland, B. 1967. Progress in consolidation in delta front and prodelta clays of the Mississippi River. In: *Marine Geotechnique*, A. F. Richards (Ed.).

McClelland, B. 1975. Trends in marine site investigation: a perspective. In: *Proceedings of Offshore Europe*, 75.

Marshland, A. 1977. The evaluation of engineering design parameters of glacial clays. *Q. J. Engng Geol.* **10**, 1.

Marshland, A. and Windle, D. 1982. Development in offshore site investigation. *Proceedings of Oceanology International* **1**, 2.7.

Noorany, I. 1972. Underwater soil sampling and testing — a state-of-art review. In: *Underwater Soil Sampling, Testing and Construction Control*. ASTM STP 501.

Noorany, I. and Gizienski, S. F. 1970. Engineering properties of submarine soils: state-of-the-art review, *J. Soil Mech. Foundn Engng, ASCE* **96**, SM5.

Norris, G. M. and Holtz, R. D. (eds) 1981. *Cone Penetration Testing and Experience*. New York: ASCE.

Quiros, G. W., Young, A. G., Pelletier, J. H. and Chan, J. H. C. 1983. Shear strength interpretation for Gulf of Mexico clays. In: *Proceedings of Geotechnical Practice in Offshore Engineering*. Austin, Texas: ASCE.

Richards, A. F. and Parker, H. W. 1967. Surface coring for shear strength measurements. In: *Proceedings of a Conference on Civil Engineering in the Oceans-I*. San Francisco: ASCE.

Richards, A. F. and Zuidberg, H. M. 1985. *In-situ* determination of the strength of marine soils. In: *Proceedings of Testing of Marine Sediments: Laboratory and* in-situ *Measurements*. ASTM STP 883.

Richards, A. F., McDonald, V. J., Olson, R. E. and Keller, G. H. 1972. *In-place Measurements of Deep-sea Soil Shear Strength*. ASTM STM 501.

Sangrey, D. A. 1977. Marine geotechnology — state-of-the-art. *Marine Geotech.* **2**.

Schmertmann, J. H. 1955. *The Undisturbed Consolidation Behaviour of Clays*. ASCE 120.

Silva, A. J. 1974. *Marine Geomechanics: Overview and Projections, in Deep-sea Sediments*. New York: Plenum Press.

Sullivan, R. A. 1980. North Sea foundation investigation techniques. *Marine Geotech.* **4**, 1.

Sullivan, R. A., Wright, S. T. and Senner, D. W. F. 1979. Evaluation of design parameters from laboratory tests. In: *Proceedings of a Conference on Offshore Site Investigation*, London.

Verruijt, A., Beringen, F. L. and de Leeuw, E. H. (eds) 1982. *Penetration Testing*. Proceedings of the Second European Symposium on Penetration Testing, vol. 2., *Cone Penetration Test*. Rotterdam: A. A. Balkema.

Vey, E. and Nelson, R. D. 1967. Environmental effects on properties of ocean sediments. In: *Proceedings of a Conference on Civil Engineering in the Oceans-I*. San Francisco: ASCE.

Winterkorn, H. F. and Fang, H. Y. 1975. *Foundation Engineering Handbook*. New York: Van Nostrand Reinhold.

Zuidberg, H. M. 1975. The sea calf, a submersible cone penetrometer rig. *Marine Geotech.* **1** (1).

Part VIII

Loading Actions

A TECHNIQUE FOR ESTIMATING WAVE-INDUCED MOTION RESPONSES OF A CRANE BARGE SEMI-SUBMERSIBLE

Xiong-Jian Wu, Shanghai Jiao-Tong University, China

Presented on **Thursday 6th March 1986 at 10.00** in the Clarence Conference Room

EMBARGOED AGAINST DELIVERY

32

A Technique for Estimating Wave-induced Motion Responses of a Crane Barge Semisubmersible

*Xiong-Jian Wu,** Ship Hydrodynamics Laboratory, Shanghai Jiao-Tong University, China

INTRODUCTION

As an extension of the current twin-hull semi-submersible, a mono-hull crane barge semisubmersible has been developed based on a concept [1] of increasing the displacement together with an increase in waterplane area and a reduction of the width in order to maintain both the motion behaviour and the operation property.

The immersed part of a crane barge semisubmersible consists of a single barge-shaped hull and struts standing on the top of the hull and supporting the work deck. Figure 1 shows the general hull configuration of a crane barge semisubmersible.

Two main theoretical methods may be considered for obtaining the hydrodynamic characteristics of such a structure. One of the methods has been suggested by Hooft. [2] This method has also been recommended by some rules and regulations. [3,4] It relies on two fundamental assumptions. One of the assumptions is the validation of the superposition principle for the added mass, wave damping and wave exciting forces of all the discretized substructures or elements. The other assumes that the diameter of the substructure is so small compared to the wavelength that a modified Morison equation can be used to calculate both inertia and damping terms, and that the interaction between substructures can be ignored. Furthermore, in most cases the damping due to wave radiation may also be negligible.

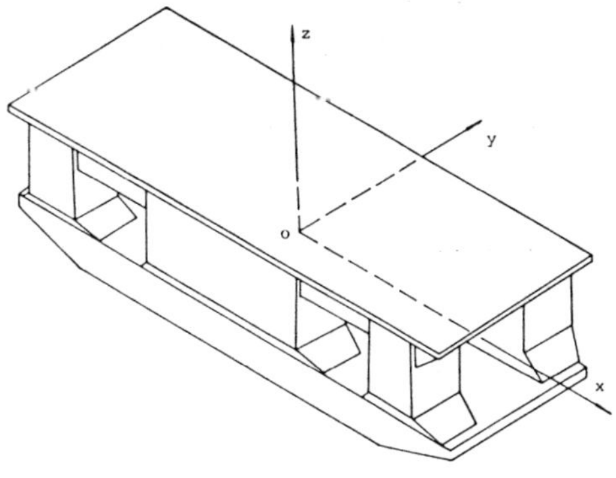

Fig. 1 General hull configuration of crane barge semisubmersible

*Present address: Department of Mechanical Engineering, Brunel University, Uxbridge, Middlesex, UB8 3PH.

The other method originates from the slender body theory, and a 2D strip theory is utilized for computation, including the interaction between massive substructures. [5] In 1970 Tasai *et al* [6] introduced viscous damping as a correction to the linear wave-damping term for predicting the heaving motion of a SWATH ship, whereas viscous correction for all the six modes of motion was proposed by Matsushima et al. [7]. A more detailed formulation of the viscous effect has been implemented by Lee and Curphey. [8] The increasing attention being paid to viscous damping is really due to its significance for predicting the motion of a semisubmersible.

The former method may be more appropriate to the semisubmersible platform composed of a great number of small members and the latter to the usual twin-hull semisubmersible with fewer substructures. For a crane barge semisubmersible with a partially truncated water region on the top of the hull, however, the conventional strip method may be unsuitable. Because of the large scale of the struts in comparison with the hull and the considerable interaction effect between struts and hull, Hooft's method may be inadequate as well. Instead, a 3D diffraction method should be adopted. Moreover, the eddy-making resistance may be significant to a certain extent for a submerged flat body oscillating in water. Therefore, a method combining the 3D diffraction theory and a viscous effect estimation is used in the present study for hydrodynamic analysis and motion predictions of the mono-hull crane barge semisubmersible.

MOTION PREDICTION METHOD

The six-degree-motion equation with viscous effect for a semisubmersible may be expressed as

$$\sum_{j=1}^{6} \{(M_{ij} + A_{ij})\ddot{X}_j + (B_{ij} + B_{vij})\dot{X}_j + C_{ij}X_j\}$$

$$= F_i + F_{vi}, \quad \text{for } i = 1\text{–}6, \tag{1}$$

where A_{ij} is the added mass, B_{ij} the wave damping, C_{ij} the hydrostatic restoring coefficient, and F_i the wave exciting force. B_{vij} and F_{vi} are the viscous damping and the 'viscous wave exciting force' due to drag. i,j = 1–6 denote the motion modes of surge, sway, heave, roll, pitch and yaw, respectively. $X_j = X_{ja}e^{-i\omega t}$ is the displacement of the jth mode of motion with motion amplitude $IX_{ja}I$ denoted by X_a, Y_a, Z_a, ϕ_a, θ_a *and* ψ_a for j = 1–6. M_{ij} is the generalized mass with M_{ii} = M for i = 1–3 and $M_{ii} = I_{ii}$ for i = 4–6, where M is the mass of the structure and I_{44}, I_{55} and I_{66} are the inertia moments of roll, pitch and yaw, respectively.

In order to solve the motion equation to obtain motion responses, the coefficients in equation (1) should be properly evaluated. With the present method, the potential terms, i.e. A_{ij}, B_{ij} and F_i, are yielded from the 3D diffraction theory [9] by using a source-dipole distribution method, [10] whereas the viscous terms, that is, B_{vij} and F_{vi}, are roughly estimated by a drag formulation. [3]

3D Diffraction Method

Within the linear wave theory the principle of superposition may be applied, and the total velocity potential due to an oscillatory body in regular waves can be expressed as

$$\Phi = (\phi^0 + \phi^7 + \sum_{j=1}^{6} X_{ja}\,\phi^j)e^{-i\omega t} \tag{2}$$

ϕ^0 is an incident wave potential and in the deep-water case can be written as

$$\phi^0 = -\frac{ig\zeta a}{\omega}\,\epsilon^{ik}(x\cos\beta + y\sin\beta)_e kz, \tag{3}$$

where ζ_a denotes the wave amplitude, ω is the wave frequency, $k = \omega^2/g$ is the wave number and β is the incident angle with β = 180° denoting a head sea case. ϕ^j (j = 1–6) denotes the jth radiation wave potential due to a unit displacement in the jth mode of motion and ϕ^7 denotes the diffraction wave potential corresponding to an incident wave potential. The Cartesian coordinate system is chosen so that the x, y axes are coincident with the still-water plane surface ($z = 0$) and the z axis is positive upwards.

The jth component potential ($j = 1$–7) is governed by the Laplace equation, the linearized free-surface condition, the seabed condition, the body surface condition and the radiation condition. [9,11] Applying Green's theorem, an integral equation in a singularity distribution form can be obtained. In the present study the source–dipole distribution method [10] is employed, and the resultant integral equation takes the form

$$2\pi\phi(P) - \iint_{S_w} \phi(Q)\frac{\partial G(P,Q)}{dn_Q}\,dS = -\iint_{S_w} v_n(Q)\,G(P,Q)\,dS, \quad (4)$$

where the superscript j is omitted from ϕ. $P(x,y,z)$ and $Q(\xi,\eta,\zeta)$ are two points on the body wetted surface s_w. $n = (n_1, n_2, n)$ is a unit normal pointing to the external fluid. v_n is the normal component of the velocity on the body surface and is equal to $-i\omega n_j$ for $j = 1$–6 or $-\partial\phi^0/\partial n$ for $j = 7$, where $(n_4, n_5, n_6) = (x - x_G, y - y_G, z - z$ n)xG with $G(x_G, y_G, z_G)$ denoting the gravity centre of the structure. $G(P,Q)$ is the Green function. In the case of deep water, the Green function expression can be found in refs 9 and 11 in an infinite integral form and in refs 12 and 13 in a finite integral form, which is much more efficient in numerical computation than the infinite integral expression. Moreover, a modified Green function form valid in all cases and used by the authors [10] can be written as

$$\begin{aligned} G &= 1/r + 1/r_1 + I_1 + iI_2, \\ I_1 &= ke^{k(z+\zeta)}\{\pi(H_0(kR) + T_0(kR)) \\ &\quad + 2ln k(r_1 - z - \zeta) + 2\int^{-k(z+\zeta)} F_0(kR,s)\,ds\}, \quad (5) \\ I_2 &= 2\pi e^{k(z+\zeta)}J_0(kR), \end{aligned}$$

with

$$T_0(kR) = \begin{cases} Y_0(kR) - 2/\pi\ ln(kR), & \text{for } kR \neq 0 \\ 2/\pi\ (\gamma - ln\ 2), & \text{for } kR = 0 \end{cases}$$

$$F_0(kR,s) = \begin{cases} (e^s - 1)/k^2R^2 + s^2)^{1/2}, & \text{for } k^2R^2 + s^2 \neq 0 \\ 1, & \text{for } R = 0,\ s \to 0 \end{cases}$$

where $R = \{(x-\xi)^2 + (y-\eta)^2\}^{1/2}$, $r_1 = \{R^2 + (z-\zeta)^2\}^{1/2}$ and $r_1 = \{R^2 + (z+\zeta)^2\}^{1/2}$. J_0 is the Bessel function of the first kind of zero order, Y_0 is the Bessel function of the second kind of zero order and H_0 is the Struve function of zero order. γ is the Euler constant.

Corresponding to the discretization of the body wetted surface into N surface panels, the integral equation, equation (4), is subdivided into N linear equations as

$$\sum_{j=1}^{N} \phi_j\{2\pi\delta_{ij} - \iint_{\Delta s_j} \frac{\partial}{\partial_{nj}}(G_{ij} - \delta_{ij}/r_{ij})dS = -\sum_{j=1}^{N}(v_{nj}\iint_{\Delta s_j} G_{ij}\,ds), \quad \text{for } i = 1,2,\ldots,N \quad (6)$$

or in a matrix form as

$$\underset{\sim}{A}\,\underset{\sim}{\phi} = \underset{\sim}{V}, \quad (7)$$

where Δs_j is the elemental area of the jth panel, $\delta_{ij} = \{\begin{smallmatrix}1 & \text{for } i=j \\ 0 & \text{for } i \neq j\end{smallmatrix}$, $\underset{\sim}{A}$ *is a* $(N \times N)$ complex matrix, $\underset{\sim}{\phi}$ is an unknown $(N \times 1)$ column matrix and $\underset{\sim}{V}$ is a known $(N \times 1)$ column matrix.

By matrix manipulation the velocity potential in (6) or (7) can be solved. Then, the added mass, wave damping and wave exciting forces can be determined by

$$\begin{aligned} A_{ij} &= \frac{\rho}{\omega}\sum_{m=1}^{N} Im(\phi^j)n_i\Delta s_m, \\ B_{ij} &= -\rho\sum_{m=1}^{N} Re(\phi^j)n_i\Delta s_m, \quad (8) \\ F_i &= -i\omega\rho\{\sum_{m=1}^{N}(\phi^0 + \phi^7)n_i\Delta s_m\}e^{-i\omega t}, \end{aligned}$$

where ρ is the water density.

A very sensitive defect of a singularity distribution technique is the existence of 'irregular frequencies' which are the eigenfrequencies of an interior eigenvalue problem defined by John. [8] At these discrete irregular frequencies the integral equation (equation (6) or equation (7)) fails to produce correct solutions. Recently, Wu and Price [14,15] derived a formulation to predict the values of irregular frequencies for an arbitrary body geometry and concluded that the irregular frequencies in a multi-hull structure are the sum of those of each individual sub-hull. Applying Wu and Price's theory, we can find that for the crane barge semisubmersible there is no irregular frequency in the frequency range of practical interest.

Both the mathematical failure due to irregular frequencies and some physical rapid changes due to resonant waves [16] can cause the integral formulation to be ill conditioned. For indicating whether or not the linear equation system (equation (6) or equation (7)) is ill conditioned, a condition number [17] is defined as

$$C_n = \frac{1}{N} \| \underset{\sim}{A} \|_E \| \underset{\sim}{A}^{-1} \|_E, \tag{9}$$

where $\| \cdot \|_E$ is the Euclidean norm. According to this definition the linear equation system is called well conditioned if C_n is close to 1.0 and is called ill conditioned if C_n is much greater than 1.0.

Viscous Effect Estimation

The viscous effect due to drag has been discussed in detail for small cylinders, [3] in very brief form for a SWATH ship [7] in six modes of motion and in detailed formulation for a slender twin-hull ship [8] in five modes of motion. For estimating the viscous effect on a crane barge semisubmersible with massive hull and struts, the following procedure is adopted. That is, such a structure is subdivided into three sets of strips along the x, y and z axes. Then suitable points on each strip contour are chosen as representative points for evaluating the relative velocity. The viscous forces and moments due to drag in six directions may be approximated as

$$\begin{aligned} \Delta F_{D1} &= 0.5\rho C_D |\dot{X}_r| \dot{X}_r \Delta s_1, \\ \Delta F_{D2} &= 0.5\rho C_D |\dot{Y}_r| \dot{Y}_r \Delta s_2, \\ \Delta F_{D3} &= 0.5\rho C_D |\dot{Z}_r| \dot{Z}_r \Delta s_3, \end{aligned}$$

and (10)

$$\begin{aligned} F_{Di} &= \Sigma \Delta F_{Di}, \quad \text{for } i = 1-3, \\ F_{D4} &= -\Sigma (z - z_G)\, \Delta F_{D2} + \Sigma (y - y_G)\, \Delta F_{D3}, \\ F_{D5} &= -\Sigma (x - x_G)\, \Delta F_{D3} + \Sigma (z - z_G)\, \Delta F_{D1}, \\ F_{D6} &= -\Sigma (y - y_G)\, \Delta F_{D1} + \Sigma (x - x_G)\, \Delta F_{D2}, \end{aligned}$$

where C_D is the drag coefficient which is Reynolds number dependent. When experimental data are not available it may be chosen from recommended values [3,18] for typical geometries. Δs_i is the projected area of a strip element in the ith direction. ΔF_{Di} denotes the viscous force on a strip element in the ith direction. Σ means summation over the total body surface. $\dot{X}_r$, $\dot{Y}_r$ and $\dot{Z}_r$ are relative velocity components in the x, y and z directions and may be written as

$$\begin{aligned} \dot{X}_r &= -\dot{X}_1 - (z - z_G)\dot{X}_5 + (y - y_G)\dot{X}_6 + W_x' \\ \dot{Y}_r &= -\dot{X}_2 - (x - x_G)\dot{X}_6 + (z - z_G)\dot{X}_4 + W_y' \\ \dot{Z}_r &= -\dot{X}_3 - (y - y_G)\dot{X}_4 + (x - x_G)\dot{X}_5 + W_z' \end{aligned} \tag{11}$$

where the velocity components of a wave particle in the x, y and z directions are

$$\begin{aligned} W_x &= W_{xo} e^{-i\omega t} = \partial\phi^0/\partial x, \\ W_y &= W_{yo} e^{-i\omega t} = \partial\phi^0/\partial y, \\ W_z &= W_{zo} e^{-i\omega t} = \partial\phi^0/\partial z, \end{aligned} \tag{12}$$

where

$$\begin{aligned} \begin{Bmatrix} W_{xo} \\ W_{yo} \end{Bmatrix} &= \begin{Bmatrix} \cos\beta \\ \sin\beta \end{Bmatrix} \omega \zeta_a e^{ik(x\cos\beta + y\sin\ )} e^{kz}, \\ W_{zo} &= -\omega \zeta_a e^{ik(x\cos\beta + y\sin\beta)} e^{kz}. \end{aligned}$$

After substituting (11) into (10) and applying an equivalent linearization procedure we obtain

$$\Delta F_{D_1} = -\frac{4}{3\pi} \rho C_D \Delta s_1 \dot{X}_{ro} \{\dot{X}_1 + (z - z_G)\dot{X}_5 - (y - y_G)\dot{X}_6 - W_x\},$$

$$\Delta F_{D_2} = -\frac{4}{3\pi} \rho C_D \Delta s_2 \dot{Y}_{ro} \{\dot{X}_2 + (x - x_G)\dot{X}_6 - (z - z_G)\dot{X}_4 - W_y\}, \tag{14}$$

$$\Delta F_{D_3} = -\frac{4}{3\pi} \rho C_D \Delta s_3 \dot{Z}_{ro} \{\dot{X}_3 + (y - y_G)\dot{X}_4 - (x - x_G)\dot{X}_5 - W_z\},$$

where $\dot{X}_{ro}$, $\dot{Y}_{ro}$ and $\dot{Z}_{ro}$ are amplitudes of $\dot{X}_r$, $\dot{Y}_r$ and $\dot{Z}_r$ and can be written as

$$X_{ro} = \omega\{[\mathrm{Re}\,(X_m)]^2 + [\mathrm{Im}\,(X_m)]^2\}^{1/2},$$
$$Y_{ro} = \omega\{[\mathrm{Re}\,(Y_m)]^2 + [\mathrm{Im}\,(Y_m)]^2\}^{1/2}, \quad (15)$$
$$Z_{ro} = \omega\{[\mathrm{Re}\,(Z_m)]^2 + [\mathrm{Im}\,(Z_m)]^2\}^{1/2},$$

where X_m, Y_m and Z_m are as follows

$$X_m = X_{1a} + (z - z_G)\,X_{5a} - (y - y_G)\,X_{6a} + W_{xo}/i\omega,$$
$$Y_m = X_{2a} + (x - x_G)\,X_{6a} - (z - z_G)\,X_{4a} + W_{yo}/i\omega,$$
$$Z_m = X_{3a} + (y - y_G)\,X_{4a} - (x - x_G)\,X_{5a} + W_{zo}/i\omega.$$

Then the viscous force can be split into viscous damping terms and a viscous wave exciting force term, that is,

$$F_{Di} = -\sum_{j=1}^{6} B_{vij}\dot{X}_j + F_{vi}, \quad \text{for } i = 1\text{–}6, \quad (17)$$

where viscous wave exciting forces may be expressed as

$$F_{v1} = \Sigma a_x \dot{X}_{ro} W_x,$$
$$F_{v2} = \Sigma a_y \dot{X}_{ro} W_y,$$
$$F_{v3} = \Sigma a_z \dot{X}_{ro} W_z,$$
$$F_{v4} = \Sigma\,(y - y_G)\,a_z \dot{Z}_{ro} W_z - \Sigma\,(z - z_G)\,a_y \dot{Y}_{ro} W_y, \quad (18)$$
$$F_{v5} = \Sigma\,(z - z_G)\,a_x \dot{X}_{ro} W_x - \Sigma\,(x - x_G)\,a_z \dot{Z}_{ro} W_z,$$
$$F_{v6} = \Sigma\,(x - x_G)\,a_y \dot{Y}_{ro} W_y - \Sigma\,(y - y_G)\,a_x \dot{X}_{ro} W_x,$$

with

$$(a_x, a_y, a_z) = -\frac{4}{3\pi}\rho C_D(\Delta s_1, \Delta s_2, \Delta s_3).$$

It is obvious that the viscous damping and the viscous wave exciting force are dependent on motion and amplitude, and therefore they will be determined simultaneously with the solution of the motion equation (equation (1)) by an iterative method.

EXAMPLE STUDY AND DISCUSSIONS

An existing mono-hull crane barge semisubmersible [1] is adopted for the sample computation. The hull configuration and main dimensions are shown in Fig. 2 and its main particulars are listed in Table 1.

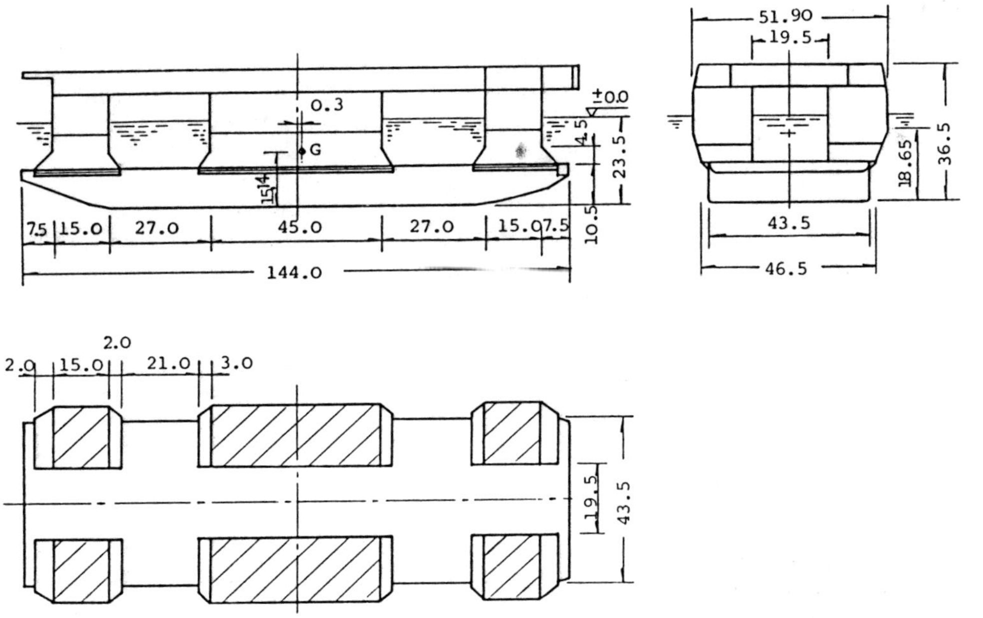

Fig. 2 Main dimensions of the crane barge semisubmersible

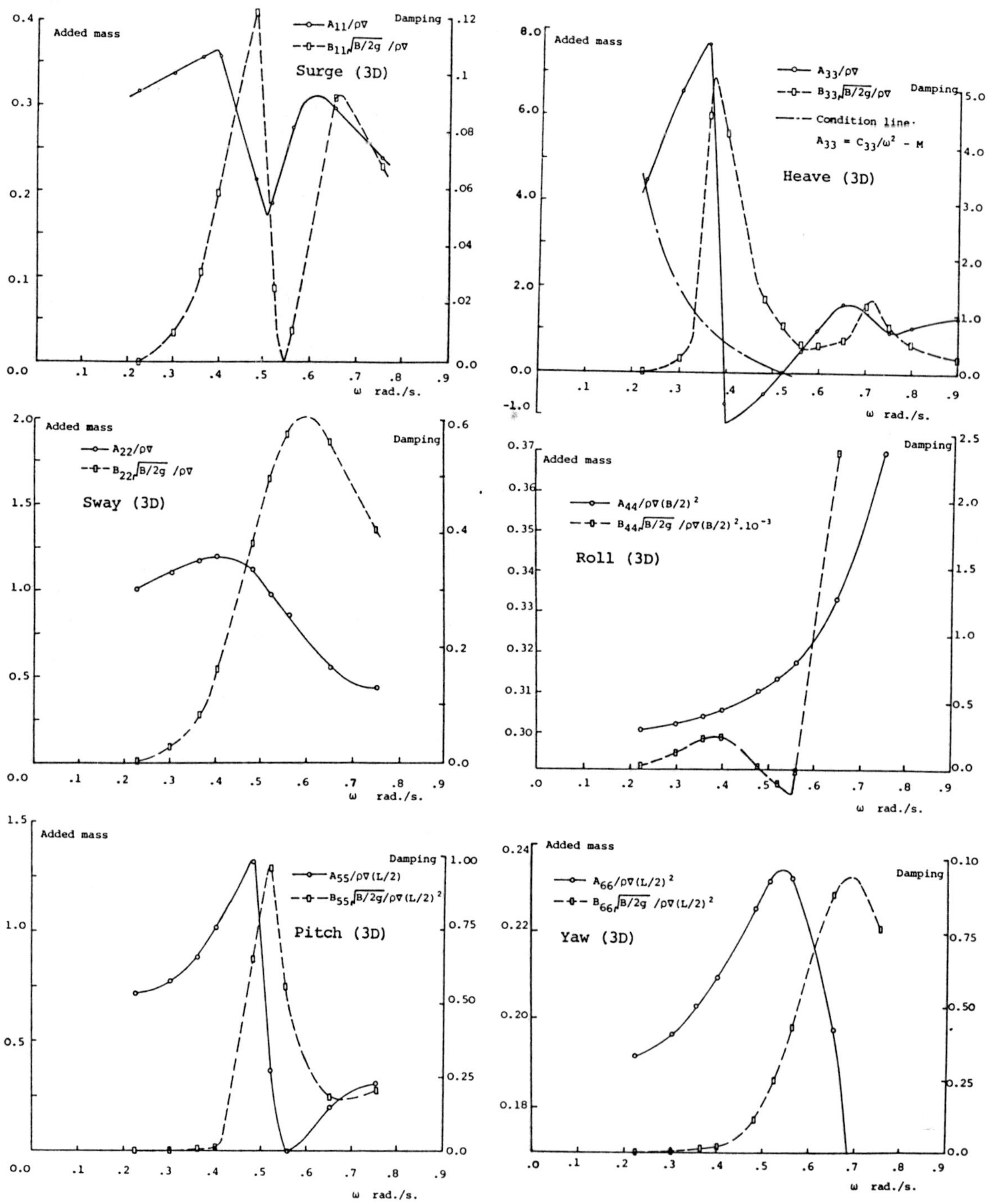

Fig. 3 3D added mass and damping coefficients of the crane barge semisubmersible for six modes of motion, that is, surge, sway, heave, roll, pitch and yaw

TABLE 1
Main particulars of a crane barge semisubmersible

Description	*Magnitude*
Hull length (L)	144.0 m
Hull beam (B)	43.5 m
Hull depth (D)	10.5 m
Draft (h)	23.5 m
Centre of gravity (X_G, Y_G, Z_G)	(0.3, 0.0, –8.36)
Displacement volume	91,196.0 m^3
Metacentric height	3.42 m
Gyradii k_{xx}, k_{yy}, k_{zz}	23.6, 41.0, 40.35 m

Taking advantage of the port and starboard symmetry, half the structure is discretized into 193 panels for the 3D diffraction method calculation. The resultant added mass and damping coefficients obtained for six degrees of freedom are shown in Fig. 3. Apparently, rapid variations occur in some calculated hydrodynamic coefficients. Especially, the heaving added mass displays sharp change with negative values in the frequency range $\omega = 0.39$–0.52 rad/s. Correspondingly the heaving damping drops from a peak value to a minimum value. These phenomena are very similar to those observed in a current twin-hull calculation and they are recognized as the interaction effect between two sub-hulls. In the present case, variations can be caused by the interaction of the struts on both sides and the submerged hull. A free-surface water region partially truncated by the struts and the hull may possess a frequency response property. The resonant motion of this water region may play an important role in some frequency ranges and induce rapid changes in the calculated hydrodynamic coefficients. By extending an approximate technique called the 'subregion matching method' [19,20] a resonant frequency of this limited water region has been predicted at $\omega \simeq 0.567$ rad/s. This resonant frequency coincides well with the frequency range of large variation in calculated hydrodynamic coefficients. A similar phenomenon has recently been confirmed by Newman, Sortland and Vinje [21] for a rectangle of small submergence.

The importance of this interaction effect in motion prediction and design considerations is the possible appearance of multiple heaving resonant frequencies. In Fig. 3 we may insert the condition lines defined by

$$A_{ii} = C_{ii}/\omega^2 - M_{ii}, \quad i = 3\text{–}5. \tag{19}$$

In basic naval architecture the intersecting points of the condition line and the added mass curve represent motion resonant frequencies. It is obvious that for a one degree of freedom oscillation the motion response is

$$X_i = \frac{F_i}{C_{ii} - \omega^2 (M_{ii} + A_{ii}) - i\omega B_{ii}}, \quad i = 3\text{–}5.$$

At an intersecting point $C_{ii} - \omega^2 (M_{ii} + A_{ii}) = 0$, and the motion response rises to a resonant peak value:

$$X_i = \frac{F_i}{-i\omega B_{ii}}$$

Because of the dominant self-induced response and small coupling effect due to the other modes of motion, a slight shift of the resonant peak value is expected. In terms of these condition lines the rolling resonant frequency is predicted at $\omega = 0.22$ rad/s, the pitching resonant frequency is at $\omega = 0.246$ rad/s and multiple heaving resonant frequencies are found at $\omega_1 = 0.22$, $\omega_2 = 0.393$ and $\omega_3 = 0.52$ rad/s, as can be seen in Fig. 3. The first heaving resonant frequency and the rolling and pitching resonant frequencies are far below the wave energy concentration frequency of a sea site and therefore are of no practical importance. However, the extra heaving resonant frequencies, in particular the third one relating to a wave period of 12.1 s, lie just within the frequency range of dominant energy concentration in a seaway. The extra resonant motion responses, if any, will reduce the operating capacity of the crane barge semisubmersible. Hence, particular attention must be paid to this problem. By combining the data from the 3D diffraction calculation and the viscous effect estimation as stated in the section on the motion prediction method, the motion responses for the crane barge semisubmersible at zero forward speed are found by an iterative procedure used to solve the motion equation

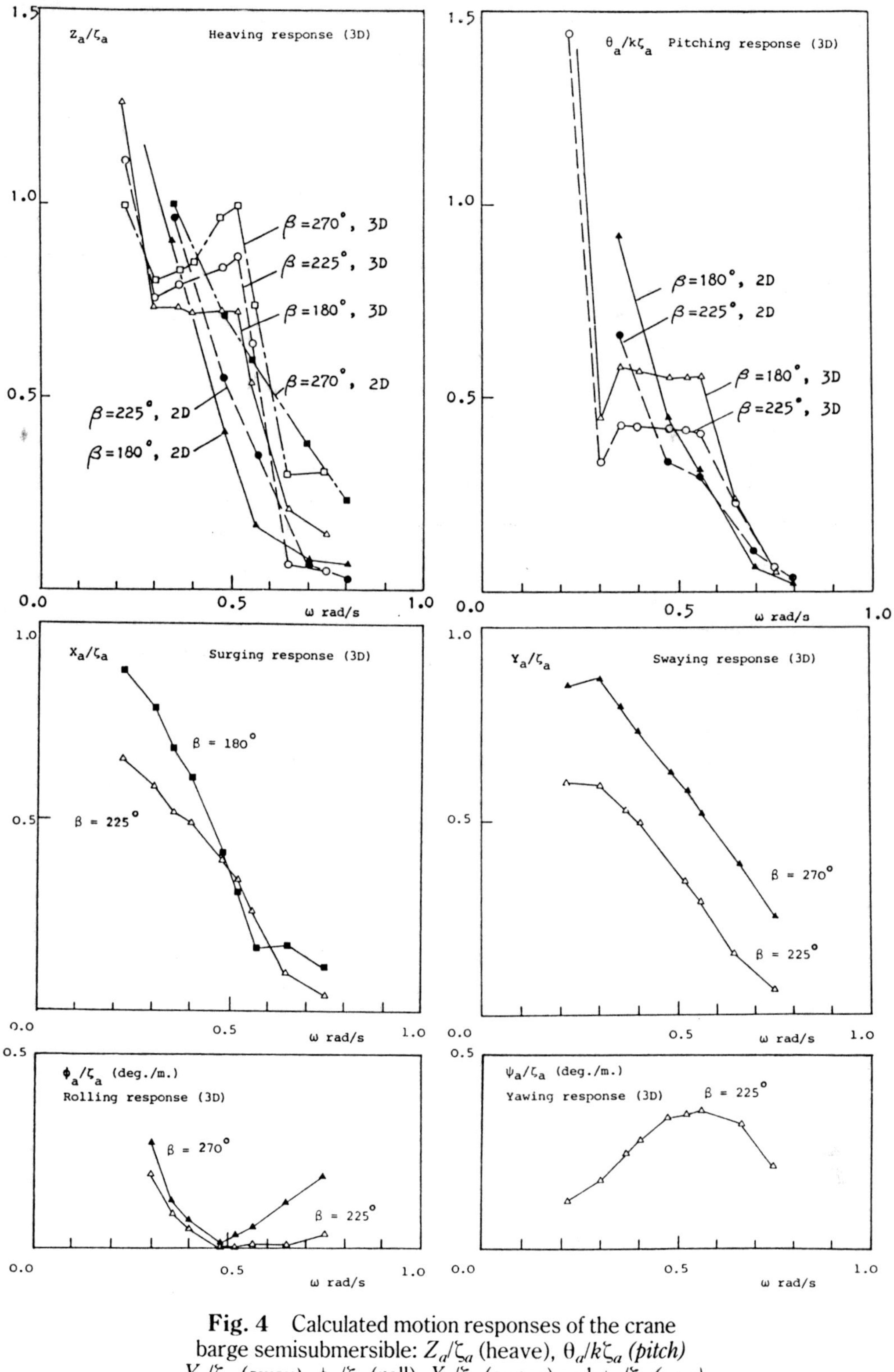

Fig. 4 Calculated motion responses of the crane barge semisubmersible: Z_a/ζ_a (heave), $\theta_a/k\zeta_a$ *(pitch)* Y_a/ζ_a (sway), ϕ_a/ζ_a (roll), X_a/ζ_a (surge) and ψ_a/ζ_a *(yaw)*

(equation (1)). Three wave headings are considered, that is, $\beta = 180°$ (head sea), 225° (bow sea) and 270° (beam sea). These results are shown in Fig. 4 and are in reasonable agreement with the experimental results [22,20] (not presented). From the heaving responses in Fig. 4 there is no response peak at the second heaving resonant frequency owing to the maximum damping value at this frequency, as can be observed in Fig. 3. But there exist heaving response peaks at the third one. Fortunately, these peak values are not serious, due to the combined action of the wave and the viscous dampings. Heaving responses under the viscous effect and without viscous influence are displayed together in Fig. 5 for three wave headings. It can be seen that the viscous effect is of importance round about the third heaving resonant frequency where radiation wave damping is small, is of no significance at high frequencies since the inertia becomes dominant, and may be significant at low frequencies when the drag becomes more important.

Now we shall investigate further the applicability of a strip theory in analysing the mono-hull semisubmersible. The structure is subdivided into a submerged hull and six struts, and then the strip method calculation is performed. The resultant heaving added mass and damping are shown in Fig. 6. Again, a condition line defined by equation (19) is inserted into Fig. 6; it intersects the added mass curve at one point, that is, $\omega = 0.22$ rad/s. This predicted heaving resonant frequency is the same as the first heaving resonant frequency predicted by the 3D calculation. There is no extra resonant frequency by the 2D strip calculation and therefore no extra heaving response peaks are expected, apart from the first heaving resonant frequency. Heaving motion responses resulted from the 2D calculation are also given in Fig. 4. In comparison with the 3D results the 2D calculation is underpredicted in the important frequency range but overpredicted at low frequencies. These may imply that for a mono-hull semisubmersible the 2D strip method is inadequate, whereas a combination of the 3D diffraction theory and an estimation of viscous effect can provide reasonable information on the hydrodynamic characteristics and motion responses.

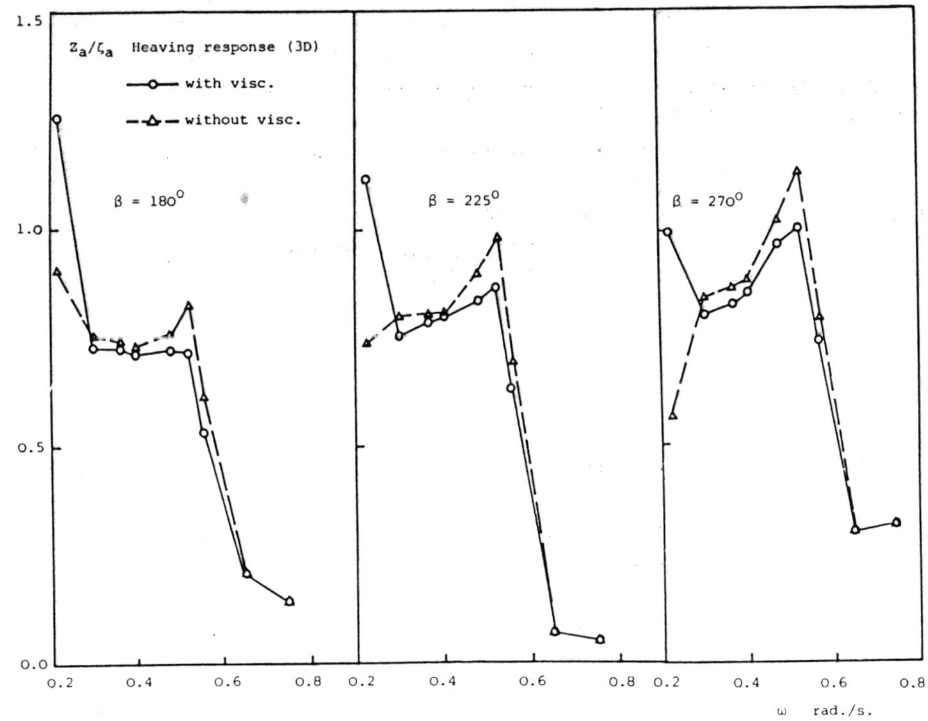

Fig. 5 Viscous effect on motion responses

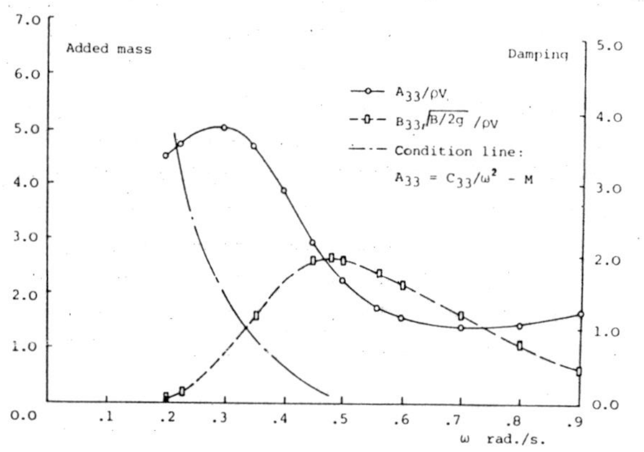

Fig. 6 2D heaving added mass and damping

Finally the condition number associated with the 3D diffraction method computation is given in Fig. 7 where serious fluctuations exist. According to Wu and Price's theory [14,15] there is no irregular frequency for the crane barge semisubmersible in the frequency range studied. There is no doubt that the fluctuation of the C_n value is induced by the mutual interaction of the substructures and perhaps by standing waves occurring in the partially truncated free-surface water region between the hull and the struts.

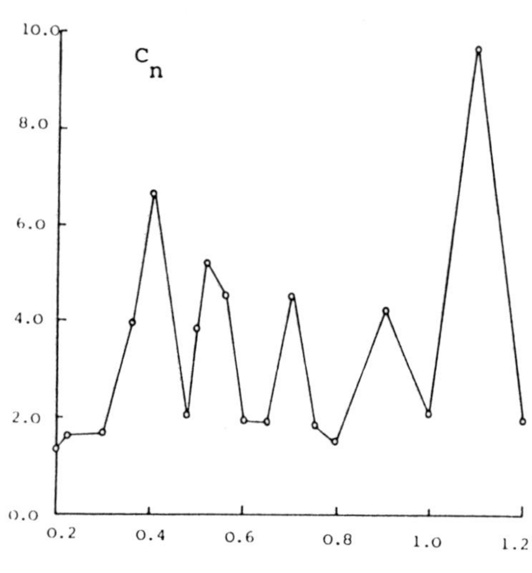

Fig. 7 Condition number

CONCLUSIONS

From the present study the following conclusions may be drawn:

(1) Due to the particular configuration of the mono-hull crane barge semisubmersible, multiple heaving resonant frequencies and extra motion response peaks may occur. This phenomenon is also found for a dry-dock in semisubmergence [23]. Therefore attention should be paid in the preliminary design stage to reducing the extra peak motion responses in order to increase the work capacity of structures of this kind.

(2) For small structures, the viscous effect predominates. For large structures, especially for large semisubmersibles with multiple natural frequencies, the viscous influence may be of some importance round about the resonant motion frequencies but may be negligible outside the vicinity of the resonant frequencies, except when a calculation at low frequencies is required.

(3) The present prediction method combining the 3D diffraction theory and a viscous effect estimation is applicable in analysing the mono-hull semisubmersible.

(4) The 2D strip method calculation may be inadequate to provide reasonable information for design consideration of the mono-hull semisubmersible.

ACKNOWLEDGEMENTS

The author is indebted to Professor J. B. Caldwell for encouragement, Mr G. E. Hearn for help and the Royal Bos Kalis Westminster Group (Holland) for support during his previous research work at the University of Newcastle upon Tyne, England. The author is also indebted to Professor W. G. Price of Brunel University for encouragement during the preparation of the present paper.

REFERENCES

1. Kaldenbach, W. P. 1978. Design consideration for heavy derrick semi-submersible. Ocean Resources Engineering, Oct. 1978, 33–41.
2. Hooft, J. P. 1971. A mathematical method of determining hydrodynamically induced forces on a semi-submersible. *SNAME 79.*
3. Bureau Veritas 1975. Rules and regulations for the construction and classification of offshore platforms.
4. American Bureau of Shipping 1973. Rules for building and classing offshore mobile drilling units.
5. Pien, P. C. and Lee, C. M. 1972. Motion and resistance of a low-waterplane catamaran. In 9th Symp. Naval Hydrodynamics, pp. 463–545.
6. Tasai, F. *et al.* 1970. A study on the motion of a semi-submersible catamaran hull in regular waves. J. Soc. Naval Archi. West Japan, No. 40 (in Japanese).
7. Matsushima, M. *et al.* 1982. Seakeeping of a semi-submerged catamaran (SSC) vessel, *J. Soc. Naval Archi. West Japan,* No. 63, 97–114 (in Japanese).
8. Lee, C. M. and Curphey, R. M. 1977. Prediction of motion, stability, and wave load of small-waterplane-area, twin-hull ships. *Trans. SNAME.*
9. John, F. 1950. On the motion of floating bodies, part II. *Comm. Pure and Appl. Math.* **3**, 45–101.
10. Wu, Xiong-Jian and Price, W. G. A source-dipole method procedure for evaluating the wave-structure interaction problem. To be published.
11. Wehausen, J. V. and Laitone, E. V. 1960. Surface waves. In *Handbuck der Physik,* Band 9, Springer-Verlag, Berlin.
12. Kim, W. D. 1965. On the harmonic oscillations of a rigid body on a free surface. *J. Fluid Mechanics, 21,* 427–451.
13. Hearn, G. E. 1977. Alternative methods of evaluating Green's function in three-dimensional ship-wave problems, *J. Ship Research,* 89–93.
14. Wu, Xiong-Jian and Price, W. G. Is the irregular frequency predictable and solvable?: an efficient numerical procedure for hydrodynamic analysis of arbitrary two-dimensional bodies. To be published.
15. Wu, Xiong-Jian and Price, W. G. An equivalent box technique to predict irregular frequencies in arbitrarily-shaped three-dimensional marine structures. To be published.
16. Wu, Xiong-Jian and Price, W. G. Resonant waves in the fluid-structure interaction problem involving a free-surface. To be published.
17. Churchhouse, R. F. (Ed.) 1981. Numerical methods. In *Handbook of Applicable Math.,* Vol. 3. John Wiley and Sons.
18. Saunders, H. E. 1957. Hydrodynamics in ship design. *SNAME,* **2p**, 420–421.
19. Wu, Xiong-Jian, 1981. Theoretical estimation and experimental method on the parameters of the flume stabilizer. *J. Shanghai Jiao-Tong Univ.,* No. 2, 37–53 (in Chinese).
20. Wu, Xiong-Jian, 1982. Motion characteristics and relative prediction method of the derrick barge semi-submersible. *Report, Dept Naval Archi. and Shipbuilding, Newcastle University.*
21. Newman, J. N., Sortland, B. and Vinje, T. 1984. Added mass and damping of rectangular bodies close to the free surface. *J. Ship Research,* 219–225.
22. de Boom, W. C. 1978. Model tests on the semi-submersible derrick barge, 'Narwhal'. *NSMB report No. 02125-2-ZT.*
23. Wu, Xiong-Jian and Price, W. G. Behaviour of an offshore structure with a partly-truncated free-surface water region. *Paper to be presented in the OMAE Speciality Symposium (Feb. 1986), New Orleans.*

33

Intercomparison Tests Related to the Application of Drift Buoy Systems for Lagrangian Current Measurements in the Yellow Sea

Paul Ferris Smith, Ferranti ORE, Inc. Falmouth, Massachusetts, USA, and
Richard Limeburner, Woods Hole Oceanographic Institution, Woods Hole, Massachusetts, USA

INTRODUCTION

Based on the success of the torpedo drifter (Cresswell et al. 1978; Fig. 1) developed and used by Cresswell of CSIRO in the Indian Ocean, the Tasman Sea, and in eddys off the east Coast of Australia, (Cresswell and Vaudrey, 1977; Cresswell and Wood, 1977) and the Ferranti ORE version (Smith *et al* 1984; Fig. 2) used by Luyten in the Agulhas current off Africa (Luyten and Smith, 1985), the ORE torpedo drifter was planned for use in the Yellow Sea Experiment, a joint USA-PRC coastal processes study to be conducted in 1986. However, a test performed at the Scripps Institution of Oceanography (p.p. Niiler, personal communication) indicated that despite the advantages of the torpedo shape in reducing wind and wave drag force errors in major ocean current areas, in regions having low velocity current regimes these errors might still be significant. Therefore, an intercomparison experiment was conducted using the ORE torpedo drifter, small spherical drifters, and the Scripps spherical drifter–drogue system. Various drogues were used with the torpedo and the small spherical drifters. While wind and sea conditions at the time of the experiment were not extreme enough to conclusively answer the question raised by the Scripps observations, it is believed an important aspect of the use of drogued drifters was revealed.

THE DRIFTER BUOYS

Of the four drifter buoys used in this experiment, three were essentially identical foam-filled fiberglass spheres 45 cm in diameter, each having a heavy duty steel pad in the lower pole for the drogue attachment point. These spheres displace 113 pounds in seawater. The ORE spherical buoy (Fig. 3) weighing 28 pounds contains within the sealed watertight shell a 12 V alkaline battery pack providing in excess of six months' operation, the ARGOS satellite transmitter (PTT Model 2101A), a sea surface temperature sensor, and a new disc-shaped 401.650 mHz antenna developed by Chu Associates. The antenna provides a full horizon-to-horizon radiation pattern ideal for a spherical buoy. The SIO spherical buoy was not instrumented for these tests since visual tracking was employed. The ORE torpedo drifter con-

INTERCOMPARISON TESTS RELATED TO THE APPLICATION OF DRIFT BUOY SYSTEMS FOR LAGRANGIAN CURRENT MEASUREMENTS IN THE YELLOW SEA

P Ferris Smith, Ferranti ORE Inc; R Limeburner,
Woods Hole Oceanographic Institution; USA

Presented on **Thursday 6th March 1986 at 10.30**
in the Clarence Conference Room

EMBARGOED AGAINST DELIVERY

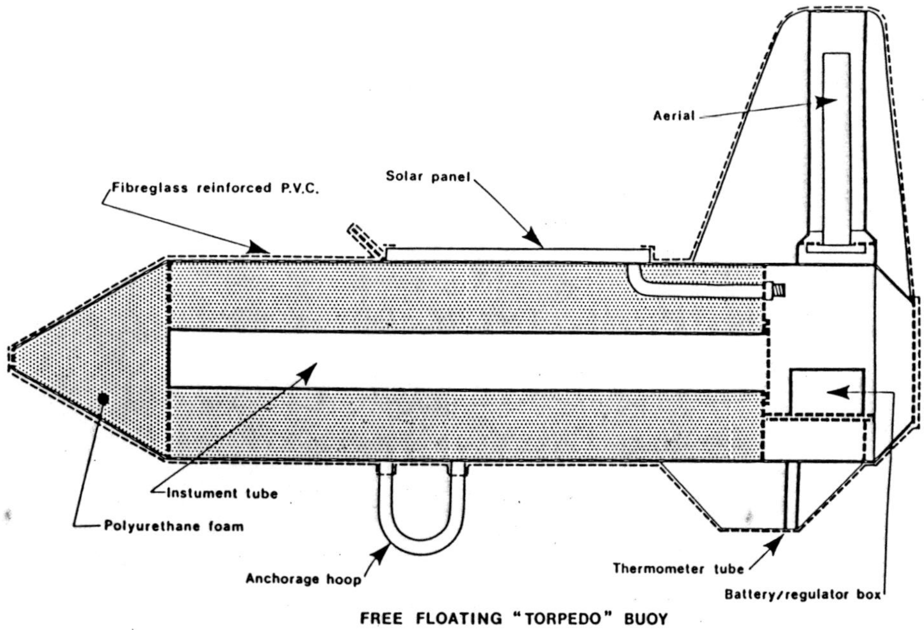

Fig. 1 CSIRO torpedo drifter buoy; overall length ca. 1.5 m. (From Cresswell *et al.*, 1978)

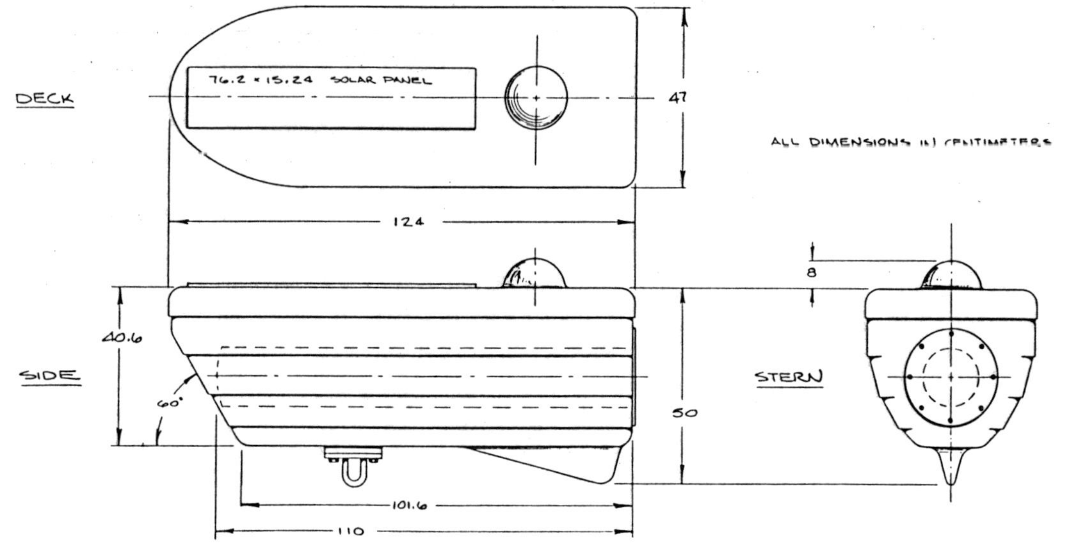

Fig. 2 Low-cost torpedo-shaped drifter, Ferranti ORE Model 830; overall length 1.24 m

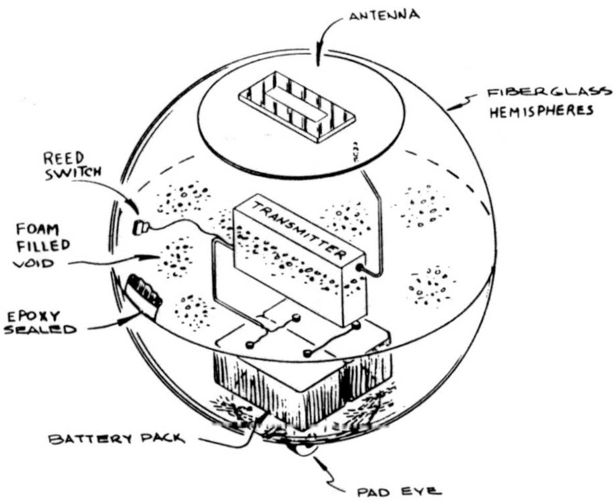

Fig. 3 Low-cost spherical drifter, Ferranti ORE Model 831; diameter 45 cm

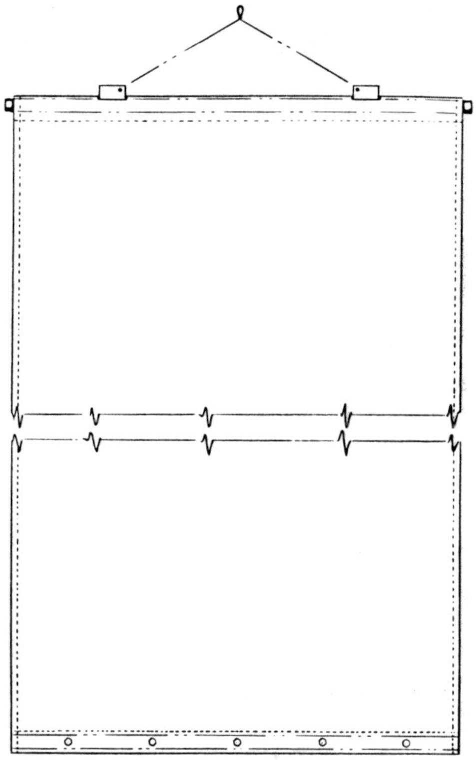

Fig. 4 Windowshade drogue; length 10 m, width 2 m

sists of a crosslinked polyethelene moulded hull filled with high-density polyurethane foam containing a PVC cylinder housing the ARGOS transmitter and an eighteen month operating-life alkaline battery pack. The ARGOS antenna and associated ground plane are sealed under a moulded blister on the upper surface. An internal steel keel supports the drogue attachment fitting and internal structural bulkheads. The sea surface temperature sensor is a waterproof probe located inside the skeg. The flat upper surface provides for battery-charging solar cells whereil long operating life in excess of eighteen months is required. Operation for over two years can be expected using these and lead–acid gell cells. This drifter weighs 90 lb and displaces 275 lb in seawater.

THE DROGUES

Three drogue designs were used: parachute, windowshade, and triplane. The parachute is a surplus 24 ft diameter military cargo parachute made of ripstop nylon ballasted by a 12 lb length of chain. The windowshade drogue (Vachon, 1975; Fig. 4) was made of nylon scrim reinforced plastic cloth 10 m long and 2 m wide, ballasted by the two upper and lower steel bars weighing respectively 15 lb and 10 lb. Two triplane drogues were used, both having the same 12 ft diagonal dimensions (Fig. 5) and both incorporating an auxiliary 12 in diameter plastic float to decouple the drogue from the motion of the surface drifter. The SIO assembly is built of PVC tubes. Its collapsible design allows it to be erected by tightening one central line. The other uses three 12 ft fiberglass tubes. It is assembled by inserting and securing the two horizontal tubes at the time of deployment. Both have sewn plastic cloth sails stretched tight to the structural members, and both are negatively buoyant in seawater. This downward force is countered in part by the 22 lb net buoyancy of the auxiliary float.

THE BUOY DROGUE RIGGING

It is proposed in the Yellow Sea studies to elucidate the Lagrangian current field for the

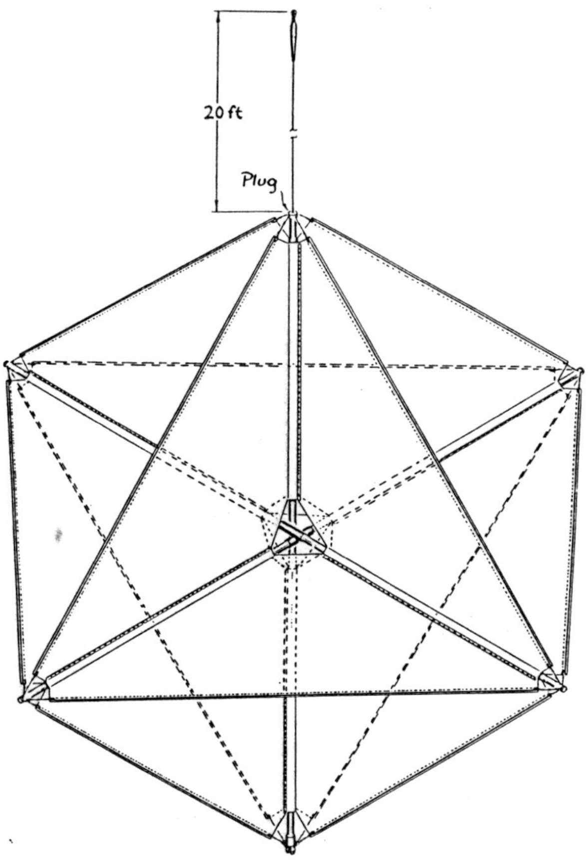

Fig. 5 Triplane drogue, diagonal members 4 m long

near surface water mass layer between 6 and 10 m deep. For this purpose, a number of drogue designs were considered. The required near-surface position of the drogue made any wave-induced motion a matter of concern in selecting the one to use. For this reason, the triplane design seemed appropriate since its horizontal plane would follow but not rectify this motion. The rigging used in this experiment was thus intended to place the bottommost part of the drogue at a depth of 10 m. In every case, this may not have been achieved for reasons discussed below. Figure 6 shows the various buoy–drogue arrangements. Note that the SIO drifter system put the drogue close to the desired depth. The torpedo–windowshade system placed the drogue over the depth range of 4–14 m. The sphere–parachute system had the effective drogue at 10 m. The ORE sphere–triplane was deployed first with and then later without a ballast chain, causing the drogue to be first over the depth range of 12–16 m and then over the range 6–0 m.

THE EXPERIMENTAL AREA AND RESULTS

All four drifters were initially deployed about 120 ft from each other in a square pattern in water approximately 85 ft deep in Vineyard Sound. This is a region dominated by a semi-diurnal tidal current of the order of 100 cm/s. The drifters were observed and positioned using LORAN-C for the following six hours. An attempt to measure drifter separation using Radar was unsuccessful due to the poor signal from these small low windage buoys. In addition a current profile was taken using a Neil Brown direct reading acoustic current meter (DRCM-2) as the drifters were passing the anchored Woods Hole Oceanographic Institution (WHOI) research vessel *Asterias*.

Twenty minutes after completing deployment, three of the drifters were lined up within 40–50 ft of each other normal to the current velocity. The sphere-parachute was somewhat behind or up stream of this group. As time (and tide) proceeded, the following pattern (details given below) evolved. The torpedo–windowshade system pulled out well in front, and the ORE sphere–triplane lagged well behind. The SIO sphere–triplane system and the sphere–parachute stayed quite closely together throughout the experiment.

After about three hours or halfway into the experiment, the leading torpedo–windowshade was recovered and reset near the ORE sphere–triplane, which was immediately recovered and reset without the lower ballast weight. The drogue was now over the 6–10 m depth range. Thereafter, both of these two drifters stayed together.

Figure 6 gives the measured current profile as well as the buoy and drogue depth combinations. It is this profile which explains the observed performance differences and indicates the key importance of drogue depth in the

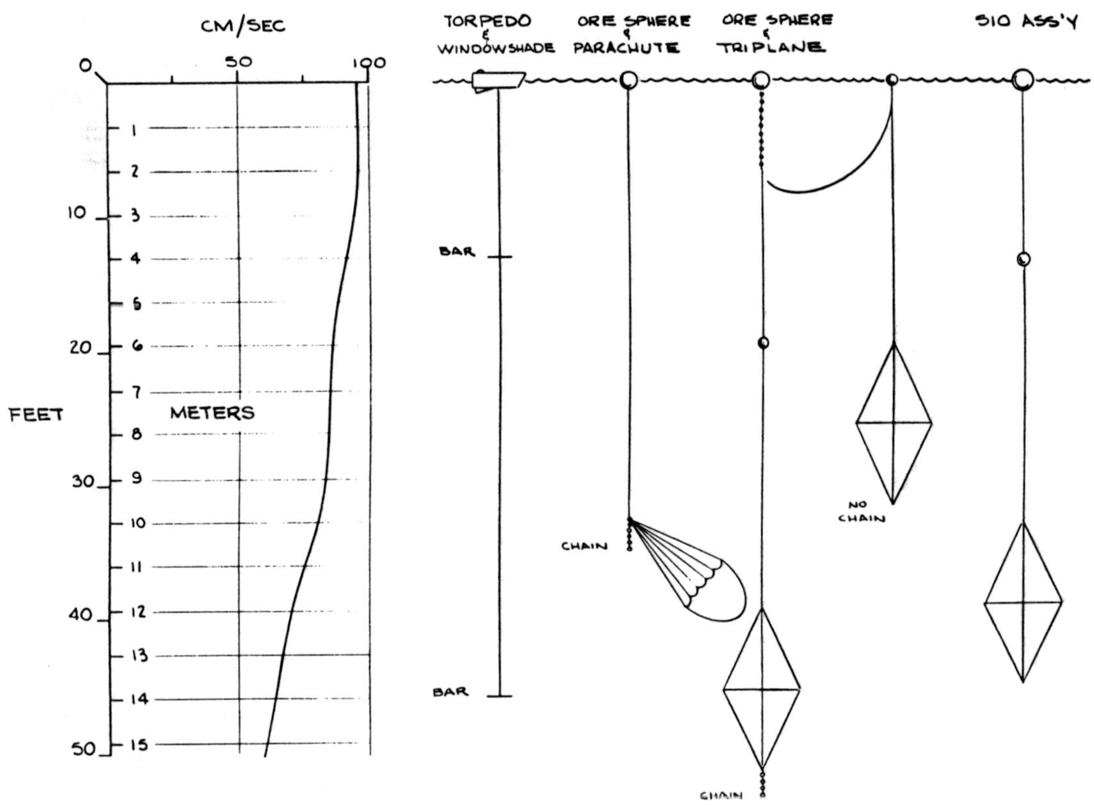

Fig. 6 Drifter buoy systems and corresponding measured current profile

presence of current sheer, rather than subtle drogue or drifter design differences.

The torpedo was out in front because the windowshade was integrating the higher near-surface currents. The ORE sphere with ballasted triplane was slowest because the drogue was deepest and thus in the lowest velocity water. The SIO sphere with triplane drogue and the sphere with parachute were coupled to the same water and thus tracked the same. When the ballast was removed from the ORE drogue it moved up to the higher velocity region felt by the windowshade and they tracked together. It is of importance to note that in the presence of current sheer small differences in drogue depth can result in considerable differences in performance.

WIND EFFECT

During this experiment the wind was 8–10 knots effectively at right angles to the current velocity. It was at first southeast and later south. The fact that all buoys continued to line up throughout the entire six hours, despite the considerable differences in exposed hulls, indicates the wind played a far less significant role than the drogue position in the water column during this experiment.

MEASURED VELOCITIES

The measured velocity magnitudes of the drifter–drogue systems were computed from the plotted LORAN-C positions and recorded time differences. The SIO system and sphere–parachute indicated a 70 cm/s velocity which corresponds to the measured value at their drogue depth as shown in Fig. 6. When reset, the torpedo–windowshade and ORE sphere–triplane indicated a velocity of 55 cm/s. This was lower than the directly measured value nearer the surface since this was at a later period in the diurnal tidal cycle. However, it does correspond to the predicted surface currents for this date and time (Eldredge, 1985). The indicated current direction was generally 240 degrees true, corresponding exactly to the tidal current charts, this being a well-defined channel. ec

ACKNOWLEDGEMENTS

The assistance of the WHOI research vessel *Asterias* skipper Steve Page during the field operations and Ferranti ORE technician Craig Apolinario both during the field work and in preparing the illustrations is much appreciated and gratefully acknowledged.

REFERENCES

Cresswell, G. R. and Vaudrey, D. J. 1977. Satellite-tracked buoy data report I: Western Australia releases 1975 and 1976. CSIRO Aust. *Div. Fish. Oceanogr. Rep. 86.*

Cresswell, G. R. and Wood, J. E. 1977. Satellite-tracked buoy data report II: Tasman Sea releases November 1976–July 1977. *CSIRO Aust. Div. Fish. Oceanogr. Rep. 91.*

Cresswell, G. R., Richardson, G. T., Wood, J. E. and Watts, R. 1978. The CSIRO Satellite-tracked 'Torpedo' buoy. *CSIRO Aust. Div. Fish. Oceanogr. Rep. 82.* Eldridge White, R. 1985. *Eldridge Tide and Pilot Book.* Robert Eldridge White, publisher.

Luyten, J. R. and Smith, P. F. 1985. Agulhas current trajectory from new ARGOS drifter compared with simultaneous shipboard measurements. In *Proceedings of Oceans-85,* November 12–14.

Smith, P. F., Frye, D. E. and Cresswell, G. R. 1984. New low cost ARGOS drifting buoy. In *Proceedings of Oceans-84,* September 10–12.

Vachon, W. A. 1975. Instrumented full-scale tests of a drifting buoy and drogue. *C. S. Draper Laboratory Report R-947,* December 1975.

34

Use of Optical-fibre Sensors for Crack Detection in Marine Environments

K. F. Hale, Senior Manager, Advanced Technology *and H. B. Boyle,* Head of Measurement (Systems and Engineering) Division, British Maritime Technology Ltd, Hythe, Southampton, UK

Standard optical-fibre crack-monitoring sensors can detect the initiation and growth of cracks with widths in the range 20 to 40 μm. They can be desensitized if required to detect wider cracks. The sensors are easy to apply to most structural surfaces and can withstand a wide range of hostile environments. Instrumentation is fully developed to allow continuous remote on-line surveillance of both fatigue test specimens or actual structural components.

INTRODUCTION

In-house development of the optical-fibre crack-monitoring sensor has now reached a highly satisfactory stage and the system is starting to find widespread use. The concept is very simple. An optical fibre is bonded to the surface of a structure by means of a high-modulus adhesive. If the structural surface cracks immediately below the fibre, then the fibre will crack also and thus light transmitted down the fibre will be attenuated. This attenuation can be detected remotely, so the system is a real-time crack-detection device. The principle of the crack-monitoring sensor is shown schematically

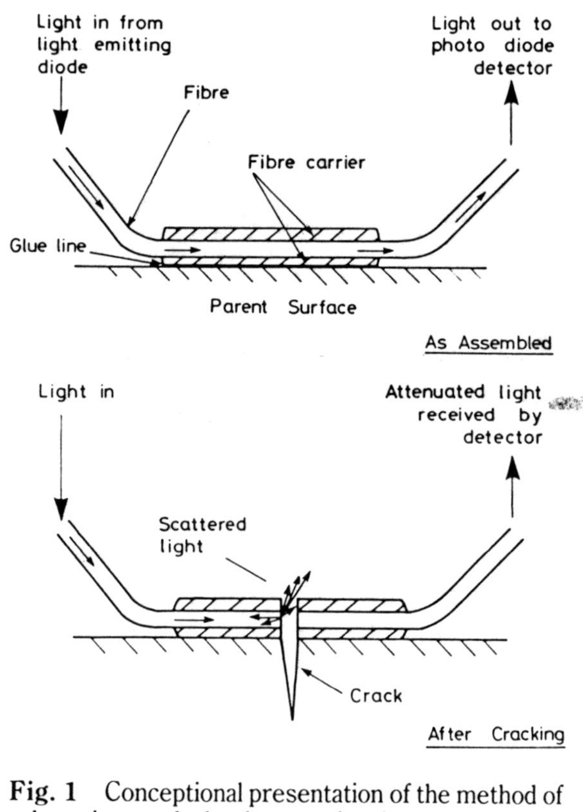

Fig. 1 Conceptional presentation of the method of detecting cracks by the use of optical-fibre crack-monitoring sensors

USE OF OPTICAL-FIBRE SENSORS FOR CRACK DETECTION IN MARINE ENVIRONMENTS

Dr K F Hale, H B Boyle, British Maritime
Technology Ltd, UK

Presented on **Thursday 6th March 1986 at 11.00**
in the Clarence Conference Room

EMBARGOED AGAINST DELIVERY

in Fig. 1. The fibres are encapsulated in translucent reinforced epoxy packages which can be made with a large range of dimensions, which both protect the fibres and also facilitate their accurate placement.

The advantages of this system over conventional witness devices, electrical crack-propagation gauges or other crack-detection systems are that it is:

- non-electrical and free from electromagnetic interference.
- non-conductive to electrical sources;
- safe in inflammable or explosive environments;
- capable of operation in more severe chemical and aqueous environments;
- capable of operation over a wider temperature range;
- capable of faithful signal transmission over long distances;
- usable with equal facility with the simplest light source or light source/detector monitoring instrumentation of the required degree of sophistication, depending upon the application;
- a precise crack locator by virtue of scattered light, even after a crack has closed up after removal of the strain that caused the crack to initiate.

The level of crack-detection sensitivity and the range of hostility of the working environment will be discussed.

DESCRIPTION

The sensors are designed to enable bare fibres of the required fracture strength to be bonded close to the surface of the structure to be monitored. The fibres must be quality-assured within tight optical and mechanical specifications

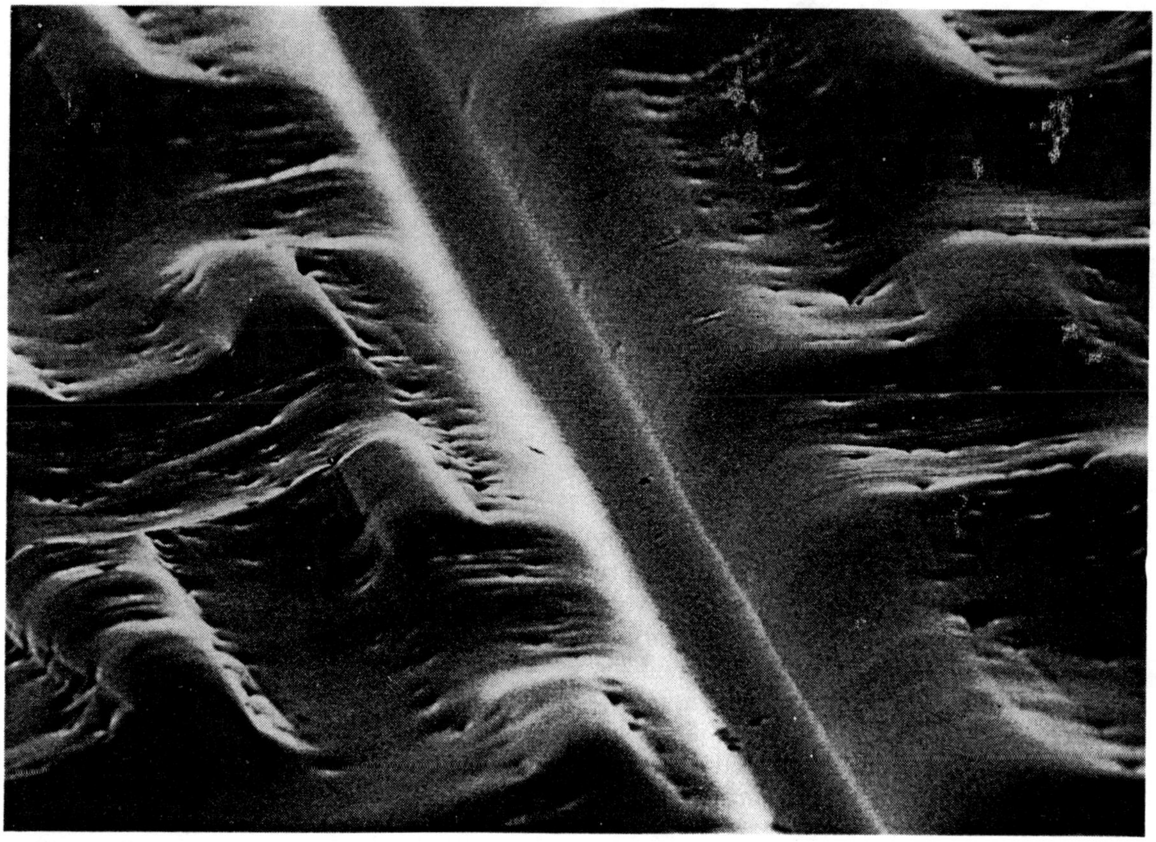

Fig. 2 Scanning electron micrograph of part of the sensing surface of an optical-fibre crack-monitoring sensor (magnification 50×)

in order to have a repeatable crack-width detection sensitivity with a reasonably narrow range of scatter and to be compatible with attenuation detection instrumentation.

Development of the sensors has benefitted from the large quantities of high-quality optical fibre, produced for the very demanding telecommunications industry, that are now readily available at low cost. This fibre is required to have closely defined optical properties for low-attenuation long-distance transmission. It is also designed to withstand abuse whilst it is pulled along narrow ducts and around small radius bends, and therefore has specified mechanical strength. It is coated with polymer to maintain its strength indefinitely.

For crack-monitoring sensors, the polymer protection is removed along a length where crack detection is to take place. Along this section of the fibre, a chemical sensitization treatment is given, and a length of backing material is provided which protects the fibre from damage and contamination, and also facilitates the application of adhesive to allow bonding of the fibre to the structure surface. The backing material has to be compatible with an essential high-modulus adhesive, to be sufficiently flexible to mould easily around a contoured surface and must also allow rapid production rates of the sensor. A suitable backing material is available as a quality-assured product for use in the aircraft industry. Quality-assured high-modulus adhesive is also readily available. It is absolutely essential to use the correct adhesive to obtain a high degree of crack-width sensitivity with a narrow range of scatter of sensitivities.

Development of the sensor was assisted by the use of a scanning electron microscope which ensured that the optimum bond was obtained between the sensing fibres (usually three) and the backing material. Electron microscopy also aided the selection of the optimum fibre stripping and cleaning process to eliminate contamination, and thereby improved the consistency of crack-detection sensitivity. Figure 2 shows a typical scanning electron micrograph (magnification 50×) of the sensing surface of a sensor.

Figure 3 is a photograph of two crack-

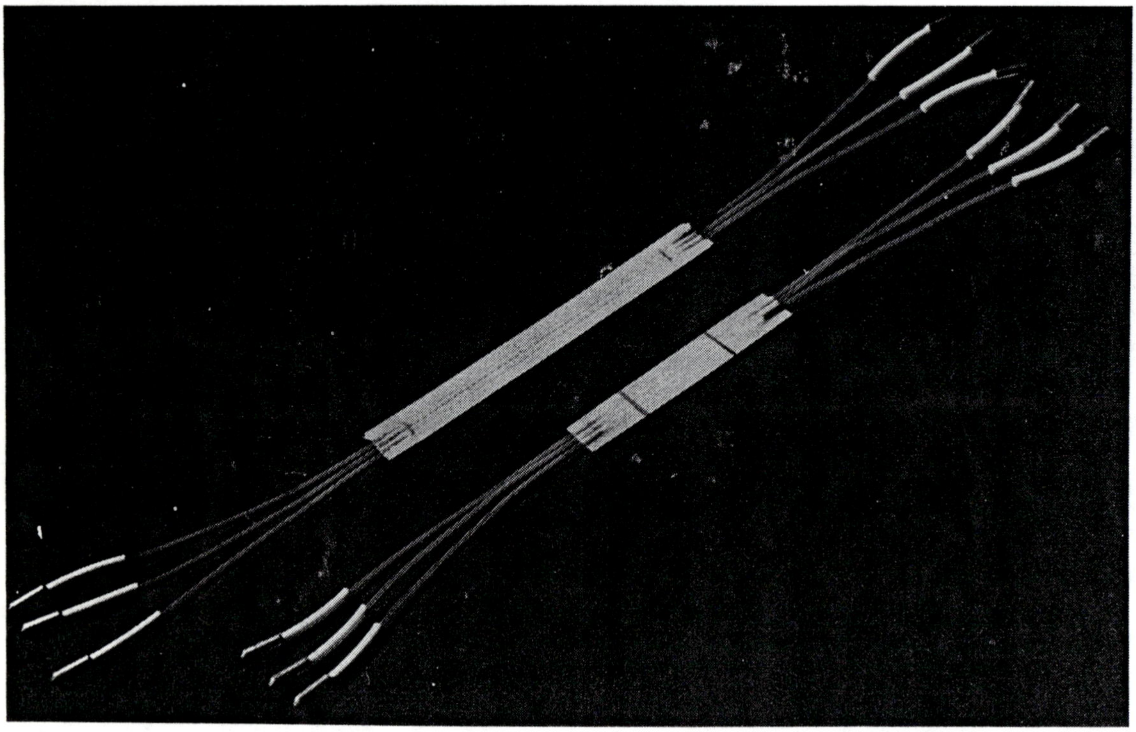

Fig. 3 Optical-fibre crack-monitoring sensors: 75 mm (top) and 25 mm (bottom)

monitoring sensors, one with an active length of 25 mm and the other with an active length of 75 mm. Each of the sensors has three parallel sensing fibres with a spacing of 2 mm. This arrangement allows crack growth rates to be determined from the times of fracture of the individual fibres.

Inexpensive optical-fibre connectors are required to allow low-attenuation coupling between the sensor fibres and optical-fibre extension leads to connect with detection instrumentation. Fortunately, these connectors have been developed for the optical-fibre industry and the technique for bonding the fibres into the connectors and producing low-attenuation joins are well developed for use in working environments. Where weight considerations are important, optical-fibre connectors can be dispensed with altogether, and low-attenuation joins between sensors and extension leads can be made by fusion splicing. Portable fusion splicing equipment for use in the field has been produced for the telecommunications industry and is commercially available from many suppliers.

CRACK-WIDTH SENSITIVITY TESTS (CRACK-SIMULATION TEST RIG)

Crack-width sensitivity tests are regularly carried out on randomly selected sensors on a specially designed crack-simulation rig. It basically comprises two demountable plates with ground edges which are supported by a base plate which may be adjusted to give effectively zero crack width. The contacting edges are immediately above a reduced thickness in the base plate, which can act as the hinge of a cantilever and which is acted upon by a screw. Deflection of the cantilever causes the ground edges of the demountable plates to move apart with a separation approximately one fifth of the depression of the end of the cantilever.

The simulated crack has the same width along the whole length of the ground edges. A dial gauge enables the separation of the plates to be measured to an accuracy of 1 μm. The area of the demountable plates is sufficient to mount four standard optical-fibre sensors using the recommended adhesive.

After bonding the sensors on to the plates, the adhesive is cured with a pressure of 0.03 N/mm^2 on the sensor, giving a glue-line thickness of 10 μm. White light is directed through the individual optical fibres and the screw is slowly turned to open the simulated crack, and converted into crack-width values. Forty-two sensors were tested giving 126 individual fibre crack-width sensitivities. The crack widths at which fibres broke ranged from <5 to 103 μm. They were grouped into 5 μm-wide bands and the percentage broken at each step is shown in Fig. 4(a). The actual number of fibres breaking within each band is shown in Fig. 4(b). It can be seen that approximately 50% of all sensing fibres had fractured at a crack width of 35 μm, 67% by 50 μm, 85% by 70 μm, 95% by 90 μm and, of course 100% by 103 μm.

Whilst it is not possible to determine the probability of false alarms with this technique, none appeared during these tests.

Crack-width sensitivities can be reduced to allow the monitoring of cracks wider than 20–40 μm by increasing the thickness of adhesive bond from 10 μm to 100 μm, or even 1 mm, by using more ductile adhesive than the standard adhesive recommended for use with the sensors, or any other method of reducing the efficiency of strain transfer to the sensing fibres.

It was hoped that the crack-simulation rig would give values of crack-detection sensitivity which would allow statistical analysis to be performed on a reasonably large sample of sensors with considerable cost and time savings, compared with the effort required for fatigue testing. However, the scatter of results from the crack-simulation rig were always inferior to the sensitivity range of 20–40 μm obtained for real cracks in fatigued specimens. The discrepancy appears to result from the unavoidable finite initial crack width between the plates of the crack-simulation rig which effectively gives the sensor a small but significant gauge length. The cracks in fatigue specimens start from a true zero width and, in addition, the plastic strain ahead of the cracks assists the breakage of the optical fibres. Thus the crack-simulation rig can only be used for quality-assurance tests; absolute values of

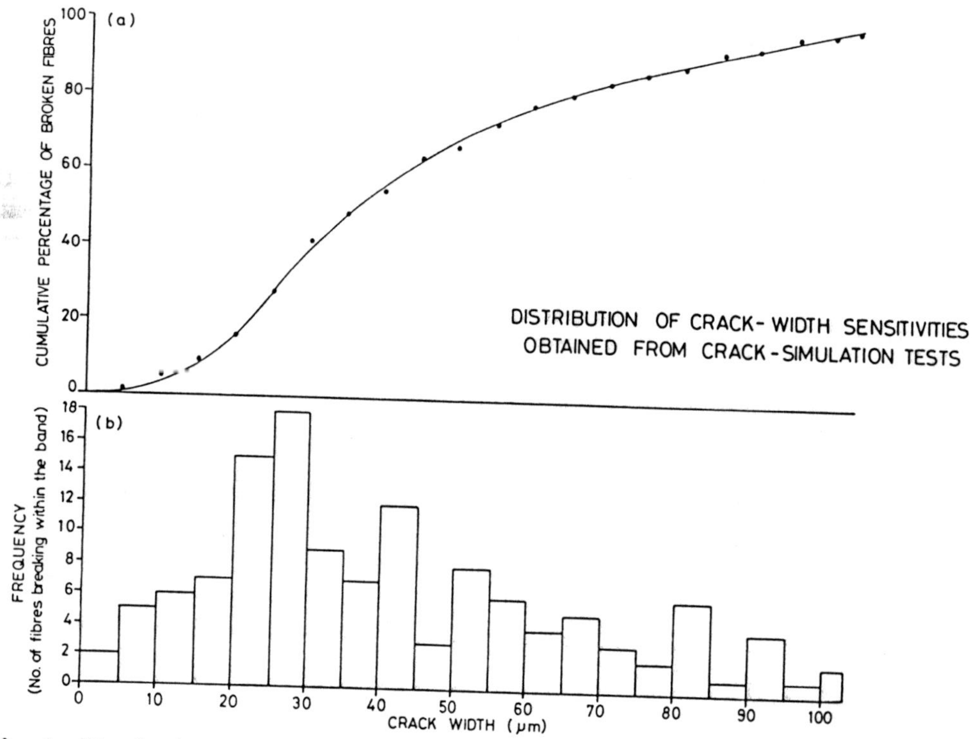

Fig. 4 Distribution of crack-width detection sensitivities obtained from crack simulation tests

crack detection sensitivity can only be obtained from actual fatigue tests.

CRACK-WIDTH SENSITIVITY TESTS (CYCLICLY LOADED SPECIMENS)

Steel Specimens

Cyclic load tests have been performed on a number of large steel specimens requiring stresses up to 160 N/mm^2 at 10 Hz and up to 770 microstrain. The fatigue tests were mainly performed in the temperature range −25°C to +60°C at relative humidities up to 100%. The specimens were of three basic types: fillet weld specimens, butt weld specimens and longitudinal stiffened plates. The total number of fatigue cycles was in excess of 70 million. The fatigue crack-width detection sensitivities were in the range 20–40 μm for optical fibres placed orthogonally to the direction of crack growth, and 20–75 μm for sinusoidal sensors used on butt welds, where the fibres were at an oblique angle to the crack length. The individual fibres in each sensor were connected to infrared source/detection/alarm instrumentation, which would switch off the fatigue testing machine when the light transmission was attenuated by more than a pre-calibrated amount.

Cyclic load tests were also performed on many L65 aluminium alloy specimens (Fig. 5) having an overall length of 340 mm, an effective length of 130 mm, an effective width of 31 mm and a thickness of 6 mm. Each specimen had two, three or five 6.5 mm diameter holes drilled along the axis of the specimen with a centre-to-centre spacing of 25 mm or 50 mm. The three sensor fibres passed within 1 mm, 3 mm and 5 mm of either side of the holes. Some of the specimens had spark-eroded crack starters at one or both sides of the holes: others were unnotched.

The fatigue stresses were up to 130 N/mm^2 at 27 Hz. The total number of fatigue cycles was in excess of 24 million. The fatigue crack-width detection sensitivities were in the range 20–40 μm. It should be noted that the true crack widths at the time of fracture of the

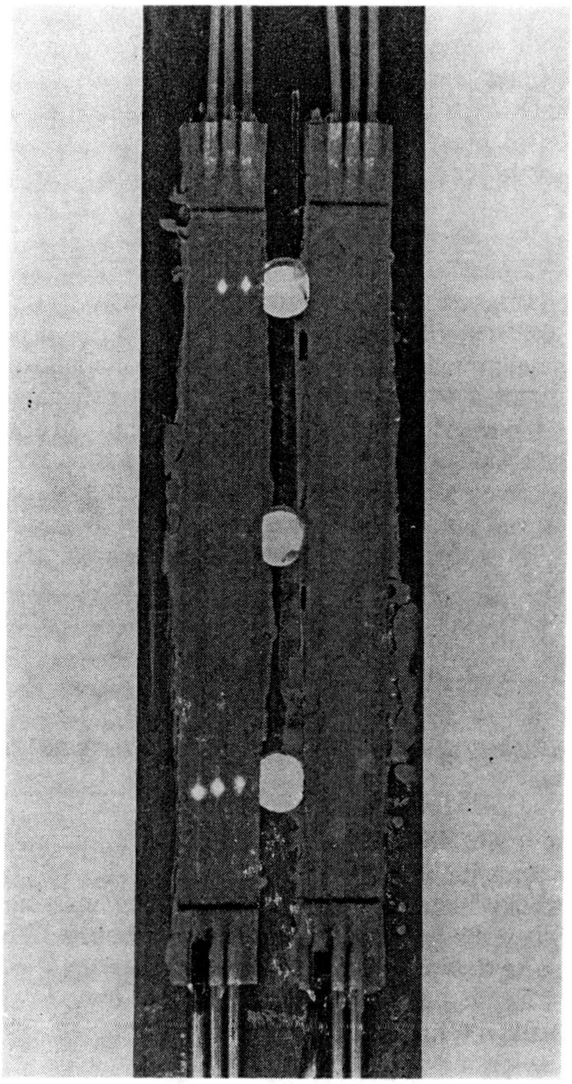

Fig. 5 Optical-fibre sensors detecting cracks initiating from holes in an aluminium alloy specimen

optical fibre immediately above the surface cracks in the specimens were determined by applying a static tensile load to the specimen equal to the maximum cyclic tensile load.

Results from all these fatigue tests have not given any false alarms. There is no evidence that crack-width sensitivities are affected by temperature in the selected range or by humidity or wetness.

Optical attenuation measurements were made on the sensor fibres before and after the fatigue tests using standard test equipment. Attenuation values usually changed from about 2 dB to between 10 dB and 15 dB, with exceptional changes to 25 dB and even 30 dB. However, attenuation due to loss of light at a fibre break, or the brightness of white light escaping at a break, cannot be used reliably as an indication of the width of the crack in the structure immediately below the fibre break. The fracture strength and mode of fracture of glass rods and fibres is a complex phenomenon involving the microstructure and distribution of surface microcracks. Thus a concentration of strain in a glass fibre because of strain transfer from cracks in the structure can result in fibre breakages, ranging from planar mirror-like surfaces which produce very little attenuation of light (~2 dB) up to explosively shattered surfaces with high attenuation (~20 dB or more). Also, inexpensive optical fibre connectors are used and, although satisfactory for the detection of incremental attenuation changes, they can lead to variable absolute values unless care is taken. Other intervening breaks in a sensor will introduce yet another variable.

Crack detection tests have also been performed on reinforced concrete beams, using optical fibre sensors one metre long, with very satisfactory results.

INSTRUMENTATION

Instrumentation has been developed for detecting cracks in optical fibres in sensors continuously and remotely. The instrumentation consists of sources, and detectors and electronics capable of detecting attenuation changes of infrared transmission with a sensitivity of about 1 dB and above. A typical in-line fracture in a fibre causes an attenuation change of the order of 2–20 dB. Extension leads have an attenuation of about 5 dB per kilometre, optical-fibre connectors cause an attenuation of 1–2 dB and fusion splices can be made routinely with attenuations of less than 0.5 dB. The maximum total loop attenuation possible with such instrumentation is approximately 30 dB.

Although it could be possible to reset the alarm instrumentation to allow more than one crack to be detected by each sensor, this is not advisable because cyclic loading can cause attenuation changes in the early fibre breakages which might be interpreted as the detection of a new crack. The attenuation change is due to escape of light at the fibre breakage. If the infrared emitter is replaced by a simple tungsten lamp, this escaping visible light provides a useful marker for the location of a crack, and remains even if the crack in the structure closes up after removal of the strain that caused the structural failure. The instrumentation is available at various levels of sophistication depending upon the type of application, from fatigue testing of specimens to complex structures operating at remote locations. Figure 6 shows two standard source/detector/alarm units housed in a case with a common power unit.

Optical time-domain reflectometers are commercially available that can determine the location of a fracture in an optical fibre with an accuracy of one metre in optical circuits up to several kilometres in length. It is therefore possible to join several crack-monitoring sensors of various lengths in series, and to record the position of structural failures remotely.

HOSTILE ENVIRONMENTS

Work was carried out to extend the temperature range of the optical-fibre crack-monitoring sensors beyond the normal temperature range of –25°C to +60°C.

Use was made of the –200°C to +200°C environmental facility available on one of the National Physical Laboratory fatigue machines. The test specimens were made of aircraft aluminium alloy, with holes to simulate rivet holes. It was found that crack-width sensitivity of the sensors remained at 20–40 μm at

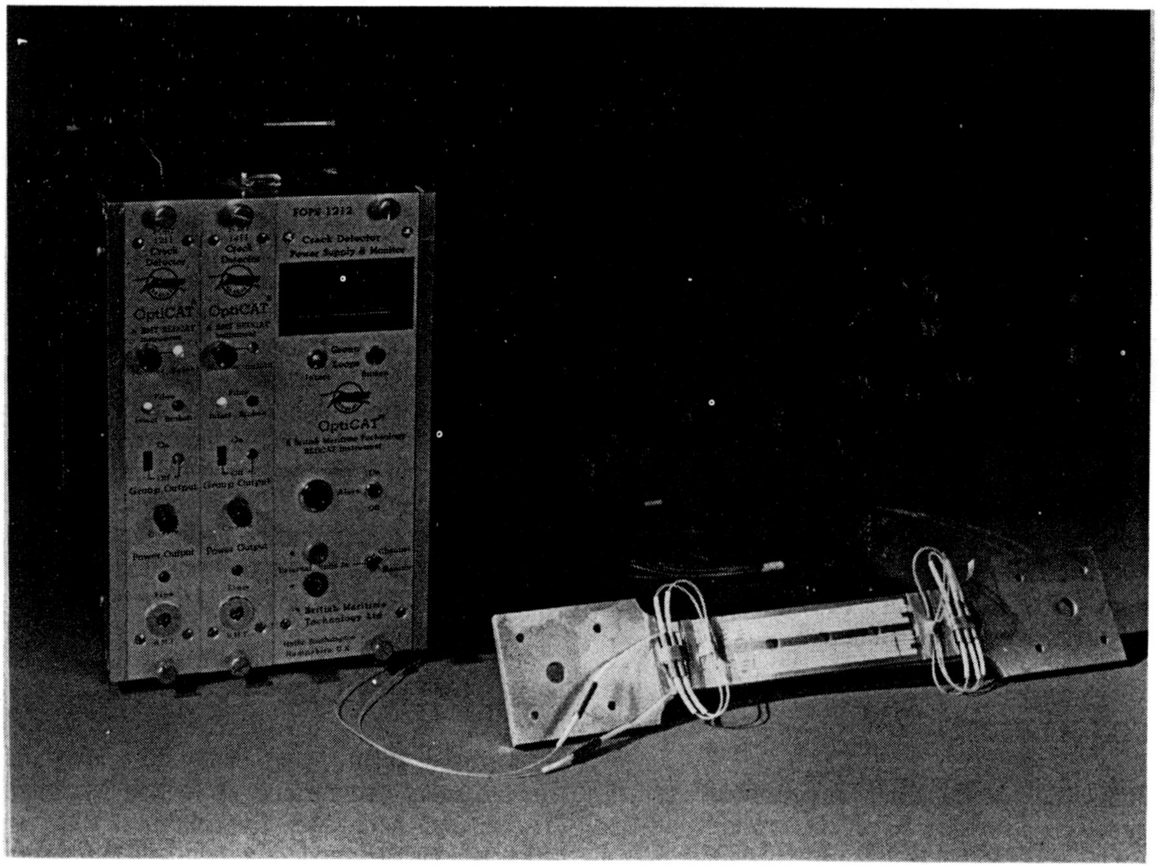

Fig. 6 British Maritime Technology Ltd optical-fibre attenuation change alarm instrumentation

temperatures as low as –196°C (liquid nitrogen temperature) and up to +180°C at normal humidities. Attenuation changes in the sensor connectors were accommodated by the detection instrumentation in a completely satisfactory manner. The maximum temperature for long-term work is about +120°C; above this temperature, operational times become progressively shorter with increasing temperature. The use of heat-resisting optical fibres, backing materials and adhesives would extend operational times at temperatures above 120°C, should they be required.

Cyclic tests were also performed on optical-fibre crack-monitoring sensors bonded onto steel specimens totally immersed in sea water. The steel specimens were shaped to give constant strain over the entire surfaces on both sides of the specimen. The sensors were covered with a thin coating of transparent water-resistant epoxy. Although the tests are in their early stages (30 million cycles at 500 microstrain during 30 weeks) there would appear to be no insurmountable problems with the seawater environment. The life of the bond between the sensors and the steel specimens is not dependent upon the magnitude of the cyclic strain experienced by the structure at normal operational stresses. Earlier experiments were conducted with sensors bonded to steel plates immersed in sea water. These plates were attached to the Christchurch Bay Tower, and were subjected to severe wave action over periods of time up to seven months. The steel plates were protected from corrosion by sacrificial zinc anodes. No disbonding of the sensors occurred. More experiments on a steel plate not protected from corrosion are in progress on the Christchurch Bay Tower.

The sensor and bonding adhesives are made from materials that are resistant to standard industrial and aviation fluids.

CONCLUSIONS

Standard production optical-fibre crack-monitoring sensors have been developed that are capable, with a high degree of confidence, of detecting crack-width increases in the range 20–40 μm in structures subjected to cyclic loading. The correct adhesive must be used, the recommended sensor bonding procedures must be followed, and post-bonding checks with a simple optical power meter must be performed on sensors, extension cables and alarm instrumentation. This ensures that false indications of crack growth are reduced to a very low level indeed. The cost of the system compares very favourably with conventional methods of detection for sensor lengths of 25 mm and 75 mm and considerable savings are effected for longer detection lengths. Prototype 3-m long sensors are currently undergoing evaluation and sensors with lengths of 20 m or more should be possible. The sensitivity of the sensors can be reduced if required.

The sensors, being optical, are completely free of problems associated with electrically operated sensors, such as insulation, interference, corrosion etc. and short-circuiting after strain removal.

The evidence to date indicates that they can withstand cyclic loading for much longer periods than the structures being monitored.

They operate satisfactorily in severe aqueous environments, as well as in the standard industrial and aviation fluids.

They operate over a temperature range of –196°C to +120°C, with the possibility of temperatures beyond these extremes.

Cracks are indicated by relatively simple instrumentation operating with infrared light. The position of a crack is readily seen by substituting white light for infrared, and the escaping light continues to indicate the crack after the strain is removed and the crack closes up. For remote location of cracks, optical time-domain reflectometry equipment is available which locates cracks to a positional accuracy of one metre along an optical fibre circuit.

ACKNOWLEDGEMENTS

This project formed part of the Research and Development programme of British Maritime Technology Ltd, and the authors wish to acknowledge the encouragement and support of the Company. Acknowledgement is also due to

Mr M. A. Halliday, Mr H. W. Duggin, Mr K. E. Henry and Mr H. J. Hatchard for their contribution to the project.

REFERENCES

Hale, K. F. 1982. The application of optical fibres to structural integrity monitoring. *Proceedings of Oceans 82 Conference*, pp. 344–348. Washington, September.

Hale, K. F. 1983. New applications of fibre optics in displacement measurement. *Proceedings of Transducer Tempcon Conference*, pp. 599–610. June.

Hale, K. F. and Boyle, H. B. 1985. Use of optical fibre sensors for crack detection. Proceedings of ERA-COMRAD Seminar on *Condition Monitoring in Hostile Environments*, June.

Hale, K. F., Hockenhull, B. S. and Christodoulou, G. 1980. The application of optical fibres as witness devices for the detection of plastic strain and cracking. *Strain*, October, 150–154.

Hughes, R. and Hale, K. F. 1984. Fibre optics monitor cracks and strain. *The Oilman*, October, 49–50.

Part IX

Geology/Geographics/Geotechnics

35

The Collection, Interpretation and Presentation of Geotechnical Data for Marine Pipeline Projects

P.T. Power, Project Engineer and *J. M. Paisley*, Engineering Geologist
Furgo Limited, Consulting Geotechnical Engineers, Hemel Hempstead, Herts, UK.

This chapter discusses the collection of higher quality soils data for pipeline projects and new equipment that has been developed for that purpose. Having been obtained, the data must be interpreted and presented in a manner that maximizes its value. Suggestions are made on how this can be achieved. Examples of the current trend towards "high technology" soils investigations for pipelines are also given.

INTRODUCTION

High technology soil investigation methods can now be used on marine pipeline route surveys as a matter of routine, in the way that they are used for the investigation of production platform sites.

The Cone Penetration Test (CPT), for many years the favoured method for *in-situ* measurement of soil parameters in offshore soils investigations, can now be performed without the need for a drilling ship or special deployment equipment.

This chapter describes some of the new equipment developed for this purpose and how, when combined with improved sampling equipment, it can lead to a much higher quality of geotechnical input to the design and construction of a pipeline.

To be of most value, the data needs to be correctly interpreted and presented in a form that can be easily assimilated by the pipeline engineer. Some guidelines on these aspects are given and some typical projects are briefly described.

THE COLLECTION OF DATA

In the past the tendency with soil sampling on pipeline projects was to add it, as an afterthought, to the specification for the "geophysical" route survey: so if the survey vessel being used could not handle a vibrocorer, the only soil sample recovered was by gravity corer or grab sampler. The end result was that the evaluation of soil conditions for all aspects of the pipeline design, installation and operation was entirely based on the interpretation of the shallow seismic profiling and handfuls of soil at infrequent intervals along the route. Little wonder that claims for additional trenching time due to unexpected conditions are common, that scour

THE COLLECTION, INTERPRETATION AND PRESENTATION OF GEOTECHNICAL DATA FOR MARINE PIPELINE PROJECTS

P T Power, J M Paisley, Fugro Ltd, Uk

Presented on **Thursday 6th March** 1986 **at 14.30**
in the Norfolk Conference Room

EMBARGOED AGAINST DELIVERY

still surprises people and that pipeline supports sink into the sea bed.

Faced with this problem, the oil industry rightly claimed that even if it did upgrade its specifications for soil investigations, there were few options available for obtaining improved data without reverting to expensive geotechnical drilling vessels. There has therefore been a concerted effort to fill that need by some companies in the geotechnical field, and a number of independent seabed testing units have now been developed for *in-situ* soil testing and sampling, including Fugro's "Seabed Wison" and "Seasprite" systems.

"Seabed Wison" CPT System

The "Seabed Wison" is a marriage of two well proven tools, the "WISON" downhole cone penetrometer test (CPT) system, that has been used for many years in soil investigation borings for offshore structures such as production platforms, and the tried and tested vibrocorer. Developed in conjunction with the vibrocoring specialists, Alluvial Mining Limited, the unit is basically a standard rectangular based "4 metre" vibrocore frame with the penetrometer unit occupying the space normally taken up by the vibrator and core barrel.

Hydraulic pressure is used to push the instrumented probe, which is the cone penetrometer, into the sea bed and, at the end of the test, to retract it. During the test strain-gauge load cells in the cone measure the resistance to penetration experienced by the tip and the frictional drag on a sleeve just above the tip. The more advanced "piezocone" can also measure the excess pore-water pressure in the soil via a porous element in the cone tip. The electrical signals from the cone are transmitted continously to a recording system on board the parent vessel via the unit's electrohydraulic umbilical cable.

The results are plotted graphically by pen recorder as the test proceeds, and can also be recorded on magnetic tape if required. As the results are immediately available, on-the-spot interpretations of the soil conditions can be made. Data that can be derived from the test results include soil type and composition, the

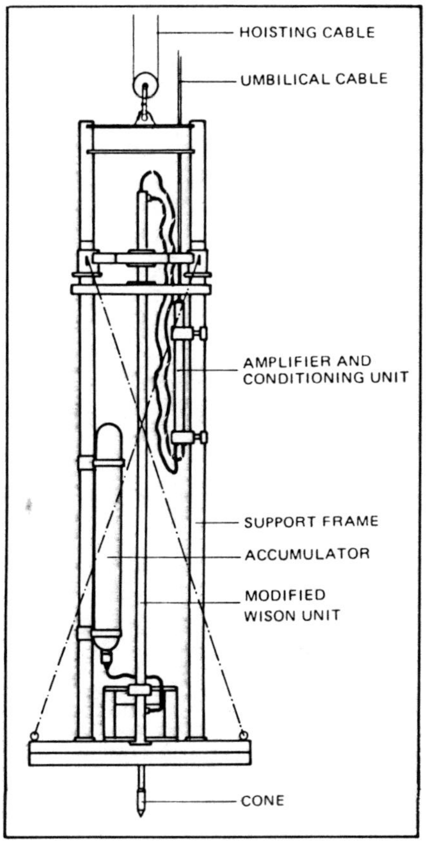

Fig. 1 (a) Schematic diagram of the "Seabed Wison".

(b) "Seabed Wison" being deployed.

relative density of granular soils, and shear strengths in cohesive soils. The interpretation of the test data is discused in more detail in a later section.

The unit, which is shown schematically in Fig. 1(a) and being deployed in Fig. 1(b) is capable of penetrations up to 2.6 m and can operate from a variety of vessels in water depths up to 300 m. This enables it to fulfil the

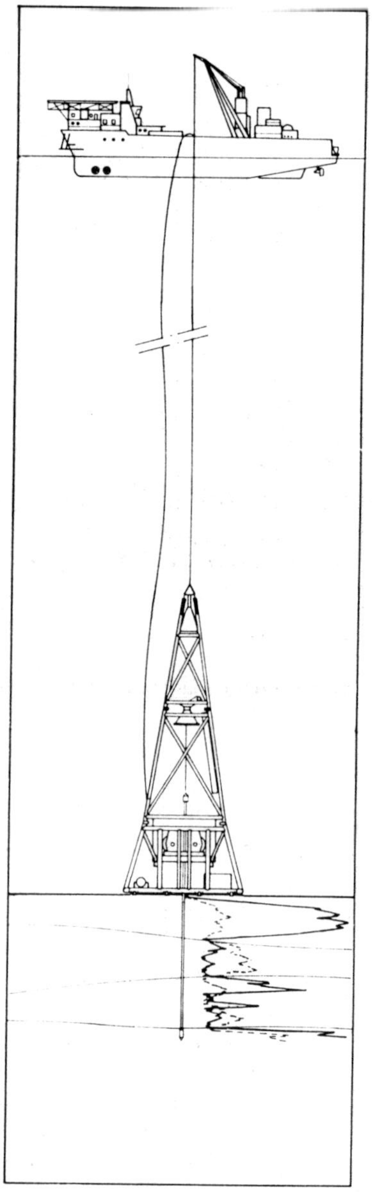

Fig. 2 Diagram showing "Seasprite" deployment method

Fig. 3 (a) "Seasprite" being lowered into the water.

(b) "Seasprite" being deployed from D.P. vessel *Polar Queen*

requirements of many pipeline soil investigations, but certain applications require greater penetration below the sea bed and operation in deeper waters. The need was therefore perceived for a tool with greater capabilities, and the "Seasprite" system was developed to fulfil this need.

"Seasprite" Seabed Testing Unit

The "Seasprite" system (the name is an acronym for Seabed Soil Penetrating Remotely-Controlled *In-situ* Test Equipment) is capable of performing tests to 5 m penetration in water depths of up to 1500 m. Like the "Seabed Wison" it is designed to be deployed by any vessel that has the lifting gear to handle it and, as no heave-compensation system is required, the choice is quite wide. The simple single-line deployment method is shown schematically in Fig. 2, and seen in action in Figs 3(a) and 3(b).

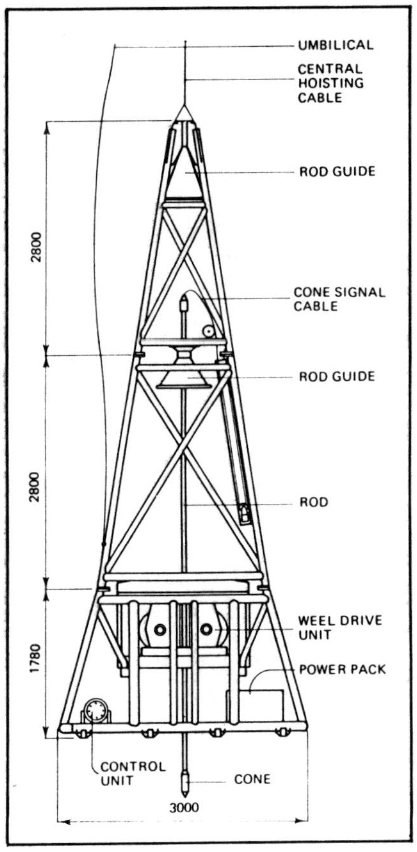

Fig. 4 Schematic diagram of the "Seasprite"

The seabed unit, detailed in Fig. 4, incorporates a number of new developments for this type of equipment. The penetration force, for example, is provided by a new "Wheel Drive" system in which four wheels grip the test rod and, when rotated, cause it to move down into the sea bed or retract it when reversed. The wheels themselves are driven hydraulically from a submersible power pack which is mounted on the seabed frame. The electrical supply for the power pack is provided via the combined power and signal umbilical, which is also used to transmit instructions to the control unit and carry back the measured data.

The control unit on the seabed frame receives the signals from the cone and converts it into digital format before transmitting to the control cabin on the parent vessel. The data is displayed on the screen of a microcomputer as the test proceeds, and recorded on floppy disk for later processing. The computer is also used by the operator to control all of the functions of the seabed unit. On completion of the test, the screen display is printed out to give a hard copy of the test result. Also recorded are other data including readings of water pressure at the sea bed, from a transducer on the frame, and the verticality of the frame from a dual-axis inclinometer.

Although designed primarily for the cone penetrometer test, the system is extremely versatile and can be adapted to perform *in-situ* shear vane tests and take "push-samples" in clay. If required, a television camera can be attached to the frame and the signals transmitted via the main umbilical.

As well as the standard cones and piezocones, two new types are currently being adapted for offshore use; these are the thermal and electrical conductivity cones, both of which have specific applications to pipeline design. The thermal cone can provide data for use in the calculation of heat loss from a pipeline and the assessment of thermal expansion effects. The electrical conductivity cone has obvious applications in the field of cathodic protection and the assessment of corrosion potential. A cone for measuring *in-situ* temperature as well as cone tip resistance has already been used on a number of occasions for investigations in permafrost regions and on a pipeline route (Kolk and Power, 1983).

As well as providing the information described above, the great advantage of the CPT over conventional coring methods is that it can generally penetrate much deeper in hard or dense soils, it does not suffer from core loss or soil "plugging" in the barrel, and it always produces a result.

Other Developments

Submarine Penetrometer (Mini-WISON)

A miniaturized "WISON" cone penetrometer unit has been developed for attachment to manned or unmanned submersibles. The system, which uses a 5 cm^2 cone tip and has a penetration stroke of about 1 m has already been used successfully in both modes (Kolk and Power, 1983).

Fig. 5 Vibro-impact corer being deployed

Improved Vibrocorers

In its basic form, the electric or hydraulic vibrocorer, in which a heavy vibrating "pot" on top of the core barrel drives it into the sea bed, has been around for a long time. However, more recently, the British Geological Survey (BGS) and Alluvial Mining Limited, amongst others, have upgraded their equipment to a higher level of sophistication. The former have incorporated functions such as penetration depth and resistance sensors and the ability to retract the core barrel, by means of a winch on the seabed frame, before lifting the device off the sea bed (Ardus *et al.*, 1982). As well as incorporating penetration sensors on some of their vibrocorers, Alluvial Mining have developed an "Impactor" attachment which improves penetration in hard strata. The impactor comprises a device which incorporates a reciprocating spring-mounted percussion weight which is activated at a certain vibration frequency. The vibro-impactor corer is shown being deployed in Fig. 5.

Box Corers

These devices, designed for scientific rather than engineering applications, have been used-probably for the first time-during 1985 on pipeline projects. Best suited to soft sediments, these devices, as the name implies, push an open-based steel box into the soil, which on retraction is sealed by a "trap-door" device. When the box is opened it is possible to visually inspect a relatively undisturbed portion of the surface of the sea bed as well as obtaining a useful bulk sample for testing. A diagram of one type of box corer is shown in Fig. 6.

"Sea Dart" Corer

This simple device provides another method of obtaining a relatively undisturbed surface sample in soft cohesive sediments. It is a small drop-corer that incorporates a 1 m long 70 mm diameter, thin-walled stainless steel tube. On top of the tube is a special one-way ball valve system that enables the sample to be retained in the tube by suction only. In this way it is

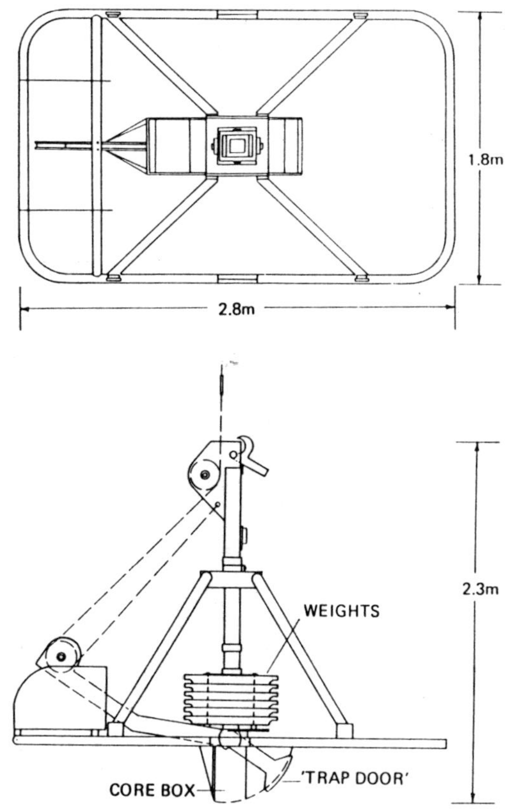

Fig. 6 Schematic diagram of "Reineck" type box corer

Fig. 7 "Sea Dart" drop corer ready for deployment

possible to overcome the disturbance to the first half metre or so of sample normally caused by standard corers, with their heavy wall thickness and "core-catching" devices. The Sea Dart is shown, ready for deployment, in Fig. 7.

Diver-operated Samplers

In certain situations, it may be practical to equip a diver with the means to obtain a soil sample during an excurison to the sea bed. A number of handheld devices have been developed that enable divers to hammer or push in sample tubes of varying sizes.

Acoustic positioning of samplers

It is now becoming common practice, when a more sophisticated vessel and sampling spread is mobilized, to attach acoustic transponders to the sampling devices to enable their actual touch-down position to be pinpointed more accurately. This is of particular value when profiling across-seabed features such as "pock-marks" or iceberg "ploughmarks".

Planning the Soils Investigation

The Programme

The scope of the soil investigation should be based upon a comprehensive review of all existing data, including local geology, for the route and general area. Ideally, this should incorporate the results of the "geophysical" route survey, including side-scan sonar, shallow seismic and bathymetric data. It is now becoming quite common to have a geotechnical engineer present during the route survey to aid this process.

The physical factors that will influence the scope and density of the sampling/testing programme will include:

- anticipated soil types;
- variability of surface soils;
- the presence of problematic features such as pockmarks, iceberg ploughmarks, sandwaves, potentially unstable slopes, and rocky landfalls.

Unstable slopes and major constructions in the area of pipeline landfalls may warrant the full weight of a major geotechnical investigation by deep boreholes from drilling ships (Marsland and Windle, 1982) or jack-up rigs (Adam and Hodgson, in press).

The possibility of trenching should also be taken into consideration, as a sufficiency of sampling stations and adequate penetration is required to ensure that unexpected conditions do not lead to inflated installation costs.

Number and frequency of samples and tests

No idealized programme can be proposed, as it will always depend on a large number of variables such as those described above. However, the tried and tested formula of one sample station per kilometre is a good starting point for the minimum programme. Consideration must then be given to increasing the frequency in areas where the stratification is changing rapidly, or where there are specific seabed features.

The pipeline route will normally deviate to avoid features such as large sandwaves, pockmarks or iceberg ploughmarks (Ice Age rather than current phenomena). If constraints such as the areal density of the features preclude this, the features-or at least a representative selection-must be investigated in detail. The traversing of such features by the pipeline will inevitably require actions such as trenching, infilling or pipeline supporting, all of which will be significantly influenced by the soil conditions. Unfortunately, these can vary considerably from feature to feature. Depressions such as pockmarks (Hovland et al., 1985) and ploughmarks may have hard shoulders and soft infillings, or soft shoulders and harder exposures in the trough. Experience has also shown that the soil conditions can vary within the depression itself.

A typical iceberg ploughmark or pockmark, of the order of 100 m across, would therefore warrant something like six investigation points covering the shoulders, internal slopes and base of the depression. A large sandwave would probably require sampling or probing at its crest in the trough and on the slopes.

The rapid profiling of such features is particularly well suited to *in-situ* testing tools such as the CPT, which can be moved quickly from point to point without being brought back on board the parent vessel. However, some samples should be taken for the purposes of correlation, visual examination and laboratory testing.

As for the ratio of CPTs or other *in-situ* tests to samples, current practice seems to favour about two or three to one in favour of the CPT.

The Equipment Spread

For a major pipeline soils investigation, a typical spread is now likely to be based upon a dynamically positioned survey vessel and comprise:

- lightweight seabed CPT equipment;
- a vibrocorer;
- a selection of gravity/drop corers;
- a grab sampler;
- an onboard soils laboratory.

The grab sampler, although of limited value, can sometimes prove useful in obtaining samples of coarse granular material which is difficult to core.

The onboard laboratory can be of differing levels of sophistication depending on the size and complexity of the investigation. It can range from simple handheld testing devices to a fully equipped containerized laboratory, capable of a range of tests including triaxial compression tests.

Where special considerations, such as a preponderance of soft sensitive soils, warrant it, other specialized items can be added, such as:

- a box corer;
- push sampling equipment;
- an *in-situ* shear vane tester.

The last two items can be deployed as adaptions to the CPT system.

Finally, in the case of the most comprehensive investigations, manpower and computing facilities would be provided on board the vessel to process and interpret the *in-situ* and laboratory test data.

In this way, the field programme can also be continuously modified to take into account the findings of the investigation and thus provide the optimum coverage. It also allows for a much earlier evaluation of the findings to be fed into the pipeline design process.

For smaller investigations, the compliment of equipment can be tailored to suit the more modest requirements and budget of the project. The range of sampling and testing equipment available for pipeline soils investigations is shown schematically in Fig. 8.

Project Examples

Most of the developments described above have only become operational during the summer of 1985, and therefore the project details and the results of the investigations are still confidential at the time of going to press. However, it is possible to give an indication of the scope and nature of some of these projects.

Pipeline Project-Norwegian Sector, North Sea

A preliminary soils investigation has been performed in areas where possible pipeline routes will encounter deep iceberg ploughmarks. The spread mobilized was very similar to that described in the previous section, with the addition of an ROV for a visual and bathymetric survey.

A number of individual ploughmarks was investigated in three separate areas. In total, 29 CPTs were performed with the "Seasprite" system and 27 cores were obtained with a

Fig. 8 Range of equipment for pipeline route soils investigations

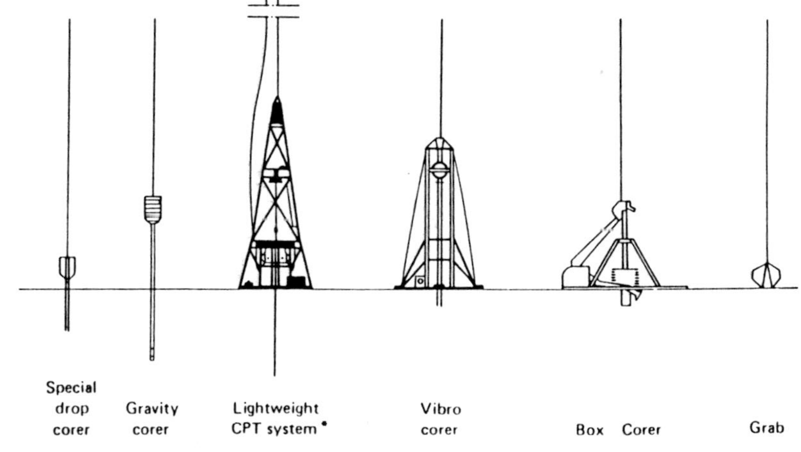

Vibro-impact corer, standard gravity corer and "Sea Dart" corer.

The investigation, performed in water depths between 200 and 300 m, was successfully completed within seven days.

Pipeline Project-UK Sector, Southern North Sea

A major pipeline route survey and soils investigation has been performed that incorporated about 100 CPTs with the "Seabed Wison" system, supplemented by 18 tests with the "Seasprite" and approximately 80 vibrocores. In addition, a number of boreholes were drilled on the shore approach and landfall sites. The investigation, which was performed in parallel with the geophysical survey, covered approximately 150 km of main trunkline and interfield pipeline routes.

Other Applications–Harbour and Coastal Surveys

Although most of the equipment described here has been developed with the pipeline industry in mind, it also has numerous applications in the areas of coastal and harbour works and predredging surveys. The "Seasprite", for example, has recently performed an investigation a few kilometres from the coast of Scotland.

The investigation programme required CPTs on a grid pattern within a 200 m diameter circle to profile the thickness of sediment over "rockhead". Using a dynamically positioned vessel, in a water depth of about 60 m, 20 CPTs were performed to between 3 m and 5 m penetration in about 24 hours.

The "Seabed Wison" has also been used in a predredging survey for a harbour expansion scheme, and the "Seasprite" on its most recent project was operated from a vessel which could, if required, deploy the equipment while beached at low tide.

INTERPRETATION OF DATA

Laboratory Testing

Before interpretation can be finalized, a full laboratory programme of tests needs to be performed in order to define the physical characteristics of the soils. The tests will include many standard soil tests as well as some developed specifically for pipeline analyses. A list of typical tests is given in Table 1, together with potential applications.

TABLE 1

Test Type	Applications
Particle size distribution	Definition of soil type and constituents, friction factors
Atterberg (liquid and plastic) limits	Indication of engineering behaviour of clay, including liquefaction/ flotation (Institute of Petroleum, 1982)
Moisture content, bulk and dry densities, specific gravity	Engineering behaviour of the soil
Mass relative density (Soil–water mixture)	Pipeline flotation (Institute of Petroleum, 1982)
Shear strength tests (fall cone, pocket penetrometer, torvane, undrained triaxial compression tests)	Strength, stability, bearing capacity, trenchability*
Shear box tests (sand)	Friction factors, trench slope angles
Oedometer (consolidation) tests	Settlements of pipeline and associated structures
Lacquer peel treatment	Determination of depth of mobile sand †

*The shear strength tests described are relatively quick and simple and are used to determine the shear strength in the undrained or rapidly loaded situation. This is adequate for most analyses. However, for complex engineering problems such as the stability of slopes or large structures, other, more complex tests, including "effective-stress" type triaxial tests may be required.

†The "Lacquer Peel" treatment of complete sand cores is a method by which the internal sedimentary structure of the core can be exposed and enhanced. This, combined with a study of biological activity within the core, can yield a very definite assessment of the depth of "mobile" material. Thus it provides a useful prediction of potential scour depth around a pipeline in sandwave areas.

Having completed the programme of testing, the results are used to formulate the engineering description of the core. This includes the strength and fabric of the material and the correct main and constituent soil types and their influence on physical behaviour. Where appropriate, data are compared and correlated to establish any continuity of strata from core to core. Finally, correlations of the results should be made with the CPT tests to improve their interpretation and aid in the interpolation of data between cores.

The Interpretation of the CPT Tests

Much has been published on the interpretation of CPT test data (Schmertmann, 1969; Sanglerat, 1972; de Vos, 1982) and so the actual mechanics of the process will not be dwelt upon here. However, the information that can be obtained from the test is considerable and this will be briefly summarized.

With a standard cone, measuring point resistance and sleeve friction, the first thing that is immediately shown is the thickness of soil layers and whether they are sand or clay. The contrast between the two soil types is significant, and is illustrated in Fig. 9. The next phase of interpretation is to refine the soil type interpretation, and this is done by considering the "friction ratio" (the ratio of cone-tip resistance to sleeve friction) in any layer. Empirical correlations (Schmertmann, 1969; de Vos, 1982) are then available to enable one to establish whether the sand is silty or the clay, sandy and so on. From the tip resistance it is also possible to estimate the *in-situ* relative density and friction angle in sand and the shear strength in clay.

With the "piezocone" the additional measurement of the excess pore-water pressure enables the interpretations to be refined even further (Senneset *et al.*, 1982; Wroth, 1984).

Correlation with the laboratory data is important as the empirical relationships described above can vary from area to area due to geological inconsistencies, and because specific parameters from a few tests may need to be extrapolated along the pipeline route. One such case is that of the sand mobility determination by "Lacquer Peel" analysis. Research is currently under way to establish a relationship between the parameters measured in the CPT and the difference between "mobile" and stable sand deposits. The material will naturally be more loosely packed, and therefore generate less tip resistance, but it is also hoped to more positively "fingerprint" such layers by means of other parameters such as the friction ratio and pore-pressure parameters.

If successful, such a correlation could be invaluable in establishing how deeply a pipeline would need to be trenched in certain areas to avoid being uncovered by migrating sandwaves, or alternatively to predict the depth and frequency of free spans that might occur should the pipe not be buried.

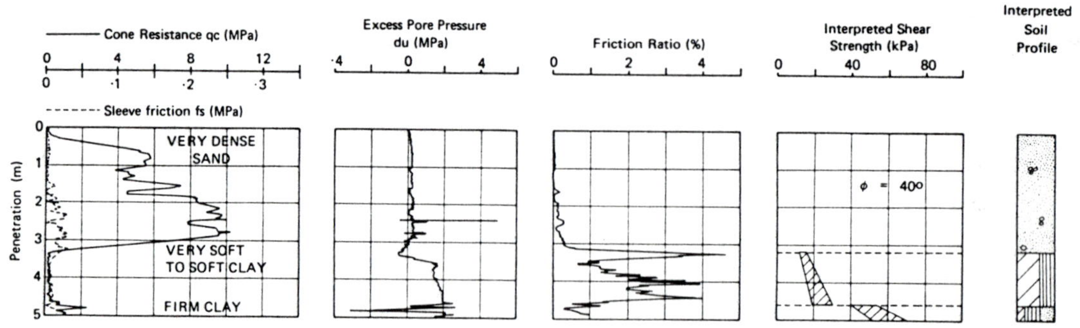

Fig. 9 Example of "Seasprite" CPT result

Integration with Geophysical Data

Wherever possible, the geotechnical data should be correlated and integrated with the geophysical data collected on the route survey. In this way the optimum value is obtained from the two operations. If, as is usual, the interpretations are proceeding in parallel, there should be maximum liaison between the two teams to ensure that:

- the geophysical evaluation has the maximum amount of core and test data input to assist in interpretation;
- the geophysical interpretation concentrates upon aspects of engineering relevance;
- the geotechnical interpretation has the necessary data to enable interpolation between test or core stations, and to plot and present the data in correct geographical orientation.

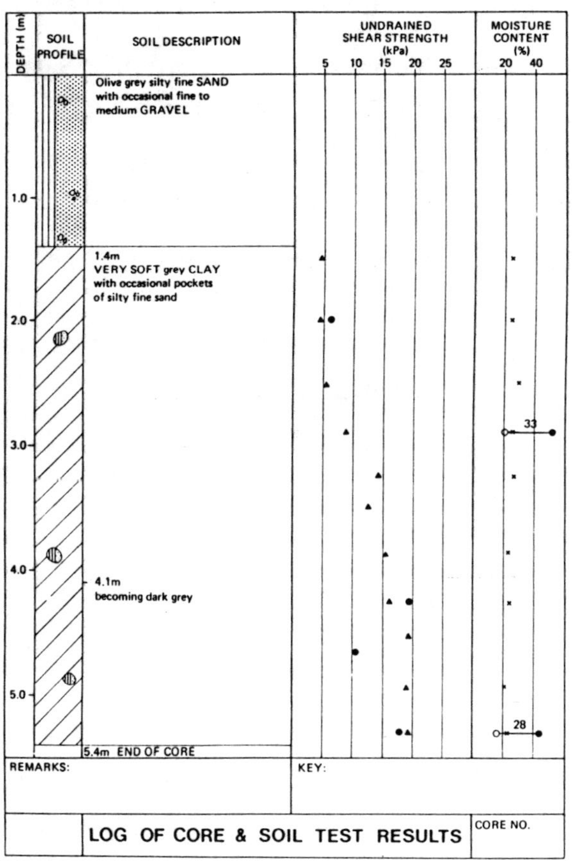

Fig. 10 Typical core log sheet

DATA PRESENTATION

Laboratory Test Results

While it is important that the results of all the individual laboratory tests are presented for specific reference when required, they should not overwhelm the soils report and, if necessary, they can be relegated to a separate volume. However, the main report should contain tabulated summaries of the test results.

Core Logs and CPT Graphs

For each core, a separate log sheet should be produced which clearly describes the most important features of the cores and presents them in an easily digested form. These are, in fact, simplified borehole logs and a typical example is presented in Fig. 10.

Equally, the individual CPT results need to be presented in such a way that the results explain themselves as far as possible, and do not require a specialist to translate them into meaningful information. The presentation of a typical "Seasprite" CPT is presented in Fig. 9.

Summary Profile

When all interpretation is completed the geotechnical information needs to be presented in a summarized form that can easily be assimilated by the pipeline engineer. For this reason an "Alignment Sheet" format, similar to that used for the finalized route survey data, is recommended. This should show the whole route in profile on one sheet or, if impractical due to length, this could be broken down into a number of separate sheets. The bathymetric profile of the route should show the location of each core and test and, if possible, a graphical representation of the soil profile. The findings of the shallow seismic interpretation can also be summarized on this profile and used, if possible, to show the continuity of layers between the sample/test locations.

Above the profile, a graphical presentation of critical geotechnical parameters can be displayed, e.g. the variation in thickness of a surface sand layer and the proportion of this which is "mobile", and the average grain size of the sand for assessing friction factors. In predominantly clay soils, factors such as shear strength and related bearing capacity should be displayed. Other data that can be shown might include the CPT cone resistance or parameters such as:

- the density of liquefied soil;
- compressibility;
- soil temperature;
- thermal conductivity;
- electrical resistivity;

and so on, the data presented being tailored to suit the predominant soil conditions and the engineering problems anticipated. Should there be any seabed features that have been investigated in more detail, they will be worthy of a separate diagram giving a similar presentation on an expanded scale.

Highlighting Problem Areas

Having provided a summary of geotechnical data, any areas that are expected to present problems — such as excessive scouring, difficult trenching or very low bearing capacity — should be highlighted and discussed in the text of the report. In this way, all of the relevant data will have been presented to the pipeline engineer, and his attention drawn to areas that might require additional scrutiny. During the detailed design and installation phases, specific problems can then be analysed in more detail with the aid of the available geotechnical data.

CONCLUSIONS

With regard to geotechnical investigations for pipeline routes, our main conclusions can be summarized as follows:

- Plan the investigation carefully, taking into account knowledge of the local geology and seabed topography.
- Choose the equipment spread to suit the expected soil conditions.
- For very soft soil sites, give consideration to specialist equipment such as *in-situ* shear vane equipment, "push-samplers" and box corers.
- For the most cost-effective collection of the optimum amount of data, a ratio of two or three Cone Penetration Tests (CPTs) to one core is suggested.
- As a bare minimum, allow for one test or core per kilometre plus additional stations at specific seabed features and changes in strata.
- Perform an adequate programme of laboratory tests to ensure that the necessary geotechnical parameters are obtained for further analyses.
- For "sandwave" areas subject cores to treatment such as "Lacquer Peel" to define the depth of mobile sand, and correlate with CPT results.
- When interpreting the results of the in-situ and laboratory tests ensure that they are correlated with each other and the results of the geophysical survey to obtain the best interpolation between sample/test locations.
- Maximize liaison between geophysical and geotechnical interpretation teams to ensure that the geotechnical data can be presented in the correct geographical framework, and that the geophysical interpretation concentrates on aspects of engineering significance.
- Present the core and *in-situ* test data in a simple, clear and easily understood format, and summarize as much of the data as possible in a graphical summary profile for the whole pipeline route.
- Highlight the areas where problems for the pipeline might occur, so that they can be subjected to more stringent analysis during the detailed design phase of the pipeline project.

With regard to that final point, if the preceding steps have been followed, sufficient high-quality geotechnical data should be available to analyse and solve such problems as they arise.

ACKNOWLEDGEMENTS

The authors would like to thank Kate Mansfield for her invaluable assistance in preparing the illustrations for this paper, other colleagues in Fugro Limited for their constructive comments, and the typing staff for their patience and perseverance.

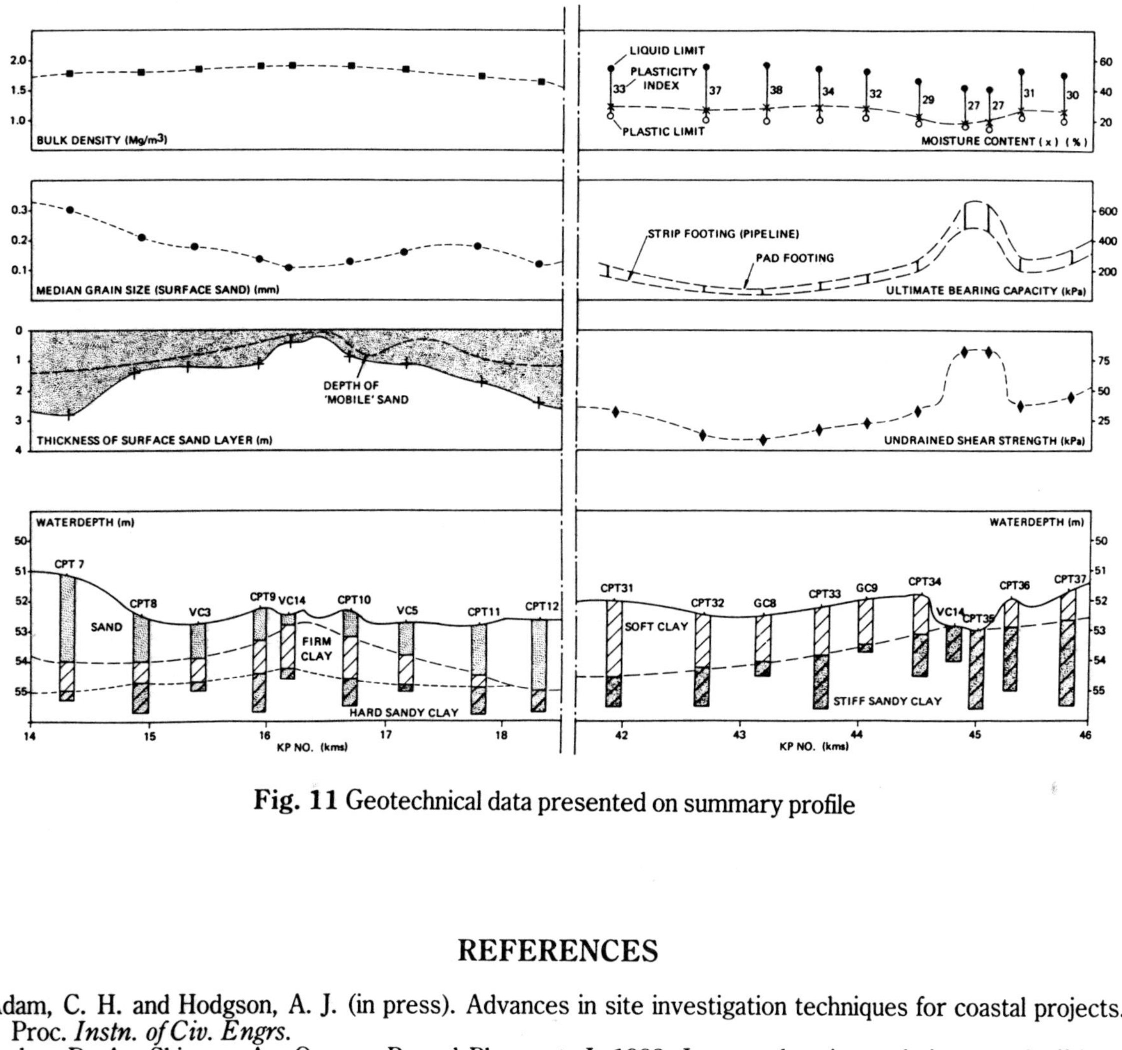

Fig. 11 Geotechnical data presented on summary profile

REFERENCES

Adam, C. H. and Hodgson, A. J. (in press). Advances in site investigation techniques for coastal projects. Proc. *Instn. of Civ. Engrs.*

Ardus, D. A., Skinner, A., Owens, R. and Pheasant, J. 1982. Improved coring techniques and offshore laboratory procedures in sampling and shallow drilling. *Oceanology International Exhibition and Conference,* Vol. 2, Paper 58, 18 pp.

de Vos, J. 1982. The practical use of the CPT in soil profiling. In: *Proceedings of the Second European Symposium on Penetration Testing, ESOPT II,* Amsterdam, Vol. 2, pp. 933–9.

Hovland, M., Judd, A., and Maisey, G. 1985. North Sea gas feeds the North Sea fisheries. *New Scientist,* August, p. 26.

Institute of Petroleum 1982. *Pipeline Safety Code, Model Code of Safe Practice Pt. 6,* 4th Edn. London: Heyden, p. 65.

Kolk, H. J. and Power, P. T. 1983. Advances in geotechnical investigations and design for offshore pipelines. In: *Proceedings, Offshore Oil and Gas Pipeline Technology Seminar,* Copenhagen, pp. 1–48.

Marsland, A. and Windle, D. 1982. Development in offshore site investigation. *Oceanology International Exhibition and Conference,* Vol. 1, Paper 2.7, 37 pp.

Sanglerat, G. 1972. *The Penetrometer and Soil Exploration*. London: Elsevier.
Schmertmann, J. H. 1969. Dutch friction cone penetrometer exploration of research area at Field 5, Eglin Air Force Base, Florida. *Contract Rept. S-69-4*. U.S. Army Engrs. Waterways Exp. Stat., Vicksbug, Miss.
Senneset, K., Janbu, N. and Svano, 1982. Strength and deformation parameters from cone penetration tests. In: *Proceedings of the Second European Symposium on Penetration Testing, ESOPT II*, Amsterdam, Vol. 2. pp. 863-70.
Wroth, C. P. 1984. The interpretation of *in-situ* soil tests (The 24th Rankine Lecture). *Geotechnique* **XXXIV** (4), 449–89.

PIPELINE INSPECTIONS: AN OVERVIEW

D L Webb, Hunting Surveys Ltd, UK

Presented on **Thursday 6th March 1986 at 15.00**
in the Norfolk Conference Room

EMBARGOED AGAINST DELIVERY

36

Pipeline Inspections: An Overview

D. L. Webb, Hunting Surveys Limited
Hunting Surveys Limited, Borehamwood, Herts, UK.

The number of submarine pipelines in the North Sea continues to grow and with it comes the need to monitor and service such systems. In most areas of the North Sea, pipeline inspections are required annually by government inspectorates. These twelve-monthly inspections must be capable of determining, as far as reasonably practical, the following:

- movement of the pipe;
- unsupported spans;
- loss of cover in buried lengths;
- damage to the pipe;
- debris adjacent to the pipe.

For the most part these conditions require visual inspection to determine the extent and magnitude of damage. However, to inspect the entire pipeline visually is slow and costly, especially when it is considered that huge sections of the pipeline show no anomalies.

A sonar and sub-bottom profiler inspection indicating possible problem areas on which a visual inspection may be planned eliminates the need to inspect visually the entire pipeline. Recent advances in sonar equipment, towing platforms and data handling combine to allow this type of reconnaissance sonar survey to be carried out swiftly and accurately, with results being available at the end of fieldwork or soon afterwards (depending on data volume). In many cases the sonar/profiler survey even eliminates the need for a following visual inspection, thus further reducing time and cost.

Obviously there are some problem areas, for example trenched sections, but it is foreseeable that continuing developments in sonar imaging and data handling will limit the need for ROV/diver inspection to points of proven damage.

This paper firstly points out the requirement for confident sonar inspection, details some systems currently in use and attempts to suggest possible future developments.

FIELD PLANNING AND PROCEDURE

In order to view conditions on both sides of the pipeline the sonar must traverse the pipeline length on both sides. With normal passive tow fish this necessitates two separate passes of the pipe.

The position of the vessel and the position of the towing frame relative to the vessel must be

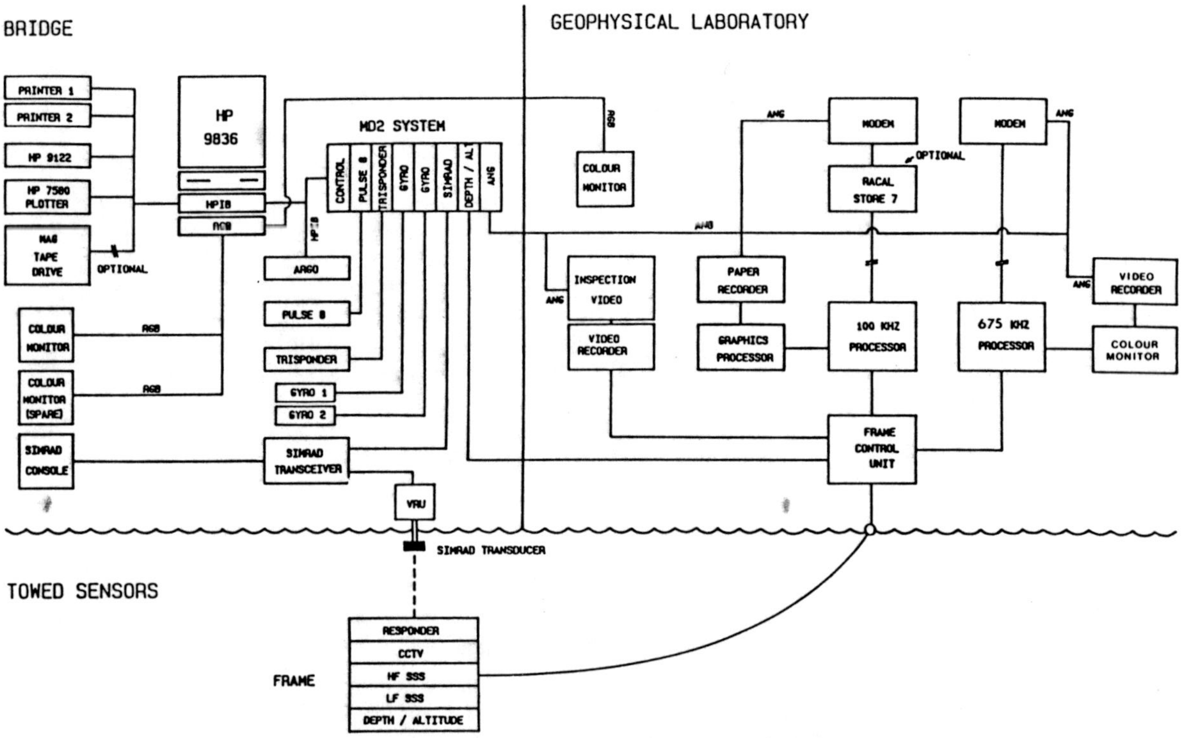

Fig. 1 Side-scan sonar survey: on-line system

accurately determined with all data acquisition systems, then integrated to the on-line navigation computer. A typical example of a vessel set-up is shown in Fig. 1.

The towing vehicle chosen should provide a stable platform capable of being maintained at an optimum position relative to the pipeline. This optimum position should be determined by the sonar frequency and beamwidths, and the desired vessel speed.

With the introduction of steerable towing frames it may now be possible to run two tow vehicles simultaneously, one on either side of the pipe, thus eliminating the need for two passes of the pipe and so cutting the field time in half.

The sonar must have a frequency and beamwidths capable of resolving actual damage or features indicating that damage may have occurred or could occur in the future. This question of resolution is particularly important when considering free spans. A span may be only a few centimetres high yet of sufficient length to cause stress on the pipeline, possibly resulting in fracture. The detection of a span is demonstrated schematically in Figs 3 and 4 and is shown on the data examples (Figs 5 and 6). In addition, Fig. 4 also shows the importance of tow vehicle position by indicating the minimum detectable span at any given angle of incidence on the pipeline. It should be noted that if the transducer platform were to yaw, pitch or roll, a feature resembling a span could be generated on the data record. This point further emphasizes the need for a stable tow body and/or the monitoring of the motion of the tow body. Were this latter information to become available it could be incorporated into the data base and used during the interpretation phase.

Problems will be encountered in trenched areas, as to detect a span we must detect the base of the pipe. In order to ensure that the pipe is not obscured by the trench wall the height of the side-scan sonar above the seafloor must be increased. This increases the angle of incidence on the pipeline and hence increases the size of the minimum detectable span.

Thought must be given to how and at what

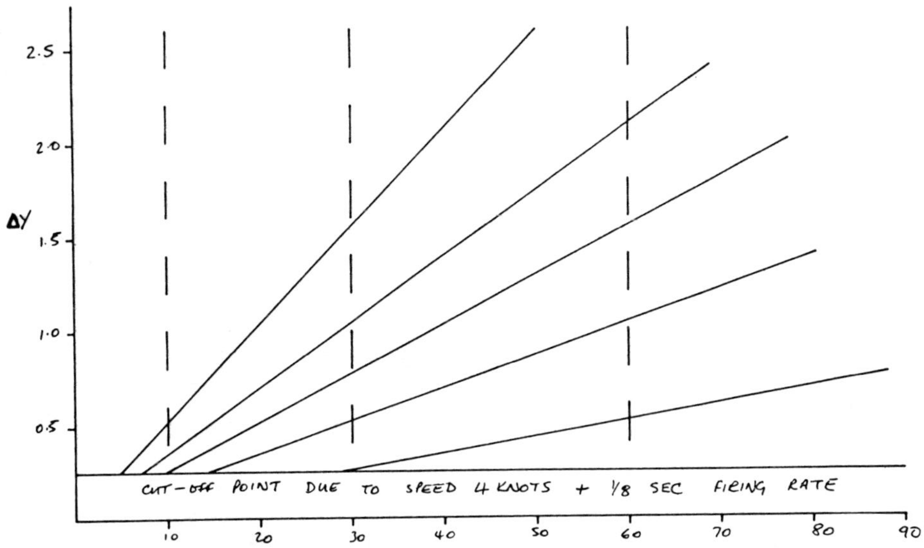

Fig. 2 Longitudinal resolution (Δy) versus slant range

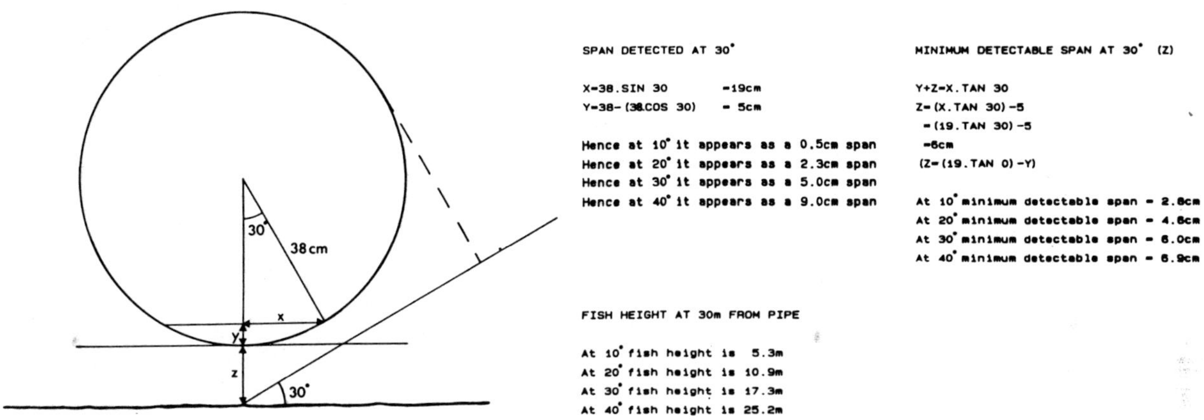

Fig. 3 Diagram showing the minimum detectable span length at various angles of incidence of the sonar beam

range the data are displayed. It is pointless utilizing a high-frequency system which has a resolution of a few centimetres when the medium and scale on which the data are displayed makes it impossible to see an object of this size. Additionally the dynamic range of the display medium is important as it governs the definition of features and the ability to resolve changes within features. Table 1 lists the dynamic ranges of a number of commonly used display media.

TABLE 1
Dynamic range of display media

Medium	*Dynamic range (dB)*
Wet paper	28
Dry paper	23
Colour monitor	40
Colour TV	46

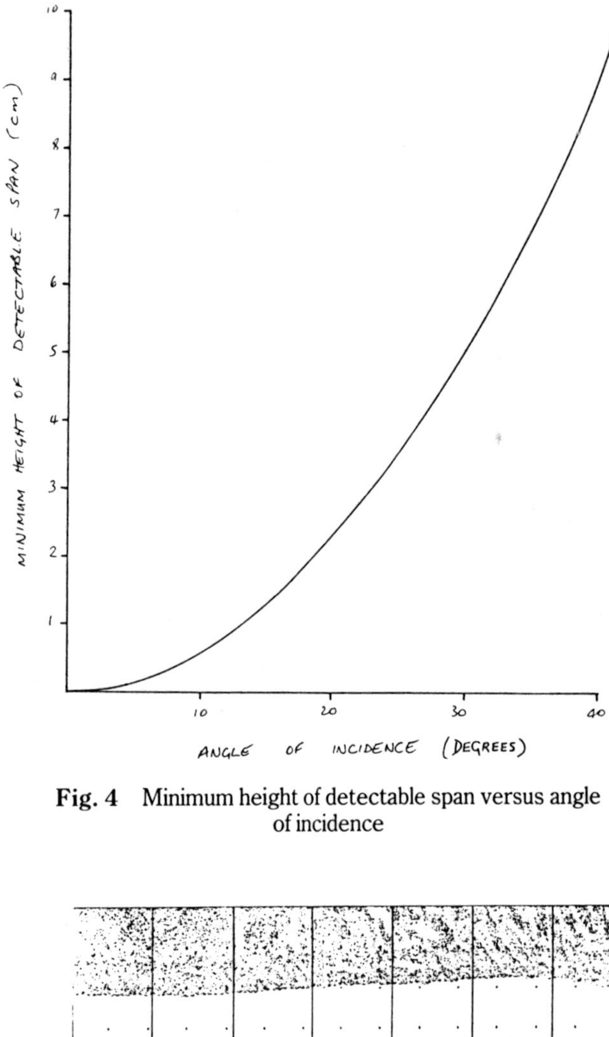

Fig. 4 Minimum height of detectable span versus angle of incidence

When considering the above it should be remembered that in addition to scrutinizing the actual pipe we also need to view the surrounding seabed for any objects which could be dragged onto the pipe. If a very high frequency system is used and the data are displayed at a usable scale, we can look closely at the pipe itself and its immediate surroundings. However, this system will not have the range to enable it to detect objects further away. If, on the other hand, we choose a low-frequency system, we lose the resolution needed to monitor the pipe (possibly even introducing some penetration in sediments under the pipe). The use of a mid-range frequency could be considered, but this is not really a satisfactory answer. The use of two systems run simultaneously and mounted on the same towing frame is probably the best route, despite the increased volume of data, providing that the systems are sufficiently different in frequency not to interfere. Data processing and interpretation should aim at correcting the various distortions in the data (slant range, water column, tow vehicle motion) and then at positioning the data on a UTM grid,

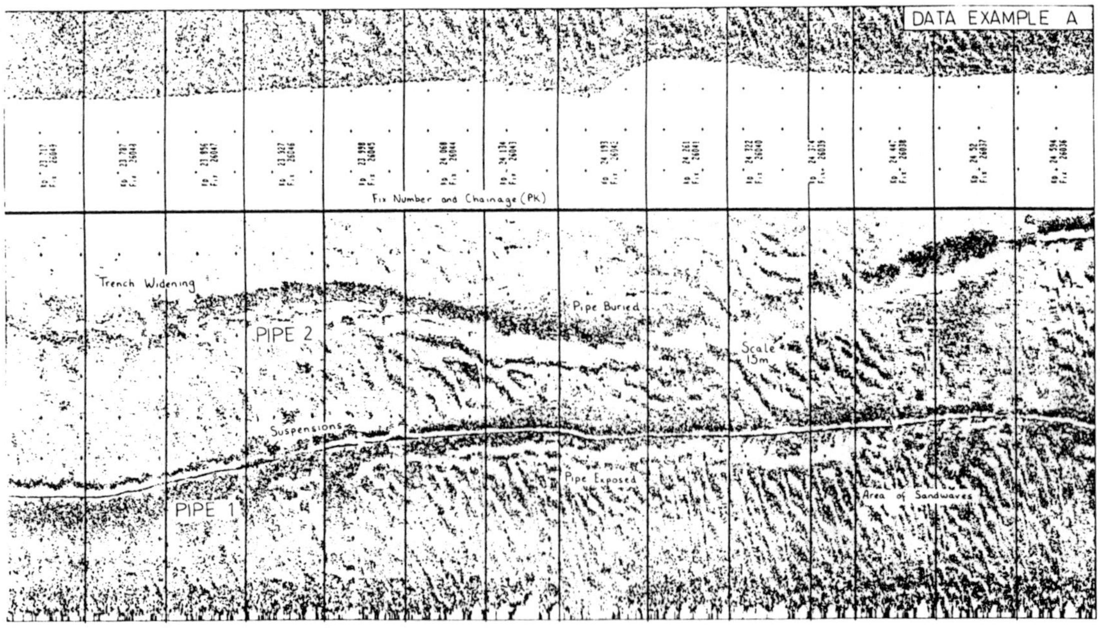

Fig. 5 Data example A

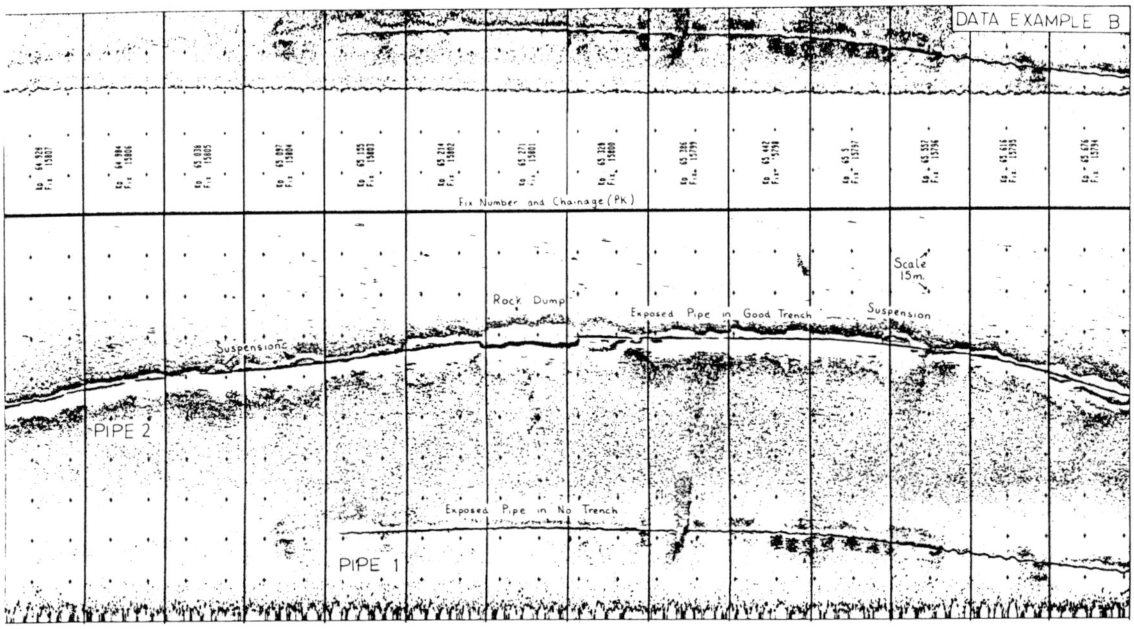

Fig. 6 Data example B

BLOCK DIAGRAM SHOWING HP9836 ACQUISITION SYSTEM USED IN CHARTING SURVEY

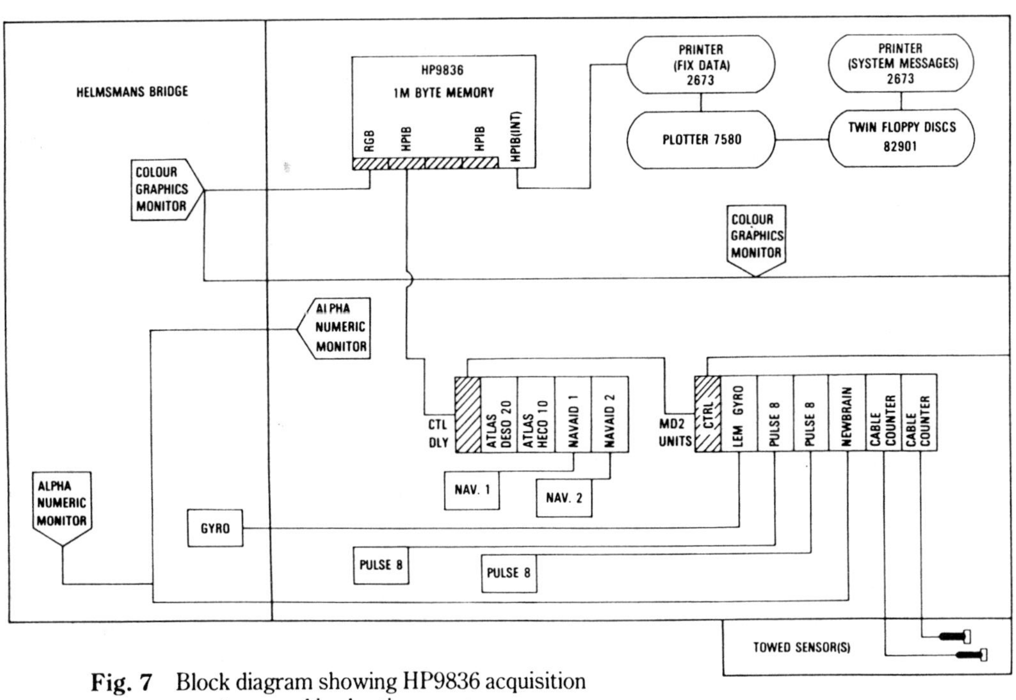

Fig. 7 Block diagram showing HP9836 acquisition system used in charting survey

irrespective of tow vehicle and vessel movement. The possibility exists to use pattern recognition techniques to detect anomalies such as spans. This processing can take place on line or post-acquisition.

In regions where the pipeline is buried, numerous seismic profiling systems may be utilized to locate the pipe by zig-zagging across the pipe route. An alternative possibility is the use of a magnetometer array. By the analysis of the sizes and distribution of the resultant magnetic anomalies the depth of burial of the pipe can be determined. Interpretation and chart production should ideally be carried out on board and be geared to enable a rapid throughput of data. An example of such a system developed by Hunting Surveys Limited is shown in Fig. 7. During interpretation, each feature is coded and stored on micro disc, with its position, before being plotted on final charts. Obviously some checking and smoothing must be involved, together with close liaison between the acquisition and processing stages of the data. In utilizing this type of on-board post-processing, results become available soon after survey work is completed, thus allowing a visual survey or remedial work to be planned and commenced quickly.

To summarize, the criteria for successful sonar inspection surveys include the following:

- Reasonable sized vessel with good working space.
- Accurate navigation of vessel and towing vehicle.
- Stable towing vehicle with motion sensors.
- Accurate positioning of the towing body relative to the pipe.
- Sonar frequency/beamwidth to give data required.
- Display medium capable of achieving the resolution requirements.
- Automated data handling to allow correction and fast production of results.

SYSTEM OPTIONS

The following sections detail each part of the on-board system and give some of the currently available options.

- Navigation
- Towing vehicle positioning
- Towing vehicle
- Sonar systems
- Profiling systems
- Data handling and processing
- Results

Navigation

It is perhaps obvious that accurate positioning of the inspection vessel is essential for both visual and sonar work. It is particularly important to consider the repeatability of the navigation system if any comparison of the pipe position and status is to be made from year to year, or if a maintenance/visual inspection is being based on the results of a sonar survey. Several systems are available, all having their advantages and problems. A number are listed in Table 2, with range, absolute accuracy and repeatability. The figures given are the makers' specifications, but these may be unattainable under field conditions.

TABLE 2
Navigation systems

System	*Maximum range* Day	Night	*Repeatability (approximate)*	*Accuracy*
Pulse 8	900	550	15	50
Argo	740	230–550	3	5
Hyper-Fix	700	250	1	1.5
Syledis	150–250	—	2	5
Trisponder	0.1–80	—	1	1
Artemis	30	—	2	15

The choice of which system to use is a question of reliability, repeatability and the predicted lifetime of the pipeline. A change in the positioning system used can create 'shifts in position' which have not occurred in reality, and hence make it impossible accurately to determine pipe movement or status change. It is possible to make comparisons of positional information and obtain some correction, but this can be time-consuming and costly. Consequently it would be preferable to use an identical navigation chain throughout the life of the pipeline.

Towing Vehicle Positioning

Once again, for information to be of practical use, the position of the towing body, be this ROV or sonar, relative to the vessel is required.

Methods range from assuming the towing body progresses along the vessel's path, but with a 'lay back' dependent on the amount of towing cable payed out. This is of course inaccurate as no serious account may be taken of the body's position off the ship's track or the exact distance from the vessel to the body.

Long baseline, seafloor responders are impracticable as in many cases the pipeline may be hundreds of kilometres in length, thus requiring a vast network of such responders or their constant relocation.

The most satisfactory method available at present is a super short baseline hull-mounted system such as the Simrad, ORE or Honeywell systems. These systems allow one or more bodies to be tracked relative to the vessel. Details of these systems are given in Table 3.

TABLE 3
Tow vehicle tracking systems

System	*Maximum range (m)*	*Range Accuracy (%)*	*Bearing Accuracy (%)*
ORE Model 460 'Trackpoint'	800	1	1°
Honeywell 'Hydrostar'	2000	1	N/A
Simrad 'HPR'	2000	0.5	0.55

The reasonably impressive errors given by the makers take no account of thermoclines which may increase the ray path from the vessel to the towing body.

Despite these errors, and given good navigation, a short baseline tracking system will give the towing body's absolute position to a high degree of accuracy and hence allow reasonable comparison from year to year.

Towing Vehicles

As pointed out previously, the positioning of the sensors relative to the pipeline and the stability of this position is vital if the data are to be

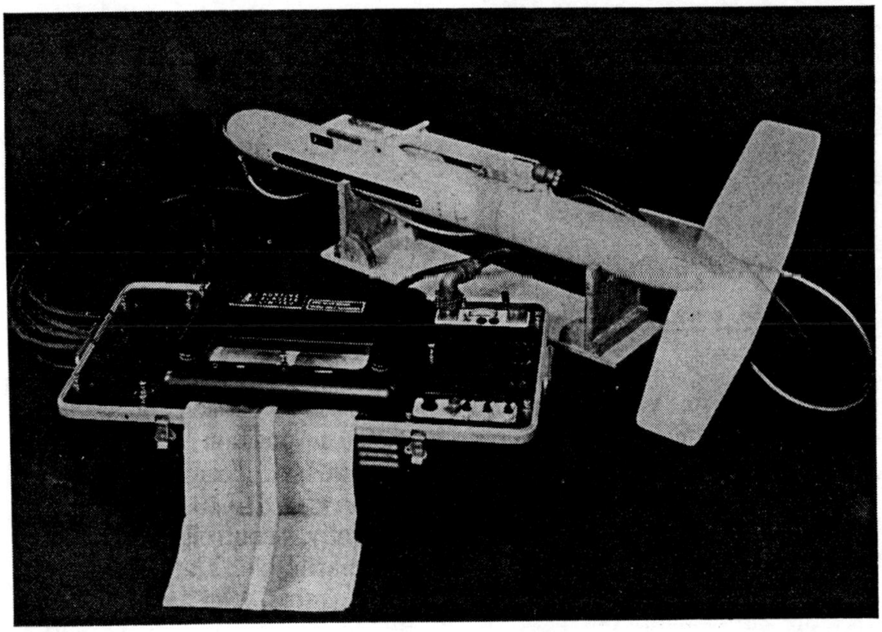

Fig. 8 Side-scan sonar system

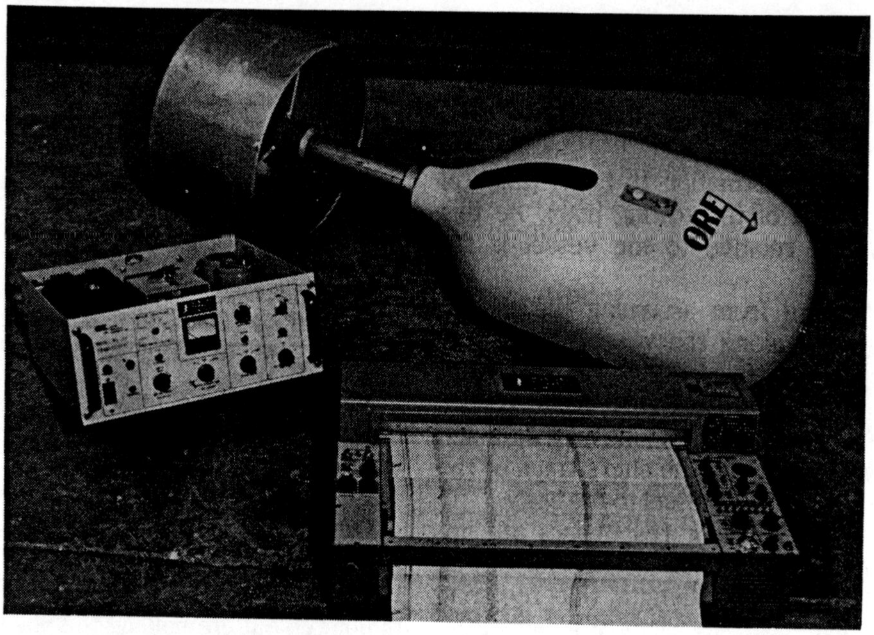

Fig. 9 EG&G pinger and side-scan sonar system

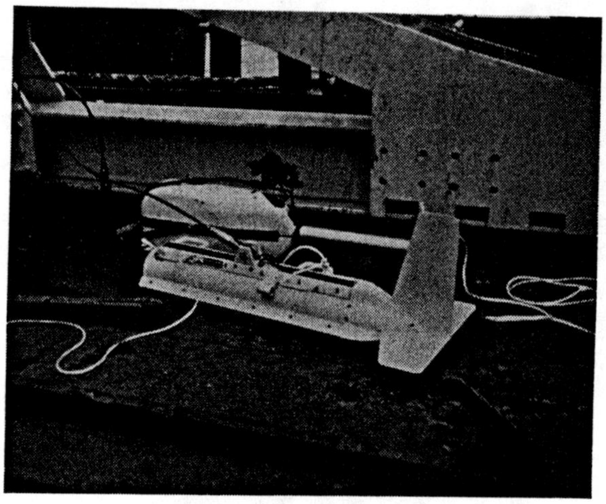

Fig. 10 EG&G sonar fish

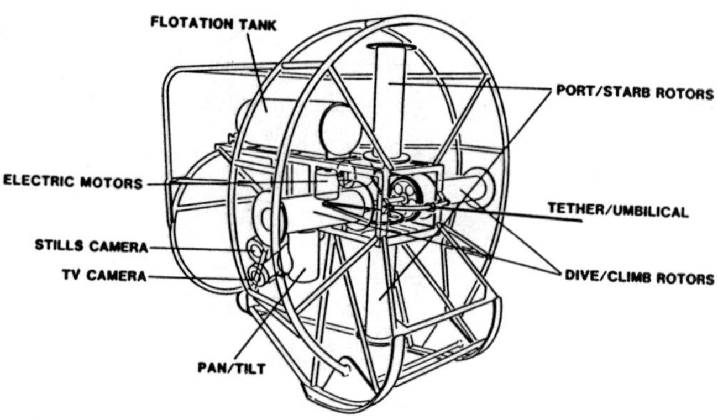

Fig. 11 The Ocean Rover towing frame

optimized. Unfortunately no vehicle can be completely stable, suggesting the need for motion sensors for later data correction. However, as much effort as possible should be made to obtain a stable platform for the sensors, which can be positioned at the required offset relative to the pipe.

Until recently, sonar towing bodies have consisted of passive hydrodynamic tow 'fish'. Despite being reasonably stable, no means is available to position the sensors independently of the vessel. Examples of various fish are shown in Figs 8 to 10.

Recently, controllable tow vehicles have become available. These are steered by hydrofoils or rotors, the latter being preferred as they provide a greater degree of control, cause little turbulence and hence give a more stable

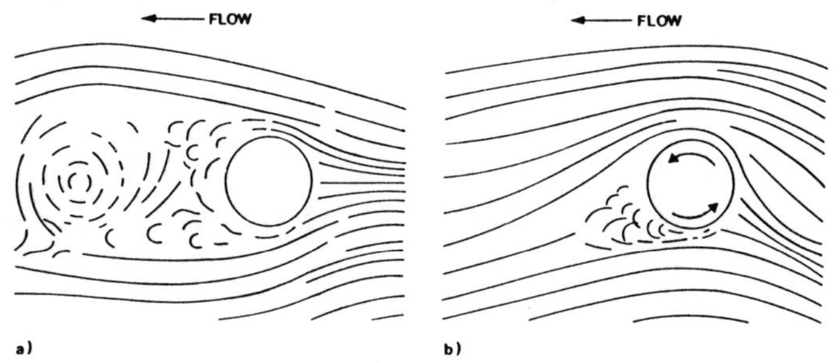

Fig. 12 (a) Diagram showing flow over rotor in the Ocean Rover. (b) Diagram showing flow over rotor as it rotates anticlockwise. Note flow pattern

platform. Figure 11 shows an example of a currently available towing frame, the Ocean Rover.

The Ocean Rover is a development from a steerable towing frame produced to monitor fishing trawls. This remotely operated towed vehicle (ROTV) uses the Magnus rotor principle (an extension of the Benoalli principle).

Figure 12 shows how a force is developed across a rotating cylinder (rotor) by setting up an uneven flow on each side of the rotor. The Ocean Rover uses four of these rotors, two controlling horizontal motion and two controlling vertical motion, giving it a capability for movement within ± 80 m of the vessel path. The vehicle has a large frame capable of taking a large payload, allowing it to carry TV cameras side-scan transducers, trench profiler and sub-bottom profiler.

Examples of data from a standard tow fish and from the Ocean Rover demonstrate greater stability of the latter and its ability to maintain the sensors in the optimum position relative to the pipe.

Sonar Systems

The range of sonar systems is vast, with frequencies and beamwidths being equally variant. The choice of system should be made carefully by considering not only the actual sensor itself but also the display medium and the towing device body.

There is little sense in choosing a very high resolution system if the display medium is unable to depict the detail available, or if effects due to the fish movement are of the same magnitude as definable real features.

Table 4 lists a few well-known systems, noting their frequency, beamwidths and pulse lengths, together with usual display medium. Data from some of these systems are shown in

TABLE 4
Sonar systems

System	*Freqency (kHz)*	*Pulse length (ms)*	*Verticle beam width (degrees)*	*Horizontal beam width (degrees)*	*Display medium*
Klein	37	0.15	0.6	20	Wet paper
ORE 160	100	0.1	1.5	55	Dry paper
Klein	100	0.1	1.0	20/10	Wet paper
EDO Western 606A	100	0.1	1.0	25	Wet paper
EG&G 272	105	0.1	0.1	20/50	Wet paper
Waverly	110	0.1	1.5	50	Dry paper
Klein	500	0.02	0.2	40	Wet paper
Mesotech 971	675	N/A	1.7	30/60	TV monitor + video recorder

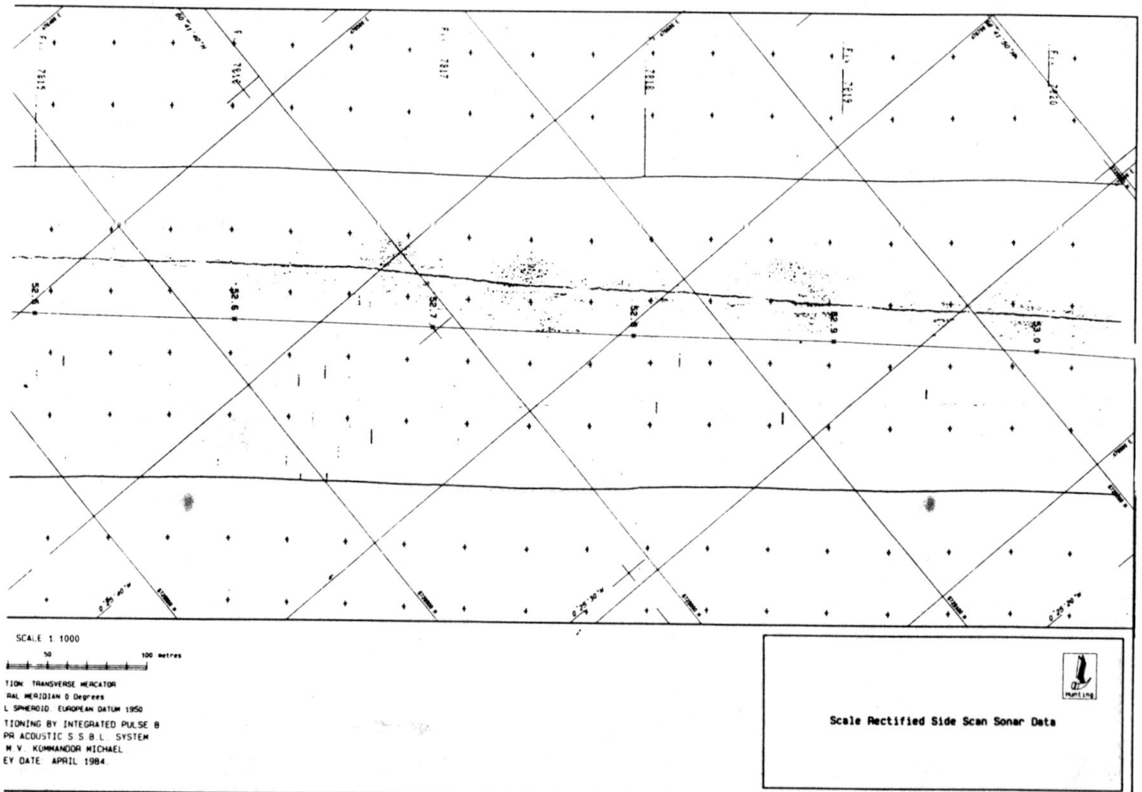

Fig. 13 Scale-rectified side-scan sonar data

Fig. 14 Scale-rectified side-scan sonar data

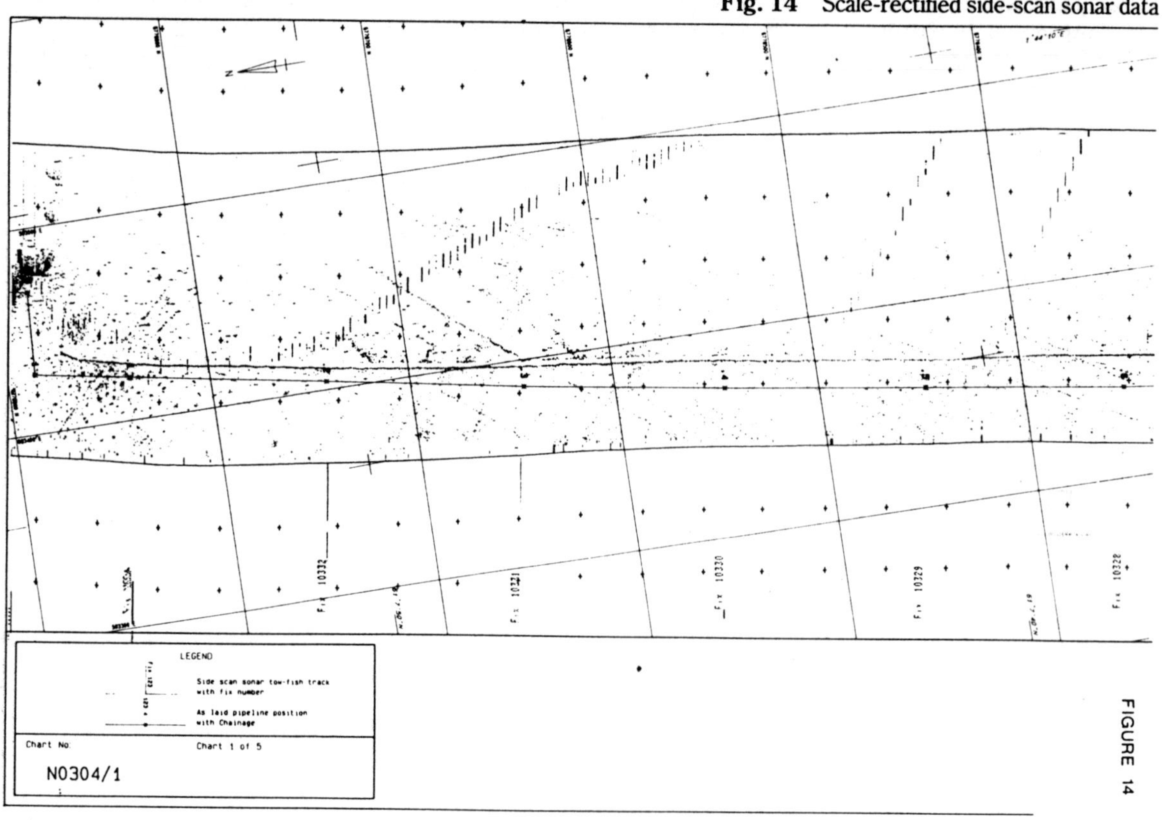

FIGURE 14

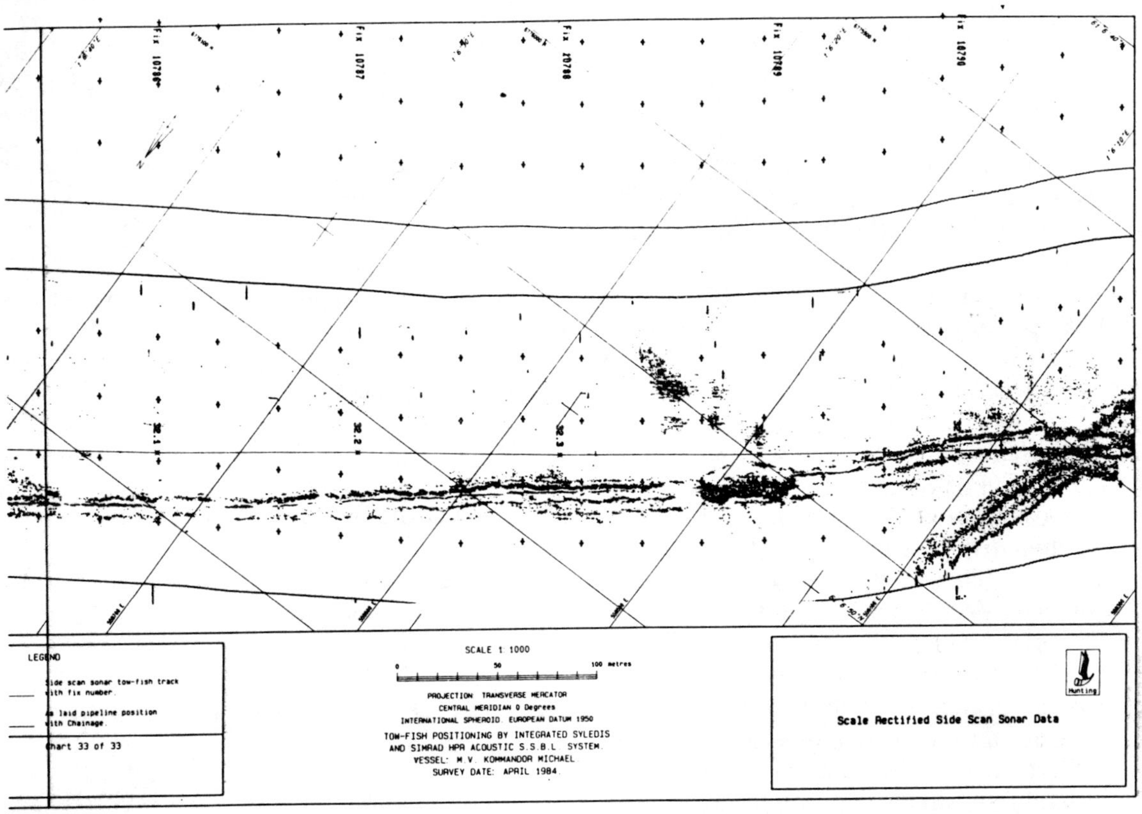

Fig. 15 Scale-rectified side-scan sonar data

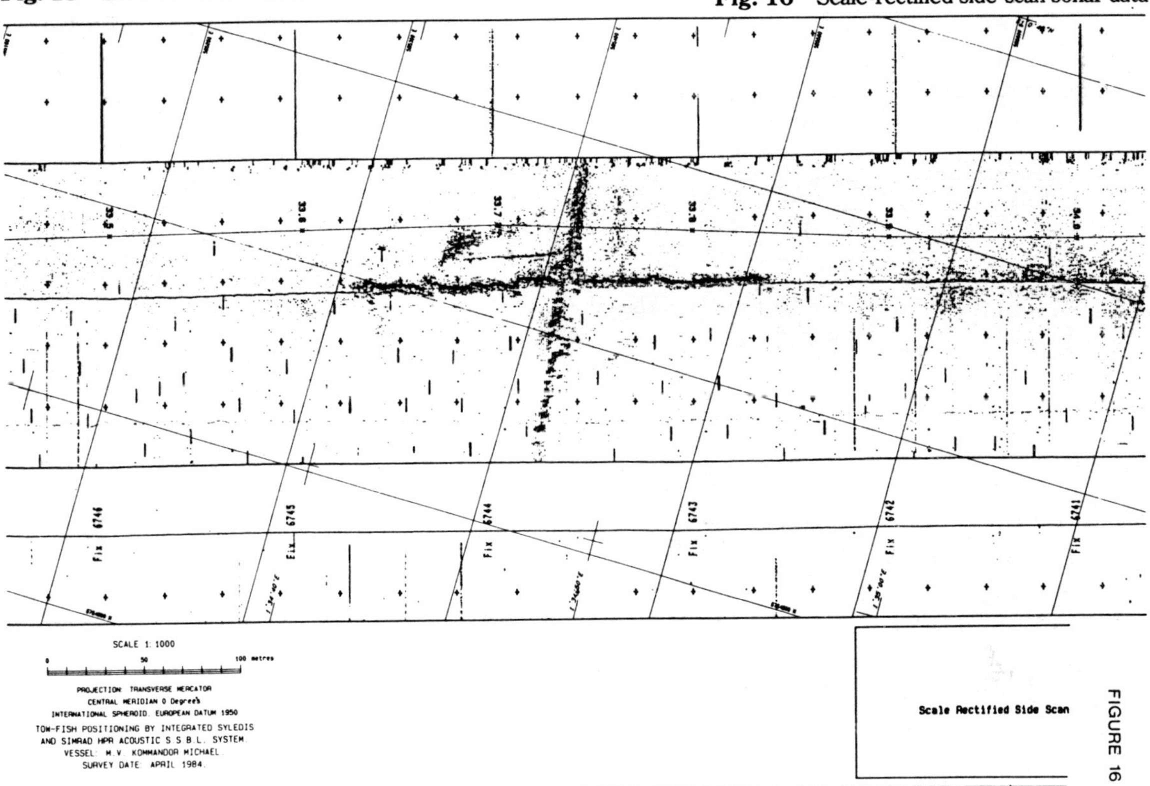

Fig. 16 Scale-rectified side-scan sonar data

Figs 13 and 16. It is worth observing how features are defined differently in different systems. Particularly note in Fig. 14 the ORE 160 (100 kHz) and Mesotech (675 kHz) data which are from the same section of pipe. These were actually run simultaneously with both sets of transducers mounted on an Ocean Rover towing frame.

Profiling Systems

In areas where the pipe is buried the depth and position may be checked by use of sub-bottom profilers. These systems can also be of use in trenched sections where the side-scan sonar has some limitations.

The most well known system is the pinger, and an example of data from such a system – the ORE pipeline system – is shown in Fig. 17.

With the introduction of more stable and steerable towing vehicles it becomes possible to utilize systems normally reserved for ROV inspections. For example, the Ultertech scanner could be mounted on the frame to give information on the pipe status in trenches.

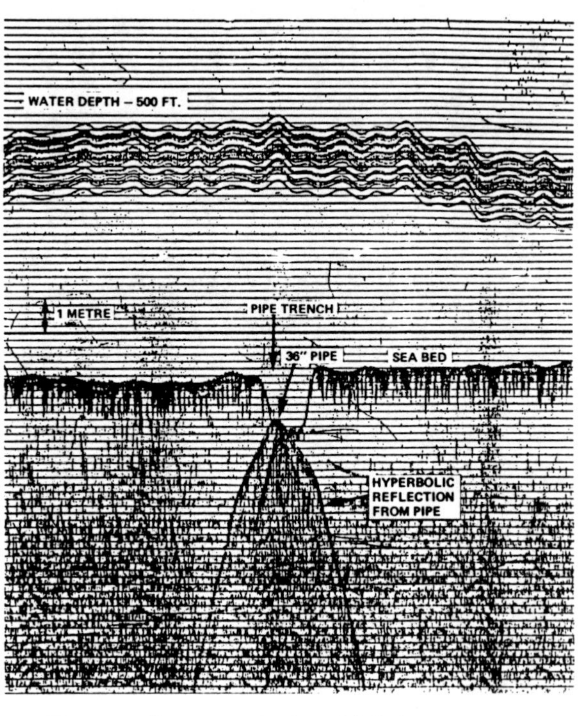

Fig. 17 Data example from ORE pinger system

Profiling is not the only answer to surveying buried sections of pipe. Given a steerable stable platform and possibly motion sensors, an array of magnetometers would enable the pipeline's depth and position to be determined by analysing the anomalies caused by the pipe. ROV-mounted systems already exist, e.g. the Innovatum maggetic tracking system, and it may be possible to mount them on a controllable tow frame.

Data Handling and Processing

Figure 18 shows the schematic layout of both the acquisition and the processing systems. It

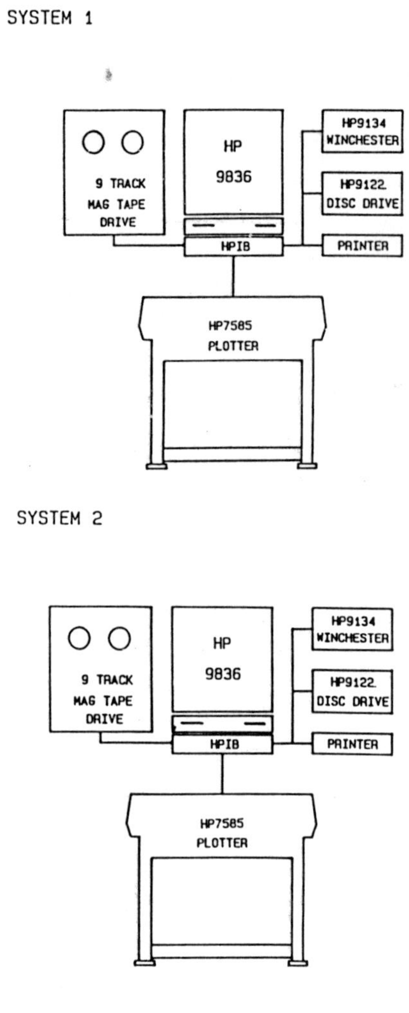

Fig. 18 Side-scan sonar survey: acquisition and processing systems

may be seen that the central on-line navigation computer is linked to all parts of the acquisition equipment, by means of a data bus system (in this case Hunting Md2). The computer, in addition to positioning the vessel, logs positional data for the vessel and tow vehicle, supplies fix marks and information to the survey sensors and vessel speed to processing units for scale correction. The data are then logged on flexible discs. These are then copied and used during post-processing.

The present manual interpretation of the sonar data must attempt to categorize the features observed. Typically the categories include debris, damage, spans, scouring, gravel dumps and trenching, amongst others. The individual features are recorded on specifically designed log sheets detailing their dimensions and positions relative to either kilometre posts or fix positions. The coded information is then transferred onto flexible disc to create data files. Each file contains details of only one feature type and may contain more detail than can always be shown on the final charts. As such they can be viewed as an exhaustive reference listing over and above the information included on the final graphical presentation.

The files should be designed readily to facilitate editing and data manipulation. This will allow, for example, reports to be completed detailing all or selected features along a specific section of the pipeline, or selected features down the whole length of the pipeline.

The navigational data can then be applied to the sonar features files, allowing preliminary alignment charts to be drawn. These are then checked for continuity, this being particularly important when more than one person is involved in data interpretation. Final drawings can then be produced. With the exception of the interpretation part of this processing sequence, the data are processed by desk-top computers. Hopefully, with the developments discussed below even the interpretation of the data may be computerized (obviously with supervision), hence further reducing the time to produce preliminary and final results.

Data processing: scale correction

The effects due to slant range, water column and vessel speed are well known. Correction to these distortions can be on line as in the ORE 158 graphic processor system, EG&G seafloor mapping system or the Klein K map system. In each of these cases the height of the fish above the seabed and the vessel's speed are monitored and used to correct the data to a true plan format at the desired scale.

Systems have been developed which enable raw data to be processed after acquisition; for example, Geoteam have utilized an array processor on a mainframe computer and Hunting Surveys the HP9836b mini-computer system (the latter destined to become an on-line system). These systems, unlike the on-line systems, position the data on a UTM grid, hence removing the effects of vessel and fish

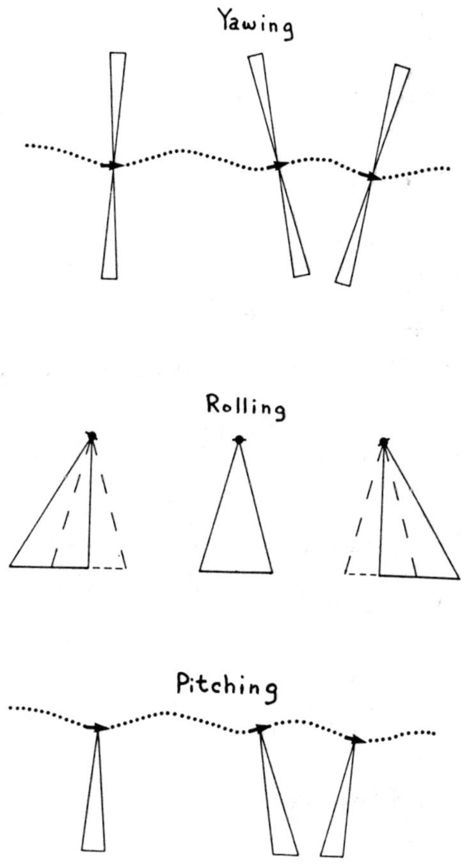

Fig. 19 Fish movement

movement. Examples of both on-line and off-line systems are given in the data example. At present none of these systems attempts to correct for yaw, pitch and roll of the towing vehicle, despite the great effect these have on the data (see Fig. 19). To correct for these effects requires three-axis monitoring of the vehicle motion. Recently miniature solid-state accelerometers have become available, allowing motion sensors to be fitted in even the smallest tow body.

Data enhancement and pattern recognition

The possibility exists for enhancement of the data and pattern recognition. However, development is at an early stage. Figures 20 to 22 show enhancement and pipeline tracking using the Hunting Surveys and Consultants' HIPAS system. Figure 20 shows the raw field data together with a single trace and the frequency content. Note on the trace profile the strong negative at around 100 ms. This is the return from the pipe. It should also be observed how the frequency content is grouped in a narrow band, and that the field has not totally removed the effects of spherical divergence and attenuation.

Figure 21 shows the effect of enhancing the data by gain averaging and increasing the frequency content. Note the increased clarity of the scars.

Figure 22 is an early attempt at tracking the pipeline using the strong negative seen in Fig. 20. It can be seen that despite the early stage of development these results are encouraging.

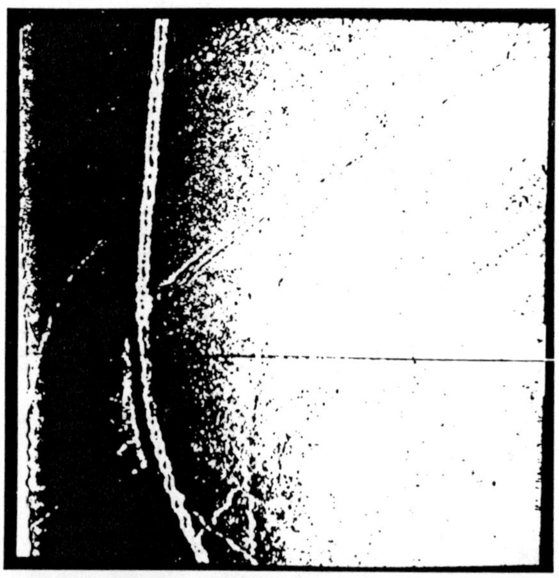

Fig. 20

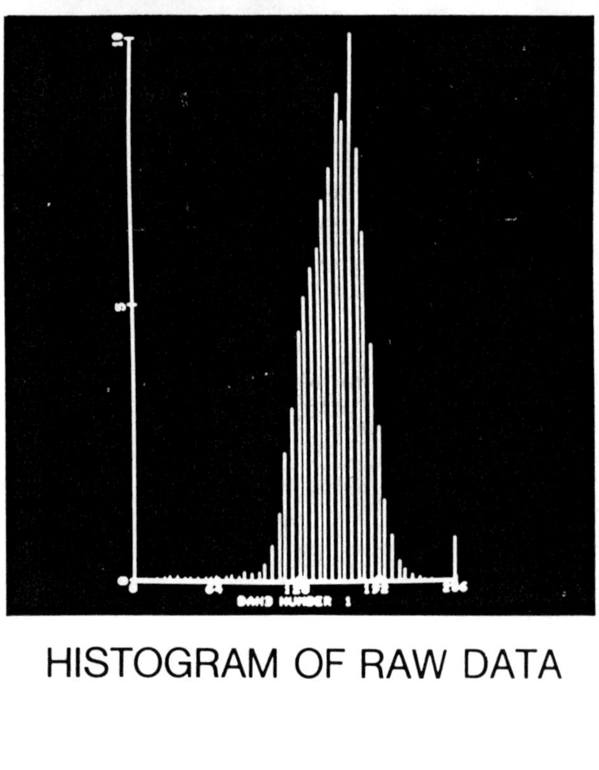

HISTOGRAM OF RAW DATA

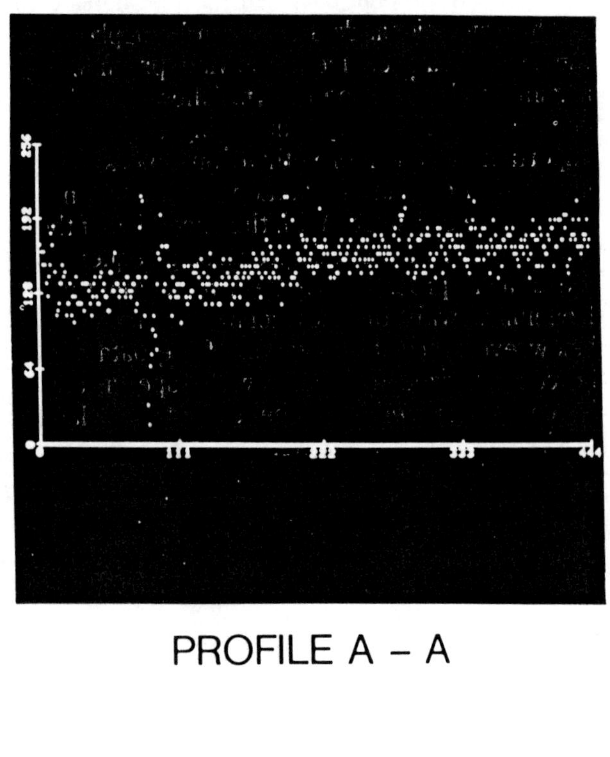

PROFILE A – A

The natural progression from this stage is pattern recognition to automatically detect spans and other anomalies. This would both speed up processing and give more accurate results than the standard manual techniques. A further advantage of this automatic interpretation is that it would remove inconsistencies over a period of interpretation and between different interpreters.

Fig. 21

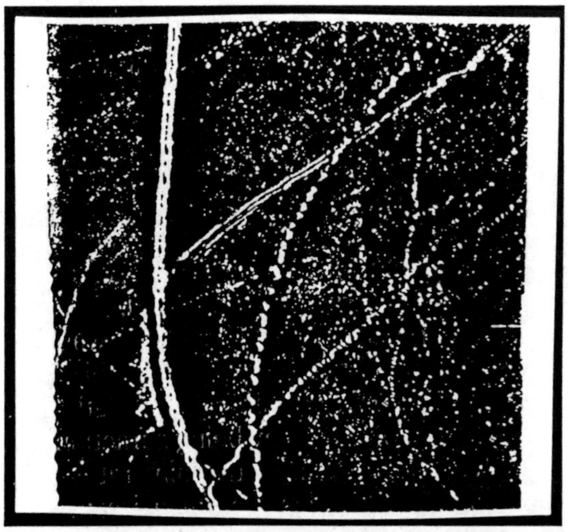

Results

An example of a chart depicting the sonar results is shown in Fig. 23. This format was designed to meet the client's requirements and to allow a visual survey and/or remedial work to be planned on the system. These final charts were computer drawn utilizing the HP 9836 system, using the data files compiled at the interpretation stage.

The use of computers to produce a high-quality final chart suitable for submission to the client is a particularly important feature of the system as a whole. By drastically reducing, or removing altogether, the need for manual draughting, a saving in cost and time is achieved. In this context the latter is much more important as it allows the results of a sonar survey to be presented soon after the completion of the fieldwork. Indeed many of the final charts would normally be available prior to completion of the survey, provided that the pipeline was examined in a modular way.

This data base technique produces an additional bonus as it rapidly allows comparison with previous inspection, hence allowing developing problems to be easily examined.

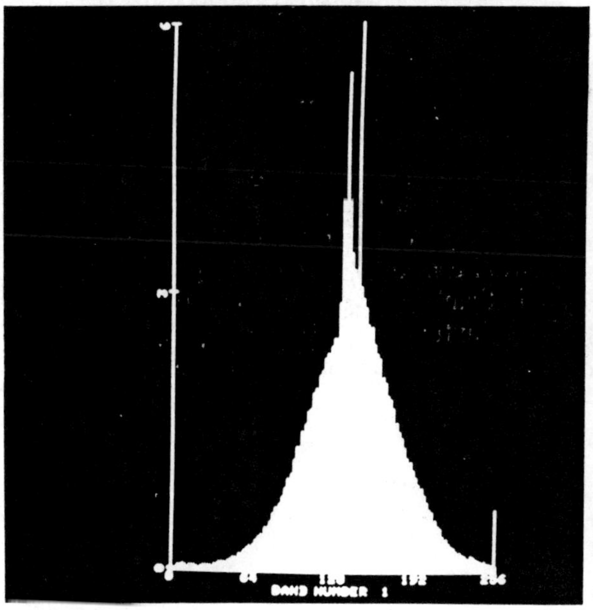

HISTOGRAM OF ENHANCED DATA

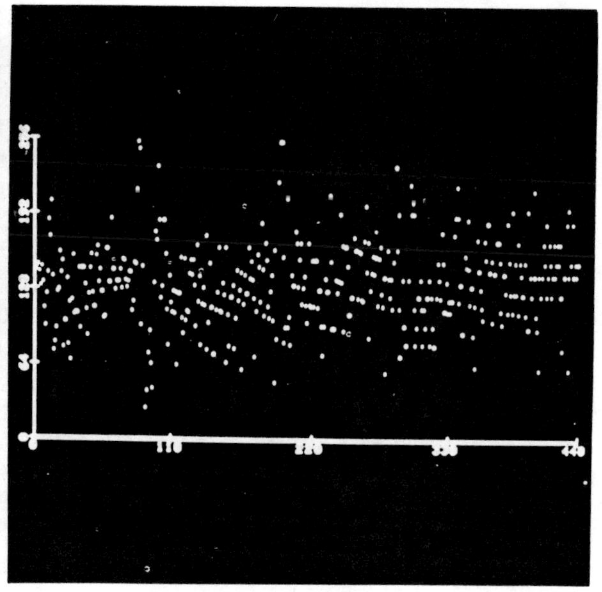

PROFILE B – B

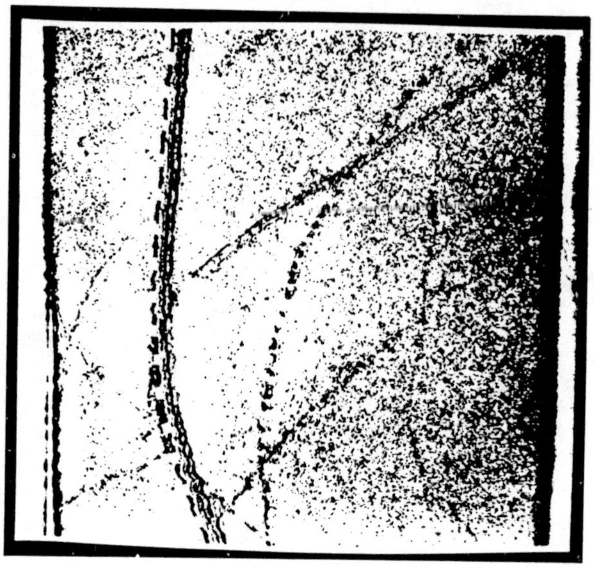

Fig. 22 Interpreted data

Vessel	MV *Master Surveyor*
Positioning system	Argo/Pulse 8/Trisponder
Computer/data handling	Hunting Surveys HP9836 and Md2 system
Tow vehicle	Ocean Rover
Tow vehicle tracking	Simrad HPR
Sonar systems	Simultaneous Mesotech 971 (675 kHz) and ORE 160 (100 kHz)
Display medium	TV monitor with video recording and dry paper
Post processing	Hunting Surveys feature logging, processing and charting systems (based on the HP9836 computer and HP7585 plotter)

Results were available at the end of fieldwork in the case of one pipeline system and soon after for the remainder, thus allowing a following ROV visual survey to be planned.

The result of this survey was a ROV inspection of the areas of possible problems rather than of the entire pipe length. With the price of a sonar survey being a fraction of that of a ROV survey, in both daily rates and time requirement, this reconnaissance sonar inspection significantly reduced the overall cost of the annual inspection.

CONCLUSIONS

A pipeline sonar inspection requires careful planning if the results are to reduce the need visually to inspect the pipeline. The choice of equipment is a case of optimizing the data through both acquisition and processing techniques. A recent sonar inspection of over 1500 line kilometres carried out by Orion Surveys A/S and Hunting Surveys Ltd is an example of modern techniques designed to meet these requirements.

ACKNOWLEDGEMENTS

I would like to thank the following for their help in compiling this paper: Shell (UK) Production and Development Limited; Statoil; Total plc; Orion Surveys A/S; Hunting Surveys Limited. My thanks are also due to Dr P. R. Thomas, M. C. Hayter, J. B. Hopwood, D. Berry, P. N. Thomas and C. E. J. Hill for their time and efforts.

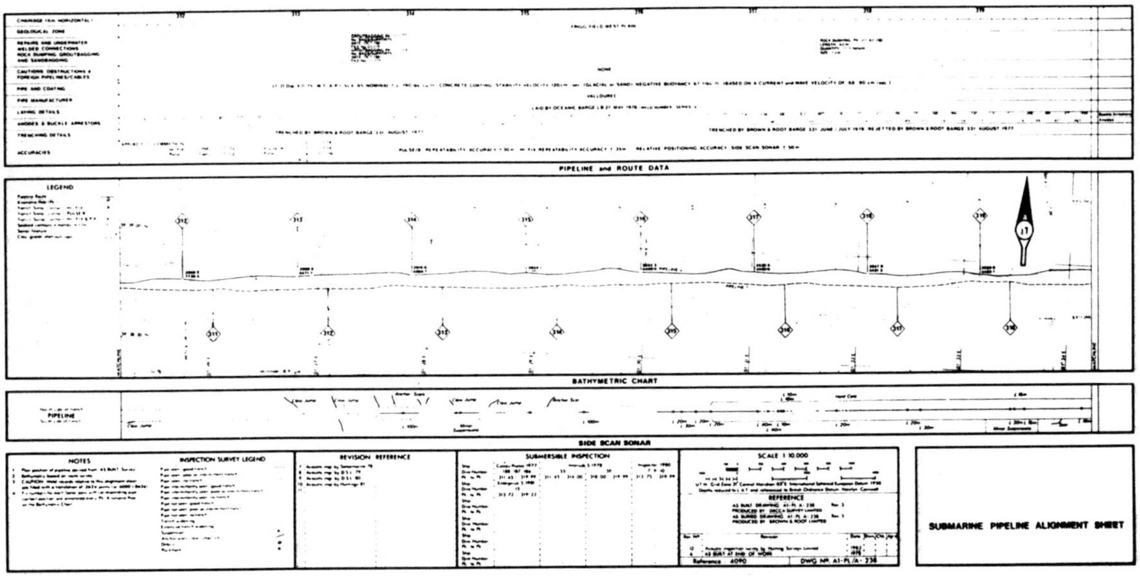

Fig. 23 Submarine pipeline alignment sheet

REFERENCES

Urick, R. J. *Principles of Underwater Sound for Engineers.* McGraw Hill, New York.

Fleming, Klein and Denbigh, *Recent Developments in Side Scan Sonar Techniques.* NGA Russel Cargill, Cape town.

Hayes, J. W. 1970. Planning and maintenance of submarine pipeline routes. *Europe and Oil.*

Preietly, Wordle and Hall 1985. *The Marine Laboratory Remote Controlled Fishing Gear Observation Vehicle.* International Council for the Exploration of the Sea.

Department of Energy 1984. *Submarine Pipelines Guidance Notes.* The Pipeline Inspectorate.

37

Analyses of Instability of Seabed and Substrate in the East China Sea

Ye Yincan, Second Institute of Oceanography, National Bureau of Oceanography, Hangzhou, People's Republic of China

INTRODUCTION

Ocean engineering has been developing with exploration and utilization of ocean resources. In the last 20 years, ocean engineering has moved forward on an unprecedented scale with the development of the offshore oil and gas industry. In most cases seafloor sediments support much more loading than on-land soils because of the dynamic environment of the ocean and the special requirement of seafloor structures. In ocean engineering construction serious accidents often happen as the result of foundation failure. Based on data for the bathymetry, shallow seismic profiling and soil sampling with laboratory investigations, we report briefly here on the instability of the seabed and substrate in this area.

MOVEABLE AND VARIABLE BEDFORMS

The development and variation of bedforms are closely relevant to environmental protection and various aspects of ocean engineering. Moveable and variable bedforms result in sand accumulation or erosion and can cause seabed pipelines to lose their bottom support and threaten the safety of shallow foundation structures on the seafloor.

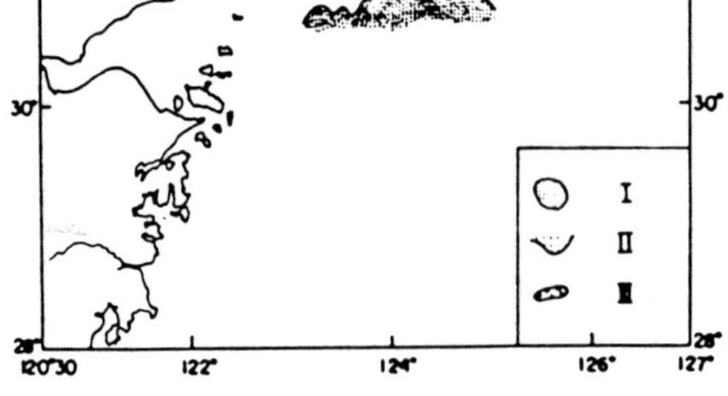

Fig. 1 Distribution of the bedform morphologies. I, well-developed megaripples; II, partially developed megaripples; III, sandwaves

ANALYSES OF INSTABILITY OF SEABED AND SUBSTRATE IN THE EAST CHINA SEA

Ye Yincan, Second Institute of Oceanography, China.

Presented on **Thursday 6th March 1986 at 16.00** in the Norfolk Conference Room

EMBARGOED AGAINST DELIVERY

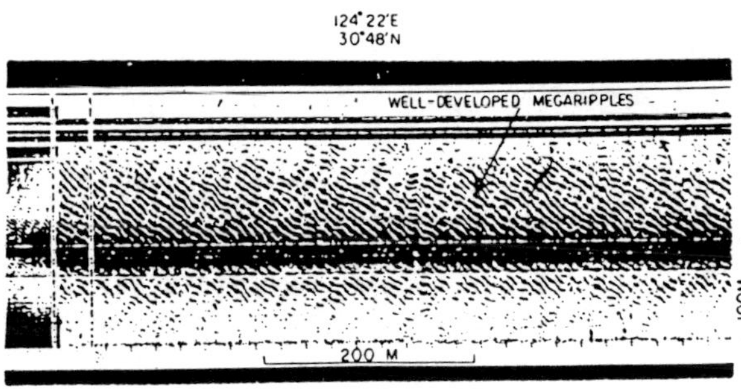

Fig. 2 Sidescan sonar record showing well-developed megaripples on the seafloor

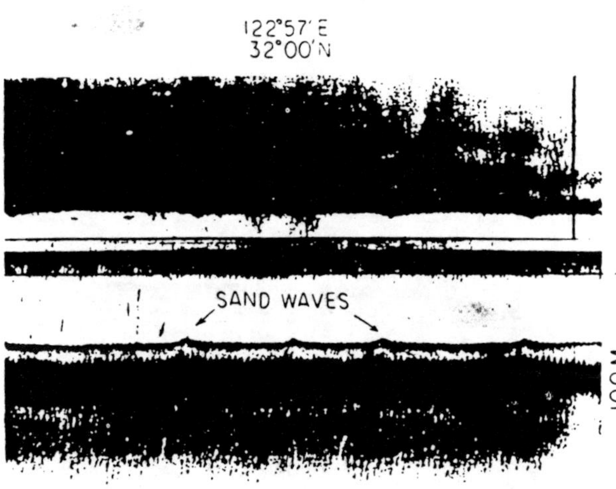

Fig. 3 Sidescan sonar record showing sandwaves on the seafloor

It has been detected that the bedform morphologies occupy an area of about 20 000 km^2 in the northern East China Sea shelf (Fig. 1) (Ye *et al.*, 1983). The transverse bedforms found on the shelf can be divided into megaripples and sandwaves (Figs 2 and 3). The megaripples are most widespread and have waveheights less than 1 m, commonly 0.3–0.6 m, and wavelengths vary from 2.3 to 13.6 m. The sandwaves are only developed in the relict sediment area off the Changjiang River mouth, with heights from 0.5 to 2.6 m, generally over 1 m, and wavelengths from 70 to 1265 m.

From the measurements in different years and analyses of the asymmetrical profiles of sandwaves, it is concluded that the sandwaves are moveable and variable. As an example, it has been estimated that the sandwaves around the area of 123° E, 32° N have migrated seaward 6.88 km in a five-year period. The migration rate is 21 times greater than the migration rates in Chesapeake Bay (Swift et al., 1972). Relief less than 1 to 2 m, such as the sandwaves and megaripples, does not constitute a real danger to structures. Nevertheless, for moveable and variable bedforms it is important to investigate in more detail their shapes and sizes, their movement and the environmental conditions, prevailing diving formation, and to find ways to deal with such situations.

OPEN AND BURIED CHANNELS

The most characteristic feature of the shallow stratigraphy in the East China Sea is the series of paleochannels, presumably representing the Late Pleistocene to Early Holocene drainage system. The paleochannels found in the continental shelf can be divided into relict and buried paleochannels. The former retain their initial V-shape valley on the seafloor and can be delineated mainly by the data from echosounding and sedimentology; the latter have been buried by subsequent sediments, and the V-shape of valleys can not be found on the seafloor but can be recognized from high-resolution acoustic profiling records.

Figure 4 shows the distribution of the paleochannels in the East China Sea shelf (Ye and Feng, in preparation). The maximum width of buried paleochannels is 3400 m. The thickness of sediment fill within the channels is commonly more than 10 m with a maximum thickness up to 22.3 m. The sediments overlying the valley range from 2 to 8 m (Figs 5 and 6).

Paleochannels may pose problems for bottom-supported structures for the following reasons:

- The sediment fill within channels is usually loose, with high water content and high compressibility.
- The character and thickness of the sediment fill can vary rapidly either vertically or transversely.
- The paleochannels are frequently associated with near-surface faults.

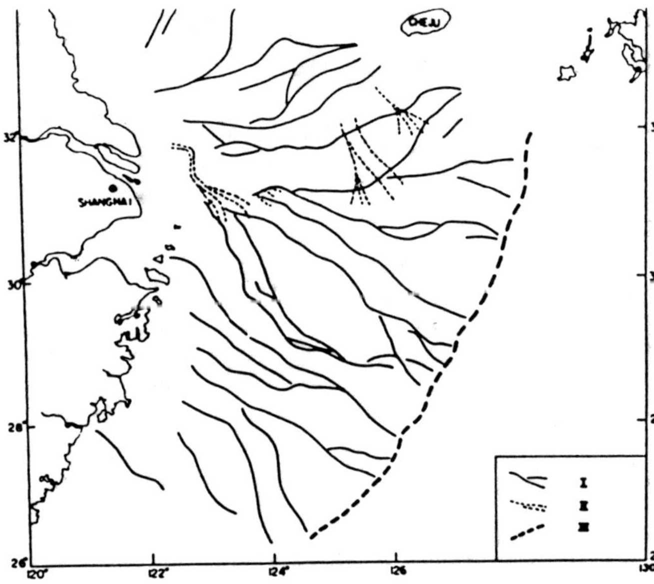

Fig. 4 Distribution of the paleochannels.
I, relict paleochannels; II, buried paleochannel;
III, shelf break line

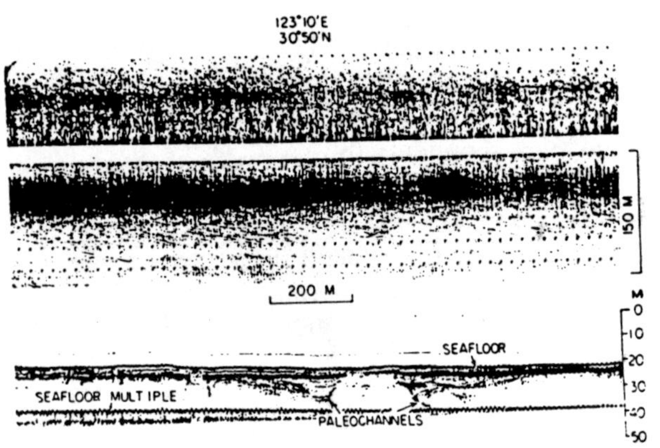

Fig. 5 Several filled paleochannels that presumably represent the ancient Changjiang River distributaries

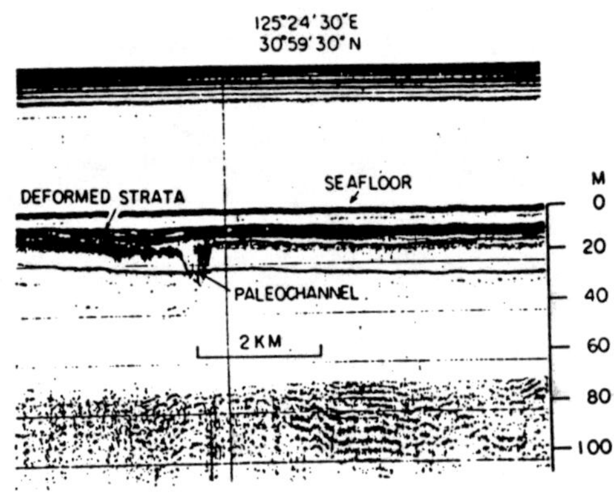

Fig. 6 Buried paleochannel on the outer shelf

NEAR SURFACE FAULTS

High-resolution acoustic profiling has revealed near-surface faults in the East China Sea shelf area. Deformed strata, offset across the fault plane, and, in some cases, seafloor scarps can be seen on acoustic profile records.

As an example, Fig. 7 shows the shallow faults associated with a filled channel. The fault in the east side of the filled channel cuts across the underlying strata and results in

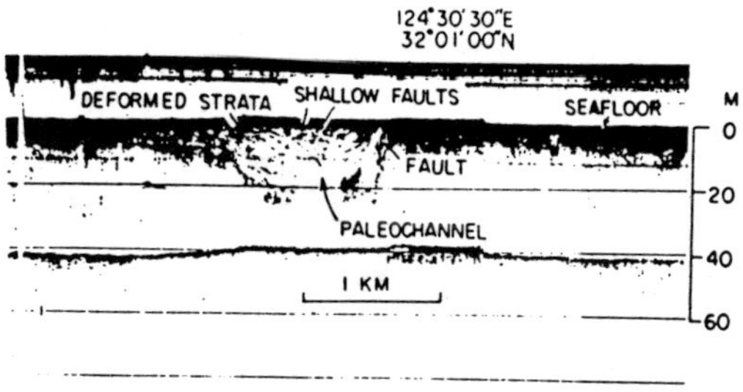

Fig. 7 Faults associated with the filled channel on the middle shelf

deformation and truncation of the sediment fill. Scarps 2 to 3 m high can be seen on the seafloor nearby. It is suggested that the fault existed before the development of the channel, and continues to be active up to the present time.

Near-surface faults are frequently associated with paleochannels and continued activity will strengthen the asymmetry across the fault plane. It is obvious that near-surface faults may endanger the bottom-supported structures.

SUBMARINE SEDIMENT SLIDES

Based on the conditions causing slides and analyses of geological and topographic data, we consider that the nearshore zone and the continental slope of the East China Sea are two areas where submarine sediment slides are widespread. In the south-east of the Changjiang River estuary there are many nearshore harbours and islands. Rock outcrops or partially buried outcrops occur on the seafloor there and the topography is relatively complex. As the nearshore zone is in a high rate deposition area it is possible that sediment overloading on some slopes and loss of its stability may occur.

Figure 8 shows a buried submarine slide which is located in the mouth of Xiangshan Bay where the initial slope is about 3° 55′. We can clearly see the slide plane, slide bed, rising dune from the sub-bottom profiling records representing approximately 100 000 m^3 in volume of translated sediments. Figure 9 shows a recently formed slide which

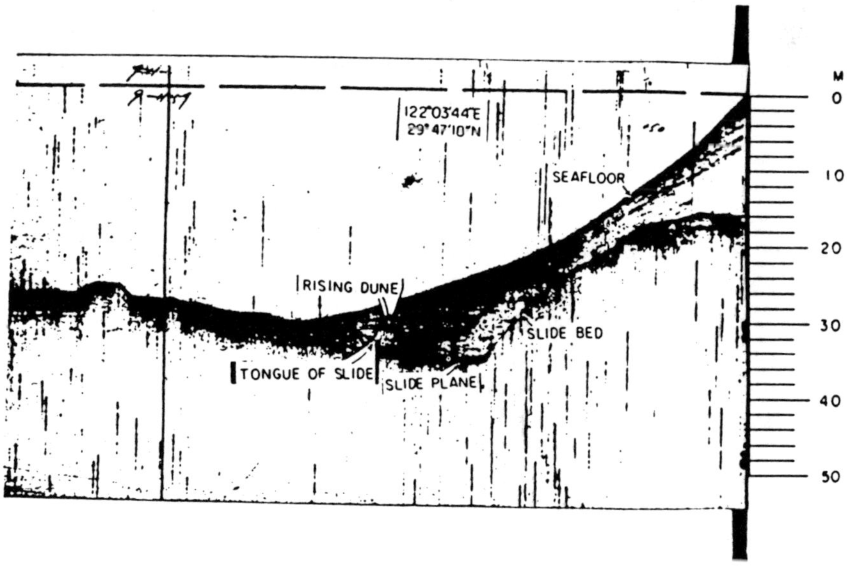

Fig. 8 Sub-bottom profiler record showing the buried submarine slide

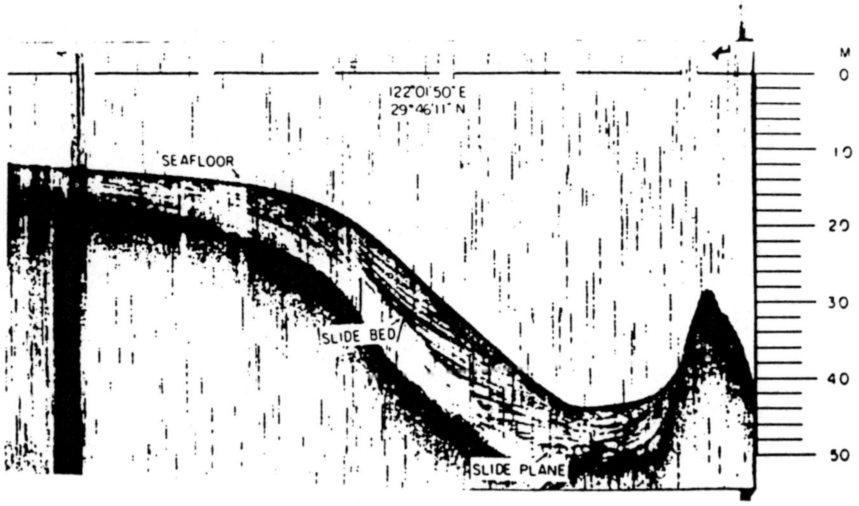

Fig. 9 Sub-bottom profiler record showing the recently formed submarine slides

consists of a series of small feather-like slides. Similar slides can also be seen in the Chejiang Province nearshore area.

High-resolution acoustic profiling data from the East China Sea continental slope is not yet available, but it is suggested from the analyses of topography and sedimentary process that the continental slope is an area where slides are widespread. Investigations by Embley and Jacobi (1977) indicated that most submarine slides on the Atlantic continental margin occur in water depths of more than 2000 m on an initial slope of less than 1.5°. The regional slope of the East China Sea continental slope is from 1 to 3° and the maximum slope up to 6–7° – far greater than that of the Atlantic continental margin. The shelf break of the East China Sea continental shelf was the position of an ancient coastline during low stands of the sea in the Late Pleistocene. At that time the shelf was subaerially exposed, traversed by large rivers which deposited their loads into the Okinawa Trough. Rapid deposition of sediments took place at the old river mouths. The pore pressure in the sediments changed with subsequent sea-level fluctuation and this has been a factor contributing to trigger slides.

The submarine slide is one of the instability factors problems to be faced by seabed constructions. The translated sediments can bury structures, or the deformed slide blocks may give rise to foundation failure.

GASSIFIED SEDIMENTS

Nearshore and deltaic sedimentary environments favour the preservation of organic matter in the sediment column. High concentrations of organic carbon are a necessary prerequisite for biogenic gas production. Because of acoustic impedance differences between gas-charged sediments and overlying deposits, strong reflectors can be seen on high-resolution acoustic profiles (Fig. 10). Based on the interpretation of acoustic data, we have delineated the distribution of the gas-charged sediments, as shown in Fig. 11. The shallow biogenic gas occurs mostly in the sediments of the Changjiang River submarine delta and the Chejiang Province nearshore zone. The depths at which gas-charged sediments occur are different from place to place, commonly shallower in the bays, but deeper in the estuaries. The blowout of gas can cause pockmarks to develop on the seafloor.

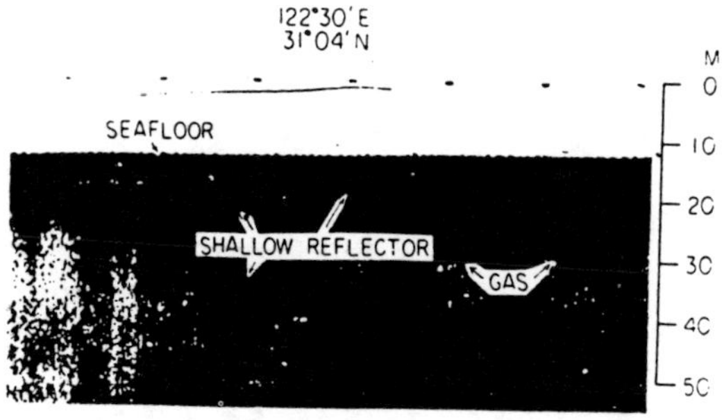

Fig. 10 Sub-bottom profiler record showing biogenic gas

The overloaded gas pressure will destroy the texture of sediment, slow its consolidation process and lessen the sediment shear strength. Studies by Whelan *et al.* (1977) indicated that in normal cases the shear strength of seafloor sediments increases with the depth, but such an increasing ratio is much less in the gas-charged sediments. The pressure of trapped sedimentary biogenic gas, coupled with bottom pressure changes associated with storm waves, may trigger subaqueous mass movement.

The geological hazards mentioned above are the unstable factors which must be considered in engineering geology reconnaissance. The interactions between different sedimentary environments and various types of seabed structures illustrate the need to conduct further detailed investigations and research to meet the demands of increasing ocean engineering activity in this area.

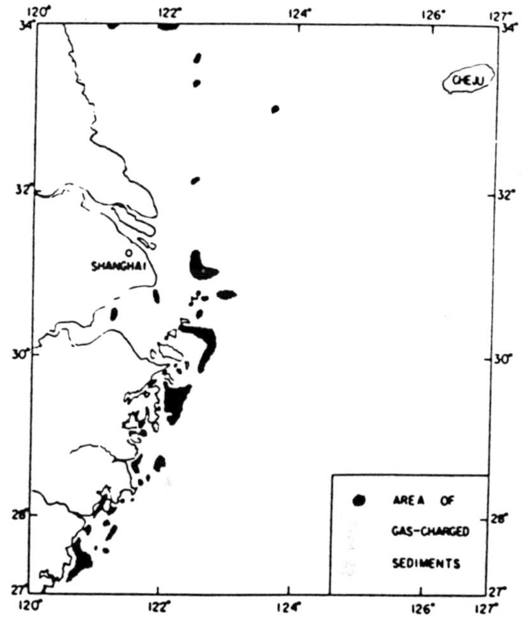

Fig. 11 Distribution of the shallow biogenic gas

REFERENCES

Embley, R.W. and Jacobi, R.D. 1977. Distribution and morphology of large submarine sediment slides and slumps on Atlantic Continental Margins. *Marine Geotechnology* **2**, 205–228.

Swift, D.J.P., Duane, D.B. and Pilkey, O.H. 1972. *Shelf Sediment Transport Process and Pattern*. Dowden, Hutchinson & Ross, Stroudsburg, pp. 377–410.

Whelan III, T. *et al.* 1977. Acoustical penetration and shear strength in gas-charged sediment. *Marine Geotechnology* **2**, 147–159.

Ye, Y.C. and Feng, I.J. Paleochannels of the East China Sea continental shelf. In preparation.

Ye, Y.C. *et al.* 1983. Bedform morphologies of the continental shelf off the Changjiang River mouth and their environmental conditions. In *Proceedings of International Symposium on Sedimentation on the Continental Shelf, with Special Reference to the East China Sea*. China Ocean Press, Beijing, pp. 762–774.

38

Channels in the North Sea: the Nature of a Hazard

D. Long and *M. S. Stoker,* Marine Earth Sciences Research Programme, British Geological Survey, Murchison House, West Mains Road, Edinburgh, EH9 3LA, Scotland

The interpretation of shallow seismic profiles has indicated that open and buried channels are present over large areas of the North Sea. These features occur at various levels within the Quaternary sequence, and their age of formation ranges from Middle to Late Pleistocene. Although the majority of the channels have been completely filled with sediment, several have remained open and have only a partial sedimentary infill. Some of the channels are of undoubted subglacial origin, but the majority were probably formed by fluvial processes, with their profiles having been subsequently accentuated by catastrophic meltwater discharges. The sediment fill is of highly variable lithological and geotechnical composition. Consequently, they may represent a hazard to structures sited above them. The variable composition of the fill is responsible for producing seismic "pull up" and "pull down" effects on the profiles, and these effects should be taken into consideration when examining geological structures at greater depth.

INTRODUCTION

Since the first shallow seismic profiles were obtained from the North Sea in the mid-1960s, features described as buried channels have been identified in many areas, at several different horizons within the Quaternary sequence. These channels, which vary from 0.5 km to 3 km wide and 50 m to 400 m deep, have been mapped during regional surveys and occur on seismic records, either singularly or as composite features. They typically extend up to 50 km in length and are usually aligned north-south in an anastomozing pattern. In one example, a buried channel was traced from the Dogger Bank to the Witch Ground Basin, a distance of 240 km (Jansen, 1976). These filled channel features have been compared with unfilled or open channels which occur throughout the North Sea and have for centuries been known to fishermen by such evocative names as Devil's Hole and Silver Pit (Fig. 1).

The stratigraphic setting of these features is

CHANNELS IN THE NORTH SEA: THE NATURE OF A HAZARD

D Long, M S Stoker, British Geological Survey, UK

Presented on **Thursday 6th March 1986 at 16.30** in the Norfolk Conference Room

EMBARGOED AGAINST DELIVERY

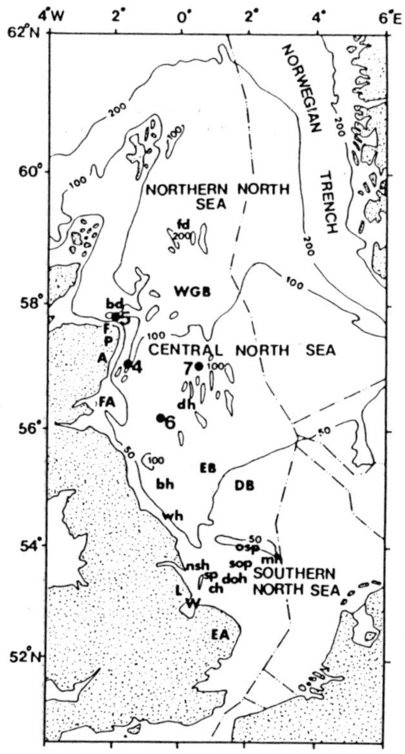

Fig. 1 Map of North Sea showing locations of areas, features and seismic sections referred to in text. WGB, Witch Ground Basin; FA, Forth Approaches; EB, East Bank; DB, Dogger Bank; W, Wash; F, Frazerburgh; P, Peterhead; A, Aberdeen; L, Lincolnshire; EA, East Anglia; fd, Fladen Deeps; bd, Buchan Deeps; dh, Devil's Hole; bg, Bayman's Hole; wh, Whitby Hole; osp, Outer Silver Pit; nsh, New Sand Hole; sp, Silver Pit; doh, Dowsing Hole; ch, Codling Hole. Numbered dots refer to locations of Figs 4–7

Fig. 2 Schematic cross-section showing the stratigraphic relationships between the three main channelled horizons in the central North Sea (modified after Fyfe, 1985). Horizon 1 is of Weichselian age; horizon 2 is of Saalian age; horizon 3 is of Elsterian age (not to scale)

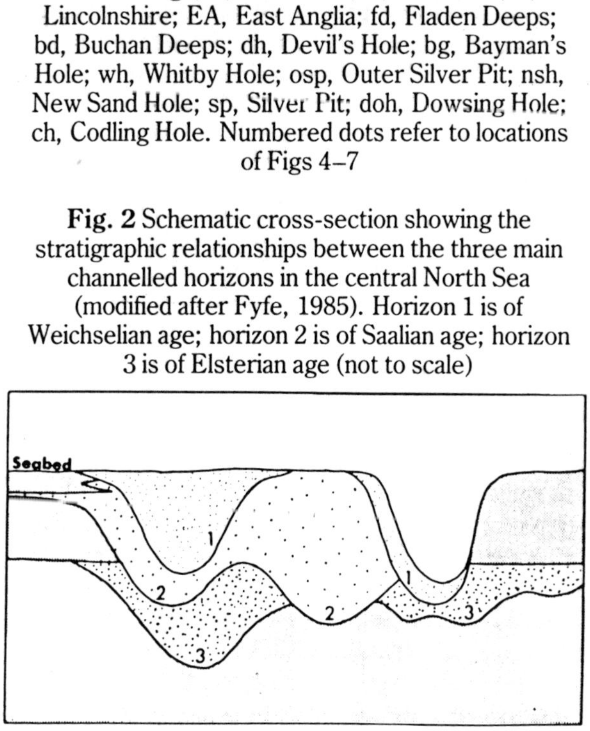

illustrated in Fig. 2. In the central and northern North Sea, the channels occur at three main horizons within the Quaternary succession and appear to correlate with the Elsterian, Saalian and Weichselian glacial periods (Stoker et al., 1985). In the southern North Sea, the channels are mainly of Elsterian and Weichselian age (Balson and Cameron, 1985).

PREVIOUS WORK

Over the past 30 years, various hypotheses for the formation of channels in the North Sea have been put forward. Most researchers have propounded the theory of glacial-related activity. Early workers suggested that the open channels of the central and northern North Sea, e.g. Devil's Hole and Fladen Deeps (Fig. 1), were formed by subglacial erosion, with the channels acting as siphons through which meltwater escaped under pressure (Valentin, 1955; Flinn, 1967). Erosion hollows formed in this way are frequently termed "tunnel valleys" and can be compared with the unfilled "tunneldale" or "rinnentaler" of Denmark and Schleswig-Holstein, the buried till-filled channels of East Anglia (Woodland, 1970), or the predominantly sand-filled channels of north Germany (Ehlers, 1981). During deglaciation, vast amounts of water are released which run off subglacially to the glacier margins. As this meltwater is confined in ice tunnels it has considerable erosive power. This mode of origin has been suggested for the infilled channels of the East Bank area (Fig. 1), offshore from Northumberland (Dingle, 1971).

Those channels which are aligned parallel to the coastline, and thus at right angles to the main direction of land ice supply, may have been created by glacier ice exploiting lines of geological weakness such as faults or boundaries between hard, old rocks and softer, relatively young rocks (Maisey, 1974). Glacial exploitation of a geologically controlled fluvial system has been proposed for hollows off the Yorkshire coast, known as Bayman's Hole and Whitby Hole (Dingle, 1970) (Fig. 1).

It has been suggested (Holmes, 1977) that the buried channels in the central North Sea may have been eroded by proglacial fluvial activity, although it was eventually concluded that they are "tunnel valleys", probably subjected to considerable glaciomarine modification prior to complete burial. More recently, these channels have been likened to shallow deltaic depositional features noted elsewhere in the world (Sissons, 1981a). It was suggested that large areas of the North Sea were exposed by lowered sea levels, during glacial periods, to form a broad coastal plain across which the major regional rivers flowed. Thus the channels were not simply erosional features but formed instead as shallow delta channels that were "carried down" and infilled as the delta subsided.

An examination of Silver Pit (Fig. 1) in the southern North Sea led to the suggestion that tidal scour eroded such channels (Donovan, 1965). As the Silver Pit has been eroded through boulder clay into chalk, it was concluded that it is a postglacial feature formed by tidal currents which were funnelled by former coastal configurations of the Wash. Similar tidal origins were suggested for other shallower channels in the southern North Sea, although it has since been proposed that the larger open channels in the south were initiated by subglacial erosion, subsequently enlarged by tidal scouring (Donovan, 1973). The hypothesis of formation by tidal scour alone has also been used to explain the origin of the unfilled channels of the Forth Approaches (Thomson and Eden, 1977) (Fig. 1).

Clearly, the origin of the channels remains problematic. In this chapter, we present new data on the morphology, infill characteristics and stratigraphic setting of the channels, in an attempt to set some constraints on the timing and mode of channel formation.

CHANNEL MORPHOLOGY

The open channels (Weichselian age) of the Forth Approaches and the central North Sea are near-parallel sided valleys, closed at both ends, varying from 0.5 km to 3 km in width and up to 30 km to 40 km in length. The base of the channels may penetrate up to 150 m below the present sea bed, more than 200 m below sea level (Fig. 3). The valley floor apparently undulates along its course and no consistent down-valley direction has been identified. The channels are usually linear, but are locally sinusoidal. Most are single features, but several, in the Devil's Hole area in particular, appear branched, giving an anastomozing pattern (Fig. 3). The channel-side slope angles vary from 27° to less than 2°, with the steepest slopes often associated with the deepest valleys in the Fladen Deeps and Devil's Hole areas. It has been noted that these channel slopes are frequently asymmetrical, with the steeper channel slope occurring more generally on the southern side of the channel, irrespective of channel orientation (Long and Stoker, in press). The southern slope on north-south profiles appears smooth, while the northern side is often hummocky.

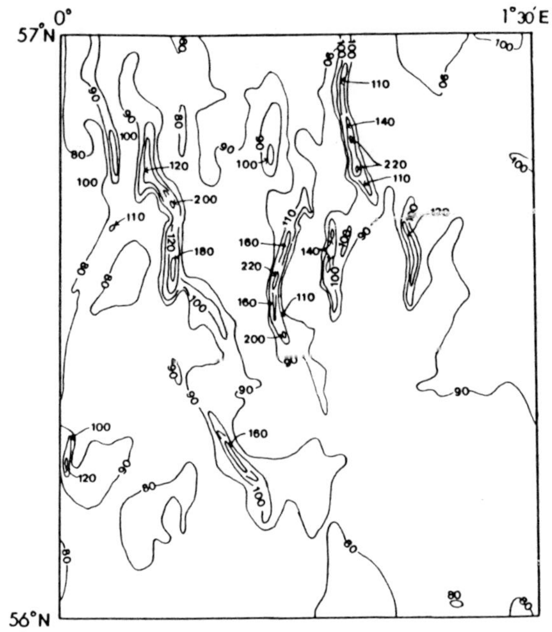

Fig. 3 Detailed bathymetric map of the Devil's Hole area, central North Sea: all depths are in metres (after Fyfe, 1985)

In the southern North Sea, the open channels are more varied in size and shape. In the south-east between East Anglia and Markham Hole (Fig. 1), a series of shallow channels (10 m deep) lie between sand-banks orientated NW-SE, parallel to the strongest tidal currents. Elsewhere in the south, the closed depressions are variable in size, ranging from small holes 1 km wide, 10 km in length and up to 25 m below the surrounding sea bed, such as New Sand Hole, Codling Hole and Dowsing Hole (Fig. 1), to large open channels such as Sole Pit which is 2 km wide, 30 km in length and 70 m below the sea bed, and Silver Pit, a large bifurcated channel 3 km wide, 40 km in length and up to 90 m below the adjacent sea bed with side slope angles of up to 6°. The largest open channel in the southern North Sea, the Outer Silver Pit on the southern edge of the Dogger Bank, is up to 25 km wide and about 100 km in length.

These open channels are cut through Quaternary sediments into Palaeozoic, Mesozoic and Cainozoic strata. Upper Cretaceous chalk crops out at the base of Silver Pit and Permo-Trias sandstones occur in the open channels offshore from Aberdeen and in the Buchan Deeps. The open channels of the Forth Approaches expose Carboniferous sediments (Thomson and Eden, 1977), and Pliocene limestone has been recovered from the walls of the Fladen Deeps (Newton, 1917). The orientation of some of the open channels offshore from Aberdeen, Peterhead and Fraserburgh appears to be controlled by Mesozoic fault lines, although similar relationships are not evident further offshore.

The buried or filled channels (Weichselian, Saalian and Elsterian age), which form the bulk of the features (Fig. 2), show similar morphological characteristics to the open valleys. These features have been described from several areas: the central North Sea (Jansen, 1976; Holmes 1977), the Forth Approaches (Thomson and Eden, 1977), south-east of Dogger Bank (Oele, 1971), the East Bank (Dingle, 1971), and offshore East Anglia (Balson and Cameron, 1985). Regional surveys (line spacing typically 10 km) show individual features to be 2–12 km in width, 20–400 m deep and up to 50 km in length. Zones of channelling are particularly common in the central North Sea; the channels are often in excess of 20 km in width, and the base of the channelled horizon may exhibit an irregular relief (cf. Fig. 6). Most interpretations of the seismic data have suggested an anastomozing pattern with an irregular channel floor of sills and depressions. Where the channels are sufficiently well defined to be traced across a network of lines, they appear to be aligned locally along a common orientation, varying from NNW–SSE in the southern North Sea to NNE–SSW in the central North Sea, although locally they trend east-west (Holmes, 1977) and are more variable in the northern North Sea.

The overall base levels of the channelled horizons show a regional linear relationship which reflects post-formational subsidence. The Weichselian channels generally have base levels of about 90 m below sea level in the Forth Approaches, which gradually increases to about 150 m east of Devil's Hole. This "slope" in base level is fairly constant (although some deviation is present) and suggests a regionally uniform, pre-subsidence base level. This relationship also seems to apply to the deeper buried Saalian and Elsterian channels. Major exceptions do occur; for example, some Weichselian channels in the Forth Approaches and in the Devil's Hole area greatly exceed depths of 150 m below sea level.

CHANNEL-FILL CHARACTERISTICS

Seismic Texture and Lithofacies

Most of the Elsterian and Saalian channels have been totally filled, but numerous Weichselian channels have only been partially infilled. On seismic profiles (sparker and boomer) there appears to be a broad similarity in the pattern of the infill of the channels. Many contain a basal unit which varies from chaotic to structureless in texture, and has an upper surface marked by an irregular-, moderate- or high-amplitude reflector. The chaotic reflection configuration occurs in two forms — symmetric channel-fill (Fig. 4), and asymmetric channel-fill (Fig. 5). The first type is characteristic of Weichselian and Elsterian channels within the known offshore limits of glaciation. The seismic characteristics are typical of till or boulder clay, an interpretation confirmed by boreholes in the Forth Approaches which have penetrated till

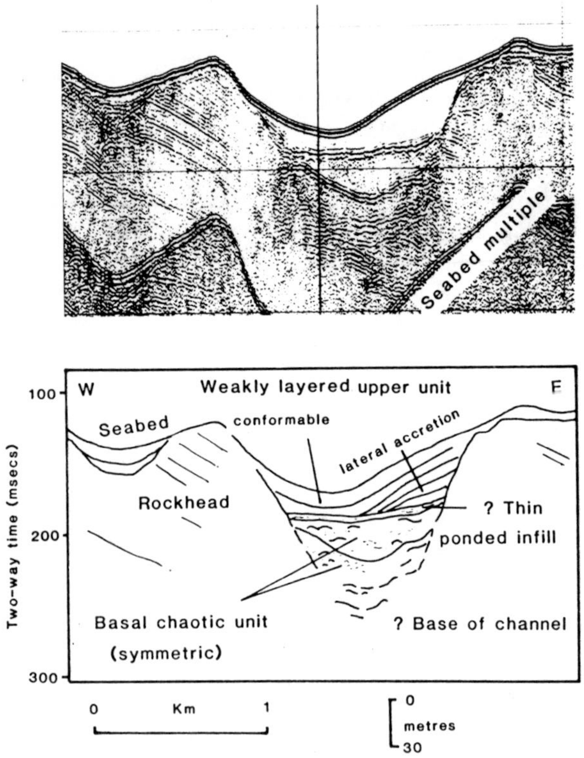

Fig. 4 Sparker profile, with interpretation, across a partially infilled Weichselian channel: profile is located offshore from Aberdeen (see Fig. 1)

and coarse sand and gravel (Stoker *et al.*, in press). This facies has completely filled some of the shallow buried channels, but comprises less than a tenth of the fill of some of the deeply incised features. The second type occurs mostly outside the known glacial limits, and has been interpreted as a slump deposit arising from slope instability in a periglacial environment (Long *et al.*, in press). This reflection pattern has, so far, been identified only in Elsterian and Weichselian channels, where it may form up to one-third of the fill. The structureless seismic facies has only been identified on high resolution boomer records across shallow Weichselian channels where it forms a thin (< 15 m) basal unit (Fig. 6). This facies forms a regionally extensive horizon in the central North Sea and a borehole in the Devil's Hole area recovered shelly, glaciomarine sands (Fyfe, 1985).

In the Saalian channels in the central North Sea, this basal unit has not been observed and these channels are generally filled by layered sediments, as described below, although inclined reflectors characteristic of lateral accretion occur locally at the base. Sediments recovered from the base of this type of fill

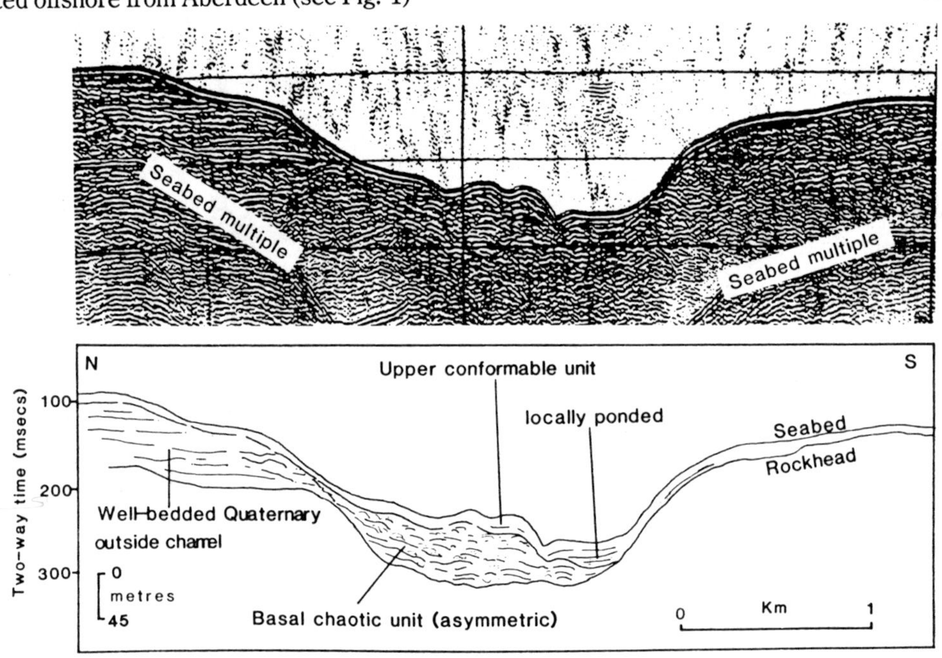

Fig. 5 Sparker profile, with interpretation, across a partially infilled Weichselian channel: profile is located in the Buchan Deeps offshore from Fraserburgh (see Fig. 1)

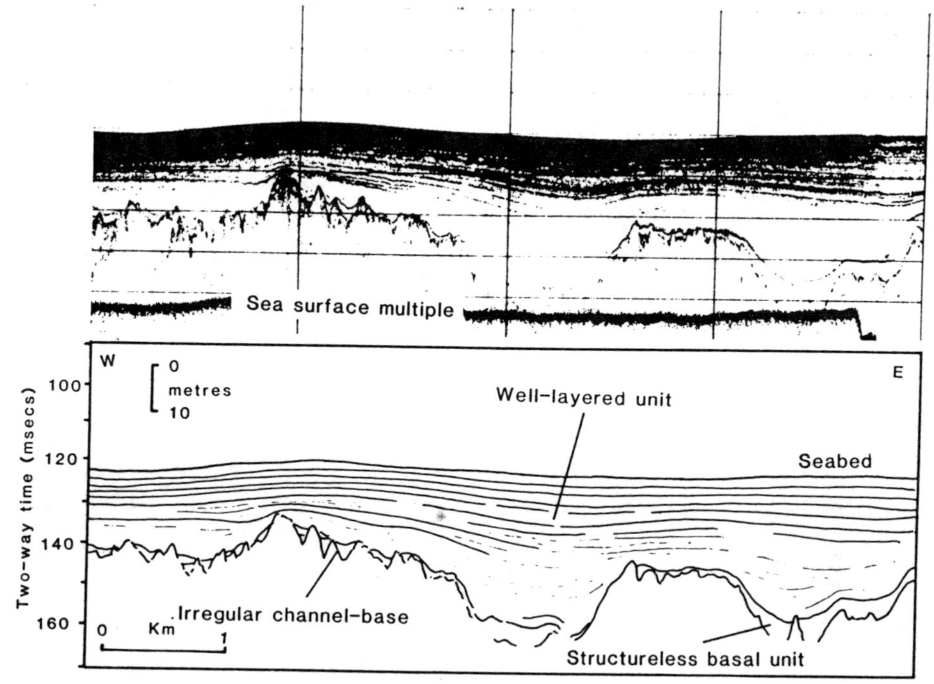

Fig. 6 High resolution boomer profile, with interpretation, across an infilled Weichselian channelled zone: profile is located in the central North Sea, south-west of Devil's Hole (see Fig. 1)

comprise glaciofluvial/glaciomarine sands and silts (Stoker *et al.*, in press).

The second unit of fill often forms the bulk of the channel fill and is characterized on spark and boomer records by a well defined, sub-parallel reflection configuration, although intraformational erosion surfaces and discontinuities are present (Fig. 7). On the margins of the channels the layering is draped over topographic irregularities on the upper surface of the underlying unit. In the central and northern North Sea, the layering comprises low- or moderate-amplitude, finely spaced, laterally persistent reflectors (Figs 6 and 7). Glaciomarine and marine argillaceous sediments have been recovered from the Weichselian channels, although localised "ponded" infills in Weichselian, Saalian and Elsterian channels may represent glaciolacustrine sediments. In the southern North Sea, the layering is less finely spaced and is interspersed with rare, high-amplitude reflectors which occur at similar intervals and depths in adjacent channels. Many of these channels formed an interconnecting network of open depressions during this phase of fill. The sediments in the southern North Sea are interpreted to have a glaciolacustrine origin (Balson and Cameron, 1985).

Many of the channels contain only two discrete seismic and sedimentary facies; however, some display a third unit of infill above an unconformity, overlying or eroded into the parallel bedded sediments (Figs 4 and 7). This upper unit is generally less than 30 m thick, and its seismic texture varies from layered, with conformable or prograding, low- or moderate-amplitude reflectors, to structureless. The prograding reflectors dip towards the west in the Elsterian channels of the southern North Sea. Similar reflectors in the Weichselian channels further north appear to dip either towards the west (Jansen, 1976), north-west (Holmes, 1977) or locally towards the east (Stoker et al., in press), implying a variable direction of sediment influx and/or lateral accretion. Sediments recovered from the Weichselian channels are predominantly shallow marine sands (Fyfe, 1985) although the presence of lateral accretion

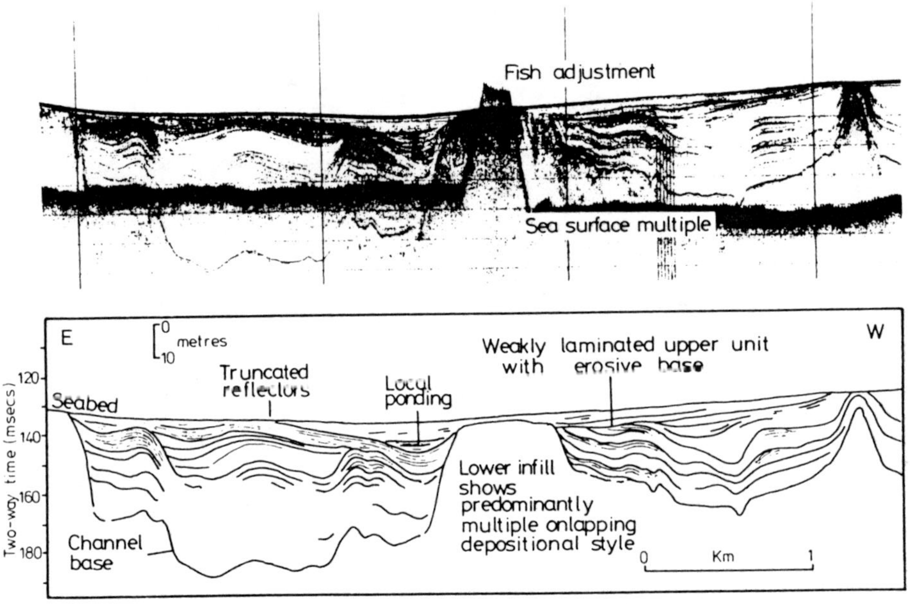

Fig. 7 High resolution boomer profile, with interpretation, across a complexly infilled Weichselian channelled zone: profile is located in the central North Sea, immediately north of Devil's Hole (see Fig. 1)

deposits suggests additional fluvial or estuarine facies.

The upper surface of the channel fill is often truncated by the present sea bed (Fig. 7). Locally, the channel has been overfilled above the general seafloor level and a mound up to 10 m above the surrounding sea floor forms positive relief over a buried channel.

The channel fill and the adjacent sediments frequently have differing seismic velocities due to their differences in lithology and geotechnical properties, and these cause seismic "pull-down" or "pull-up" effects on the underlying material. Significant "pull-down" effects are noted beneath open channels.

Age of the Sediment Fill

The Elsterian channels in the southern North Sea are filled mostly with subglacial and glaciolacustrine deposits of Elsterian age, although marine sediments from the succeeding Holsteinian interglacial stage may be represented within the uppermost part of the channel (Balson and Cameron, 1985). In the central North Sea, the situation is different; thin Holsteinian interglacial marine sediments occur at or near the base of the Elsterian channels in the Devil's Hole area, overlain by Saalian glaciomarine sediments which form the bulk of the fill (Cameron *et al.*, in press).

Saalian channels in the central North Sea show similar characteristics to the Elsterian features. A borehole in the Devil's Hole area recovered Eemian interglacial marine sediments from near the base of a channel, overlying a thin basal zone of Saalian glaciomarine sands. The Eemian sediments are, in turn, overlain by Weichselain glaciomarine deposits (Stoker et al., in press). The channels southeast of Dogger Bank are filled with Eemian shallow-water and littoral sediments and are thought to be related to the Saalian glacial meltwater valley of the Hunze in the Netherlands (Oele, 1971).

The Weichselain channels have been the most extensively sampled. In the Forth Approaches, channels cut into the pre-Quaternary bedrock have subsequently been filled by subglacial till, sand and gravel, overlain by glaciomarine and marine muds of Late Weichselian age. Renewed channelling occurred during the latest Weichselian times and these features were then backfilled by early Holocene fluvial and estuarine sands and muds (Stoker, in press). Further offshore, in the central North Sea, subglacial sediments are lacking and the fill generally comprises Late Weichselian glaciomarine and marine sediments, locally eroded and overlain by early Holocene shallow marine sands (Fyfe, 1985). In the southern North Sea, the open channels generally have a thin cover only, of Recent sediments except where bedrock is exposed (Donovan, 1973). In the buried channels of the East Bank, subglacial sediments are similarly lacking, and their fill comprises late- to postglacial deposits (Dingle, 1971).

CHANNELS OUTSIDE THE NORTH SEA

Similar channel features have been identified on other continental shelves around the world, giving rise to various interpretations of their origins. Observations from former glaciated shelves show that both glacial and subglacial erosion has produced channels from 1 km to 20 km wide and up to 400 m deep. These channels are characteristically steep walled with a till or diamictite infill, suggesting predominantly subglacial deposition, though often with proglacial or glaciomarine sedimentation near the top of the channel fill. These subglacial meltwater erosional features are found on most parts of the North American continental shelf where glaciation is known to have occurred, for instance producing the archetypal fjord erosion on the Labrador coast (Barrie and Piper, 1982), and ice-scoured valleys in steep walls of rock on the southeastern Alaskan shelf (Armentrout et al., 1979) and British Columbian shelf (Herzer and Bornhold, 1982; Luternauer and Murray, 1983). Extreme examples of subglacial erosion have been reported from the Mackenzie Canyon area of the Beaufort Sea, where a channel 90-100 km wide but ony 80 m deep has been eroded (Shearer, 1971, 1972), and from Shelikof Strait, Alaskan Peninsula, where a trough in excess of 800 m depth has been gouged out (Hampton, 1985). In contrast, several different stages of channelling off the north coast of Alaska, with features up to 1 km wide and 12 m deep, occur within marine, glaciomarine and non-marine sediments, and are thought to have been formed by high-gradient streams flowing from the nearby Brooks Range (Dinter, 1985).

Continental shelves in tropical and temperate areas of the world also show channelling on seismic records at many different levels within the Quaternary sequence. There are many channels on the Great Barrier Reef, north-east Australian continental shelf, varying from V-shaped depressions, 100 m across and 2 m deep, to large channels 1 km wide and up to 20 m deep occurring, in some cases, in zones of extensive channelling up to 6 km wide (Johnson *et al.*, 1982). The channels contain two units of fill, a lower, well bedded, typically dipping unit, and an upper unit of parallel bedding, which fills and occasionally forms a positive feature above the channel. These have been interpreted as fluvial channels formed by rivers of glacial to early postglacial age, and can be traced for distances of up to 50 km across the shelf. Tidal scour has been discounted due to the size of the features and the fact that the region is, at present, only mesotidal and lacks strong erosive tidal currents in open areas. The lower unit is interpreted as containing fluvial or estuarine sediments. Progressive backfilling during transgression by fluviomarine sediments is thought to have filled the channels.

Fluvial channels, up to 25 m deep and up to 10 km wide, have also been identified on the North Carolina continental shelf of the south-eastern United States. These fluvial channel systems have been traced shorewards to established rivers of the present day (Hine and Snyder, 1985). In other areas, where the continental shelves would have been exposed during the glacial maximum, e.g. the New Jersey Shelf, north-east United States, small channels up to 17 km wide and up to 20 m deep have been identified throughout the Quaternary sequence and reflect global sea-level lowering during glacial periods (Barrier and Piper, 1982; Swift *et al.*, 1980). Where this type of shelf was influenced by nearby glacial limits, buried channels have been recorded. For example, the former course of the Hudson River tracked south of its present course and is preserved as a large buried channel (Knebel *et al.*, 1979). The cause of river migration has been attributed to glacial loading of the shelf and a consequent change in shelf gradient (Knebel *et al.*, 1979).

Some channels identified on non-glaciated continental shelves are thought to have had a cataclysmic origin. It has been suggested that the open and filled channels of the English Channel were caused by a catastrophic flood which modified an existing fluvial drainage system, leaving an area with an anastomozing pattern of channels with irregular thalwegs (Smith, 1985). This theory is based on comparisons with the well documented Missoula (or Spokane) Flood of the Columbia River system of the northwestern United States, where it has been postulated that an ice-dammed lake of 2500 km^3 emptied in only a few days (Bretz, 1928). The sudden release of vast quantities of water is a powerful erosive agent. The Missoula Flood was shown to have eroded channels 20 m deep through basalt and to have produced channels crossing previous watersheds (Baker and Bunker, 1985), and it has been stated that "The unusual features observed in the channelled scablands of the Spokane Flood of anastomozing channel patterns and deep rock basins could not have been eroded by 'normal' streams" (Allison, 1941). Recent studies have suggested that the Missoula Flood was not an isolated event but has occurred 35–40 times during the Late Wisconsinian (Late Weichselian) alone, by repeated flotation of the ice dam (Waitt, 1980; Craig and Singer, 1984). The effects of these floods have been recorded at distances of up to 1400 km from their source (Griggs *et al.*, 1970).

A series of troughs up to 80 m deep, which are cut into postglacial deposits in the Sound of Jura (west coast of Scotland) is thought to have been created by tidal erosion (Evans and Ruckley, 1980). These features appear as closed, U-shaped basins cut into very soft sediments of postglacial age (11 000–8000 yr BP), and occur outside the known limits of the Younger Dryas (Loch Lomond) stadial.

DISCUSSION

The traditional view of the North Sea, during glacial periods, is that it has been totally covered by glacial ice, and that the channels have been eroded by glacial scouring or by subglacial meltwater flowing under hydrostatic pressure. Channels cut by glacial scouring probably occur where a geological weakness could be exploited; this implies that the deep channels should not have been formed in the middle of large expanses of relatively uniform, poorly consolidated Quaternary deposits. Subglacial meltwater has been known to erode deep channels when strong hydrostatic pressure occurs due to the vertical fall of a glacier, but is this likely to have occurred over the relatively flat North Sea plain? Meltwater falling through a glacier may cause subglacial erosion in the form of strudel holes, but this would probably be only a localized phenomenon. A major problem regarding the theory of a glacial or subglacial origin for the channels of the North Sea is the lack of evidence for a glacial ice cover across the North Sea. Marine geologists have frequently questioned the Quaternary land geologists' presumptions of glaciers extending across from Scandinavia to Britain, in particular commenting on the lack of glacial tills far from the coast. Subglacial tills have only been noted up to about 75 km offshore of the coast of Scotland, while boreholes located in the middle of the North Sea have recovered predominantly glaciomarine sediments. Although microfabric analyses have identified evidence of permafrost, they have failed to find conclusive evidence of subglacial loading (Derbyshire *et al.*, in press). These data have led some workers to conclude that the glacial ice limits were very close inshore, and that large expanses of the North Sea were exposed and were subject to periglacial processes during the glacial periods (Cameron *et al.*, in press), although shallow marine basins, in the deeper parts of the North Sea, may have experienced a sea-ice cover (Stoker and Long, 1984).

The variety of channel-forming processes is highlighted by the occurrence of channels on both glaciated and non-glaciated continental shelves. In a distinctly glacial or non-glacial setting, the number of potential mechanisms for channel formation is less than in an environment subject to both glacial and proglacial processes, such as is envisaged for the North Sea basin during Middle and Late Pleistocene times. A detailed understanding of the stratigraphic setting, morphology and infill characteristics of the channels is therefore required, and these must be considered in relation to the wider regional palaeoenvironmental setting.

In the Forth Approaches area and offshore from East Anglia, the available evidence supports a subglacial origin for the channels. They are predominantly singular features (e.g. Fig. 4), 2-12 km in width and up to 400 m deep with a subglacial fill, although proglacial and interglacial marine, fluviomarine and lacustrine sediments may occur near the top of their fill. They occur within the known glacial limits as deduced from regional offshore mapping. In many cases they have been cut into pre-Quaternary strata, and their base levels are variable and do not display any regional trend.

In the central North Sea, the channels tend to occur in wide zones (e.g. Figs 6 and 7) forming more laterally extensive sedimentary units. These channels occur outside the known glacial limits and their base level characteristics suggest that there may have been a regional downward limit to channel deepening. The infilling sediments comprise proglacial and interglacial marine and fluviomarine deposits, and the sequence often displays evidence of reactivation, with internal erosion and/or discontinuity surfaces forming in response to changing sea levels, and local shifts of the axis of sedimentation within the channelled zone. Lateral accretion bodies have been noted locally at the base of the infilling sequence, particularly in Saalian channels, but also in Weichselian channels in the Forth Estuary where fluvial and estuarine sediments have been recovered. The available evidence from these channels does not support a subglacial origin, as favoured previously by numerous workers, therefore some other mechanism is required. A study (Donovan, 1973) of channels formed by tidal scour concluded that large channels only formed in narrow straits bordered by resistant rock, although tidal scour concentrated in estuaries may produce small closed basins. While some of the shallow (10 m deep) channels off East Anglia and Lincolnshire, in the southern North Sea, may be attributed to tidal scouring, it is not regarded as the major mechanism for the central North Sea features. The concept of a subsiding delta (Sissons, 1981a) is similarly rejected as it fails to explain the stratigraphical, lithological and geotechnical differences between the channel fill and the adjacent sediments, and the presence of open channels dismisses the implication, inherent in the deltaic model, that deposition kept pace with subsidence. However, it does highlight the possibility that major fluvial activity occurred in the North Sea basin. We therefore suggest that the channelled zones of the central North Sea formed in response to the periodic expansion of the northern European river system across a partially exposed sea floor, and these rivers drained into or beyond the northern North Sea or into the Norwegian Trench during periods of glacio-eustatically lowered sea level. Examples of fluvial channels from non-glaciated areas of the world indicate that these are generally shallower than the North Sea features. Additional processes may therefore have been active; this is probably to be expected at a glacial/proglacial interface.

Some of the fluvial channels may have been altered in shape and size by the catastrophic release of large volumes of water from ice-dammed or intra-ice lakes associated with the Scandinavian, British or even Alpine ice sheets, as well as the annual release of glacial meltwater from the ice-front. Such events are not alien to this area; during the Younger Dryas (Loch Lomond) stadial, the sudden release of water from Glen Roy in the Central Highlands of Scotland caused extensive erosion up to 70 km from its source (Sissons, 1981b). The relatively unconsolidated Quaternary sediments of the North Sea could have been easily and deeply eroded by such catastrophic events, with the flood waters pouring along existing river courses and flowing over shallow watersheds into neighbouring river systems, producing an anastomozing pattern of channels with hollows and plunge-pools. This may account, in part, for some of the local deviation from regional base-level trends.

Another possible variable to be considered, but for which we have, at present, no quantitative control, is the effect on base levels that may be caused by differential uplift and subsidence associated with North Sea neotectonic deformation (Muir Wood, 1984). Evidence for crustal extension during the Quaternary has been found, although the full regional implications for this in terms of Quaternary sedimentation patterns have still to be established.

These modified fluvial channels were subsequently filled and buried over quite a long timespan, often encompassing an interglacial period. Their protracted history of fill is also applicable to the filling of subglacial features where interglacial sediments often occur towards the top of the sequence. It seems plausible to assume that the present-day situation of open channels co-existing with filled channels also occurred during previous interglacials. This would explain why, over large areas of the North Sea, interglacial deposits are rare, if not absent, outside channels but are preserved within channels. The implication is, that when the sea floor became exposed during the early part of each glaciation, the deposits of the preceding interglacial were eroded away except in areas of continuing sedimentation, i.e. in the channels.

The reason why some channels stayed open while the majority were filled remains problematical. In some cases, open features occur within an otherwise filled channel zone, and are merely depositional hollows overlying an exceptionally deep channel. In most examples, however, the channels appear to have remained open in response to some physical process, the nature of which remains uncertain. Possible explanations include tidal current scour preventing sedimentation, or topographic restrictions preventing total infill during seabed exposure, followed by a rapid rise in sea level.

CONCLUSIONS

The channels in the North Sea formed in an environment subject to both glacial and proglacial processes, and the problem of their origin is not related to a single cause but to how several processes may have interacted. Channel formation has been attributed mainly to subglacial and fluvial processes although tidal scour may have been important locally. The fluvial channels may have been modified to some extent by cataclysmic outbursts of glacial meltwater resulting in local overdeepening of channel profiles. Conversely, subglacial channels may have also acted as pathways and depositional traps for later fluviomarine sedimentation.

The high preservation potential of channels and their fills make them of great importance to both scientists and engineers alike. They are important to the Quaternary geologist as they contain extensive sedimentary sequences which have not been preserved outside the channel. For the geotechnical engineer, the channels are of interest because of their variable lithological fill, the different stress history between the fill and the adjacent sediment, and the importance of channel side-wall instability in determining foundation conditions. Moreover, the petroleum engineer may be interested in the effects such features might have on velocity contrasts, and the apparent displacement of adjacent and underlying seismic reflectors.

ACKNOWLEDGEMENTS

We are grateful to Dr D. Cameron for his comments on an earlier draft of this chapter. This chapter is published with the permission of the Director, British Geological Survey (NERC).

REFERENCES

Allison, I. S. 1941. Flint's fill-hypothesis for channelled scabland. *J. Geol.* ´**9**, 54–73.

Armentrout, J. M., Rosenmeier, F. and Rogers, J. 1979. Glacial origin of Megachannels of Upper Yakataga Formation (Pliocene-Pleistocene), Robinson Mountains, Gulf of Alaska. *Am. Assoc. Petrol. Geol. Bull.* **63**, 411 (abstract).

Baker, V. R. and Bunker, R. C. 1985. Cataclysmic Late Pleistocene flooding from glacial late Missoula: a review. *Quat. Sci. Revs* **4**, 1–41.

Balson, P. S and Cameron, T. D. J. 1985. Quaternary mapping offshore East Anglia. *Modern Geol.* **9**, 221–39.

Barrie, C. Q. and Piper, D. J. W. 1982. Late Quaternary marine geology of Makkovik Bay, Labrador. *Paper 81–17.* Geological Survey of Canada, 37 pp.

Bretz, J. H. 1928. The channelled scabland of eastern Washington. *Geog. Rev.* **18**, 446–77.

Cameron, T. D. J., Stoker, M. S. and Long, D. in press. The history of Quaternary sedimentation in the UK sector of the North Sea Basin. *J. Geol. Soc. London.*

Craig, R. G. and Singer, M. P. 1984. A theoretical study of Lake Missoula jokulhlaups in the Late Pleistocene. *Geol. Soc. Am. (Abst. with programs)* **16,** 276–77.

Derbyshire, E., Love, M. A. and Edge, M. J. in press. Fabrics of probable segregated ground-ice origin in some sediment cores from the North Sea basin. In: Boardman, J. (Ed.) *Soils and Quaternary Landscape Evolution.* Chichester: Wiley.

Dingle, R. V. 1970. Quaternary sediments and erosional features off the North Yorkshire coast, western North Sea. *Marine Geol.* **9**, M17–22.

Dingle, R. V. 1971. Buried tunnel valleys off the Northumberland coast, western North Sea. *Geol. Mijnbouw* **50**, 679–86.

Dinter, D. A. 1985. Quaternary sedimentation of the Alaskan Beaufort Shelf: influence of regional tectonics, fluctuating sea levels, and glacial sediment sources. *Tectonophysics* **114**, 113–61.

Donovan, D. T. 1965. Holocene origin of erosion hollows in the North Sea. *Rep. Challenger Soc.* **3,** 43–44.

Donovan, D. T. 1973. The geology and origin of the Silver Pit and other closed basins in the North Sea. *Proc. Yorks. Geol. Soc.* **39**, 267–93.

Ehlers, J. 1981. Some aspects of glacial erosion and deposition in North Germany. *Ann. Glaciol.* **2**, 143–6.

Evans, D. and Ruckley, N. A. 1980. The geology and origin of narrow closed channels in the Sound of Jura. *Scott. J. Geol.* **16**, 65–72.

Flinn, D. 1967. Ice front in the North Sea. *Nature* **215**, 1151–54.

Fyfe, J. A. 1985. *Devil's Hole, 56°N 2°W: Quaternary Geology.* British Geological Survey 1 : 250 000 map series.

Griggs, G. B., Kulm, L. D., Waters, A. C. and Fowler, G. A. 1970. Deep-sea gravel from Cascadia Channel. *J. Geol.* **78**, 611—19.

Hampton, M. A. 1985. Quaternary sedimentation in Shelikof Strait, Alaska. *Marine Geol.* **62**, 213–53.

Herzer, R. H. and Bornhold, B. D. 1982. Glaciation and post-glacial history of the continental shelf off southwestern Vancouver Island, British Columbia. *Marine Geol.* **48**, 285–319.

Hine, A. C. and Snyder, S. W. 1985. Coastal lithosome preservation: evidence from the shoreface and inner continental shelf off Bogue Banks, North Carolina. *Marine Geol.* **63**, 307–30.

Holmes, R. 1977. Quaternary deposits of the central North Sea, 5. The Quaternary geology of the UK sector of the North Sea between 56 o and 58 oN. *Report No. 77/14.* Institute of Geological Sciences, 50 pp.

Jansen, J. H. F. 1976. Late Pleistocene and Holocene history of the Northern North Sea, based on acoustic reflection records. *Netherlands J. Sea Res.* **10**, 1–43.

Johnson, D. P., Searle, D. E. and Hopley, D. 1982. Positive relief over buried post-glacial channels, Great Barrier Reef Province, Australia. *Marine Geol.* **46**, 149–59.

Knebel, H. J., Wood, S. A. and Spiker, E. C. 1979. Hudson River: evidence for extensive migration on the exposed continental shelf during Pleistocene time. *Geology* **7**, 254–8.

Long, D. and Stoker, M. S. in press. Valley asymmetry: evidence for periglacial activity in the central North Sea. *Earth Surf. Proc. Landforms.*

Luternauer, J. L. and Murray, J. W. 1983. Late Quaternary morphologic development and sedimentation,

Central British Columbia Continental Shelf. *Paper 83–21*. Geological Survey of Canada, 38 pp.
Maisey, G. H. 1974. Geologic properties of the Norwegian Continental Shelf relating to pipelines. *North Offshore* **3**, 37–49.
Muir Wood, R. 1984. North Sea neotectonics. *Geol. Soc. Newslett.* **13** (6), 42 (abstract).
Newton, R. B. 1917. On a fossiliferous limestone from the North Sea. *Q. J. Geol. Soc. London* **72**, 7–22
Oele, E. 1971. The Quaternary geology of the southern area of the Dutch part of the North Sea. *Geol. Mijnbouw* **50**, 461–73.
Shearer, J. M. 1971. Preliminary interpretation of shallow seismic reflection profiles from the west side of Mackenzie Bay, Beaufort Sea. *Paper 71–1*. Geological Survey of Canada, pp. 131–8.
Shearer, J. M. 1972. Geological structure of the Mackenzie Canyon area of the Beaufort Sea. *Paper 72–1*. Geological Survey of Canada, pp. 179–80.
Sissons, J. B. 1981a. The last Scottish ice-sheet: facts and speculative discussion. *Boreas* 10, 1–17.
Sissons, J. B. 1981b. Lateglacial marine erosion and a jokulhlaup deposit in the Beauly Firth. *Scott. J. Geol.* **17**, 7–19.2a
Smith, A. J. 1985. A catastrophic origin for the palaeovalley system of the Eastern English Channel. *Marine Geol.* **64**, 65–75
Stoker, M. S. in press. *Tay-forth, 56°N 4°W: Quaternary Geology*. British Geological Survey 1: 250 000 map series.
Stoker, M. S. and Long, D. 1984. A relict ice-scoured erosion surface in the central North Sea. *Marine Geol.* **61**, 85–93.
Stoker, M. S., Long, D. and Fyfe, J. A. 1985. The Quaternary succession in the central North Sea. *Newslett. Strat.* **14**, 119–28.
Stoker, M. S., Long, D. and Fyfe, J. A. in press. A revised Quaternary stratigraphy for the central North Sea. Report. British Geological Survey.
Swift, D. J. P., Moir, R. and Freeland, G. L. 1980. Quaternary rivers on the New Jersey shelf: Relation of seafloor to buried valleys. Geology 8, 276–80.
Thomson, H. E. and Eden, R. A. 1977. Quaternary deposits of the central North Sea, 3. The Quaternary sequence in the west-central North Sea. *Report No. 77/12*. Institute of Geological Sciences, 18 pp.
Valentin, H. 1955. Die Grenze der letzten Vereisung im Nordseeraum. *Verhand. dt. Geogrlags* **30**, 359–66.
Waitt, R. B. 1980. About forty last-glacial Lake Missoula jokulhlaups through southern Washington. *J. Geol.* **88**, 653–79.
Woodland, A. W. 1970. The buried tunnel-valleys of East Anglia. *Proc. Yorks. Geol. Soc.* **37**, 521–78.

POCKMARKS OF THE NORTHWESTERN ARABIAN GULF

J Ellis, W McGuinness, Arabian American Oil
Company, Saudi Arabia

Presented on **Thursday 6th March 1986 at 17.00**
in the Norfolk Conference Room

EMBARGOED AGAINST DELIVERY

39

Pockmarks of the Northwestern Arabian Gulf

Jeffrey P. Ellis and *William T. McGuinness*, Hydrographic Survey Section, Arabian American Oil Company, Ras Tanura, Saudi Arabia

Pockmarks are natural bottom features, unrelated to production activities. Physical evidence for the origin of pockmarks from three sites in the northwestern Arabian Gulf is examined. Three related features are also discussed. Two sources of gases are suggested by the physical evidence.

INTRODUCTION

Since King and MacLean (1970) first applied the term "pockmark" to crater-like depressions found on the Scotian Shelf, similar features have been described in the North Sea (Hovland, 1981a, b; McQuillin and Fannin, 1979; Green *et al.*, 1985), the East China Sea (Milliman *et al.*, 1985) and off Alaska (Sandstorm *et al.*, 1983). Pockmarks seem to be associated with hydrocarbon-bearing areas, although a legitimate question to be asked is whether this is a function of geology or survey coverage. Pockmarks are of interest to oil companies as they may pose a threat to seabed installations.

It is now recognized that most pockmarks result from the erosion of the sea floor by venting fluids, which in most cases is presumed to be gas. The gas has two possible origins: petrogenic (thermogenic) and biogenic. In Norton Sound (Alaska), for example, gases contained in the sediments of a pockmarked bottom have been shown to be biogenic (Sandstorm *et al.*, 1983). Gas-charged sediments need not lead to the formation of pockmarks. Gas associated with peat deposits in Long Island Sound (New York/Connecticut) has not led to the formation of any recognizable pockmarks (Bohlen, 1985). Little is known of the role of bottom dwellers and feeders in either the creation or maintenance of these depressions. A key question to be answered then, is why pockmarks form where they do.

Aramco conducts two types of surveys: site-specific and regional. In both cases, bathymetric side-scan sonar and Uniboom data are collected. Regionals, or field-wide surveys have a 100 m line spacing. Site-specifics, which are one kilometer squares centred on some point of interest, have a 50 m spacing for Uniboom and side-scan lines. The operational areas shown in Fig. 1, which encompass 1300 km^2, are regions in which surveys have been conducted. Except in deeper waters, where seismic penetration can be in excess of 40 m, shallow seismic data is

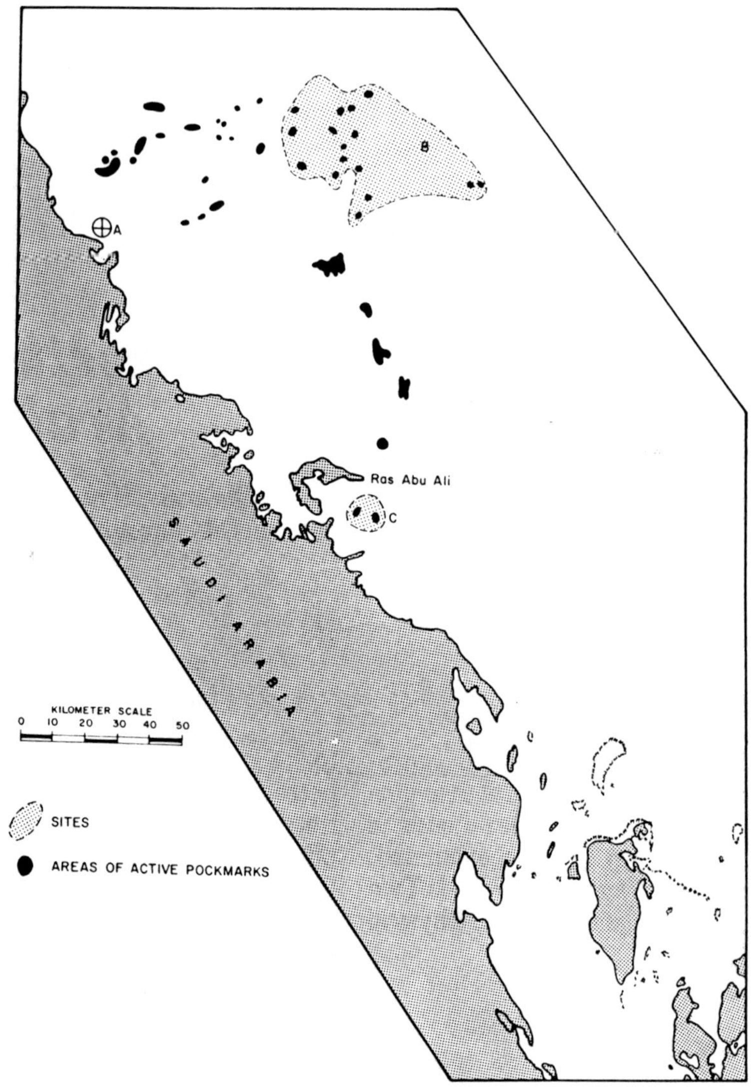

Fig. 1 The Arabian Gulf: operational areas are zones in which Aramco has undertaken surveys

to depths of less than 30 m.

A large number of engineering soil borings have been drilled within the Aramco concession area. Generally, the borings are about 40 m deep. Surface sediments range from very soft, lime mud to beachrock (locally called "faroush") and coral. Subsurface sediments are typically carbonate sands and clays. Cemented sands, mudstone and limestone are quite common, occurring in thicknesses ranging from thin partings to thick bed sequences. Evaporites are present in some of the borings from deeper waters in the form of gypsum and anhydrite. The presence of H_2S is often noted in bore logs. It is usually associated with surface and near-surface deposits. At some locations, sulfur has been observed to be a cementing agent. Using available survey data and stratigraphic information from engineering soil borings, four types of pockmarks and possible reasons for their formation will be discussed. The discussion will be limited to three specific sites (Fig. 2).

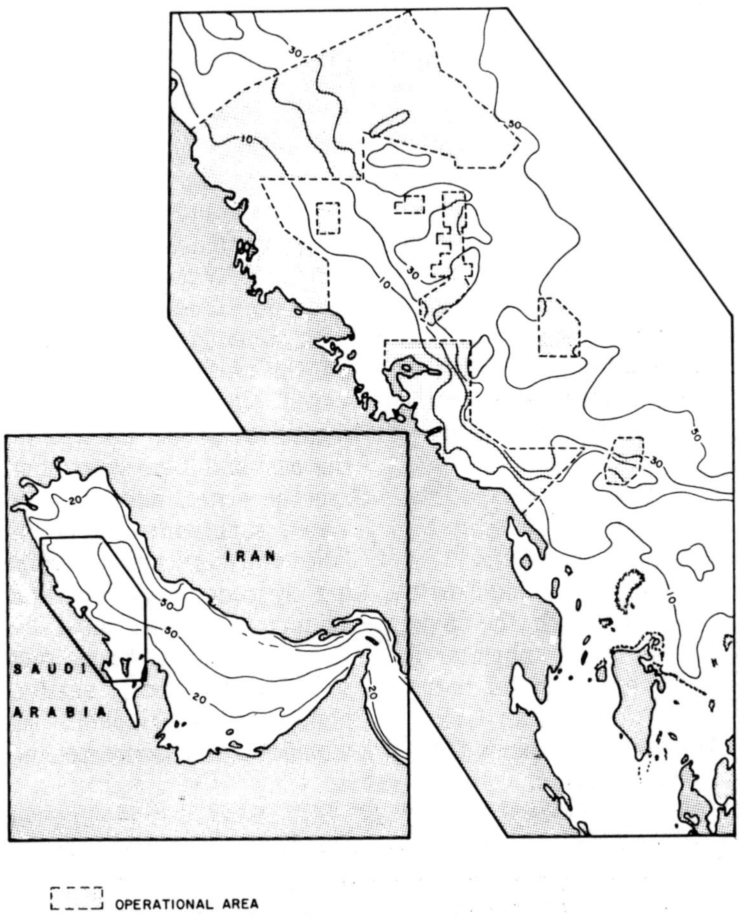

Fig. 2 Arabian Gulf: Locations of specific sites and zones of pockmark activity

REGIONAL SETTING

The deepest water within the operating area is 55 m; depths over much of the region, however are 30 m or less (Fig. 1). Off shore, surface water temperature ranges from 18.5°C in late winter to 33.2° in early fall. Salinities range from 38‰ to 40‰ (Basson *et al.*, 1981). Currents are primarily tidal; however, longshore currents generated by the prevailing northwesterly winds and the especially strong shamal winds control sediment movement along the coast. Modern sediments are primarily carbonate; however, sands of aeolian origin are present in the northern part of the Saudi Arabian Shelf.

The tectonic history of the region has been outlined by Kassler (1973) and Johnson (1978). During the maximum of Wurm glaciations, the Arabian Gulf was a river valley crossed by an ancient Shatt-Al-Arag. After the initial post-Würm rise in sea level, a short regression followed (±11 000 years BP). Marine waters probably reached the northern end of the gulf about 8500 years BP (Stoffers and Ross, 1979). The transgression continued in stages into the middle Holocene (±4000 years BP) when the sea level was higher than at present. A slight regression followed, with the gulf reaching its present level 1000 years ago (Felber *et al.*, 1978). A regional unconformity left by this last regression forms the base of the upper Holocene sediments. The regional unconformity is represented in borings by a change in color from gray to reddish brown and by a sharp rise in soil strength (blow count). There are often non-marine deposits, silica and dune sands for instance. In shallow seismic records the unconformity appears as a truncation of beds, subsurface channels or an increase in reflectivity. In deeper waters there are also sandwaves and dunes covered by a thin layer of very recent silt. Structure is a function of variation in the depositional environment. With the exception of site "C" (Figs 2 and 12), there is no evidence of faulting, folding or diapirism in shallow seismic records. Pockmarks are found only in the very recent sediment overlying the unconformity.

POCKMARKS

Since pockmarks are seemingly linked to the presence of hydrocarbons at depth, the obvious first step is to look for some physical link between the two. The existence of natural, submarine oil seeps in the Arabian Gulf has been known for some time (Emery, 1956). Evidence of near-surface hydrocarbon deposits is found in an engineering soil boring shown as "A" in Fig. 2. Water depth at the site is 4 m. Surface sediments are described as "loose, calcareous, silty sand with shell fragments." The regional unconformity is found at a depth of 10 m. "Oil saturated" sands were recovered at 15 m, and between 25.5 and 28.5 m below the sea floor. Between these "oil saturated"" zones is a "very dense, light reddish brown, clayey silt." From 28.5 m to the bottom of the hole at 34.6 m an "oil saturated" limestone was recovered. No other recoveries like this have been made. Additional borings were taken 3 km to the east and 3 km west of site "A". Lithologies are the same, although the thickness of individual sediments differs. In both cases, the holes were drilled to 45.5 m without encountering hydrocarbons. Near-surface oil deposits, leading to seeps, must be a very localized phenomenon. Side-scan records from site "A" show several widely scattered pockmarks in a highly mobile (ripples), sandy bottom. This would suggest some recent activity. To date,

Fig. 3 Side-scan sonar: "normal" case pockmarks. Scale lines are 15 m; distance between fixes is 150 m; the apparent elongation is a function of scale distortion

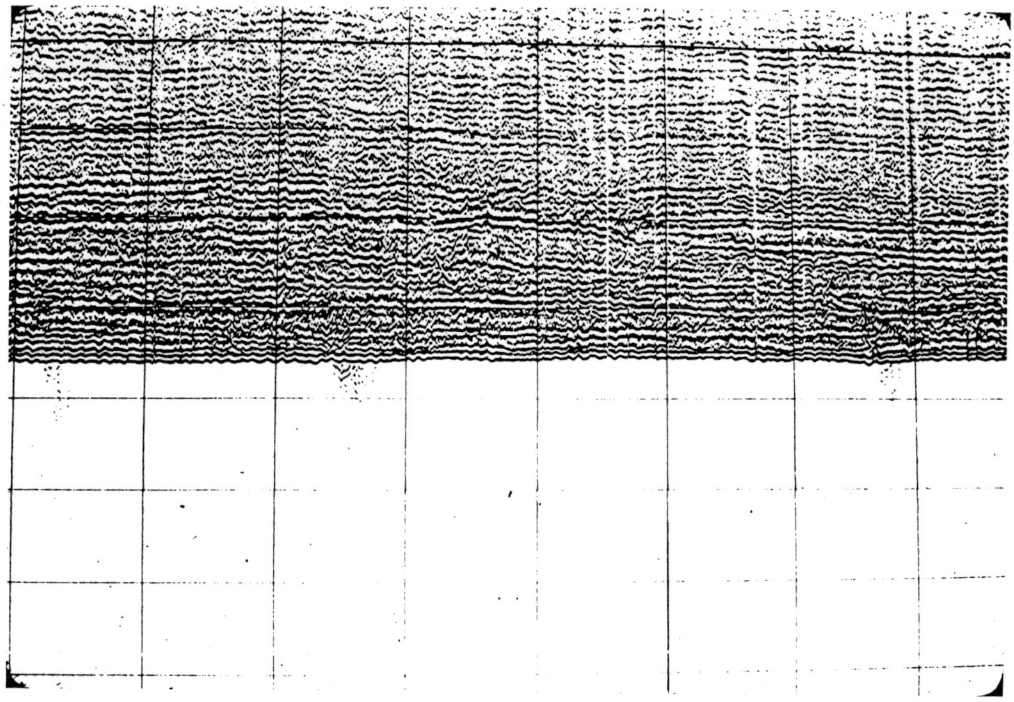

Fig. 4 Uniboom: "normal" case pockmarks. Time lines represent 12.5 ms (TWT); distance between fixes is 150 m

340 active pockmarks have been identified. This figure represents 5% of all pockmarks so far found. The distribution of pockmark activity has been shown in Fig. 2. Based on the physical evidence, it has not been possible to identify which of these active pockmarks are oil seeps.

The "normal" case for pockmarks in the northwestern Arabian Gulf is for them to be rounded in plan view (Fig. 3) and asymmetric in cross section (Fig. 4). Area "B" (Fig. 2) has been chosen to represent this normal case. Water depths vary from 40 to 55 m. Modern sediments are typically silty sands. Pockmark densities average 29 per km 2. Area "B" contains approximately 20% of all the pockmarks so far identified.

"Normal" case pockmarks are found in flat-lying sediments (Fig. 4). The top of the regional unconformity can be clearly seen in Fig. 4 at a depth of 4 m. In the cross section presented in Fig. 5, it is represented by the "stiff hard clay", at a depth of 6 m. As can be seen from Fig. 4 there are typically no bright spots associated with active pockmarks. There will be masking of reflectors and what Hovland (1981a) termed "columnar disturbance," as shown in Fig. 5. Both of these features have been interpreted as indicators of gas-charged sediments. In soft sediment the vent tends to be a depression. On sandy materials a low cone will be built up.

The modern 30 m bathymetric contour (Fig. 6) encloses the surface expression of water, at sea levels between –25 and –55 m, would have been a restricted basin. The extent of the basin (area "C", Fig. 2) is unknown at this time; however, there is some evidence that it extends to deeper water. The basin has been partially filled by up to 20 m (Fig. 7) of gas charged mud (Fig. 8). Surface samples are described as "very soft, silty clay, olive to pale olive" (5Y, 5/3 to 5/Y, 6/3; Munsell soil color chart) in color. Except where gas-charged, these deposits are acoustically transparent or finely laminated in Uniboom records (Figs. 8 and 12). In soil borings they are described as "very soft silts" with a strong H^2 odor, and

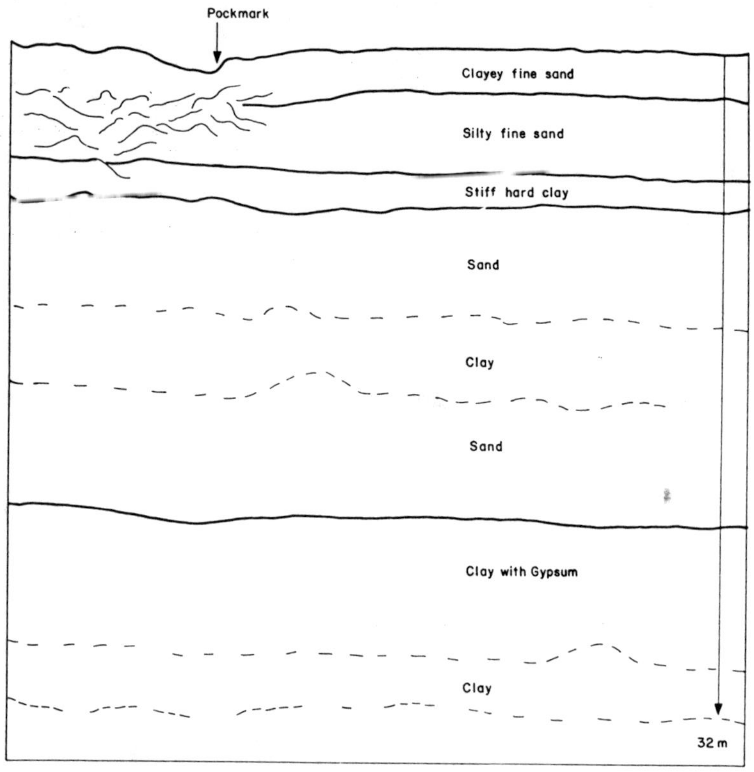

Fig. 5 Geologic cross section of area "B", based on Uniboom and soil boring data: Assumed velocity of sound 1700 m/s

having no appreciable shear strength. Water content is 80% at the surface, decreasing to 40% at 10 m.

A typical, "normal" case, pockmark, in Area "C", is a rounded feature about 13 m in diameter at the base of the cone, with a central crater 4 m in diameter. The cone tends to be a low affair not more than half a meter high (Fig. 9). "Normal" case pockmarks are restricted to the periphery of the basin, where bottom sediments tend to be silty to sandy, and are found in very large numbers, up to 240 per km^2. The distribution of pockmarks have diameters greater than 10 m generally conforms to the 30 m contour (Fig. 6). It is at this depth that the edge of the "cap rock" (faroush) is reached and the alignment of pockmarks becomes linear. Available shallow seismic data suggests that basin sediments continue under the outer lip of the "cap rock". These basin sediments could then supply the gas for the pockmarks which are found along joints in the cap rock. It is unclear at this point whether there is a deep-seated structure which forms the shoreward edge of the basin under the "faroush". In addition to what we have called "normal case pockmarks", three additional features which are products of venting gas are found within the basin. We have distinguished them from "normal" case pockmarks for purposes of description.

"Bubble marks" are very small features which can occur in very great numbers. On side-scan records they appear as light spots (craters) and adjacent reflection spots (lumps). Their size is difficult to judge, but it is thought that they are one meter or less in diameter. They have wide distribution in the flatter portions of the basin and in smooth floored pockets between highs, which are capped by a loosely cemented limestone. "Bubble marks" are found in many of the deeper parts of the Aramco operating area. In all cases they are found in areas where the sea floor is smooth and

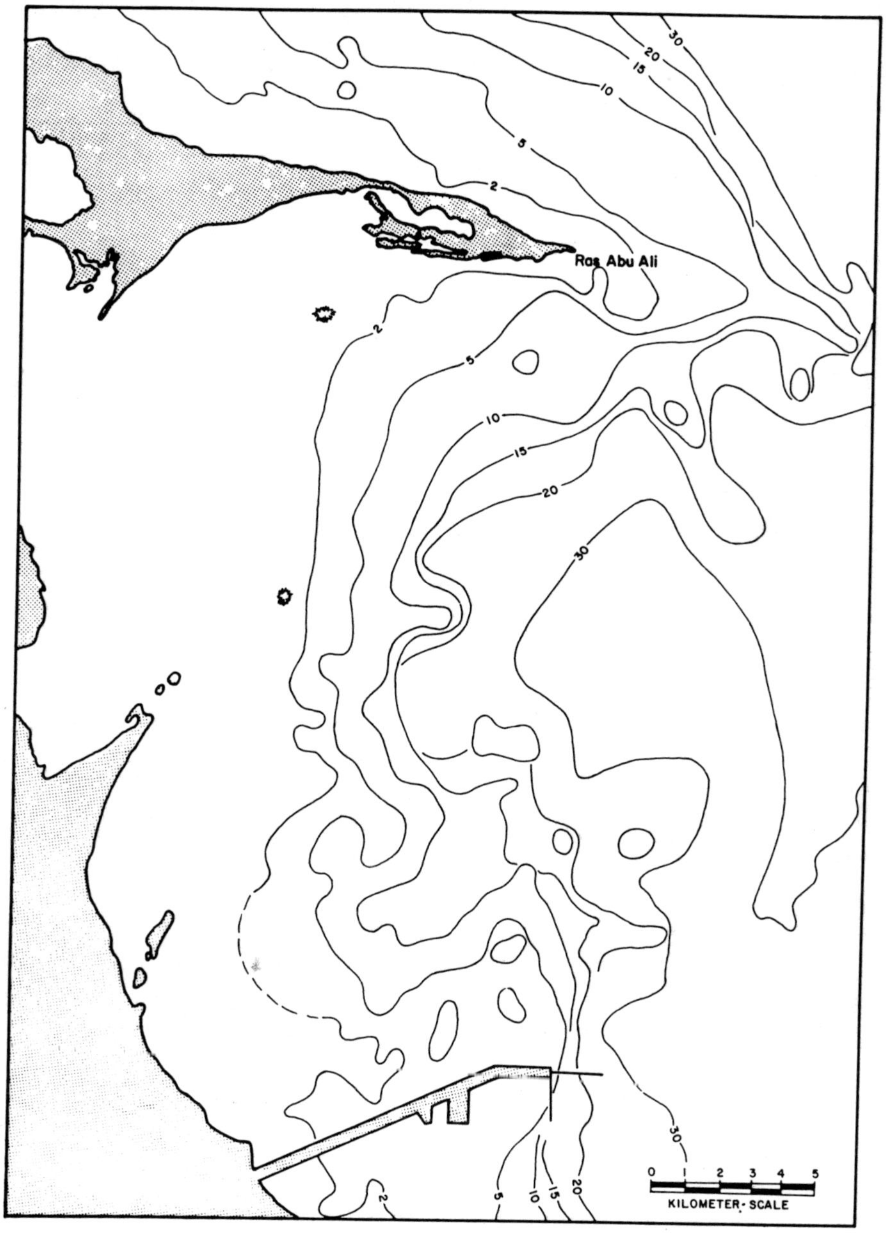

Fig. 6 Bathymetric map of area "C": contours in meters

of low to moderate reflectivity. In some cases the bottom appears to have effervesced. As can be seen in Fig. 10, there are usually larger pockmarks nearby.

"Craters" fit the description of pockmarks in the North Sea (McQuillin and Fannin, 1979). They are large, diameters up to 80 m, oblate depressions with depths to 1.5 m. Often they are compound features having multiple rims and cavities (Fig. 11), "Craters" may also have a simple profile, representing a large hole around a single vent. "Craters" can be found throughout the basin. Densities are about 20 per km 2. The perimeter of the crater is difficult to detect on side-scan sonar records, since they rarely have a cone and because the slope within the

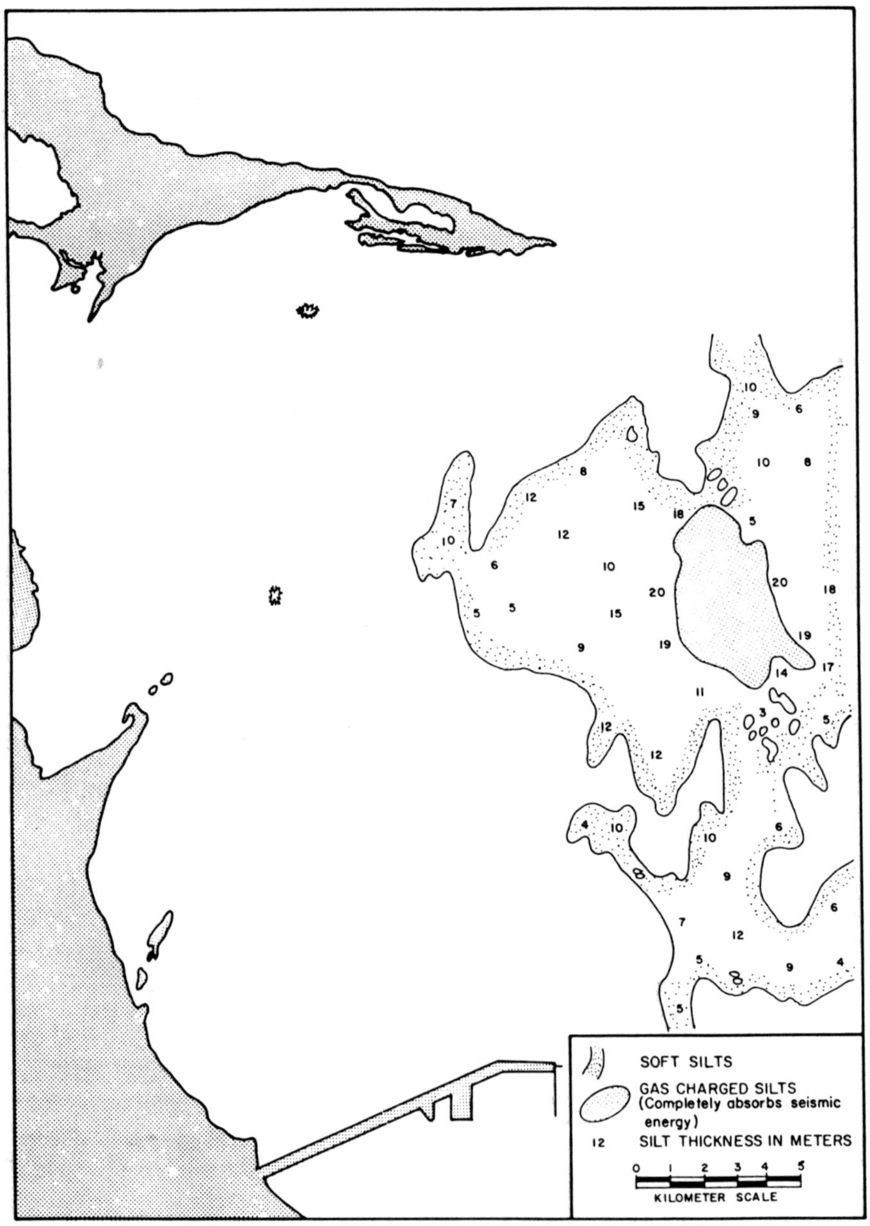

Fig. 7 Area "C": distribution of soft silts and gas-charged sediments

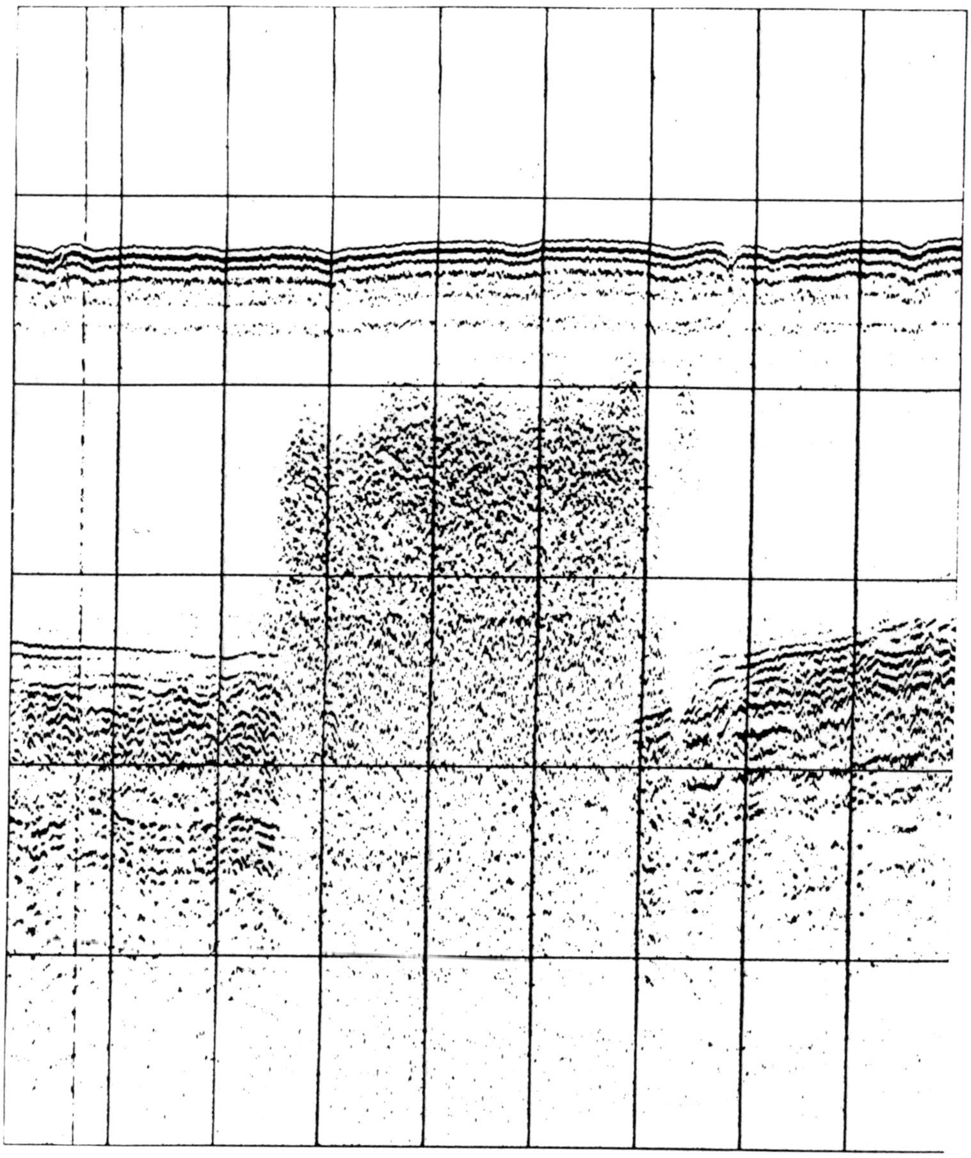

Fig. 8 Uniboom: gas-charged sediments. Time lines represent 10 ms (TWT); fixes are 100 m apart

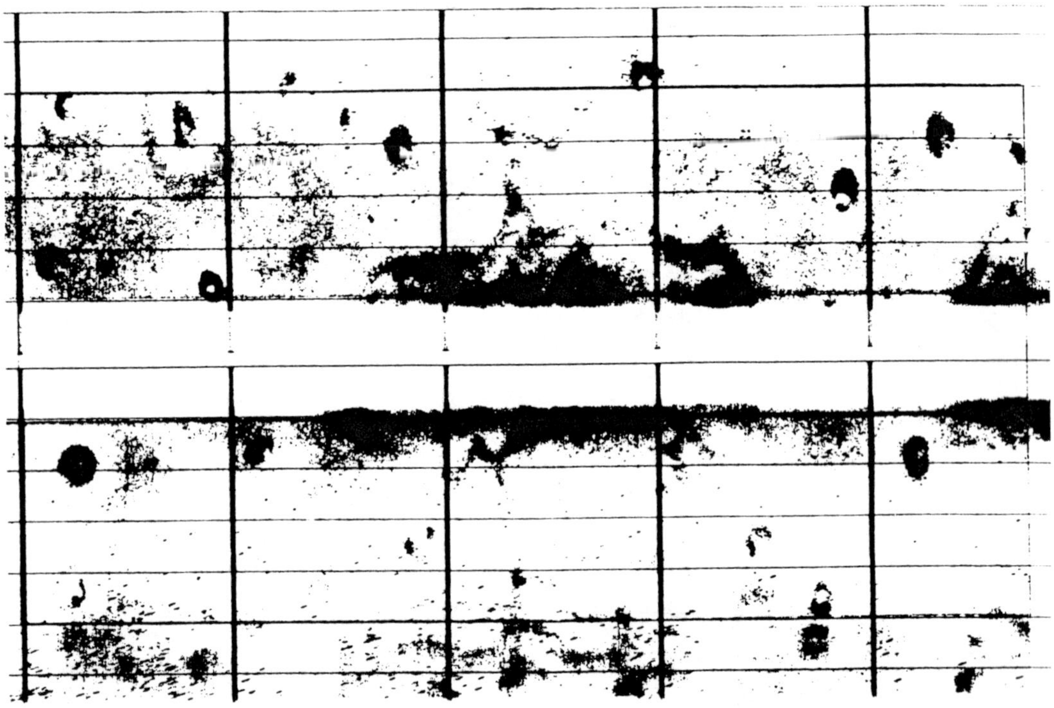

Fig. 9 Area "C" pockmarks. Reference lines are 15 m; distance between fixes is 100 m

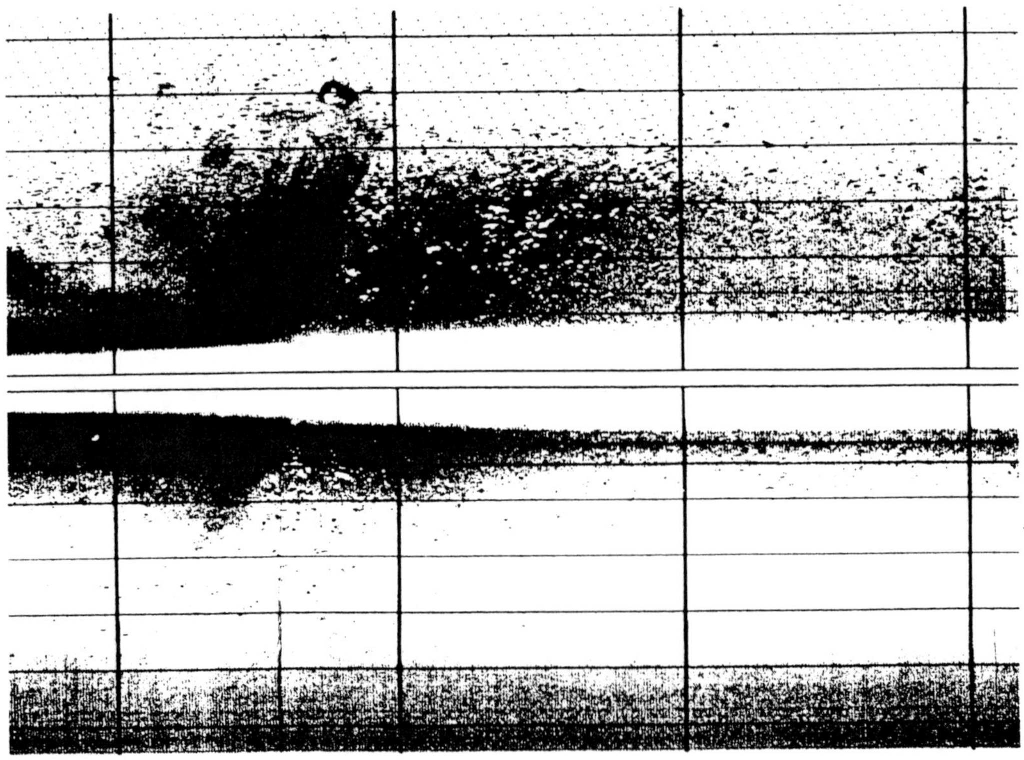

Fig. 10 Side-scan sonar: bubble marks. Reference lines are 15 m; distance between fixes is 100 m

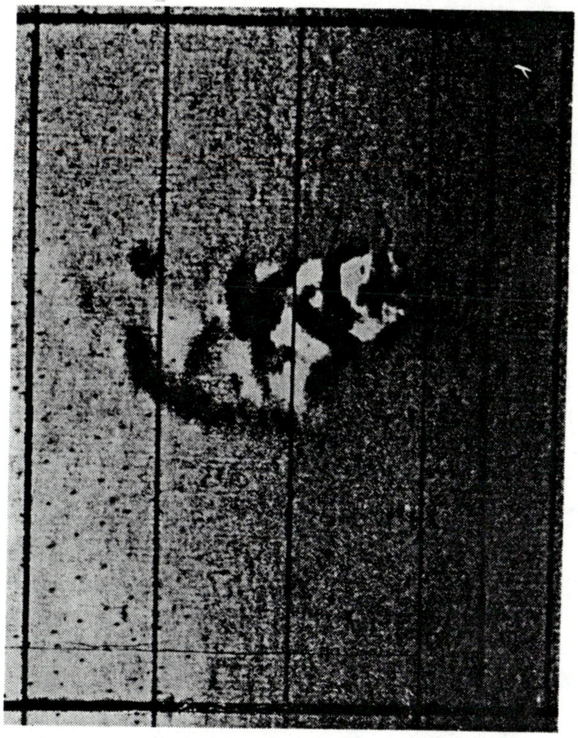

Fig. 11 Side-scan sonar: compound crater. Reference lines are 15 m; fixes are 100 m apart

cavity is gentle (3%). Features of this size have been found in only one other place in the northwestern Arabian Gulf. At this time, however, Uniboom data is not available from that location, so it is not known if the geologic conditions at the bottom are similar; water depths are.

"Outcrop mounds" and their associated depressions may be the most interesting geologic features on the Saudi Arabian Shelf (Fig. 12). They are pockmark-like, it is believed, because these features are, at least in part, the product of degassing. In engineering soil borings, the folded sediments (Fig. 12) lying beneath modern acoustically transparent muds are described as "hard grey clay" which have a water content of 30%, is plastic and has high shear strength (as measured by blow count). Between the overlying muds and deeper clays is a sequence of black to grey sands and grey clays which varies in thickness from 1 to 10 m. The sequence is characterized by the presence of shell fragments, weather limestone and H_2S. Some of the sands are partially cemented. In plan view the "outcrop mound" will be either a rounded depression with the clay outcroping at the center (Fig. 13) or an elongate depression with the outcrop at one end (Fig. 14). They are a common feature within area "C" (Fig. 2), but have been found nowhere else within the Aramco operating area.

DISCUSSION

The occurrence of biogenic gases in recent marine sediments, and the seepage of thermogenic gases migrating from deeper reservoir, which are natural phenomena unrelated to the production of hydrocarbons, undoubtedly both play a part in the formation of pockmarks in the northwestern Arabian Gulf. Their respective roles, however, remain unanswered. In some areas both sources may be active. The distribution of pockmarks may, therefore, be linked both to a reservoir for biogenic gases within the sediments and with migratory pathways through them. The size and distribution of pockmarks are controlled not only by the volume of the causative fluids available, a quantity which has not been measured to date, but by the characteristics of the containment sediments. Sandy and silty, non-cohesive sediments, produce elevated, cone-shaped mounds, while in very soft fine silt to clay muds (cohesive) depressions are the common form. Water depth (hydrostatic pressure) does not appear to influence the size of pockmarks. In two adjacent areas with similar water depths and bottom conditions, it has been found that the distribution of pockmarks can be very different. At one location, 54% of all pockmarks were between 10 and 20 m in diameter; with an additional 34% having diameters between 20 and 30 m. At the second location, 50% had diameters of less than 5 m, with 48% having diameters between 5 and 10 m. Pockmarks in areas of firm sediments are often found in geometric patterns indicating subsurface control. At the contact between silty sediments and "cap rock" along the western side of area "C", 25 pockmarks are found in a 4 km long arch. The arch must represent a joint in the thin cap;

Fig. 12 Uniboom: area "C". Time lines represent 10ms (TWT); fixes are 100 m apart

the source of the fluid could then cover a wide area under this cap. "Bubble marks" may result from the slow migration of small volumes of gas through the bottom. Bubbling of methane from sediments at Cape Lookout Bight, North Carolina has been shown to produce tubes in the sediment (Martens, 1982). A similar mechanism at work could help explain these very small features.

In general, there appears to be a conservative mechanism at work, the interplay between supply, geology and hydrostatic pressure, which regulates the formation of pockmarks. We tend to find large numbers of small pockmarks (bubble marks) and relatively smaller numbers of larger features (craters).

"Outcrop mounds" present a number of possibilities. Some of the deeper structure associated with these features resemble diapars (Fig. 12). There is also the suggestion of some faulting of clays underlying the basin. Shallow seismic records indicate that movement of these clays continued well past the erosional stage that truncated some of the beds. These data also suggest that these sediments are in the process of degassing. Something is obviously keeping the depressions open (Fig. 12). A deep source of gas, perhaps petrogenic, could be supplying the fluid which appears to be coming from them. This gas may also have a role to play in maintaining the plasticity of the clays.

Two issues which have not been addressed are: the role of bottom dwellers and feeders on the creation or maintenance of pockmarks, and what Green *et al.*, (1985) call "return period". There is known to be one location where a pockmark appears to turn on and off with the tides. The only way to adequately measure the "return period" of a pockmark would be to instrument one more location which is known to

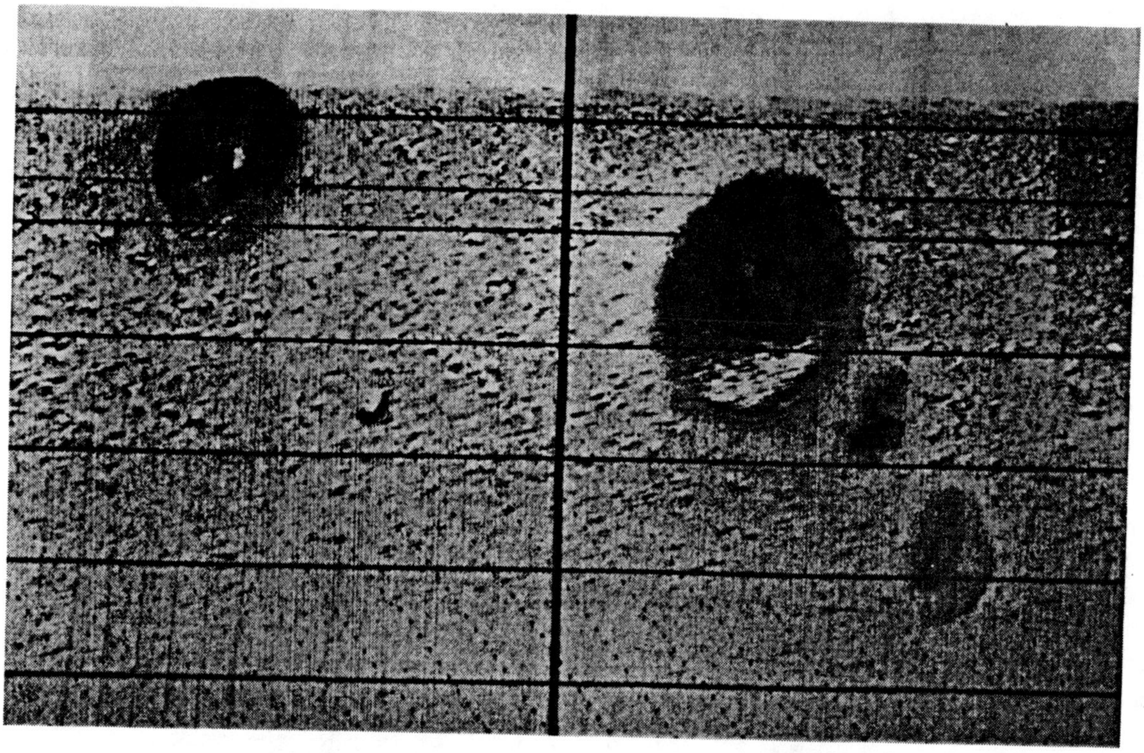

Fig. 13 Circular outcrop mound. Scale lines are 15 m; fixes every 100 m

be active. However, this has not been done as yet. There is evidence that any activity, anchoring, rigs jacking up, or fish feeding, which disrupts bottom equilibrium can initiate pockmark activity. At one location a pre-construction survey showed no pockmarks. A post-construction survey done a year later found seven pockmarks, five of which were active. Observations made from the NR-1 off the coast of Florida suggest that the size of a depression in the sea floor may be species specific, certain varieties of fish occupying specific size holes (Milliman, 1985). To establish a similar relationship in regard to pockmarks of the north-west Arabian Gulf would require long-term, on-site observation. One possible, biologic explanation for at least some "bubble marks" are "sand volcanoes" which occur in large numbers, more or less completely covering an area of the bottom. The features are conical mounds up to a meter high, which are believed to be produced by large ghost shrimp or thalassinid (Basson, *et al.*, 1981).

The major questions which still must be answered are:

- What is the source of fluid coming from these bottom features?
- What activates the pockmarks?
- What role, if any, do bottom dwelling and feeding creatures play in pockmark development?

This chapter is based on data available through July 1985. Surveys of the region are continuing and will undoubtedly change our understanding of the nature of pockmarks in the northwestern Arabian Gulf.

ACKNOWLEDGEMENTS

The authors wish to thank the management of the Arabian American Oil Company for granting permission to submit this contribution.

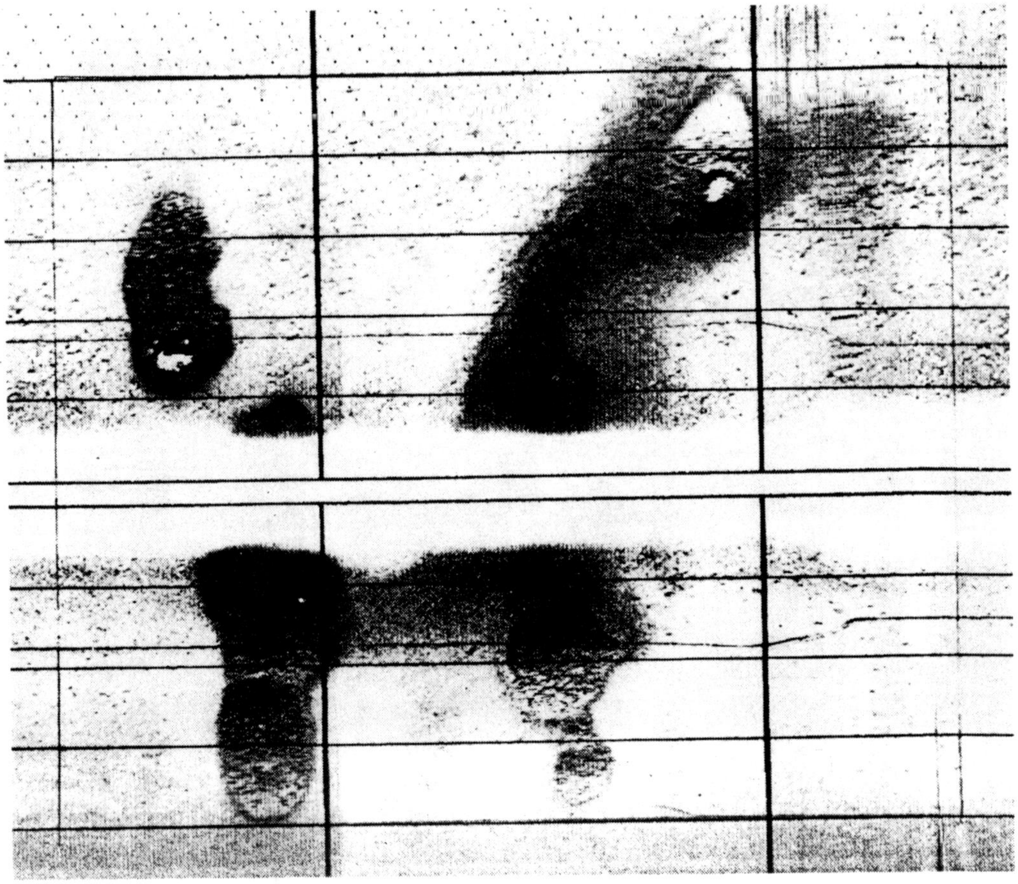

Fig. 14 Elongate outcrop depression. Scale lines are 15 m; fixes every 100 m

REFERENCES

Basson, P. W., Burchard, J. E., Jr, Hardy, J. T. and Price, A. R. 1981. *Biotopes of the Western Arabian Gulf.* Aramco Department of Loss Prevention and Environmental Affairs, Dhahran, 284 pp.

Bohlen, N. F. 1985. Personal communication (University of Connecticut).

Emery, K. O. 1956. Sediments and water of Persian Gulf. *Bull. Am. Assoc. Petrol. Geol.* **40**, 2354–83.

Felber, H., Hotzel, H., Maurin, V. *et al.* 1978. Gulf coastal region and its hinterland — sea level fluctuations during the Quaternary Period. In: *Quaternary Period in Saudi Arabia — 1: Sedimentological, Hydrogeological, Hydrochemical, Geomorphological and Climatological Investigations in Central and Eastern Saudi Arabia,* S. S. Al-Sayari and J. G. Zotl, Eds. New York: Springer-Verlag, pp. 50–57.

Green, C. D., Heijna, B. and Walker, P. 1985. An integrated approach to the investigation of new development areas. In: *Proc. Offshore Site Investigations Conf.*, London, in press.

Hovland, M. 1981a. Characteristics of pockmarks in the Norwegian Trench. *Mar. Geol.* **39**, 103–17.

Hovland, M., 1981b. A classification of pockmark related features in the Norwegian Trench. *Publication No. 106.* Trondheim, Norway: Continental Shelf Institute.

Johnson, D. H. 1978. Gulf coastal region and its hinterland — general geology. In: *Quaternary Period in Saudi Arabia—1: Sedimentological, Hydrogeological, Hydrochemical, Geomorphological, and Climatological Investigation in Central and Eastern Saudi Arabia,* S. S. Al-Sayari and J. G. Zotl, Eds. New York: Springer-Verlag, pp. 45–58.

Kassler, P. 1973. The structural and geomorphic evolution of the Persian Gulf. In: *The Persian Gulf*, B. H. Purser, Ed. Heidelberg: Springer-Verlag, pp. 11–32.
King, L. H. and MacLean, B. 1970. Pockmarks on the Scotian Shelf. *Geol. Soc. Am. Bull.* **81**, 3141–8.
McQuillin, R. and Fannin, N. 1979. Explaining the North Sea's lunar floor. *New Scientist* **1163**, 90–92
Martens, C. S. 1982. Methane production, consumption, and transport in the interstitial waters of coastal marine sediments. In: *The Dynamic Environment of the Ocean Floor,* K. A. Fanning and F. T. Manheim, Eds. Levington: Levington Books, pp. 187–202.
Milliman, J. D. 1985. Personal communication (Woods Hole Oceanographic Institution).
Milliman, J. D. Qin, Y. S. and Butenko, J. 1985. Geohazards in the Yellow Sea and East China Sea. In: *Proc. Offshore Technology Conf., OTC 4965,* Houston, Texas, pp. 73–6.
Sandstorm, M. W., Meredith, D. and Kaplan, I. R. 1983. Hydrocarbon geochemistry in surface sediments of Alaskan outer continental shelf: part 2. Distribution of hydrocarbon gases. *American Assoc. Petrol. Geol. Bull.* **67**, 2047–52.
Stoffers, P. and Ross, D. A. 1979. Late Pleistocene and Holocene sedimentation in the Persian Gulf — Gulf of Oman. *Sed. Geol.* **23**, 181–208.

Part X

Hydrography — Instrumentation

40

Computer-aided Mapping of Sonar

Sigurd Ruud, Geoteam, Norway

INTRODUCTION

Sidescan sonars have been important tools in offshore surveys for many years. Traditionally the sonars produce paper records on-line. These range from simple systems displaying only the raw sonar signal to more sophisticated systems. The latter correct the sonar signal for slant range and adjust the paperfeed according to vessel speed information. These sonar images come close to a scale-corrected picture, provided that profile lines are completely straight and speed data input to the system is of a high accuracy. The paper sidescan sonar record can then be interpreted by a skilled surveyor, geologist or geophysicist who can 'translate' his or her interpretation into a mapable form. In this process it is normal practice to define areas of common seabed features, identify major elements and, after correcting for geometrical distortion, manually plot a seabed feature map. A basic weakness in the seabed feature maps that are produced is the lack of fine detail that can be successfully displayed — for example the gradational nature of sediment boundaries. On complex areas this transformation of information from sonar paper records to seabed feature maps is a arduous job.

BASIC AIMS AND LIMITATIONS

The first basic aim was that the computer-aided mapping of sonar (CAMOS) should present the true track of the sonar towfish for survey lines that are not absolutely straight due to survey conditions, or for survey lines following an intended curved path — for instance while tracking a pipeline. The second aim was that the CAMOS software should map the true sonar image into a seabed features map. The final aim was to develop software to present several adjacent sonar profiles on one map as a mosaic. Traditionally sonar mosaics have been created using scissors and glue.

These goals can not be achieved on a real-time on-line system. A post-processing system requiring computer resources was needed in order to deal with the huge amount of sonar data. The concept should be flexible with new opportunities to control and present sonar information. On the other hand, running the software becomes expensive, due to operator

COMPUTER AIDED MAPPING OF SONAR

S Ruud, Geoteam A/S, Norway

Presented on **Thursday 6th March** 1986 **at 16.00**
in the Clarence Conference Room

EMBARGOED AGAINST DELIVERY

Fig. 1 CAMOS example of a sonar image following an artificial profile track

time and computer consumption.

The software was designed to work with the company's existing computers and sonars. It is therefore not at this stage a generalized software product that will work with any computers or sonars.

Figure 1 is intended to illustrate a CAMOS end product. An artificial survey track line following a 270 degree curve has been created. A sonar profile was applied to this line to show the response utilizing CAMOS software. Note that this is not a true survey line. The intention is to show the ability of CAMOS to display scale-corrected sonar images of any predicted or non-predicted curved survey paths.

NECESSARY EQUIPMENT

This chapter will discuss the equipment used in the development of the CAMOS software.

Geoteam operates survey vessels suitable for undertaking a broad spectrum of marine site evaluation projects. A computer-processing department in Oslo performs post-processing of navigation, position fixing, hydrographic survey data and seismics. It was therefore natural to base the development of CAMOS on already existing computers and interface to existing sonars.

Two EG&G SMS 960 sidescan sonar systems are in operation. This is one of several systems which perform on-line seafloor mapping with slant-range correction and adjustable paperfeed. This system performs a digital sampling of the sonar data and stores it on a 9-track magnetic tapes. With a formatter unit that can be incorporated into the system, the result is an IBM-compatible tape that can be easily accessed by several computer systems. During early trials, the formatter was not part of the on-board sonar system. This had several disadvantages. The tapes had to be reformatted by a subcontractor with the necessary equipment. This was both time-consuming and costly, and there were cases where sonar information was lost due to a vulnerable on-line recording system.

Summarized, this means that the CAMOS input of, or interface to, sonar information is a digitally sampled sonar tape from an EG&G system in an IBM-compatible format. The software has not been tried on other sonar systems and would require a digitally formatted tape with sonar information from these systems in order to be tried.

The actual CAMOS software is developed on a VAX 11/780, a 32-bit computer with 4 Mbytes of primary memory and a VMS operating system allowing further virtual memory to be allocated. Programs are written in Fortran and Pascal. All CAMOS images are today presented on a black and white Benson AO electrostatic plotter with 254 dots per inch resolution.

CAMOS FLOWCHART DESCRIPTION

Figure 2 shows a flowchart of the programs involved in CAMOS processing. This flowchart will be discussed later.

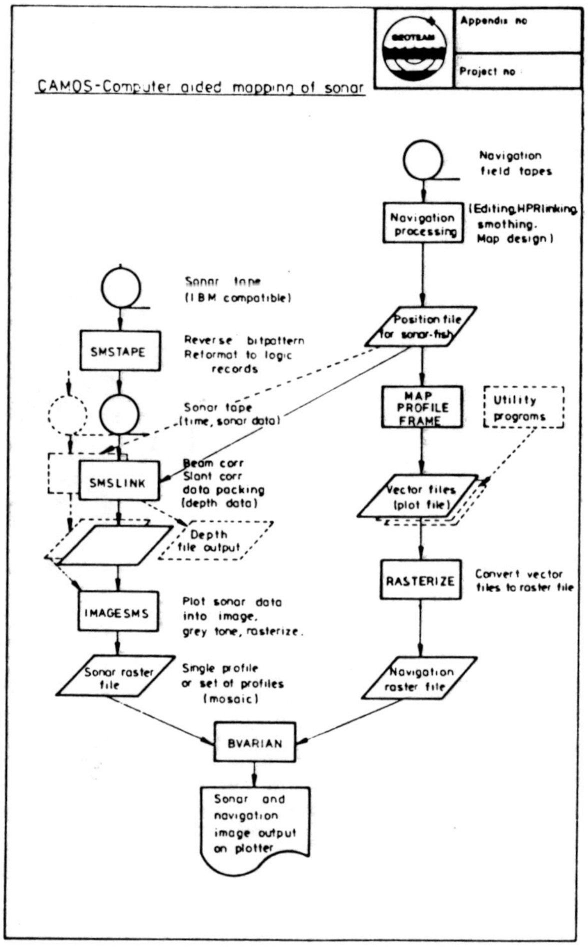

Fig. 2 CAMOS flowchart

Navigation

The CAMOS end product is dependent on overall navigation quality. Any noise or inaccuracy of the navigation data is clearly revealed on a CAMOS mosaic. Three factors have a major influence on the navigation quality.

(1) *Frequent logging of raw navigation data on tape.* The longest logging interval used is 20 s. In many cases 10 or 5 s are used.
(2) *Accurate radioposition systems.* CAMOS techniques based on Syledis, Argo, Trisponder or other accurate local navigation systems produce higher quality results than, for instance, Pulse/8.

(3) *Hydroacoustic navigation systems for sonar towfish positioning.* A Simrad HPR 309 hydroacoustic system measures range and bearing from the vessel to the sonar towfish, and this information is linked to the primary navigation. The overall towfish position quality becomes far better than can be obtained by the old, but still used, method of predicting the towfish position based on the tow cable length added to vessel position.

A CAMOS mosaic is planned prior to the processing. Map area and map scale is decided, and the navigation data are processed and presented on this map. Quality control at this stage is important. Sonar paper records are checked against the navigation plot for control, and any navigation corrections necessary are performed at this stage. When the navigation data are finally accepted, they are linked to the sonar data. From each navigation record, four parameters are used: profilename, timefix, UTM-northing, UTM-easting.

Rasterization of Navigation Data

One of the problems faced in development was the rasterization of the final navigation plot for output on an electrostatic plotter. Normally navigation plots are presented on a pen plotter or a raster plotter. Plotting programs produce a vector file, advising the computer as to the next pen position and if this position should be reached with pen up or down. A pen plotter will simply move the pen according to the vector file information. But a sonar image cannot be plotted on a pen plotter. A raster plotter is usually equipped with rasterizing software which converts the vector information to raster information (black and white dots or pixels) before presenting a top-to-bottom version of the navigation plot. The rasterization itself is usually performed by special dedicated software and is visible to the user. This is normally a great advantage, but not in this case. CAMOS requires a two-step operation. First the raster program converts the navigation vector file to a navigation raster file. Then the raster file is added to a sonar image raster file and dumped on the raster plotter.

SMSLINK Program

SMSLINK is the first program in the CAMOS concept to deal with sonar data.

Identifying sonar data

The EG&G SMS 960 sonar system performs the digitizing and multiplexing of sonar data in the towfish, transfers the data on a coaxial cable to the on-board system and stores the raw data digitally on tape.

The sonar operator may switch between a selection of ranges: 50, 100, 200, 300, 400 or 500 m for each port and starboard channel. The sonar fires one shot and starts data recording simultaneously. As soon as the signal from the outer range limit is received, the next shot is fired. The raw signal received from a single shot contains amplitude reflection information from a narrow strip of the seabed, perpendicular to the towfish direction. The analog signal is amplified with a TVG, envelope detected, sampled and converted to binary words ranging from 0 to 64 dB. Each sample corresponds to 12.5 cm of seabed. The sampling frequency is 6000 samples per second for each channel, resulting in a total of 12000 samples per second being recorded. This figure indicates the huge amount of sonar data involved in the CAMOS process.

A single record on the tape contains a header block of 32 bytes plus all the sampled sonar data belonging to one shot. The header block contains different types of status information. For the CAMOS purpose the timefix is the most important part of the header block. The record length is proportional to the range selected, such that a sonar range of 100 m gives a single record of 1800 bytes in length.

Linking navigation position and direction data to sonar information

The SMSLINK program reads each record of the sonar tape and extracts the timefix. The navigation data file is investigated, and UTM-northing, UTM-easting, profile direction for the sonar timefix is calculated. The port and starboard channel data is kept in computer memory.

Detecting fish and computing its distance from the seabed

The fish height (vertical distance between fish and seabed) is a figure needed for sonar signal corrections. Unfortunately the EG&G sonar towfish does not contain an echosounder. Therefore the fish height must be extracted by investigating the sonar signal.

Transducer beam form correction

On the final sonogram, a flat amplitude response independent of range for an even seabed is desired to give a uniform picture. Unfortunately the response is not flat due to spread loss, absorption loss, strike angle effect and transducer beam form. A TVG in the towfish is supposed to correct for spread loss and absorption loss. The CAMOS software is programmed to correct for the effect of transducer beam form. If this is not corrected, ripples from the transducer lobe overlay the seabed feature information and consequently reduce the overall sonar image quality. This correction is carried out in SMSLINK, which is to a large extent able to correct for this ripple effect.

Slant range correction

After beam form correction the signal is slant range corrected. This is a simple but time-consuming operation.

A sonar detects amplitude reflection data as a function of time. This two-way reflection time is then proportional to the distance (radius vector) from the towfish to the reflection point on the seabed. It is important to note that a sidescan sonar cannot measure the direction from the reflection point. In order to obtain a 'true scale' image, amplitude as a function of the radius vector must be converted to amplitude as a function of the horizontal vector. This conversion is called 'slant range correction'. Having already detected the fish height, the horizontal distance is calculated by the theorem of Pythagoras, that is, the horizontal distance is calculated assuming a flat seabed. Slant range correction implies that the data below and adjacent to the towfish are 'stretched'. This inevitably implies that the image closest to the fish is of a lower quality.

Finally, after slant range correction, the sonar data are written to an output file together with position and direction information.

IMAGESMS

The IMAGESMS program performs the sonar image creation. The overall resolution is usually limited by the plotter, not the sampled sonar data. CAMOS images are presented on a Benson black and white electrostatic plotter with a resolution of 254 pixels per inch (1 pixel = 0.1 mm × 0.1 mm). The sonar image is not plotted in black and white only, but in grey tones varying from black to white. A grey-scale code is created by allocating a 4 × 4 pixels set and filling this set with different numbers of black pixels. In this way a grey-scale effect is obtained, but the image resolution is reduced to 0.4 mm × 0.4 mm per image element.

The CAMOS map size is defined at an early stage of processing, as described earlier. The computer now allocates memory for all the image elements in this map, and initializes the value of each image element to zero. The SMSLINK program outputs a file with sonar and position data. The IMAGESMS program now reads this file and fills in the image elements with sonar amplitude information. One or more profiles can be processed simultaneously to produce a sonar image. Finally the image is converted to a raster file and the sonar amplitudes translated to grey tones.

This sonar raster file is added to the navigation file and dumped onto an electrostatic plotter.

Depth Data from sonars

A sonar survey of an offshore site is often combined with a detailed bathymetric mapping of the site. Bathymetric data are today usually recorded by a ship-mounted echosounder. Research and development is currently being performed by several companies on advanced equipment like multiscan echosounders, topographic sidescan sonars and multibeam sonars in order to improve bathymetric mapping. Howev-

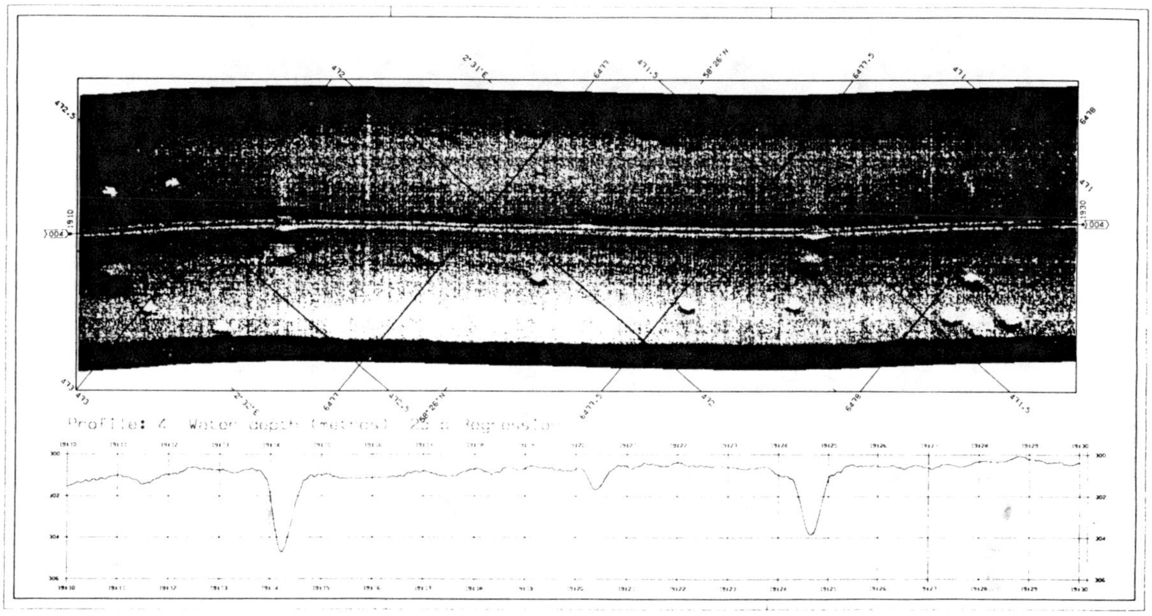

Fig. 3 Sonar image with corresponding depth profile

er, this subject is beyond the scope of this paper and will not be discussed further.

Depth recordings from a sonar can, however, be useful, and computer processing opens new opportunities for utilizing a combination of sonar and depth information. The EG&G 990 towfish contains a depth transducer which measures the distance from the sea surface to the towfish. This information is one of the parameters stored on tape in the header block. The depth transducer information added to the fish height (distance from towfish to seabed) gives the total depth from the sea surface to the seabed along the track of the towfish. If the slant range discussion is borne in mind, it can be seen that a depth profile can be obtained on the area with the lowest sonar quality. This depth data set is therefore ideal for use in combination with sonar information.

Figure 3 shows such an example of a computer-processed CAMOS sonar image combined with the corresponding fence diagram of the depth along the sonar track. The example is a pockmark area in the Norwegian Trench. The object of the survey was to find the best route for a future pipeline, avoiding obstacles such as pockmarks. The depth profile shows that the sonar towfish track has passed through several pockmarks. The depth and the vertical shape of the pockmarks can easily be detected. The same pockmarks are hardly seen on the CAMOS image due to a vague shadow effect, and low resolution, below the towfish.

A comparison of the pockmarks detected on the depth profile, against the general pockmark shape seen on the sonar image, supports the opinion among geophysicists that pockmarks are usually much bigger than the sonar impression seems to indicate.

Sonar Mosaics

One of the advantages of the computer processing of sonar data is the possibility of presenting a scale-corrected mosaic of several sonar profiles on a single map. This CAMOS mosaic application is probably not cost effective for areas with few seabed features and which require limited reporting. There are, however, areas in the North Sea which are of an extremely complex nature.

A CAMOS mosaic is shown in Fig. 4. The CAMOS mosaic is a combination of eight parallel sonarlines from a ploughmark investigation in the Haltenbanken area, offshore mid-Norway. A full picture of ploughmarks is

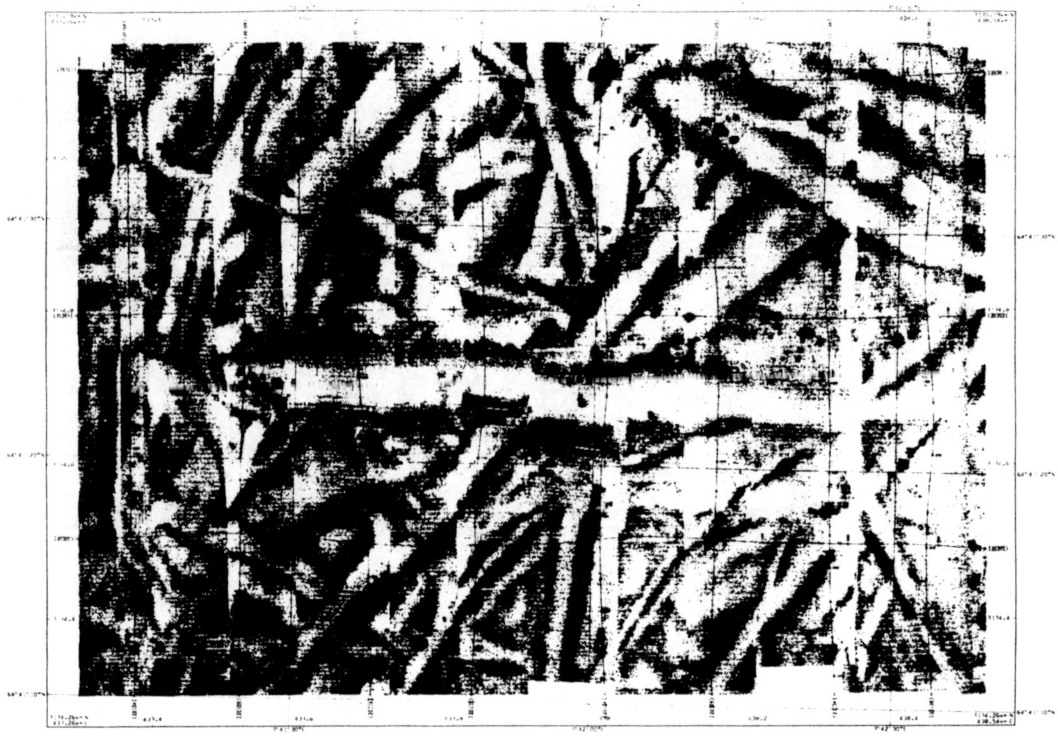

Fig. 4 Mosaic of ploughmark area (with permission from Saga Petroleum, Norway)

difficult, if not impossible, to achieve by the application of the normal interpretation methods of sonograms. A CAMOS mosaic gives the user much more information, which is, furthermore, precise about such an area than a manually interpreted seabed map could possibly achieve. The area was also bathymetrically mapped, primarily from echosounder information. The CAMOS mosaic was an important tool for the contouring of the bathymetric map.

Pipeline and platform inspection surveys are an annual requirement to ensure the operator has good control on the pipeline and platform condition. The immediate visual effect of a scale-corrected sonar mosaic gives the final user a far greater appreciation of the actual seabed and pipeline condition than has been possible in the past with interpretation maps. A base year, created from a sonar mosaic, will help the user with the evaluation of future inspection reports.

Figure 5 shows a CAMOS sonar mosaic, with a four-legged steel jacket platform, from the Alwyn oilfield in the British North Sea sector. An exposed pipeline is visible on the left of the image. The seabed consists mainly of sand, with darker spots indicating a harder seabed material, probably outcropping boulder clay.

SONAR QUALITY CONTROL

In order to improve the quality of the final sonar image, software has been developed for analysing and presenting the sonar signal. The program presents a diagram of the sonar reflection amplitude as a function of distance from the towfish. Based on paper sonograms, areas of interest can be selected and checked thoroughly against this diagram. The diagrams are important tools when optimizing slant range and transducer beam form corrections. It serves also as a quality control check of the raw sonar data.

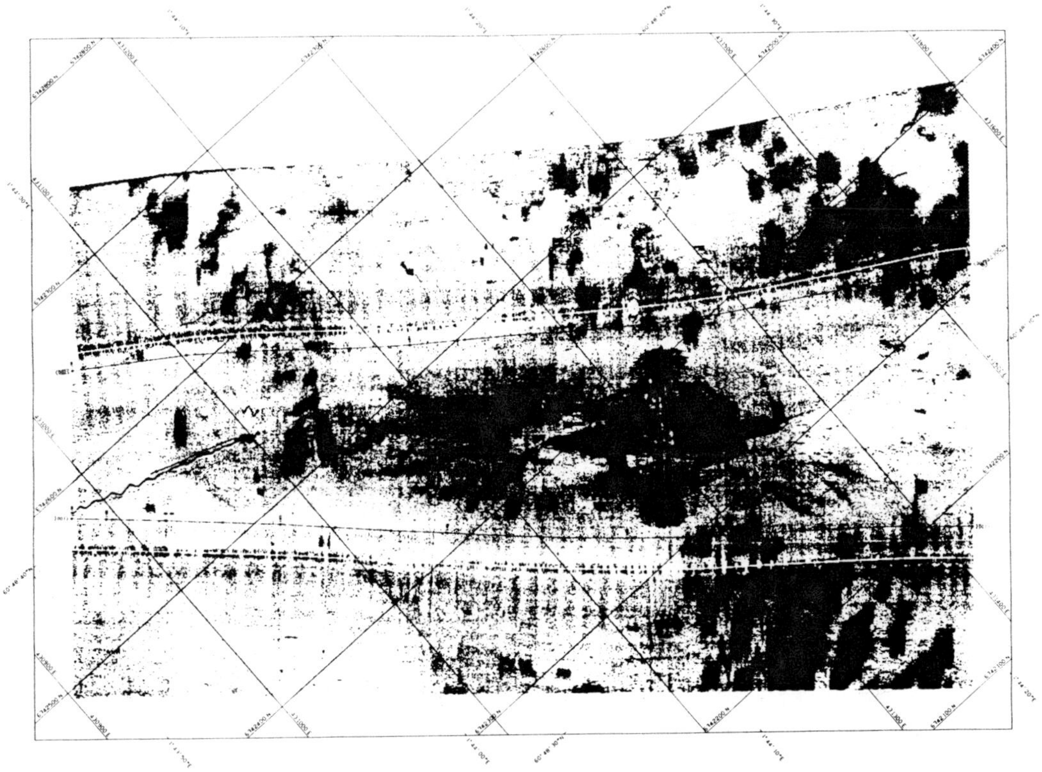

Fig. 5 Mosaic of platform site (with permission from Total Marine plc)

Case Study: Sea Surface Reflection

Theoretically sonar image quality can be improved by utilizing different sorts of image enhancement algorithms. However, these algorithms are costly in computer resources. Therefore a better way, if at all possible, is to improve the quality of the actual on-board sonar.

Reflections from the sea surface and multiples from the sea surface and seabed may seriously disturb the sonar image under certain circumstances. These reflections then may blank out part of the sonar area of interest. This has been a problem for the sonar on which CAMOS processing has been performed in the past. However, the software developed under CAMOS has been able to support important feedback to improve the actual sonar construction.

Figure 6A shows an amplitude-range diagram from the sonar with the original set of transducers. The data are collected from an absolutely flat seabed. A transducer beam form correction has been applied to the signal. This is an extreme example of how surface reflection and multiples from the sea surface and the seabed may disturb a sonar signal. Figure 6B shows the identical seabed, but the data is recorded by the same towfish with a new set of test-transducers mounted. The figure shows that the sea surface reflection and consequently all the multiples are efficiently removed due to an improved transducer construction.

FUTURE ASPECTS

An overall improvement of sonar image quality can be expected in the future. Firstly, the actual sonars can be improved. Today one sonar system is very often used for different applications. The future will perhaps show more use of different towfishes with different transducers for different types of survey. For instance, would it not be better to use a short-range, high

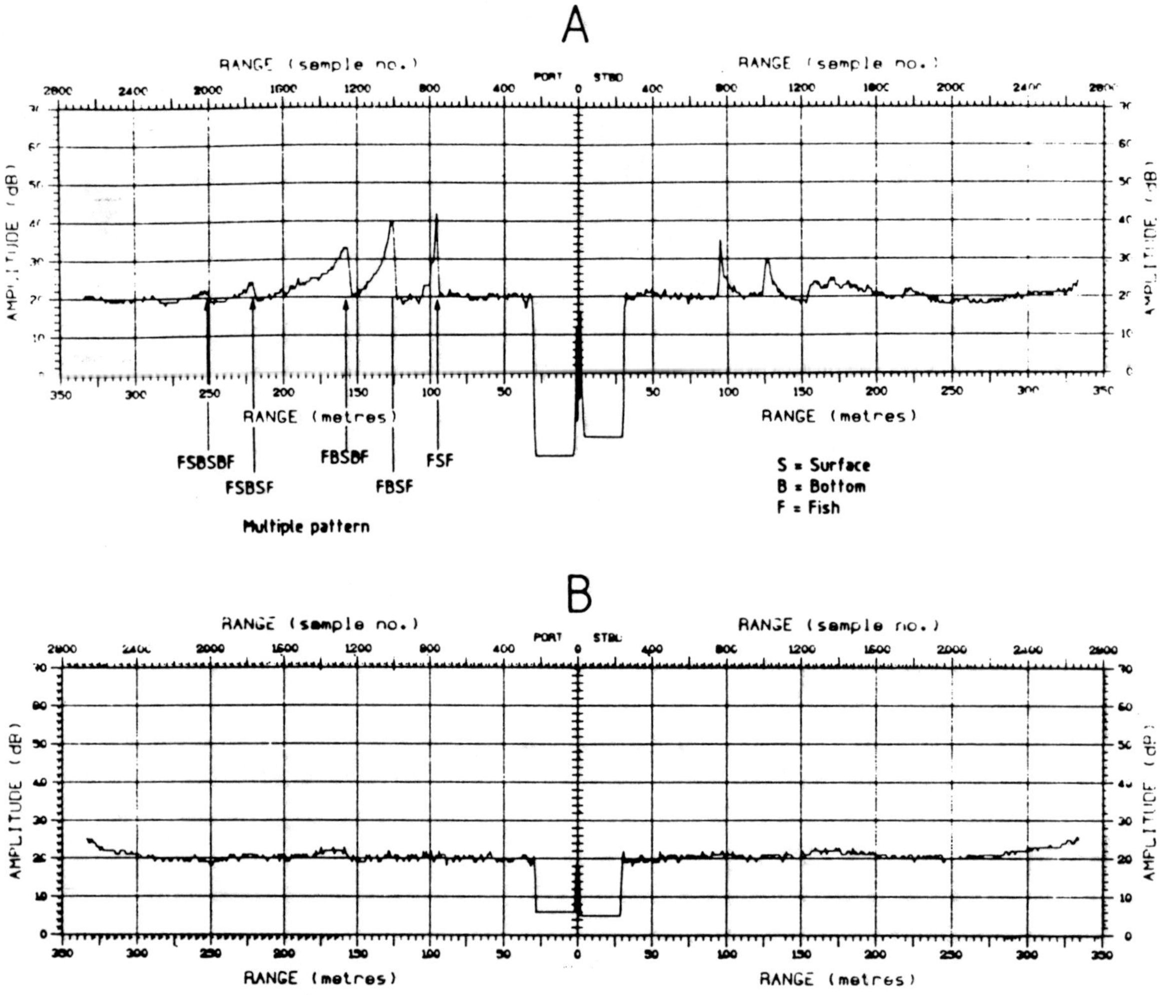

Fig. 6 Amplitude-range diagram for two different transducer sets

frequency (300–600 kHz) sonar for high-resolution applications like pipeline inspections, and a medium-frequency (50–100 kHz) sonar for general seafloor mapping?

The construction of the sonar towfish is important in order to ensure stable behaviour and minimum yaw under difficult survey conditions. Yaw can seriously reduce the sonar image quality. If yaw cannot be avoided, then the yaw effect should at least be recorded. If a sonar towfish had a gyro or compass that could record the towfish course with high resolution, then a computer-aided system for sonar mapping could utilize this information and minimize the effect of yaw on the final sonar image.

The processing of sonar data is power-consuming with respect to computer resources for CPU, memory and disk. As the prices of these computer devices keeps falling, sonar processing becomes more attractive. This will further open the possibility for the application of image enhancement algorithms from the science of image processing to the sonar data.

Colour graphic presentation is another aspect. Colour graphic screen terminals have been available for many years. But the end user needs a sonar hard copy for practical applications. Hard-copy presentation of computer colour images has traditionally been a problem. Coloured-ink jet and colour electrostatic plotters are now being introduced at lower prices but with higher quality and higher resolution. Today the sonar amplitude is presented on a grey-tone plotter. These devices offer the possibility of presenting the sonar amplitude on colour plotters.

One of the most advanced fields linked to image processing is pattern recognition. More research on pattern recognition is expected to follow in future years. Sonar processing might well gain from this research. This opens the possibility of recognition and detection of free spans on pipelines. Pattern recognition could be used to search for seabed features. This would be beneficial both for constructing sonar mosaics and for extracting information for the user about the seabed features.

ACKNOWLEDGEMENTS

The author wishes to thank Total Oil Marine plc for permission to present the sonar mosaic from the Alwyn platform site and Saga Petroleum Norway for permission to present the sonar mosaic from the Haltenbanken ploughmark investigation.

41

SONARMAP: A Convenient Integration of Positioning Information and Seafloor Mapping Sonar Data

M. Sarginson, Comap Geosurveys Ltd, Manchester, UK, *W. N. Ferwerda,* Intersite Surveys BV Haarlem, The Netherlands

INTRODUCTION

The capability to remove the water column from a sidescan sonar record and at the same time geometrically correct the seabed image such that it is a correct scale representation of the seabed is now an established part of marine hydrographic survey procedures. Most of the manufacturers of sidescan sonar now offer at least one so-called seafloor mapping sonar recorder in their equipment prospectus. The use of the term 'seafloor mapping' for the commercially available recorders is, however, a major euphemism since the scale-rectified images produced have no direct or obvious two-dimensional correlation with position. The production of a map still requires a laborious procedure of mosaicing these images using a compromise between best-fit techniques and direct measurement and plotting, or estimation of the towfish position at each data scan. Once this has been achieved it is also still necessary to transfer a positioning grid onto the mosaiced data to create a map in the true sense.

A development program was initiated in late 1984 to produce a rectified seafloor image and a UTM grid, generated on the same graphic recorder. The grid was to be annotated and would be printed in its true position and orientation directly onto the sonar image from a knowledge of the position of the sidescan sonar towfish at the time of recording.

At an early stage in the planning of this development it became obvious that in order to achieve the primary objective, the production of strips of sonar data which could be laid alongside each other and precisely positioned to form accurate seabed mosaics, it was necessary that the grid generated should be rectilinear. No matter how careful the survey crew may be it is extremely optimistic to expect any sidescan sonar towfish to traverse precisely along straight survey lines; therefore it became necessary to devise a method whereby the sidescan sonar data could be slewed relative to the grid. This immediately restricted the range of seafloor mapping recorders that could be used, both because of technical difficulties in slewing the sonar data and the need for sufficient display space on the recorder to allow some movement without corrupting other data.

The concept of merging sidescan sonar data and position information in this manner is not new, but previous successful attempts have involved extensive computer-aided processing on large office-based mainframe systems, incurring substantial expense and time. The new SONARMAP system operates in real-time using a commercially available sidescan system and Hewlett Packard 9800 series computers, which are in standard commercial use by most marine survey companies.

The ability to create efficiently in real-time a scaled rectilinear grid printed directly over a good geometrically corrected sidescan sonar image of the seabed makes the previously extremely laborious task of patchwork mosaicing of sonar images of areas of seabed very efficient. The superimposition of the grid also makes the exact positioning of features such as seabed pockmarks, iceberg scours or debris a very simple process.

BACKGROUND TO THE DEVELOPMENT

Figure 1 illustrates the principle features of a standard sidescan sonar display and Fig. 2 demonstrates the basis on which this image is transformed to a scale-correct seafloor mapping display. As stated in the introduction, this step is now an established feature of many commercially available systems.

The development required to merge this corrected image with the positioning information had been under discussion for several years, but actual fabrication commenced in November 1984. Three primary controls existed when preparing the project.

- use of existing commercially available sonar hardware;
- the need for a real-time solution to the merging of data but retaining a capability to post process;
- user-friendly operation with system flexibility and general reliability.

The first control was resolved by the choice of the Edo Western 706 Seafloor Mapping recorder as the basis of the development system. The second control required that the solution be based on firmware and software development using predictive algorithms. The third

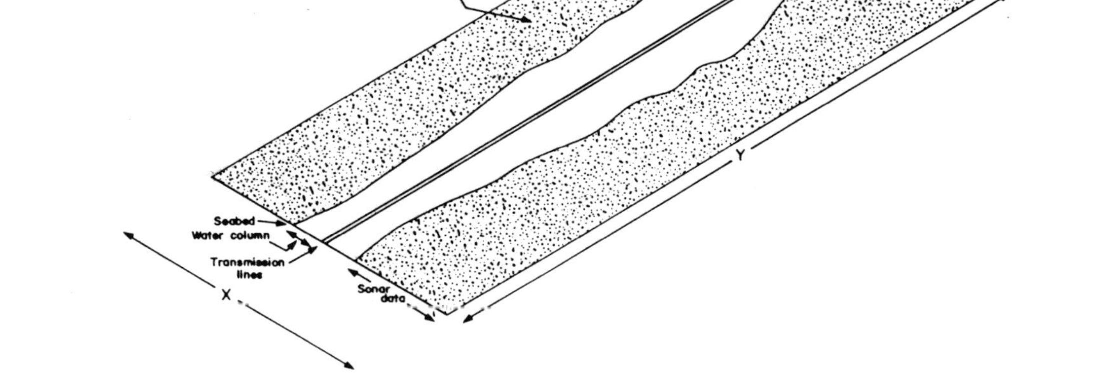

Fig. 1 Conventional sidescan sonar.

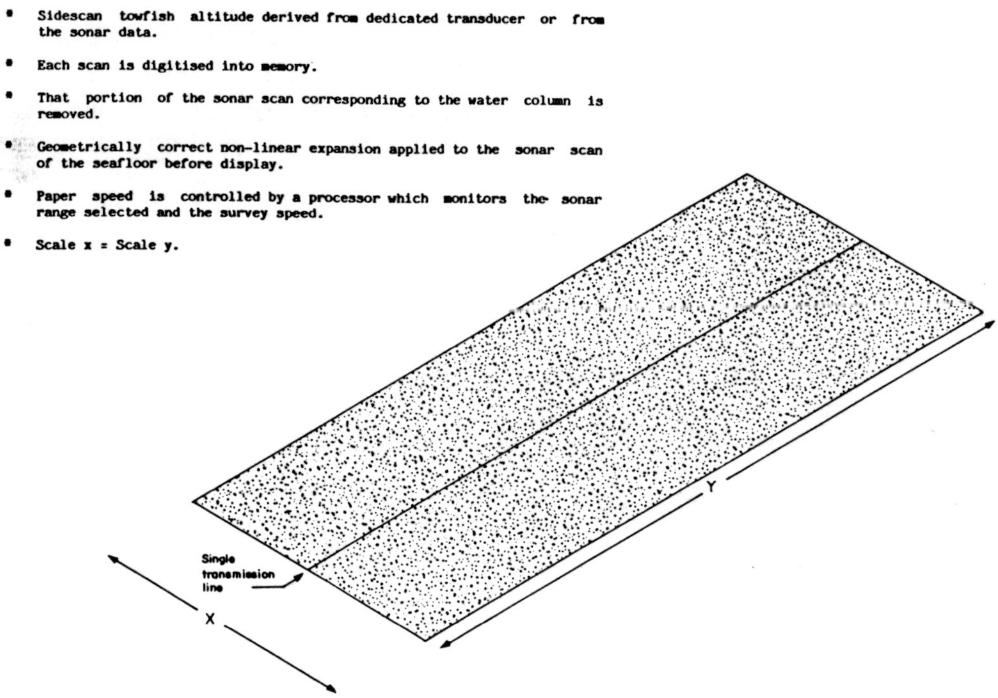

Fig. 2 Seafloor-mapping sidescan sonar.

control implied that the fabrication be based around a 'black box' device to which is linked a computer with a screen display which prompts the operator to select system parameters and supplies quality control information on the position information and synchronization features.

The position of the sidescan sonar towfish is obviously a fundamental requirement of the (U)TM grid computation; therefore this development is only one part of a shipborne package comprising ship positioning aids, gyrocompass and an acoustic short or ultra-short baseline system for tracking the towfish (Fig. 3). Any instability in this derived position can lead to problems in the real-time solution to the grid generation, but post process techniques do allow for the recovery of this data in most circumstances.

One final stage in the development preparation is illustrated by Figs 4 and 5. As mentioned in the introduction, a non-rectilinear grid can result from lateral movement of the towfish about a survey line if the grid is generated relative to the towfish position. This is not desirable and can be easily remedied by presuming the towfish is sailing down a straight line, constructing the grid relative to this line, then slewing the sonar data proportionally to the offtrack measurement.

The different tasks of the main hardware components and the software are discussed in the next section.

SYSTEM HARDWARE AND CONFIGURATION

The SONARMAP system consists of the following major hardware components:

- Edo Western 706 recorder
- SONARMAP processor
- Interplot 200 navigation and data logging system
- Racal Store 4 tape recorders

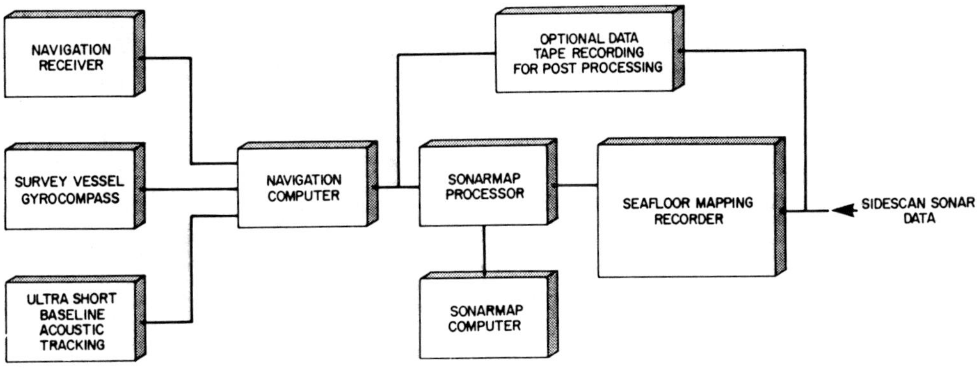

Fig. 3 Primary system components

The U.T.M. grid is generated relative to the known position of the sidescan towfish therefore any change in heading of the survey track line will result in distortion of the U.T.M. grid.

* The grid lines diverge and converge as the sidescan sonar deviates from its course.

* Sub-bottom profiler data or an expanded section of the sidescan sonar can be displayed on the auxiliary channel.

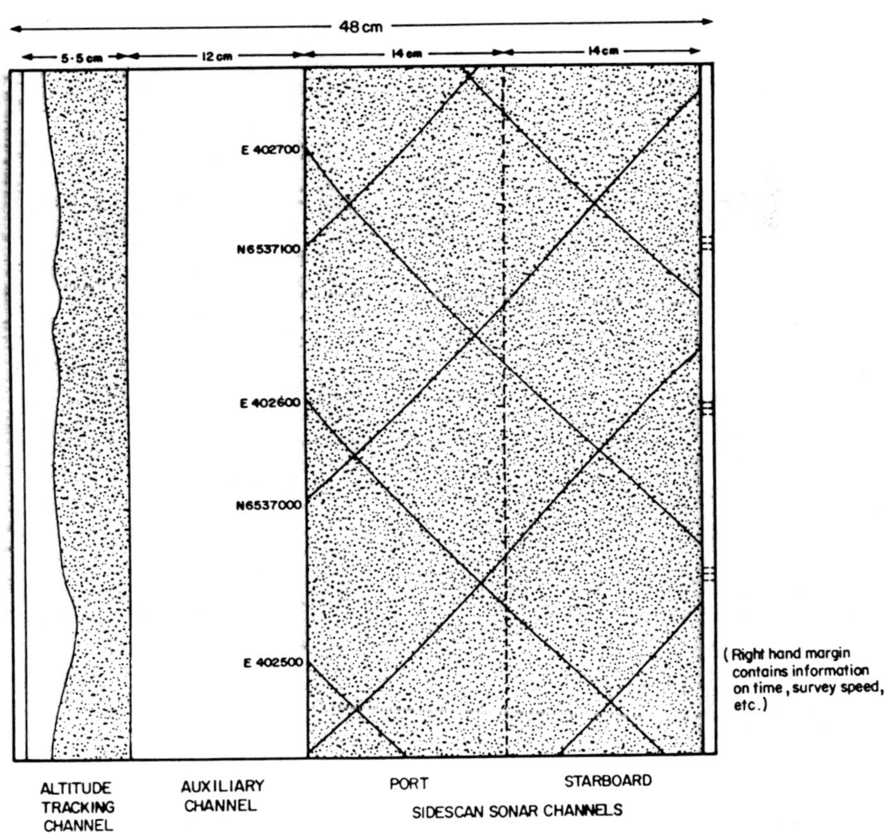

Fig. 4 SONARMAP display on the EDO 706 seafloor mapping recorder (no offset correction applied).

(OFFSET CORRECTION APPLIED)

The U.T.M. grid generated is totally rectilinear. Seabed mosaics can be constructed directly by grid matching.

* The Auxiliary Channel allocation is reduced to allow for offset movement of the main sonar display.

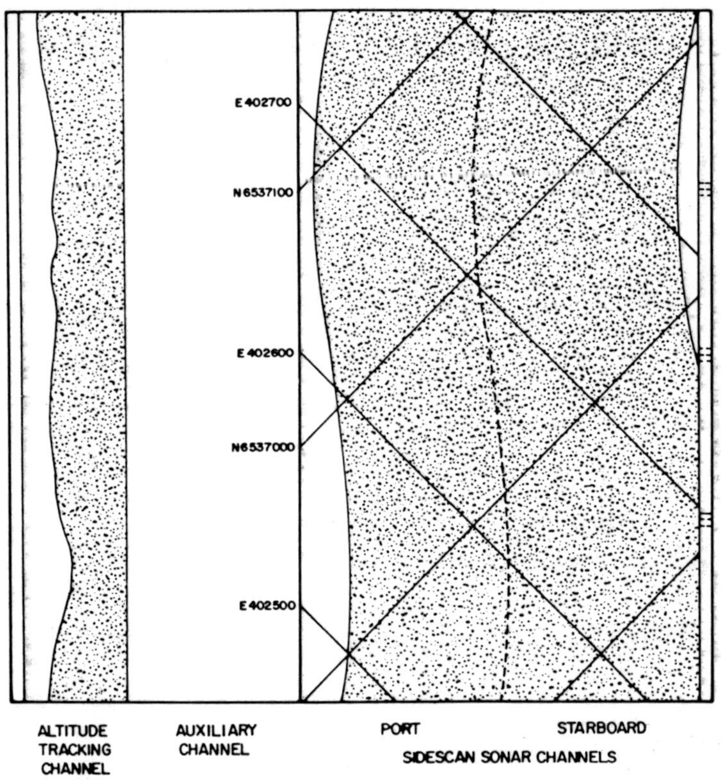

Fig. 5 SONARMAP display on the EDO 706 seafloor mapping recorder (offset correction applied).

Edo Western 706 Recorder

The most practical choice of recorder for this purpose was the Edo Western 706. This displays on 48 cm wide paper and has a programmable multichannel display capability, an RS 232 input port for external annotation, and the print mechanism is a single stylus; therefore selective time shifting of the displayed data is a relatively simple operation. The only modification required to the recorder was the addition of a direct access line to the final print stage. The 706 recorder has a fully interactive menu-controlled keyboard entry of display and signal conditioning parameters. The format of the display is selectable between 2, 3 and 4 channels, and the range and data delays for each channel are individually adjustable in 1 m units giving exceptional flexibility.

SONARMAP Processor

The SONARMAP processor has three main functions:

- generating the (U)TM grid over the SSS-records;
- correcting the SSS-records for the offtrack distance of the fish;
- controlling the tape recorders.

The SONARMAP processor is built in a Eurorack 19 in frame and is controlled by a dedicated computer of the Hewlett Packard 200 series via a GPIO interface. This computer receives the basic data from the navigation computer and sends the required information to the SONARMAP processor after manipulating this data. The way of manipulating will be described in the section on the Interplot 200 navigation and data logging system.

The SONARMAP processor contains the following major hardware components and functions:

- Two Z80 microprocessors plus peripheral components for:
  — grid annotation,
  — offtrack distance correction;
- tape control circuitry;
- connectors and cabling;
- power supply.

Five special control signals are needed for the generation of the SONARMAP. These are synchronization, tachy pulses, annotation, analog-in and analog-out. The synchronization pulse is generated by the Edo recorder at the beginning of each sweep of the stylus, and is used to synchronize the SONARMAP processor with the recorder. The tachy pulses are also generated by the recorder; the absolute position of the stylus as it moves across the paper is always known by counting the number of tachy pulses since the last synchronization pulse. The grid is generated by using tachy pulses to mark the paper. Annotation is an input to the recorder to mark the chart paper. Analog-in is the rectified sonar data transmitted to the SONARMAP processor, and Analog-out is the adjusted signal shifted proportional to the towfish offtrack distance to the intended line.

Generating the UTM Grid

The system will be set up to sail a particular line. For the sake of clarity it is assumed initially that the fish follows exactly this line. The information on the fish position along the line is transmitted to the SONARMAP processor from the navigation computer.

From the fish position along the line the points of intersection of sweep and the grid lines are computed. The computer sends to the SONARMAP processor the offset from the start of the sweep and the distances between the points of intersection of the gridlines. The system sends the data for the easting and northing lines independently. In fact as soon as a start on the grid is made the velocity along the proposed line should be sufficient to ensure the drawing of the correct grid. However, for control purposes the system is built in such a way that the mentioned values will be sent with

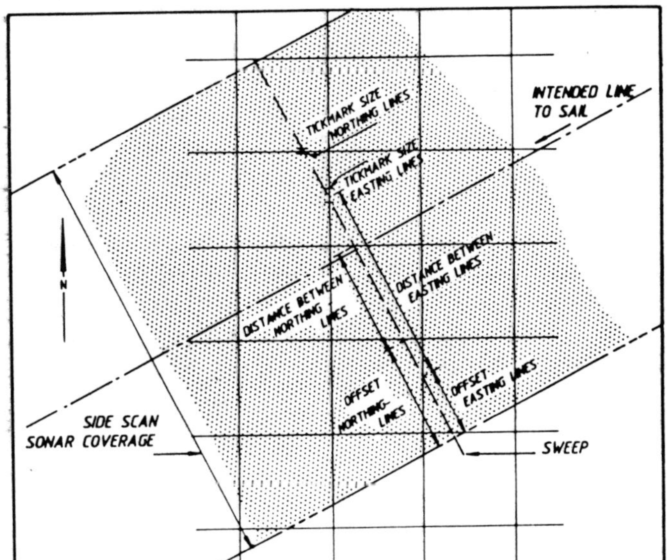

Fig. 6 Grid construction of the SONARMAP display

every update. Filter routines implemented in the Interplot 200 take care of the quality of the transmitted values.

A stylus sweep of the EDO recorder consists of approximately 4600 tachy pulses. For the SONARMAP sidescan sonar records, a maximum of 3850 pulses are used. The records are annotated by using five or more tachy pulses to generate a grid tick in a sweep; a number of these grid ticks in a row are used to generate a line. By experiment it has been found that a minimum number of five pulses will give the best results for a tickmark. To ensure that the grid has full lines in all different directions, the grid ticks per sweep are defined as a function of the bearing of the grid and the towfish speed. For example, for a 45 degrees grid line, one tickmark, i.e. five pulses per sweep, is sufficient and for a 90 degrees grid line a full line, i.e. all tachy pulses are required.

The navigation software outputs the following parameters to the SONARMAP processor:

- start offset for easting and northing grid lines;
- distance between the easting and northing lines;
- size of the grid ticks for the easting and northing lines;
- number of sweeps to go before gridline intersects edge for headings around 0, 90, 180 and 270 degrees;
- polarity of the grid.

The velocity, required to control the paper speed, will not be sent through the SONARMAP processor but will be fed directly into the Edo Western 706 recorder.

Offtrack Distance Correction

In the previous section it was assumed that the towfish would follow exactly the proposed line, but it is obvious that this seldom happens. Therefore a system was implemented for delaying or advancing the sidescan sonar signal proportionally with the offtrack distance. (Both positive and negative offtracks are possible.)

SYSTEM SOFTWARE: INTERPLOT 200 NAVIGATION AND DATA LOGGING SYSTEM

The successful use of the SONARMAP system depends largely on the quality of the navigation data. It is obvious that the use of an unstable positioning system will affect the velocity and thus the scaling of the grid and that jumps in the position will cause jumps in the offtrack calculation and thus an unrealistic movement in the lateral corrections to the sidescan sonar data. To cope as much as possible with all the effects of instability of the positioning systems and with the update speed required to service the SONARMAP processor, the Interplot 200 is split up into a two-calculator set-up, one calculator for the standard navigation and data logging and one dedicated computer for the SONARMAP system.

This Sonarmap computer has the following tasks:

- communication with the standard navigation computer;
- transmitting positioning data to the SONARMAP processor;
- remote control of the Edo recorder, such as speed, annotation, etc.;
- control over the tape drives and tape changeover;
- synchronization of the navigation data and the storage of the sidescan data on the tape;
- text annotation via a proprietary record annotator.

The standard navigation computer of Interplot 200 performs the normal tasks of a navigation and data-logging system; it is interfaced to a positioning system(s), a north-seeking gyro compass and an acoustic positioning system to track the position of the towfish. The system outputs the data to a mass storage medium, a printer, a video for the helmsman and optionally to a plotter.

The positions are filtered for navigation purposes by a Kalman filter and are displayed on the mentioned media. All raw data, such as lines of positions and gyro reading, are stored together on the micro discs. To be able to reprocess the data after the mission, it is essential that at least each five seconds a position is recorded.

The consequence of this approach is a tremendous amount of data and a substantial number of discs. To manage this data flow the discs are copied every day to a 16 or 64 MByte cassette. The administration of the data is better and easier to maintain and the transfer speed of the data storage is also guaranteed.

The raw unfiltered positions of the towfish are transmitted to the SONARMAP computer together with the exact time of the fix and the standard deviation of the fix.

In the standard navigation computer a Kalman filter can be used to get a smooth track of the vessel using the positions and the gyro information. The filtering of the towfish position is done independently from the navigation computer for two reasons: firstly, there is no heading information available on the towfish, and secondly — a more important reason — the filtering of the towfish position makes higher demands than filtering for navigation purposes only. The filtered fish position has to be such that the output to the SONARMAP processor is smooth, stable and shows no jumps in velocity and position. This means that the filter setting may be different from the navigation computer. However, a very heavy filter setting eliminates the dynamics of the fish entirely; therefore other additional steps are taken to eliminate unrealistic changes in position and velocity. The outputs of the velocity and heading by the computer are protected by using a window of a maximum/minimum change. In general, the combination of the optimal filter setting and this protection gives acceptable results; however, in the case of large instabilities the data may be too corrupted, so that post-processing is unavoidable.

The data sent to the SONARMAP processor and thus to the Edo recorder are predicted data. There is a time span of several seconds between reading the lines of position, doing the computation, output to the SONARMAP computer, filtering the data and output to the SONARMAP processor. The time difference between the fix time and the final output is measured and used for the prediction. All analog data from the sidescan sonar recorder is time-stamped every five seconds by the SONARMAP computer.

The time synchronization between the two computers is done by transmitting the exact time from the navigation computer to the Sonarmap computer. The adjustment of the time is operator controlled and in practice only required at the start-up of a survey or when one of the computers has been switched off.

The Control Functions of the SONARMAP Computer

The SONARMAP computer has full control over the tape drives. Via the special function keys of the computer, the tapes can be switched over to recording, stopped, and forward and reverse wound. As soon as one tape is full an automatic change-over from the first to the second drive, and vice versa, will be performed. The clock can be synchronized with the time of the navigation computer as described before. All filters can be set to their optimal settings; for example the Kalman filter can be set from light filtering (value = 1) to heavy filtering (value = 10) and all values in between. The heading filter and speed filter can be set on the exxpected average value, maximum deviation and the maximum change per update. The filters can be reset individually or simultaneously.

Post-processing

The SONARMAP system is designed for real-time applications, but unfortunately the long-range positioning systems do not always provide an acceptable real-time end product. For such occasions an extensive post-processing system has been developed which allows the positioning data to be separately manipulated prior to replay.

The procedure adopted is to transfer the positioning data from the cassette to a Winchester disk. At this stage the data can be recomputed using revised station coordinates or calibration values. The main function of the reprocessing is, however, to smooth the positioning data by manipulation of the complete data set. This has many advantages over real-time filtering using only historic data and can eliminate many of the lag problems associated with predictive computations.

SONARMAP APPLICATIONS

The primary application of the system will be for the generation of regional seafloor maps or sonar mosaics. The fact that the sidescan sonar image is two-dimensionally correct and the (U)TM grid is orthogonal does mean that mosaics can be easily and efficiently generated by matching the grids. Minor errors will result due to false headings on the sonar towfish, and topographic distortion of the sonar image, but both these effects are inherent deficiencies in sidescan sonar and no attempt has been made to correct for them in this development. The assembly of such mosaics does create pictorial-scale representations of the seafloor from which information on the distribution of artefacts, areas of natural or man-made disturbance, and geological features or lithological variations, can be derived. Specific and topical applications include the areal distribution of pock marks, iceberg plough marks and mudslide features.

A similar application is in the production of pipeline route planning charts where the pipeline-laying contractor and design team require a detailed presentation of seabed features, including topographic variation and lithological variation, on which to plan the route and barge-anchoring operations. In complicated areas this form of sonar presentation may be the only satisfactory way of supplying sufficient data on which to base a decision.

The production of scale-correct sidescan sonar data with the grid information is also very valuable in the sonar investigation of seabed installations, particularly pipelines. The positions of the features can be derived with a high level of confidence. Progressive burial or emergence of the pipeline can be monitored by analysing a historic sequence of such displays. A similar use of historic sequences of SONARMAP displays can be of great value when studying dynamic seabed conditions such as sandwave and megaripple fields, pockmark fields and mud slides.

ACKNOWLEDGEMENTS

The authors gratefully acknowledge the substantial contribution of the technical staff of Intersite Surveys BV, who made this development possible. Special thanks go to J. J. Alders, A. M. Jongsma, Drs A. Lubbes, R. J. Ridsdill and Ir. H. G. Zwaan.

42

NOT SUBMITTED

43

Multibeam Echo-sounder System for Seabed Mapping

Magne Glittum, Øivind Heier, Freddy Pøhner and *Karstein Vestgård, SIMRAD* Subsea A/S, Horten, Norway

INTRODUCTION

Single-beam echo-sounder sidescan sonar is currently the main sensor type being used for seabed mapping and surveying. In the future, new developments within high-resolution wide-swathe covering sensors will be available, and will to a large extent replace existing instrumentation.

A wide-swathe bathymetric system is described, viz. the Multibeam Echo-sounder EM 100. The first part of this paper presents ideas about how new developments within high-resolution swathe covering systems will influence instrumentation for seabed mapping and surveying in the future. The second part describes the Multibeam Echo-sounder.

SEABED MAPPING AND SURVEYING WITH EXISTING INSTRUMENTATION

The current seabed mapping technique is based on the use of the single-beam echo-sounder with sidescan sonar for visual inspection between the sounding lines.

The echo-sounder provides a high-density sampling of the topography along survey lines and no information of the seabed between the lines. For detailed mapping a line density down to 25 m is required, which is a time-consuming and costly investment. Production of maps from these data requires highly skilled staff and advanced interpolation techniques.

Sidescan sonar provides a means for visual inspection of the seabed between the sounding lines. Despite the many favourable qualitites of sidescan sonar, the use of this technique for bathymetric mapping has been limited by the inability to provide quantitative depth information.

The sidescan sonograph shows echo level as a function of slant range from the towfish, giving a distorted picture of the seabed. This makes the interpretation of the sonograph time-consuming and costly.

Sidescan sonar provides useful results in areas with homogeneous seabed materials, while the results from areas with non-homogeneous materials may be of very little use. In areas with rugged topography, iceberg plough marks or pockmarks, the sonograph from sidescan sonar gives a very distorted picture of the seabed. In such areas it is very

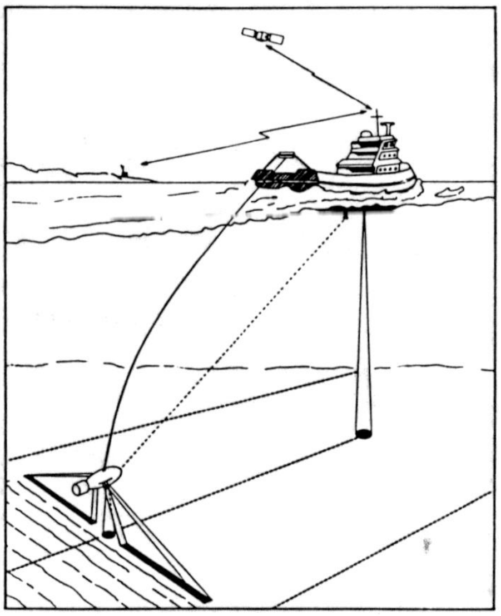

Fig. 1. Existing instrumentation for mapping and surveying. Vessel: navigation, echo-sounder, towfish tracking. Towfish: sidescanning sonar, depth sensor, attitude sensor. Scale 1 to 10 000–50 000. Contours 2–20m.

difficult to distinguish between sidescan effects and material effects.

Figure 1 illustrates current instrumentation for seabed mapping and surveying.

FUTURE SEABED MAPPING AND SURVEYING WITH WIDE-SWATHE COVERING SYSTEMS

Two main factors will push forward a change from conventional echo-sounder and sidescan sonar to a new generation of wide swathe covering systems.

(1) The demands for reliable and accurate seabed mapping is increasing. This requires a shorter distance between the sounding lines using a conventional echo-sounder. Wide-swathe systems have to be used in order to keep surveying costs within reasonable limits, and also to meet the required level of detail and reliability of the final maps.

(2) Current technology makes it possible to bring forward wide-swathe systems within acceptable dimensions and cost limits.

The Multibeam Echo-sounder (EM 100) will to a large extent replace single-beam echo-sounder and sidescan sonar techniques. The Multibeam Echo-sounder will have accuracy comparable with the single-beam echo-sounder, and the need for visual inspection between the sounding lines is no longer present.

Fig. 2. Future instrumentation for mapping and surveying. Vessel: navigation, multibeam echo-sounder. Scale 1 to 500–50 000. Contours 0.5–20 m.

Figure 2 illustrates the use of future wide-swathe instrumentation systems. As mentioned above, use of the wide-swathe systems will reduce survey costs. The saving in survey time for 100% coverage compared with a single-beam echo-sounder is proportional to the number of beams and their aperture.

In addition to the saving in survey time, the survey cost also is reduced by a reasonable use of real-time map-processing systems. By using such facilities the data quality can be checked before leaving the area to be surveyed.

Compared with earlier wide-swathe systems, the dimensions of newly developed wide-swathe systems are small. This makes the mounting of such systems on ships of length about 50 ft possible, indicating that the user can incorporate these systems on already-existing survey vessels.

DATA PROCESSING

Automatic Contouring

The introduction of wide swathe covering systems will have a significant impact upon the way in which bathymetric data are processed. The trend towards computer-assisted map production will be speeded up, since

- 100% coverage of the sea bed makes computerized contouring the appropriate and satisfactory technique. Only a small amount of interpolation is required, and straightforward techniques will give a satisfactory result.
- the amount of data produced by wide-swathe covering system is such that manual data processing is impracticable.

Real-time Mapping

Wide-swathe covering systems will be able to produce seabed maps in real time, as the survey progresses. While most applications will make use of a real-time map for quality control, it may well be that some users will accept the real-time map as a final result of the survey.

Post-survey Processing

For most applications, it is feasible to store the bathymetric data together with time tags, position data and environmental data on magnetic storage medium, for post-survey analysis and processing. Such processing will include tidal corrections, combining data from several survey lines, and elimination of redundant and bad data.

Standardization

A number of users — chart institutions and oil companies — are planning or have started to build up databases containing total information about a large area of the sea bed of interest. Data from area-covering sensors will fit directly into these activities. File format standards for bathymetric and other types of data, together with standardization of storage media, would make it easier and more economical to combine data from several sources, and also to exchange data between user groups and institutions.

EM 100 MULTIBEAM ECHO-SOUNDER — SYSTEM DESCRIPTION

The SIMRAD EM 100 is a medium-frequency multibeam echo-sounder working at 95 kHz with 32 receiving beams. The beam aperture can be switched between 2° × 3° or 2.5° × 3° (athwartships by fore and aft). Maximum coverage is 80°, corresponding to 1.7 times the water depth. The beam fan is electronically stabilized in roll and mechanically stabilized in pitch.

Depth values from the beams are shown at the operator display, giving a transverse profile together with an along-track depth profile for one of the beams for the previous 1, 2 or 5 min.

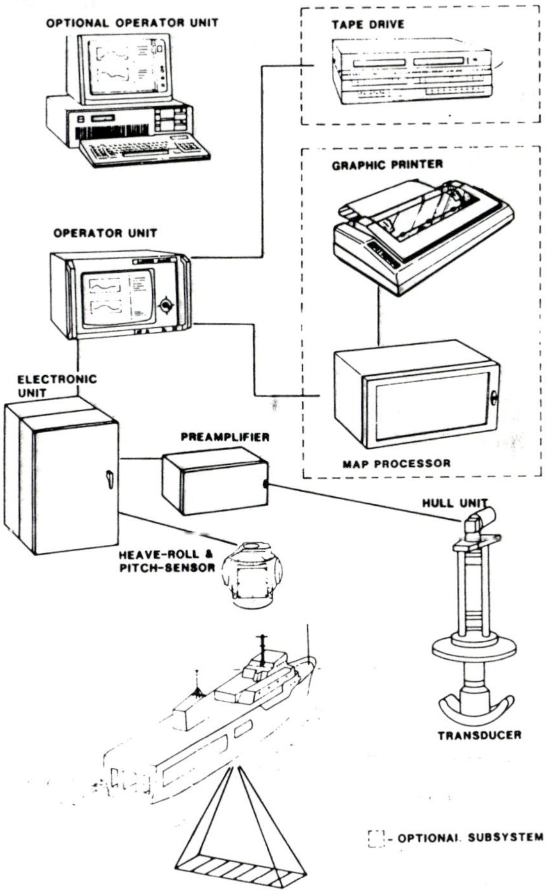

Fig. 3. EM 100 system diagram.

Depth values, together with position and control parameters, are recorded on tape and used for real-time drawing of maps, or for off-line map processing.

The system comprises the following main units (see Fig. 3):

hull unit,
preamplifier unit,
transceiver unit,
operator unit,
tape drive (optional) and
real-time map processing system with graphic printer (optional).

The hull unit is bolted to a cylindrical installation trunk. The transducer is mechanically stabilized against pitch, and can be retracted into this trunk.

The transceiver unit contains the power amplifiers, multiplexers, beam former, detection circuits, interface and control processors. The operator unit comprises a graphic display with its own processor.

The optional subsystems are self-contained units with their own processing power. All of the units are connected to and controlled by the operator unit.

Operator Unit

The operator unit consists of the following units:

main system processor,
graphic display,
joystick and
keyboard.

The operator unit is the centre of the EM 100 system. It controls the other subsystems, based on operator input via joystick and keyboard. The main system processor executes these control commands and processes the incoming data. The kinds of processing are:

calculation of relative position of the depths for each beam

calculation of data for presentation on the operator display and

preparing data for storage on tape and/or for input to the real-time map processing system.

Data are presented on the graphic display, which is divided into two sections: graphic section and text section. The graphic section is mainly used for the presentation of depth data and the text section for menu-lists for controlling the system via joystick/keyboard.

Graphic Section

For viewing depth data, the depth of each beam showing the transverse profile for one ping will be presented, and in addition the depth profile for one of the beams for the previous, 1, 2, or 5 min. Also, the signal strength of each beam is shown in both of the profiles (see Fig. 4).

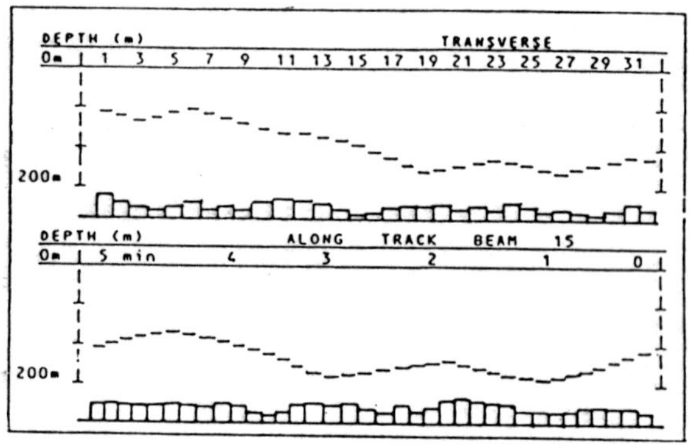

Fig. 4. Presentation of transverse depth profile and along-track beam no. 15.

Text Section

The text section displays the operator control functions and system parameters. The control functions are organized in main menus with corresponding sub-menus.

Figure 5 shows an example of the text section.

EM 100 85.02.03 15.50

MAIN MENU

Echo Sounder
Displlay
Pos. estimator
Type
Map
Annotation
Date/Time

Fig. 5. Menu example.

Data-logging

The tape system is a tape cartridge system, which uses the DC 300 tape cassette standard. The tape cartridge storage capacity is 11.5 Mbytes and the system may continuously log data on tape for more than 5 h (450 ft tape length).

The data stored on tape are:

time, depth and relative position of depth referred to the ship's roll/pitch centre and heave, roll and pitch data,

time, UTM position and antenna offset referred to the roll/pitch centre,

time and operator annotations,

time and echo-sounder parameters and

file name, data and other data defining the survey.

All data stored on tape may also be available on RS 232 interface standard. These depth data are corrected for ship motions in the same way as for the EM 100 storage system.

Real-time Map Processing System

The mapping subsystem consists of a cabinet with a multimicroprocessor system, connected to the operator work station on one side, and to a graphical printer on the other side.

Maps may be produced in real time while the echo-sounder is in operation, or "off-line", using recorded data as input. Each map will present the results of one survey line. For a water depth of 200 m, this will be a chart of width about 300–400 m, but of arbitrary length.

The main purpose of the on-line map is to display the information collected by the multibeam echo-sounder in an orderly and complete manner, so that the quality of the data for contouring and chart production is ascertained. However, in some cases the on-line map may itself be a desirable product.

If there is no navigation system connected to EM 100, the map is always centred around the ship, and the scale of the map will only be approximately correct. If a navigation system is connected and active, the map will have a defined orientation, it will be to a specified scale, and a UTM co-ordinate grid will be marked. The projection will be UTM.

Map heading. The map will start with a heading, displaying initial values of parameters and settings. These are:

scale factor	selectable from 1:500 to 1:10000
contour interval	selectable from 0.5 to 20 m
tide correction	selectable from –20 to +20 m
map orientation	relevant only if navigation system is connected
beam filtering	filter constant for each beam
two-dimensional smoothing	parameter for two-dimensional smoothing
navigation	on/off
contouring	on/off

Distance marks. On the left-hand side of the map area there will be an indication each 100 m.

Ship's track and extension of map. The ship's track will be marked by a dotted line. To each side of this line there will be a dotted line marking the extent of coverage of the echo-sounder.

UTM position marks. UTM positions are marked in a regular grid. The position is marked by a cross, and to the side of this is printed the UTM eastings and northings in metres.

Depth of centre beam. On the right-hand side of the map the depth directly under the ship is drawn as a continuous curve, with a scale selected according to the contour interval setting.

Noise. A separate information field is allocated to display the number of beams marked as non-valid measurements.

Uncovered areas. Areas of the map that are within the extent of coverage, but without acceptable depth data, are marked with a special raster pattern.

Depth contours. If specified, depth contours will be drawn. Each contour line corresponds to a certain depth, and the depth spacing between each line is given by the parameter "contour interval".

Depth numbers. As an alternative, or supplement, to the contours, depth numbers can be printed. A depth number will increase by one when the depth increases by one contour interval. Only one digit is displayed, so the next number above 9 is 0 (last digit of 10). The depth is referred to the centre of the number.

When specified as a supplement to the

contours, the depth numbers are placed in a sparse grid, the ease the understanding of what are hills and what are valleys.

When specified as the only depth indication, the depth numbers are printed in a dense grid. It is then possible to drawn contours manually.

Figure 6 illustrates an example of a typical printout of a map computed in real time.

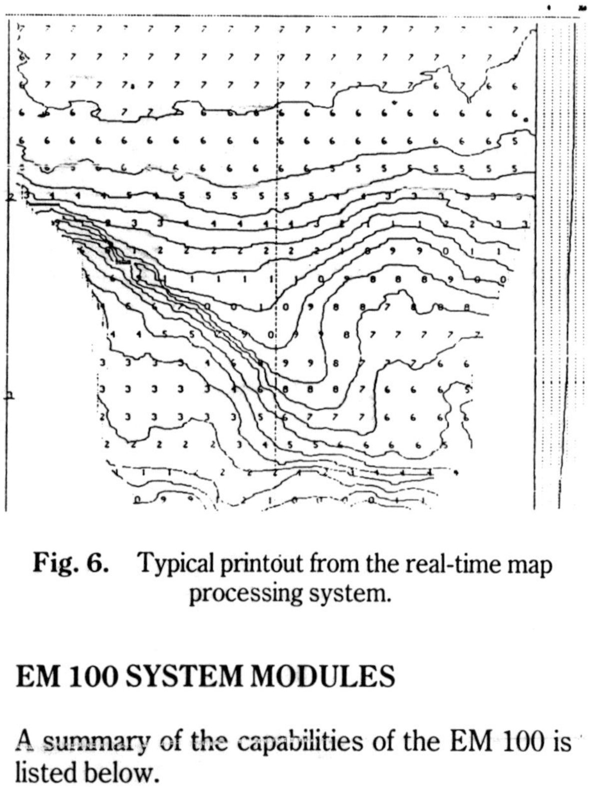

Fig. 6. Typical printout from the real-time map processing system.

EM 100 SYSTEM MODULES

A summary of the capabilities of the EM 100 is listed below.

Basic system
hull unit with raise/lowerable transducer,
transducer,
transceiver and amplifier units, and
operating unit with display and keyboard.

Subsystems

data-logging,
real-time/off-time map processing,
interface to ship motion sensors and
interface to positioning systems.

Basic system — hull unit, transducer, transceiver

raise/lowerable transducer,
electronic roll stabilization,
mechanical pitch stabilization,
95 kHz operating frequency,
32 beams
$2°/2.5° \times 3°$ beam widths,
beam spacing 1.25°/2.5° and
depth coverage of max. $1.7 \times$ depth ($\pm 40°$).

Basic system — operator unit

display divided into graphic and text sections,
display of depth of each beam,
signal of amplitude of each beam,
alongship depth profile from operator-selected beam,
error and status reporting,
menu-based operation (joystick/keyboard),
standard 16-bit processor,
real-time operating system,
input from position estimator system or radio navigation system
output to external computer or data-logging system and
output to real-time contour map processor.

Basic system — data processing

Input of
range from each beam,
heave, roll and pitch data,
gyro,
sound velocity and
relative position of antenna (of positioning system) referred to ship's reference centre.

Calculate
depth and relative position of each beam referred to ship's reference centre.

Ouput of
depth and relative position of each beam,
position data with antenna offsets and
attitude data — heave, roll, pitch and gyro.

Subsystem — data-logging

Logging of
time, 32 depths, relative position of depths, attitude data,
time, position, antenna offsets,
time, annotation,

time, echo-sounder parameter changes and
file/survey data.

Subsystem — real-time/off-line map processor
- real-time map processing,
- map presented on graphical printer,
- contouring of map by lines and/or numbers,
- annotation of geographical co-ordinates,
- operator changes and text on contour map,
- print-out of data quality parameters directly on contour map and
- depth profile of centre beam presented.

THE BATHYMETRIC SWATHE SOUNDING SYSTEM

Dr R L Cloet, Dr C R Edwards, Bathymetrics Ltd, UK

Presented on **Thursday 6th March 1986 at 15.00**
in the Clarence Conference Room

EMBARGOED AGAINST DELIVERY

44

The Bathymetric Swathe Sounding System

R. L. Cloet and *C. R. Edwards,*
Bathymetrics Ltd, c/o Thornton Baker,
7 Old King Street, Bath, UK

INTRODUCTION

Users of depth data have felt for some years now that echo-sounder surveys are inadequate both for navigation with close tolerances and for various engineering works below the waterline.

Sidescreen sonar helped enormously in increasing the awareness of the degree of inadequacy of line scanning echo-sounding surveys. However, sidescan has never quite become a survey equipment for a number of reasons. It is difficult to relate sidescan, in anything but the broadest sense, to a depth survey. It is selective in what it shows, depending on the scan aspect to a feature or target. Sidescan sonar, used with caution, is a diagnostic tool, i.e. it can indicate the presence of a something which requires further examination. Therefore, in one respect, it has tended to increase the time spent on depth surveys, although as a consequence the latter were improved.

Interferometric swathe sounding now adds depth values to the shadowgraph information of sidescan sonar. The Bathymetric swathe sounder, operating at 300 kHz, succeeds in producing depth data at a density which is about equal both down-track and cross-track, whereas the echo-sounding down-track to cross-track ratios are almost invariably appreciably in excess of 1:100.

The swathe sounding system has proved capable of coping with gentle and steep slopes, and has already operated successfully in inclement weather conditions from small and large vessels. It does not need shadow information to delineate seabed features, and has been described as a "one-pass" system.

THE PRINCIPLE OF THE SYSTEM

The system has been developed on the basis of an acoustic interferometer which was the subject of a SERC-funded research project from 1979 to 1985, at the University of Bath.

The concept was described in a paper first presented at the 1982 Oceanology Conference, and subsequently published in the *Journal of the Royal Institute of Navigation* (Cloet *et al.*, 1982). The aim was to achieve depth data, laterally from the survey ship's track, of a quality equal to that which could be attained by means of an echo-sounder along the ship's track, i.e. to fill-in the space between survey sounding lines with soundings rather than sidescan sonar shadowgraphs.

The system was designed to be mounted in a towed vehicle which needed to be sufficiently small to be deployable from small vessels. Towing was adopted for the sake of transferability from ship to ship, and to enable it to be lowered below the seasonal thermocline when necessary. The ability to operate from a small vessel was to exploit to the maximum the possibility to measure depths in navigationally hazardous areas from a stand off position without endangering the vessel.

The interferometers operate at 300 kHz. A beam 1° wide in azimuth and at 60° elevation is transmitted on one transducer and received, so that two interferometers are generated. The transducers forming the interferometers are spaced slightly differently so that a difference phase ramp is produced which allows the normal phase ambiguity to be resolved. This ambiguity is caused by the physical dimensions of the sidescan transducer arrays, which prevent them from being placed close enough to each other to cause a single-phase ramp to occur.

Ranges in excess of 100 m are obtained, and the swathe sounder transmits 5 pulses/s. In each profile about 200 depth values are measured. Thus, when operating at a normal sidescan towing ship's speed of 5 knots, soundings are recorded at approximately 0.5 m spacing along each profile, and each of these is about 0.5 m from its neighbour. It follows that the soundings lie roughly on a 0.5 m square pattern.

THE BENCHMARK ECHO-SOUNDING SURVEY

The question of the accuracy of a hydrographic survey is not easily resolved because such a survey is composed of many measurements. The major difficulty is one of replication, i.e. if the same measurement is repeated a number of times, what is the range of values which results? The inability to repeat the same measurements is the most widely recognized problem of surveying, and this recognition has resulted in major advances being made in the development of positioning systems, of which there are now quite a number, several of which are very good.

What is, perhaps, rather disappointing is that the process of surveying the shape and depth of the sea bed has not received the same attention. Even today we still have to consider such basic questions as, for example, is it safe to come to within 1 or 2 m of the charted depths, even in our busiest shipping lanes? We cannot answer the question: does the sea bed move, in an area only 10 miles offshore and, if so, by how much? Yet not knowing the answers to these questions can prevent the full economic use of such shipping lanes, unless expensive dredging is carried out, or unwarranted risks are taken. It amounts to an expensive ignorance.

An area of sea bed was chosen in Plymouth harbour, which has depths ranging from about 5 m to 30 m, and which contains gradients ranging from 1:2 to 1:200 (Cloet *et al.*, 1985). The texture varied from gravel and stone to silty peat.

The principle of the benchmark survey was to carry out two surveys of line spacing 5 m, at right angles to each other so that the effect of proximity of a sounding line to a feature could be taken into account as well as the depths actually measured (Cloet and Duncan, 1984). The proximity effect was tested by interpolating the data onto a 5-m grid. It showed, not surprisingly, that sounding lines, run in the direction of the steepest slope, represented this slope more accurately than lines, even as little as 5 m apart, parallel to the slope. The accuracy of an echo-sounding survey is therefore also aspect-dependent, because of the orientation of sounding lines relative to that of bed features. Because the "footprint" of a sounding is rather wide, covering about 10 m^2 or more, it cannot discriminate height differences within the footprint which may amount to about 20 cm, and this could well be exceeded on occasion.

When the two orthogonal surveys were compared it was concluded that, on average, and on the basis of very close line spacing, the replication accuracy of such an echo-sounding survey could achieve 10 cm, with a standard deviation, on the steeper slopes, of about 40 cm.

THE PLYMOUTH HARBOUR SWATHE SOUNDING SURVEY

In October 1983 the research model of the swathe sounder was used to repeat the survey, previously made by echo-sounder, in the benchmark area. As in the echo-sounder case, two surveys were carried out at right angles to each other. One of the main aims of the surveys was to obtain complete coverage in each survey. Since it was uncertain how much lateral range would be achieved consistently because processing was off-line, a line spacing of 50 m was adopted. Normally ranges in excess of 100 m were expected.

The vessel used was the *Catfish*, an 11 m catamaran kindly made available by the School of Maritime Studies of Plymouth Polytechnic. The towfish, about 1.5 m long, was suspended from the ship's A-frame over the stern at a depth of 2 m below the water-line. There was ample room in the cabin for the recording and test equipment. To assist in the assessment of the performance of the equipment the output from one of the transducer arrays was displayed on a moving window VDU in normal sidescan mode, and it was also possible to view the interference fringes whenever desired.

Two tape-recorders were used. One, the digital recorder, receives every 0.2 s:

2000 phase angle and range measurements,
four position ranges,
one ship's echo sounding and
three rotational and three translational atti-

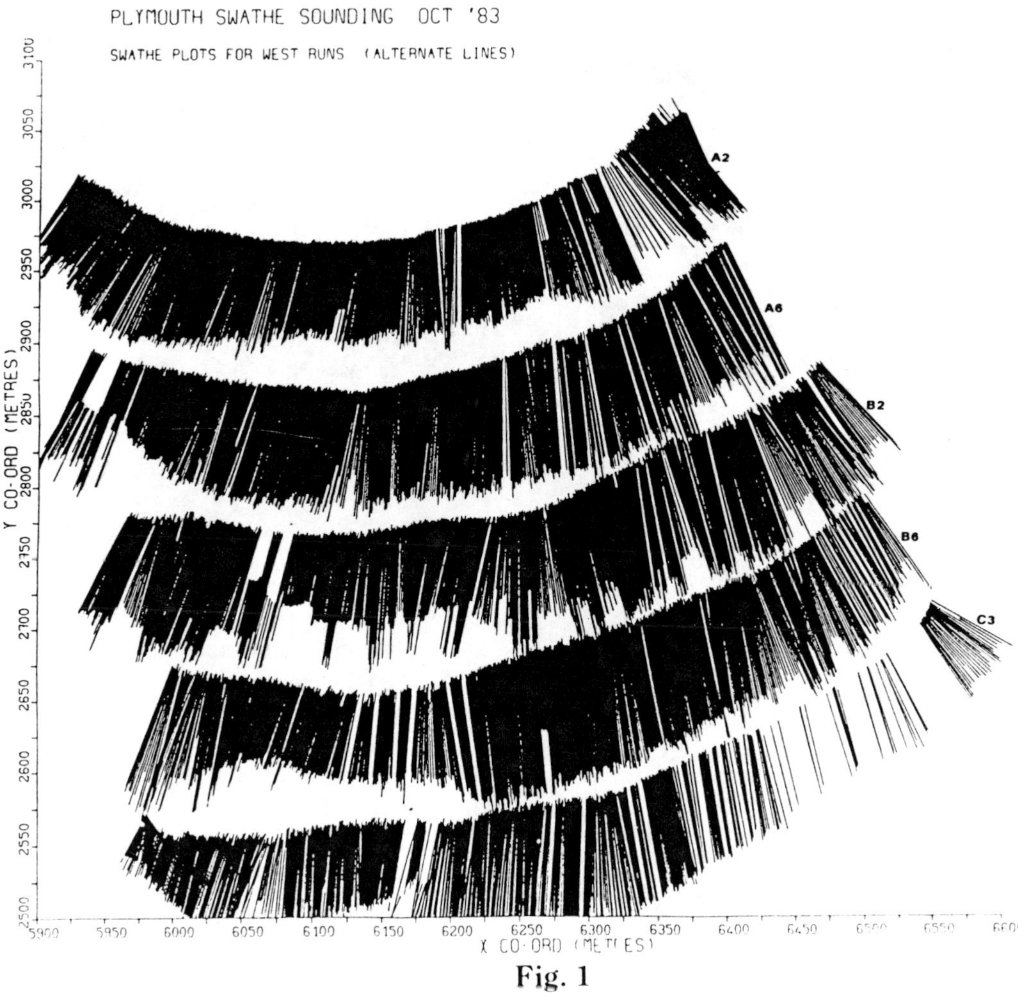

Fig. 1

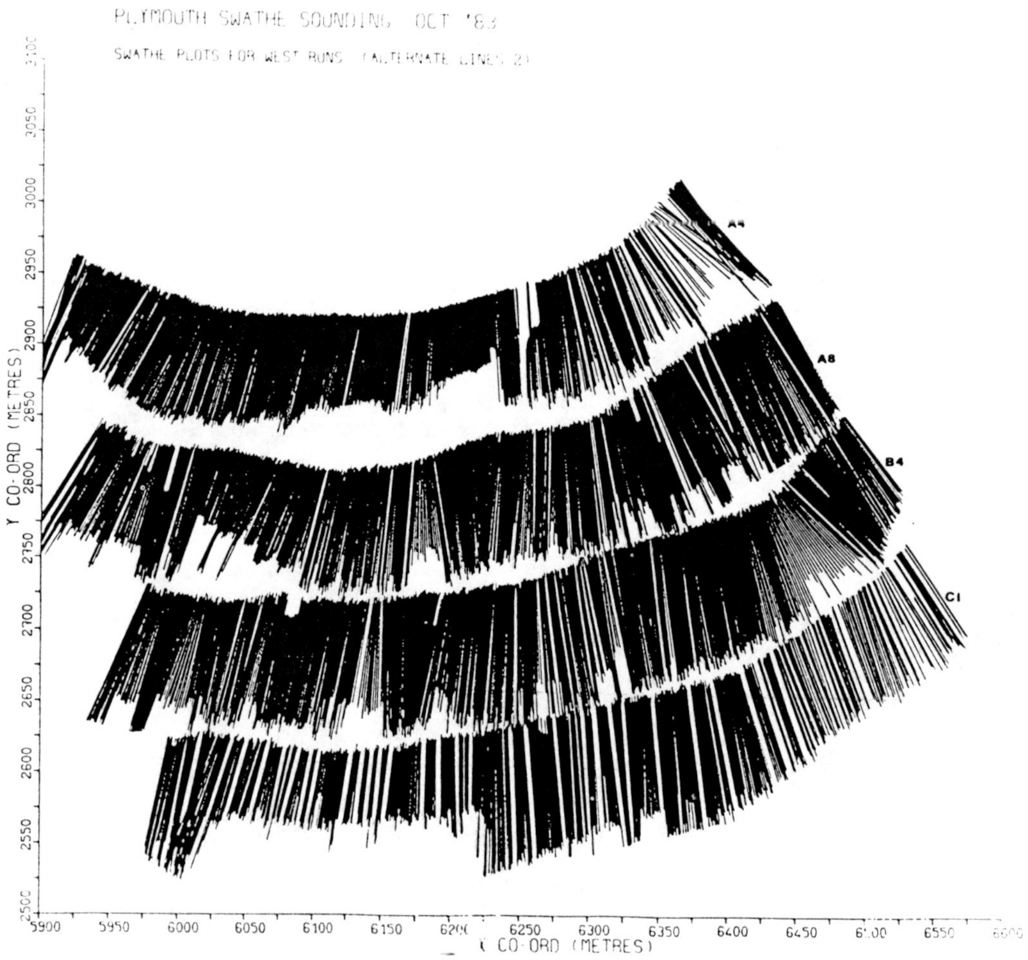

Fig. 2

tude measurements.

The second recorder obtained the ordinary sidescan amplitude signal.

For research purposes, only one side was instrumented with acoustic transducers, and it was seen that this had the effect of producing a permanent list of 3°.

In order to be able to visualize the survey data it was felt that several graphics will normally be needed. Although all of the data was processed off-line, an attempt was made to organize the data in the order which is most helpful; first, while surveying, and secondly, as the equivalent of a completed survey fair sheet. The analogy with echo-sounding surveying is that a track plot is needed on the bridge to establish the adequacy of the coverage achieved, and an analogue echo trace to provide immediate information on any signs of noteworthy features. The fair sheet will then collect the tidally corrected and edited soundings, as well as selected contours.

When swathe sounding, coverage is no longer adequately shown by a track plot. Instead we can now produce a "footprint" plot, which shows the position of each lateral profile along each track. For the sake of illustration, the plot has been divided into odd and even line plots to avoid generating virtually a black sheet of paper (Figs 1 and 2). It will be seen that yawing caused some crossing over and splaying of profiles. Also, in this first survey, there are occasional losses of profiles. These were generally software-induced and not the result of inability to generate a profile from the acoustic data. Where the rate of roll exceeded 2° per ping, it was felt that the profile datum was no longer a straight line, but had become curved. Since roll was only being sampled every 0.2 s that curve would possibly not be reproduced correctly to recover the profile datum. Because of the superabundance of data it was simpler to reject those profiles. Similarly, in Fig. 1, swathes A6, B2 and B6, several profiles are truncated. This is the area of a 1:2 to 1:4 cliff with a large shadow area. Again, in the initial euphoric state of mind, the largest segment of valid profile was retained in each case, and no attempt was made to recover the portion beyond the shadow zone. It will be shown that in a later survey more data could be retained.

In Fig. 2, swathe C1, a large segment of reduced range profiles occurs. Bottom samples were taken in the area by Plymouth Polytechnic, which revealed a very soft peaty silt. The resulting low signal level had caused the correlation value between the two interferometers to be less than the stringent 0.85 correlation demanded in the software. Much of this can be recovered by a slight relaxation of the correlation value.

Figure 3 shows a profiles cascade. This is obtained by offsetting the datum of the next profile, below the previous one, by a small amount. The illustration is of a set of profiles uncorrected for heave. Even though this was less than 0.5 m, the bunching and gaping of the profiles is very noticeable.

Both the footprint plot and the profiles cascade need to be available on-line because they are basic decision information. There is no problem in adding to this an uninterrupted output of sidescan sonar if desired. Indeed, since there are several sidescan-like transducers there is a choice of which to display or record.

It will already be obvious that there is considerable redundancy in the data for purposes of mapping the sea bed. In each profile depths are produced approximately every 0.5 m, and the profiles themselves are about the same distance apart. Assuming we have to deal with natural unconsolidated features it was felt that a 5 m depth matrix would be sufficient to represent faithfully all features whose amplitudes could exceed 0.5 m. Normally, when it is

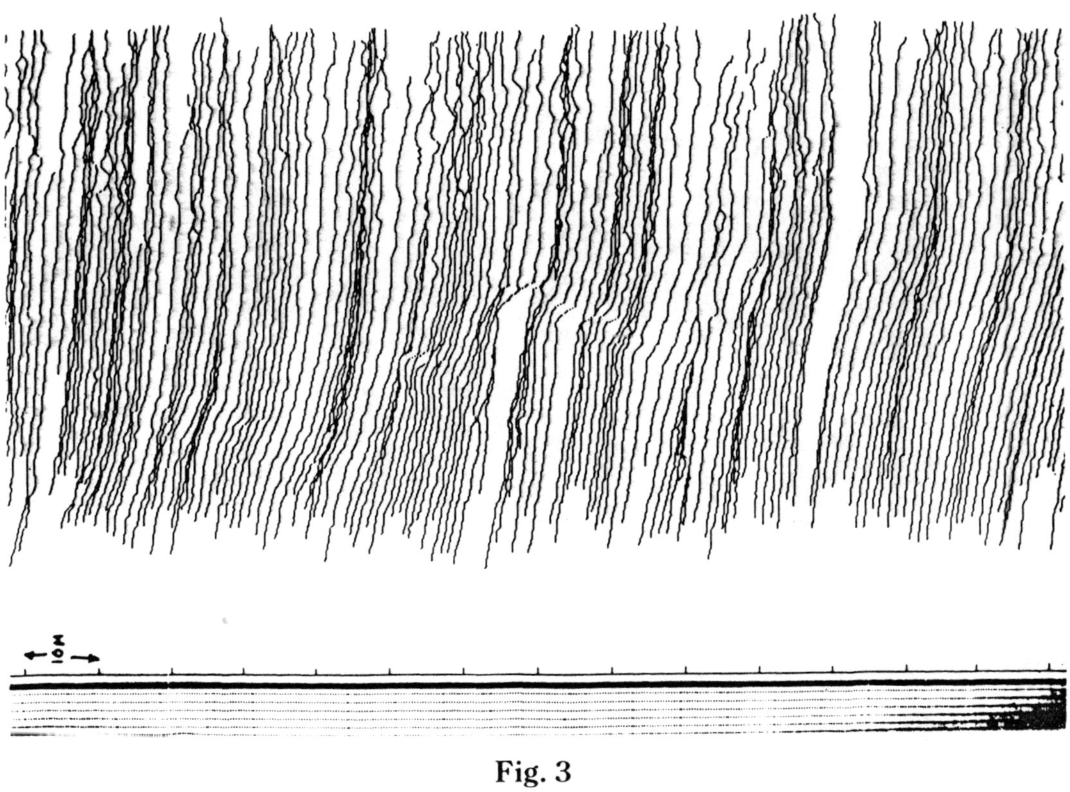

Fig. 3

necessary to produce a grid of seabed data, recourse must be made to one or other ingenious interpolation routine, because the data density is less than the grid density. More precisely, the data from echo-sounding is very unevenly distributed, being 100–1000 times more closely spaced along the track, than from track to track. In the case of swathe sounding the data density is very similar in both directions, and is well in excess of what is required for a 5 m matrix. Therefore there is no interpolation, since there are always a number of depth measurements to justify each matrix depth. Figure 4 was compiled as a density plot. If each 5 m square is considered as a "bin", then the diagram indicates the population density of each bin, i.e. the number of depth values used in the computation of each matrix depth. The count was stopped at 32 because the range of available symbols using a single colour was insufficient to extend beyond it.

Even though we had been rather profligate in the way some profile data was discarded, it was evident that with 50 m line spacing the sounding density was far in excess of what had been achieved by the echo-sounding benchmark survey, which had a line spacing of 5 m.

Data presentation, from echo-sounding data is still principally by means of a "fair sheet", or "smooth sheet". This graphic normally doubles as the authoritative data store, although increasingly the survey data is also retained on magnetic tape. In the case of swathe sounding the graphic is only one of a number of possible presentations and may only contain a selection of the mass of stored data. Indeed, the issue of how much of this data is to be archived is not trivial, and this *embarras de richesse* may well help to concentrate the mind on the cost-benefit of data storage compared with the possibility of having to commission new surveys of an area for different applications. Figure 5 illustrates soundings on a 50 m matrix, whereas the contours are generated from the 5 m matrix

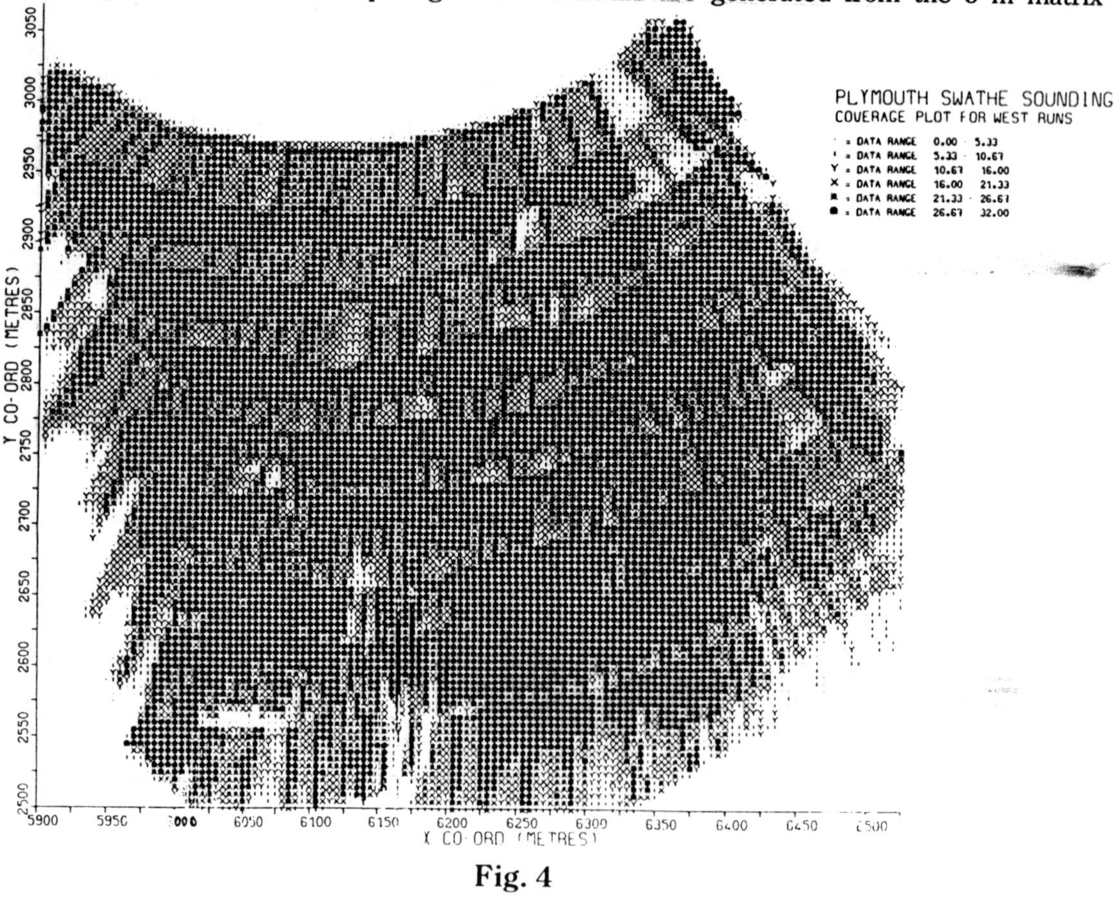

Fig. 4

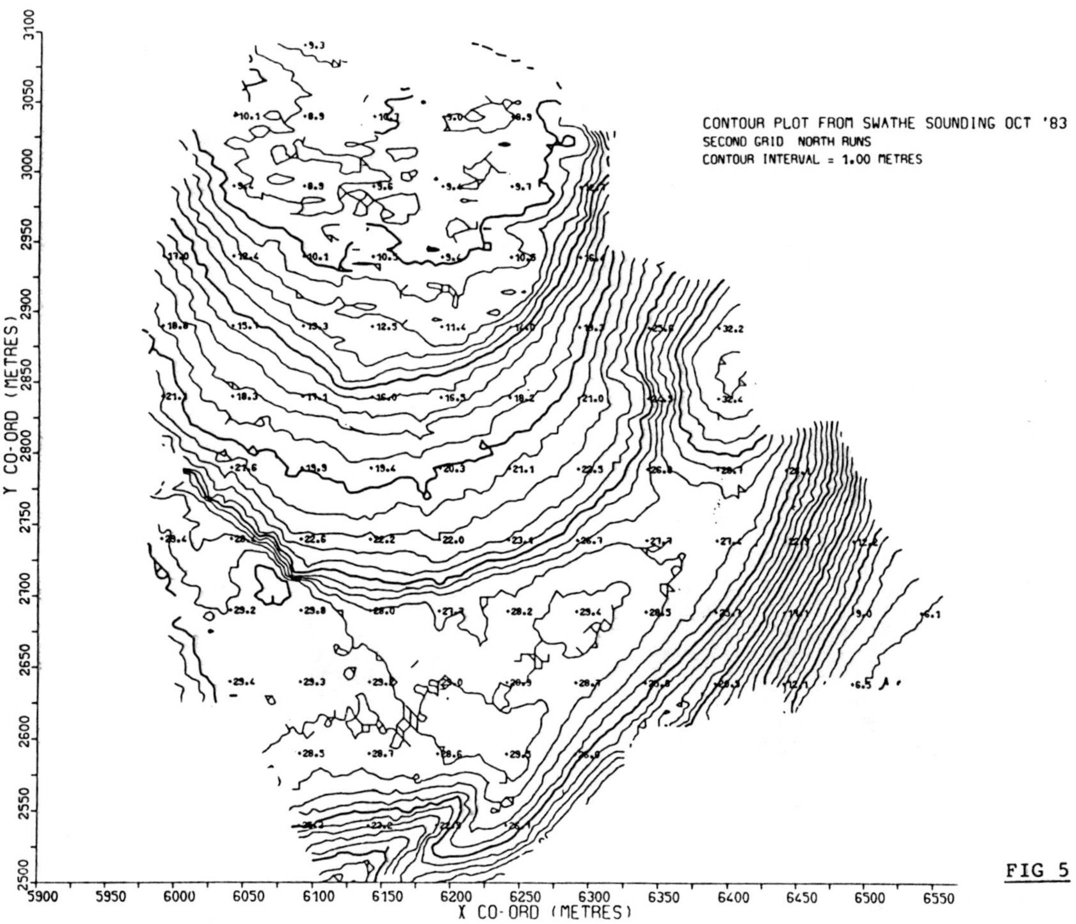

Fig. 5

without any massaging for cosmetic reasons. Alternatively Fig. 6 shows a pseudo 3-D model of the area, looking down-channel. The model can be rotated if desired.

COMPARISON WITH THE ECHO-SOUNDING BENCHMARK SURVEY

When attempting to establish the accuracy of the swathe sounding system, the same difficulty arises as when comparing two echo-sounding surveys, viz. the limitations of positional accuracy.

Depending on which of the two echo-sounding surveys is used as baseline, the mean depth differences vary from zero, in the level sea bed areas, to 15 cm, or 40 cm in the steeper parts. This compares with a mean depth difference between the two echo-sounding (E/S) surveys of 15 cm (Cloet and Duncan, 1984).

More meaningful is the standard deviation between E/S and E/S, and between E/S and swathe sounding. Whereas in the former the s.d. varies from 10 cm to 40 cm with increasing slope, in the latter the s.d. ranges from 25 cm to 45 cm with increasing slope.

It should be appreciated that the surfaces generated from E/S data tend to be smoother because they are based on depth matrices derived from far fewer measurements. Also, depth values obtained by the swathe sounder looking up-slope, are bound to be more accurate than those coming from down-slope profiles. Both are produced from seabed areas which are smaller than those covered by the acoustic beam of an echo-sounder. The latter is well known to produce too shoal a depth value on slopes, irrespective of positional accuracy.

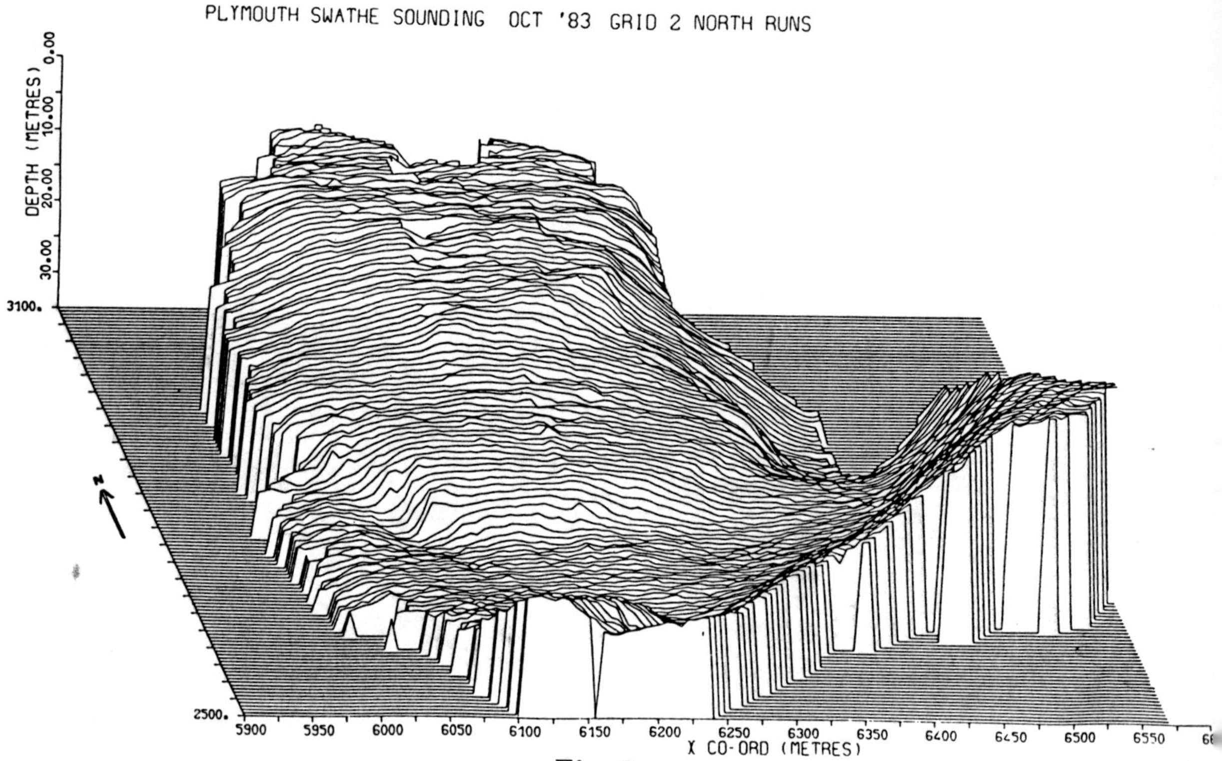

Fig. 6

THE CLYDE APPROACHES SURVEY

Early in December 1984, a demonstration survey was carried out at a site chosen by the Hydrographer of the Navy. It lay in an area, due for re-surveying, just east of the Isle of Arran in the approaches to the Clyde.

The ship was HMS *Hydra,* and the towfish was suspended from the crane on the foredeck. To avoid the risk of damaging the towfish when launching and recovering, and because of the rather inclement weather, the captain chose a shoal off Brodick Bay. The water depth ranged from 30 m to about 80 m, and the fish was towed at 10 m down, although a greater depth would have been preferred.

Positioning system problems, also aggravated by the weather, finally limited the available survey time to one night.

On this occasion also it was decided to run swathe tracks 50 m apart, because processing was still off-line, and the effect of underfish depths of 70 m on the attainable horizontal ranges, normally expected to be of the order of 100 m, was unknown. Figure 6 shows the track plots for the lines run. Normally some of these would have been rejected by the surveyor and re-run, but pressure of time forced their retention here. Figures 7 and 8 illustrate the odd and even footprint tracks which should normally be combined into a single plot. The loss of profiles seen in some swathes is caused by loss of position data. For much of the time only three ranges were obtained, and where these resulted in error circles of more than 4 m long axis, the fixes and accompanying profiles were rejected.

The acceptance or rejection of segments of each profile is carried out in range increments of 2 m. The data were much more noisy because of the weather and low signal level due to excessive underfish depth. In spite of this, usably long profile segments were obtained in depths of 65 m underfish, even when looking down-slope. The holiday in the footprint plot, Fig. 8, between swathes 4 and 6, due to the ship being blown off track, was filled in by a pair of lines, run in the opposite direction.

As a demonstration of the performance of the system, two adjacent swathes, both looking

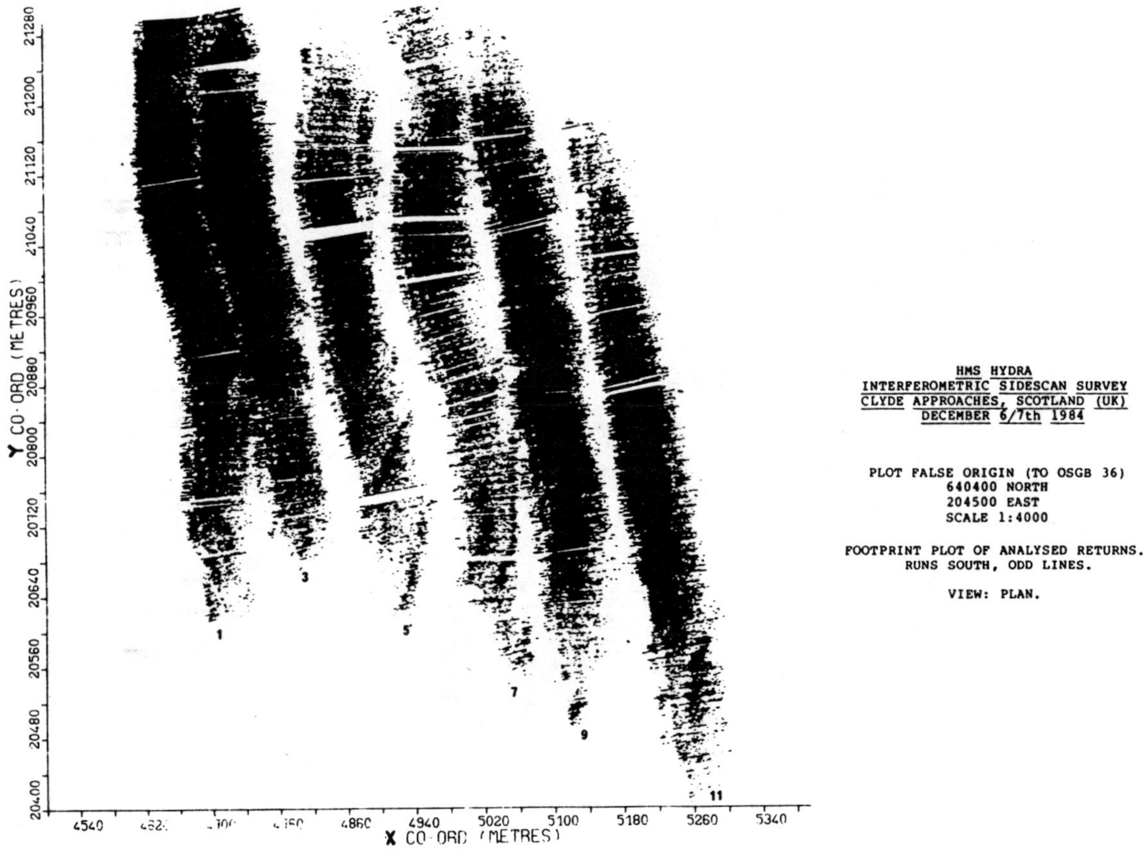

Fig. 7

down-slope, were independently contoured, and plotted together in Fig. 9. The misclosure of the contours, in the overlapping area, is of the order of 25 cm. Since this comprises the sum of errors in fish positioning, gyro readings both of roll and heading, and the azimuthal setting up, the result is considered to be quite satisfactory.

Contours were required at 1 m intervals, and these are shown in Fig. 10, together with a 50 m depth matrix and two least depths.

This survey was subjected to a detailed analysis by the Hydrographic Department, based on the Hydrographer's own survey, made at the same time, but not rendered until after the swathe sounding survey report was presented. The report of the comparative analysis concluded:

> 'Taken in isolation, either survey would be considered satisfactory for the correction of Nautical charts. The interferometric data produces small details between survey lines not discernable by conventional echo sounding. Indeed, the actual minimum depths identified by BATH lie off the survey lines.'

A COMPARISON OF SIDESCAN SONAR AND SWATHE DEPTH SOUNDING

On both occasions, in the course of the Plymouth Harbour survey, and of the Clyde Approaches survey, the amplitude signal received on one of the transducers was recorded. This signal is a normal sidescan sonar operating at 300 kHz.

Because we now possess depth information which corresponds exactly with the sidescan, a close comparison can be made.

In the case of Plymouth Harbour, Figs 11(a) and (b) illustrate two sidescan lines which were run parallel to a 1 : 2 sloping cliff, located in position 6050–6100E, 2700–2750N (see Fig. 5).

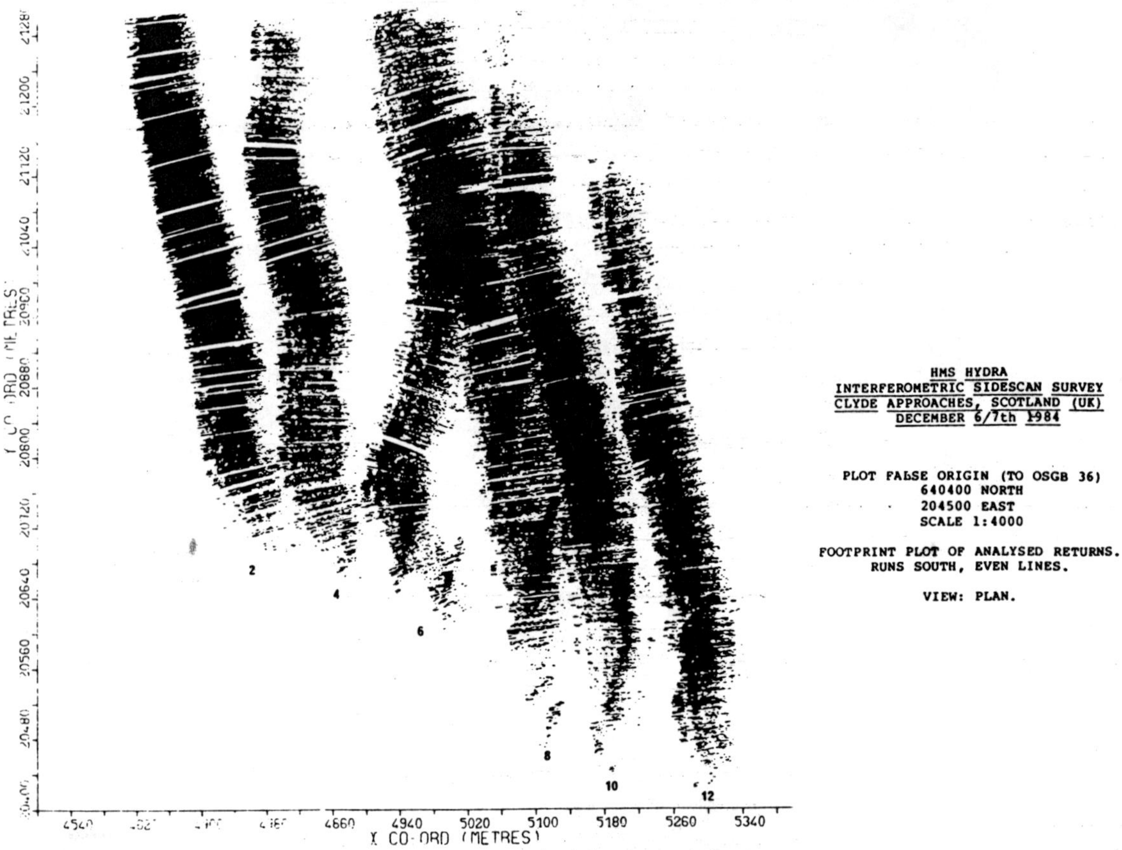
HMS HYDRA
INTERFEROMETRIC SIDESCAN SURVEY
CLYDE APPROACHES, SCOTLAND (UK)
DECEMBER 6/7th 1984
PLOT FALSE ORIGIN (TO OSGB 36)
640400 NORTH
204500 EAST
SCALE 1:4000
FOOTPRINT PLOT OF ANALYSED RETURNS.
RUNS SOUTH, EVEN LINES.
VIEW: PLAN.
2
4
6
8
10
12
4940
5020
5100
5180
5260
5340
X CO-ORD (METRES)

Fig. 8

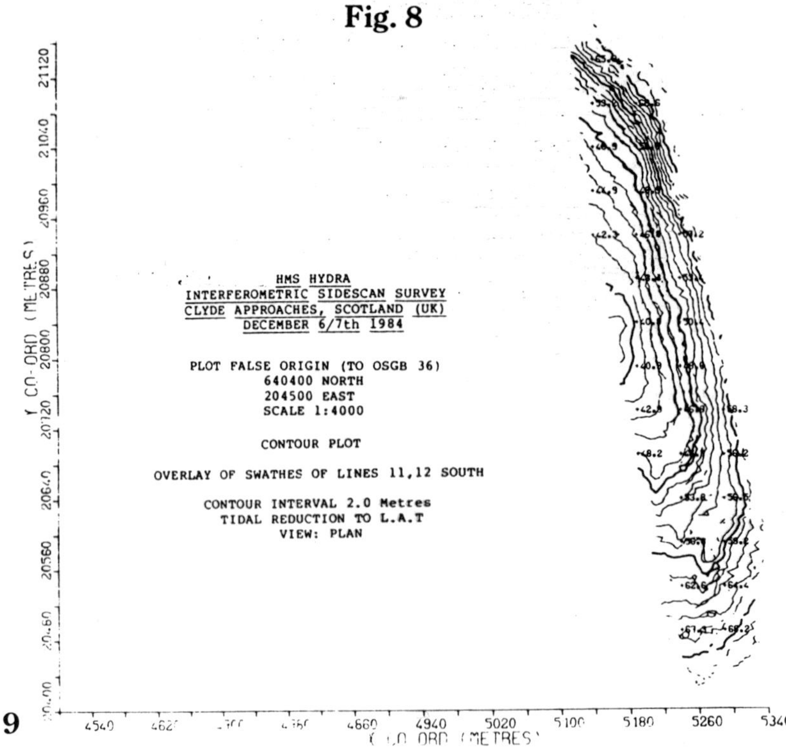
HMS HYDRA
INTERFEROMETRIC SIDESCAN SURVEY
CLYDE APPROACHES, SCOTLAND (UK)
DECEMBER 6/7th 1984
PLOT FALSE ORIGIN (TO OSGB 36)
640400 NORTH
204500 EAST
SCALE 1:4000
CONTOUR PLOT
OVERLAY OF SWATHES OF LINES 11,12 SOUTH
CONTOUR INTERVAL 2.0 Metres
TIDAL REDUCTION TO L.A.T
VIEW: PLAN
21120
20880
20800
20560
4540
4940
5020
5180
5260
5340

Fig. 9

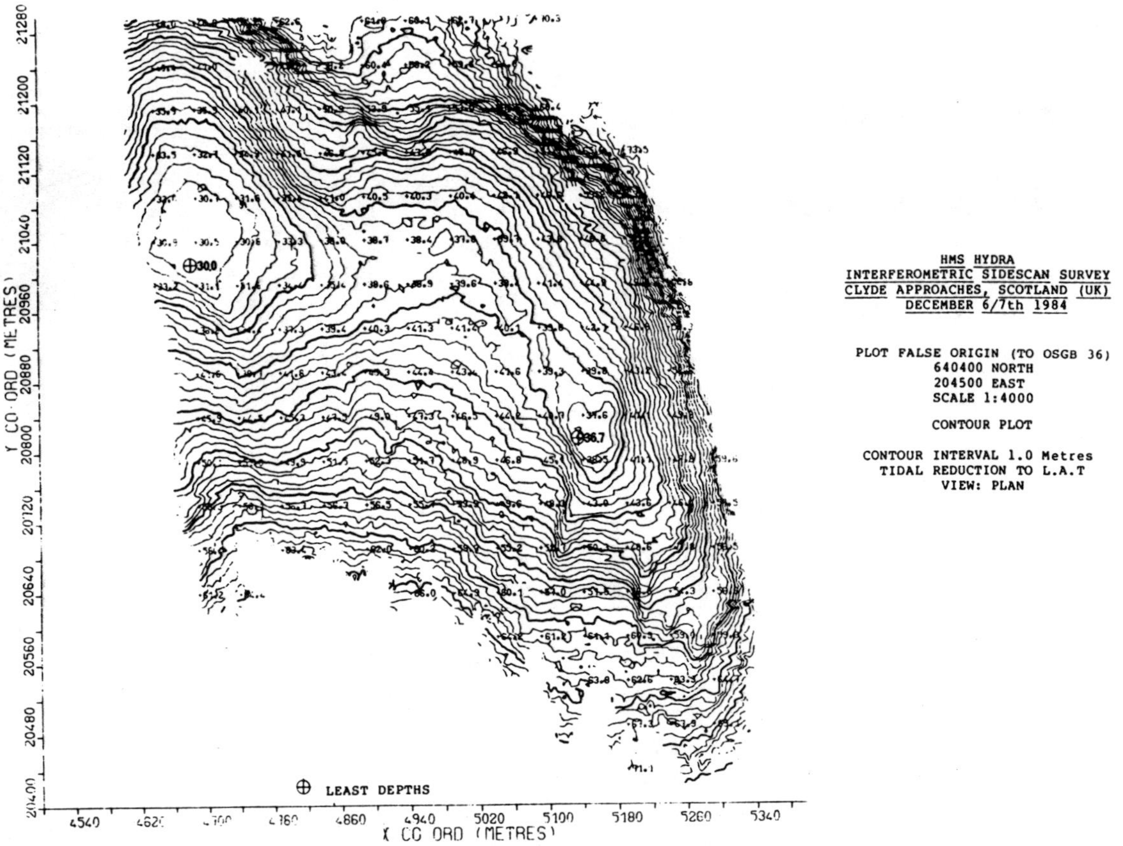
HMS HYDRA
INTERFEROMETRIC SIDESCAN SURVEY
CLYDE APPROACHES, SCOTLAND (UK)
DECEMBER 6/7th 1984
PLOT FALSE ORIGIN (TO OSGB 36)
640400 NORTH
204500 EAST
SCALE 1:4000
CONTOUR PLOT
CONTOUR INTERVAL 1.0 Metres
TIDAL REDUCTION TO L.A.T
VIEW: PLAN
LEAST DEPTHS
Y CO ORD (METRES)
X CO ORD (METRES)
21280
21200
21120
21040
20960
20880
20800
20720
20640
20560
20480
4540
5020
5100
5180
5260
5340

Fig. 10

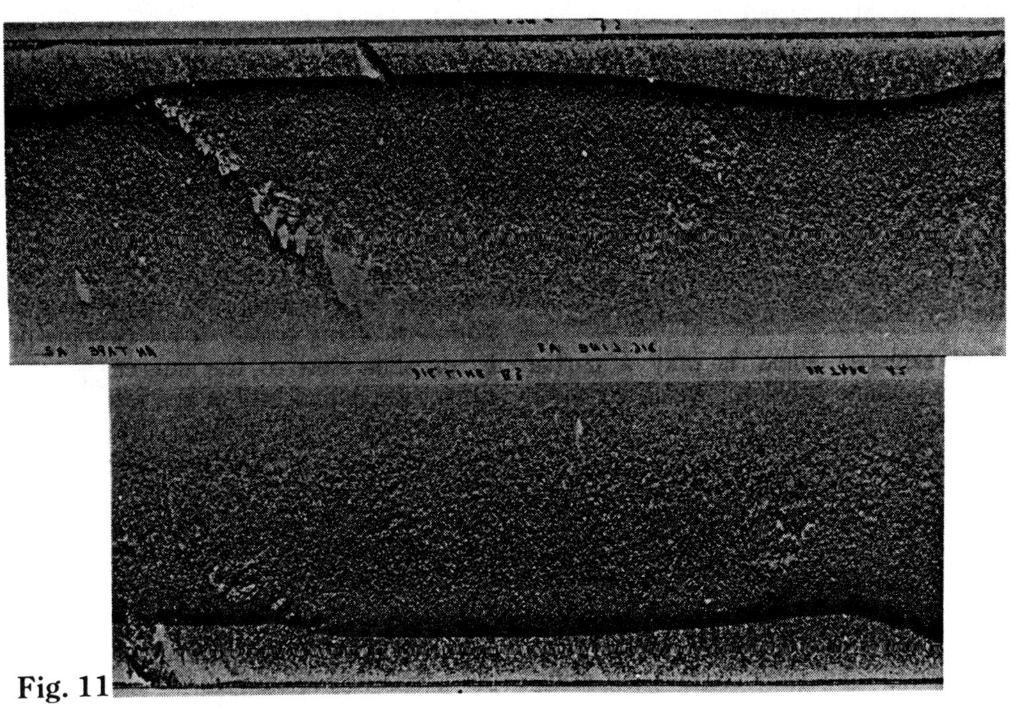

Fig. 11

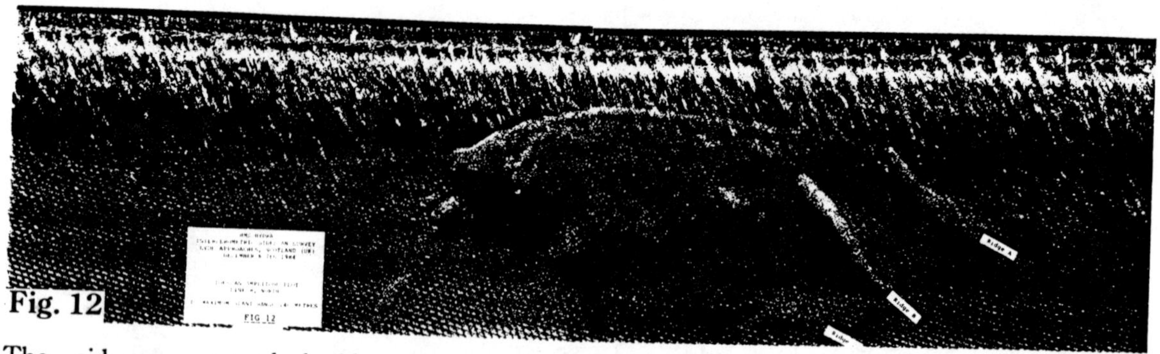

Fig. 12

The sidescan record looking south (Fig. 11(a)) entirely unambiguously indicates the presence of a major feature, but one could be foregiven for not paying undue attention to the same feature in Fig. 11(b), where it is viewed from the south.

The shoal surveyed in the Clyde Approaches showed some startling features on the sidescan (Fig. 12). It would have been difficult for a surveyor to ignore these without sounding over them, if the general depth were anywhere near critical. Figure 13 shows the same features on the 3-D display. Measuring their dimensions, they are found to be about 0.5 m on ridge A, 2.5 m on ridge B and 0.75 m on ridge C.

These two examples help to illustrate the reasons why sidescan sonar images are increasingly looked at with caution. They are:

(a) sidescan can be strongly aspect-dependent, i.e. now you see it, now you don't,

(b) it can virtually ignore large features and exaggerate the importance of relatively minor ones and

(c) because it is not always easy to discriminate between the meaningful and the trivial, it can cause survey time to be expended in precautionary, if unnecessary, investigations.

In short sidescan sonar is only a partial solution to a genuine problem, which can now be solved by swathe sounding.

THE IMPLICATIONS OF SWATHE SOUNDING

A first reaction could well be that an excessive amount of data is produced. However its ready availability should dispose of the fear surveyors cannot altogether get away from when using existing instrumentation; that some hazard of significance has remained undetected. Even in well surveyed areas old wreckage can take years to be discovered. We are still uncertain whether certain features do or do not exist on our major shipping lanes. Available records of depths in areas suspected to be mobile do not contain sufficient data to establish whether they are mobile, and if so how large that mobility may be.

The high density of data generated by a swathe sounder, from a ship travelling at 5 knots, is equivalent, if seen in line miles of data, to that of ship surveying at 1000 knots, and having the ability to steer tracks only 0.5 m apart. Evidently merely for the sake of legibility, it would be impractical to produce a graphic archival data store, such as is the present survey fair sheet, even though it would be easy enough to select data down to that density. A distinct advantage of computer-readable data storage is its easy and cheap accessibility for post processing and analysis, unlike the costly and error-prone method of digitizing graphic media. The advent of the electronic chart will necessitate the production of computer-readable data stores, and render obsolete the notion of "chart scale". It will become practical to present data to the degree necessary to reflect the complexity of bed features, or other navigational detail, without the need to compromise because of the inflexible dimensions of the paper chart.

The swathe sounder is capable of producing all the seabed data needed by the widest range of potential customers. Because of its ready accessibility it will be more fully used, and the cost-benefit of hydrographic surveys may well be significantly enhanced.

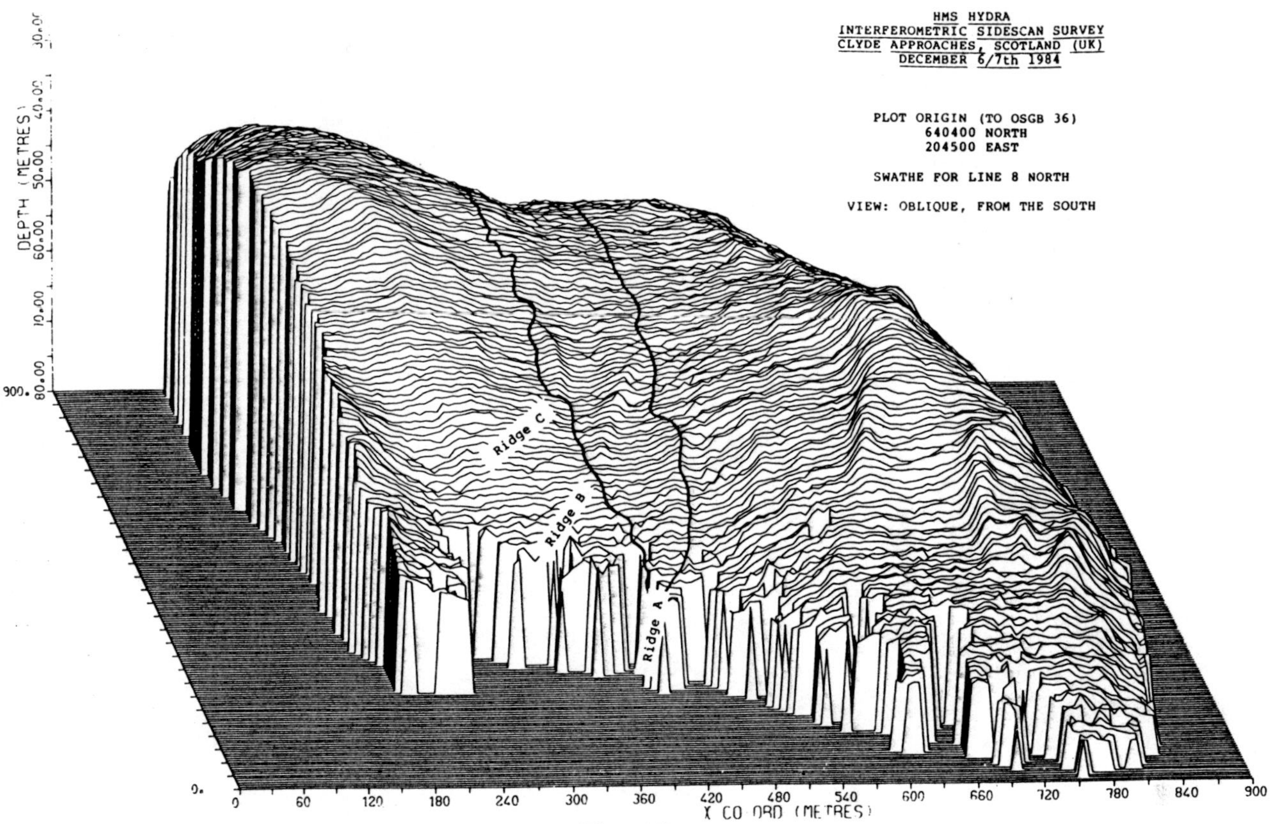

Fig. 13

REFERENCES

Cloet, R. L. and Duncan, A. J. 1984. The Plymouth echo sounding bench mark survey. *Hydrol. J.* **34**, 1–4 (October).

Cloet, R. L., Hurst, S. L., Edwards, C. R., Phillips, P. S. and Duncan, A. J. 1982. A sideways-looking towed depth-measuring system. *J. R.* Inst. Navig. 411–420 (September).

Cloet, R. L., Edwards, C. R. and Duncan, A. J. 1985. The University of Bath swathe sounding system. *Proc. Underwater Ac. Conf.*, Portland, UK, 117–125.

Part XI

Hydrography — Dredging/Mud/General

45

High-accuracy Dredging Surveys at the Port of Hamburg

H. F. Wentzell, Krupp Atlas Elektronik GmbH, Bremen, West Germany

INTRODUCTION

The Hamburg Port Authority is among the first to have adopted a digital hydrographic survey sweeping configuration incorporating a position-fixing capability designed to meet required continuous positional accuracies of ±2 m.

With the exception of a shore-based positional tracking station, all equipment is installed on the Authority's own survey vessel, *Deepenschriewer 11.* The equipment comprises a bottom-mapping system, a precision survey echosounder, gyrocompass, a laser range-azimuth position-fixing unit and, finally, a fully automated navigation and survey processing configuration for both pre- and post-survey operations.

Factors governing overall survey requirements and performance are discussed, and the various systems used are described.

HYDROGRAPHIC ASPECTS AND APPLICATIONS

When discussing survey requirements, it is, of course, the morphology of the bottom which should be considered first. Of equal importance, though, is the prescribed manner of positional determination, which is primarily dependent upon topographical considerations, as well as the economic factors affecting the rationalization of survey operations.

Morphology

Example fair sheets (Fig. 1 (a) and (b)) show an area where the River Elbe divides into a northern and southern arm. In Fig. 1, (a) shows depth contour lines generated from cross-profile data while (b) illustrates the same area, but from a survey carried out with an Atlas Boma 20 sweeping system. Comparison of (a) and (b) shows clearly that assessment of morphology dictates whether or not cross-profile surveys can usefully determine a bottom structure.

A further typical practical example would be the River Rhine, near Bingen, where deepening or correction of the fairway can only be carried out with very precise blasting. Without the use of a sweeping system, it would be extremely costly and time-consuming to ascertain if and where boulders or shelves remained after

HIGH ACCURACY DREDGING SURVEYS AT THE PORT OF HAMBURG

H F Wentzell, Krupp Atlas Elektronik GmbH,
Federal Republic of Germany

Presented on **Friday 7th March 1986 at 10.00**
in the Clarence Conference Room

EMBARGOED AGAINST DELIVERY

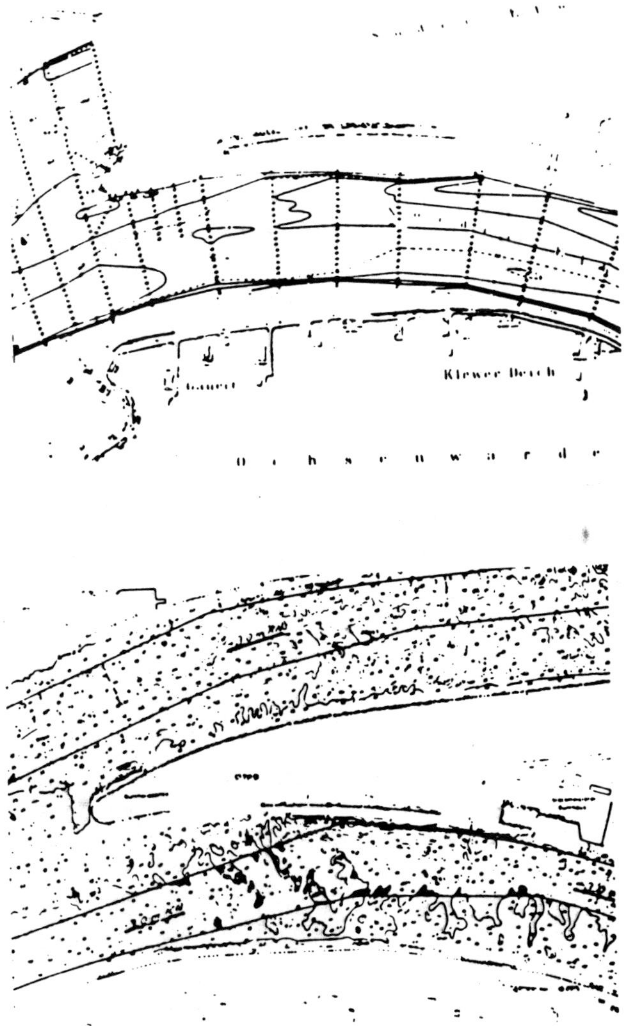

Fig. 1 Comparison between cross-profile method (a) and sweeping survey method (b).

blasting operations.

The morphological conditions of a sea bed or river bed, together with the current velocity of the water, may influence the design concept of a survey vessel. That is, whether the required ultrasonic transducer can be installed in booms for the vessel to move in the direction of ship's axis along the river, or whether transducers should be installed alongside the hull or in the keel in order for the vessel to move perpendicularly to its axis and across the river.

Economic Factors

Given an average vessel speed of 4 m/s for longitudinal surveying and 2.5 m/s for cross-profile surveying, it would typically be assumed that a river area comprising a length of 25 km and a width of 100 m would have to be surveyed. The question of which of the following two survey methods would be the less time-consuming often arises.

(a) A cross-profile survey with a parallel track-distances of 30 m.
(b) A gapless sweeping survey with longitudinal tracks and 50 m sweeping width.

The answer is self-evident. A survey vessel equipped with a sweeping system is able to carry out the same survey operation in a considerably shorter time than would be possible by cross-profile methods; thus accurate, sweeping systems are gapless. In specific regions where, because of morphological conditions, the spacing between tracks must be closer, or where additional longitudinal profiles have to be run, the bias is even more in favour of the sweeping system.

Further, it is a truism that dredging work is expensive. Inaccurate surveys in areas such as the Port of Hamburg can add significantly to overall dredging costs if only because of the relatively large safety margins which have to be applied. Thus an accurate and gapless survey sweeping system can soon pay for itself.

Position Determination

A sweeping system can basically fulfil a combination of two differing tasks. Firstly, safety surveillance: a survey vessel may typically run extended distances along a given survey area, indicating instantaneously all hazardous depth areas for traffic warning or dredge guidance management. Secondly, hydrographic surveying: highly accurate and gapless information is obtained, together with details of bottom morphology, in an operation comparable only to aerial photogrammetry.

For the first application, safety surveillance, only moderate positional accuracy is required. It need ony be adequate enough to identify shoal areas, since the primary concern is not one of accuracy but, rather, one of safety where it is important not to overlook any hazardous areas within the traffic channel.

The second concern is the efficiency of the vessel itself in terms of covering a long distance in a short time. Positional determination by optical means or via radio systems, such as Mini-Ranger or Trisponder, are in many cases not practicable, since these will demand either people or reference stations onshore with all the associated logistical costs that they imply. A further problem with these systems is that their area of actual survey coverage is very much dependent upon channel width, since this dictates the length of the baseline used.

In built-up areas in harbours and ports, where much dredging work is carried out, problems with radio-positioning systems may be further accentuated by unwanted signal reflections from nearby buildings or structures. The Port of Hamburg Authority, for example, had made extensive tests and acquired considerable experience of various radio-positioning configurations, such as Radiolog, Doppler Navigator, MRD 1 and Mini-Ranger. But with none of these could a required positional accuracy of ±2 m be realized in excess of 50% of the survey area. In part this led to the development by Krupp Atlas Elektronik of an entirely new concept of dynamic position fixing, a combined range-azimuth laser-based unit using only a single shore-based automatic tracking station, Polarfix. As its name implies, by combining range and azimuth measurements, the system essentially provides polar coordinates for position fixing.

Fig. 2 Sweeping survey vessel with booms

Layout of Survey Vessels

Broadly, two types of survey vessel are usually deployed for bottom-sweeping system applications. The first (Fig. 2) is designed to carry out longitudinal surveys only. It is equipped with booms mounted on both sides which can be swung out and in quickly by means of hydraulic support drives. This is the type of vessel now in operation at the Port of Hamburg.

The shape of the booms is hydrodynamically neutral, so that virtually no vertical momentum is generated at any speed of the vessel. Residual forces due to fabrication tolerances are compensated for by means of small fins at the tips of the booms. This design concept offers distinct advantages in that not only are booms horizontally stabilized in the water but they also stabilize the vessel in its roll axis.

Ultrasonic transducers are installed in the booms and between them in the vessel hull. Each boom can be as much as 20 m in length, so that the total sweeping width can be as much as 50 m, depending on the overall beam length of

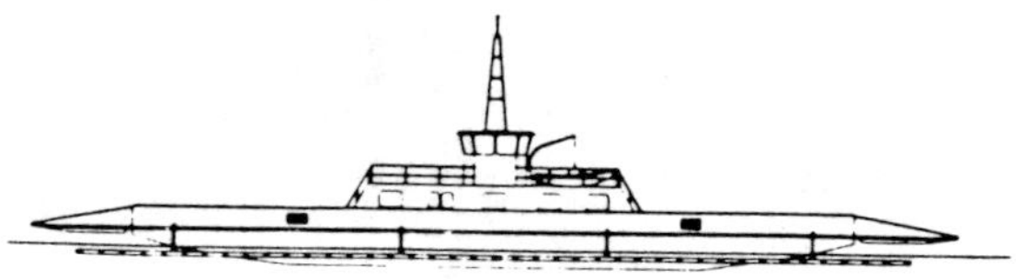

Fig. 3 Sweeping survey vessel with transducers installed along the side

the vessel itself. With booms extended, the speed of such vessels can be as high as 10 kn, with vessels typically having a length of around 30 m.

The second type of sweeping survey vessel (Figs 3 and 4) is designed to carry out surveys either along or across the longitudinal axis of a survey area. The axis may either be in the keel or housed in a special support bar along-side the vessel. Vessels of this type are used on the Rhine and in the Albert Canal in Belgium for detailed hydrographic surveying. Although admittedly their sailing speed is relatively low, it should be borne in mind that their initial conversion into sweeping craft was realized at a fraction of the cost of constructing new purpose-built vessels.

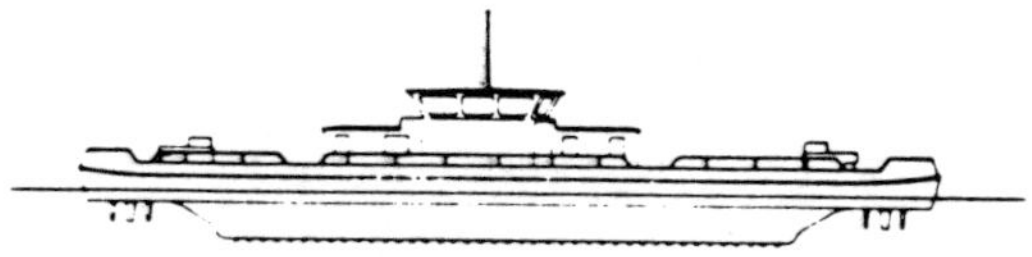

Fig. 4 Sweeping survey vessel with transducers installed in the keel

Principle and Data Processing of the Boma 20 Survey Sweep System

Here we look at the various constituents of a total survey sweep system as used at the Port of Hamburg, beginning with a basic Atlas Boma 20 bottom mapper and then defining the most sophisticated configuration.

Echosounding equipment (Fig. 5)

Discounting comprehensive technical details, the equipment consists of a number of ultrasonic transducers connected to the transmitter/receiver. A central computer controls the

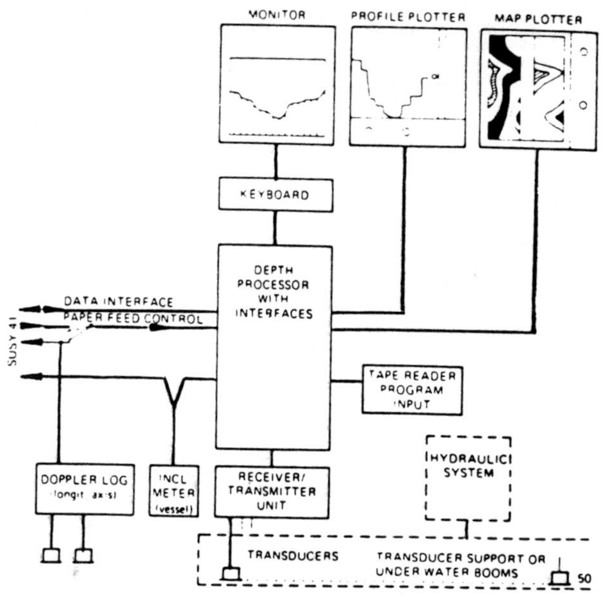

Fig. 5 Boom 20 Bottom Mapping System, standard configuration

transmission and receives the depth information approximately twice per second over the entire sweeping width. Three outputs are provided, each displaying data metrically, viz.

(*a*) *Cross-section display (Fig. 6)*. This is the main interactive medium for the operator. The upper two lines are used for entry of system parameters which can be modified by simple cursor address and without interrupting measurements. The lower part shows the instant cross-section, updated in 1 s intervals and showing:

the shoalest depth (asterisk, 101 cm)
reference depth (= = =, 1 m)
deepest depth (396 cm)
unreliable soundings (inverted column)

(*b*) *Cross-section plot (Fig. 7)*. This can be generated either automatically in fixed-time increments, or only upon initiation by the operator. Output is basically a plot of the cross-profile visible at the CRT at specific time of event, but it may also be plotted in different scales of both axes. This type of plot is particularly useful for subsequent evaluation of the third type of output, the bottom map.

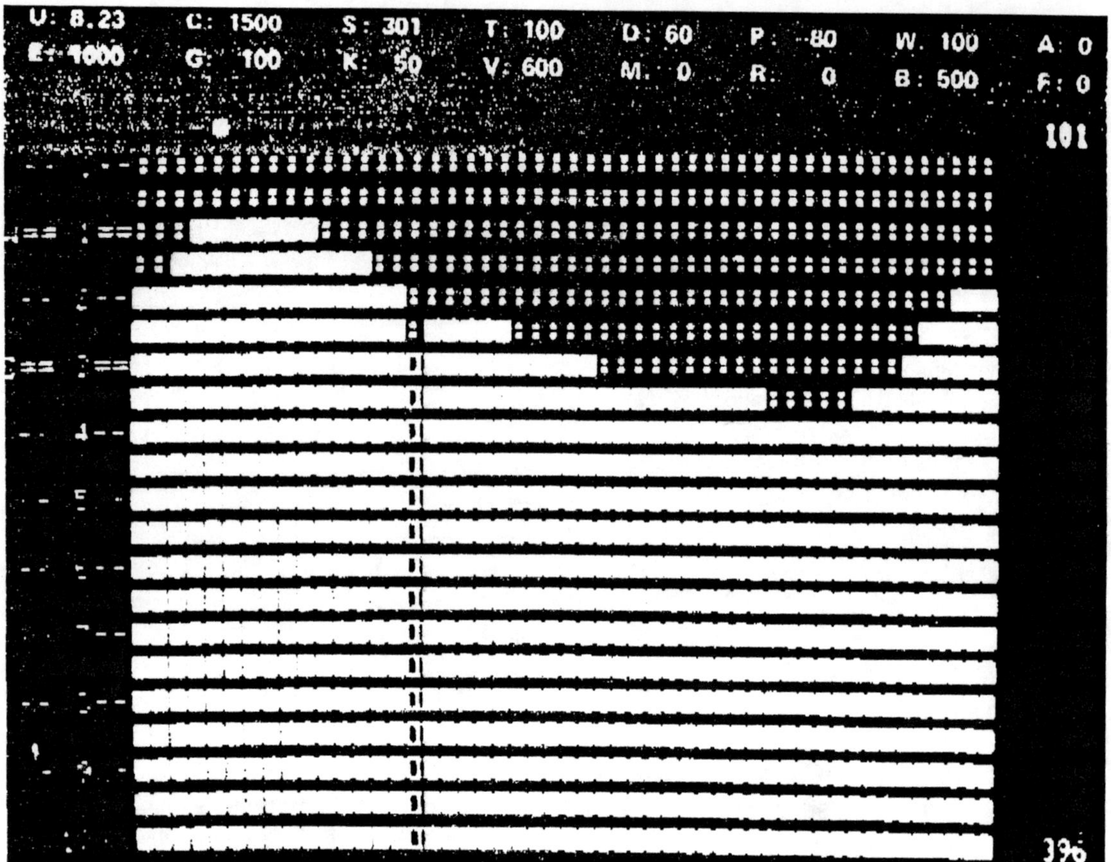

Fig. 6 CRT monitor: cross-section display

Fig. 7 Cross-section plot

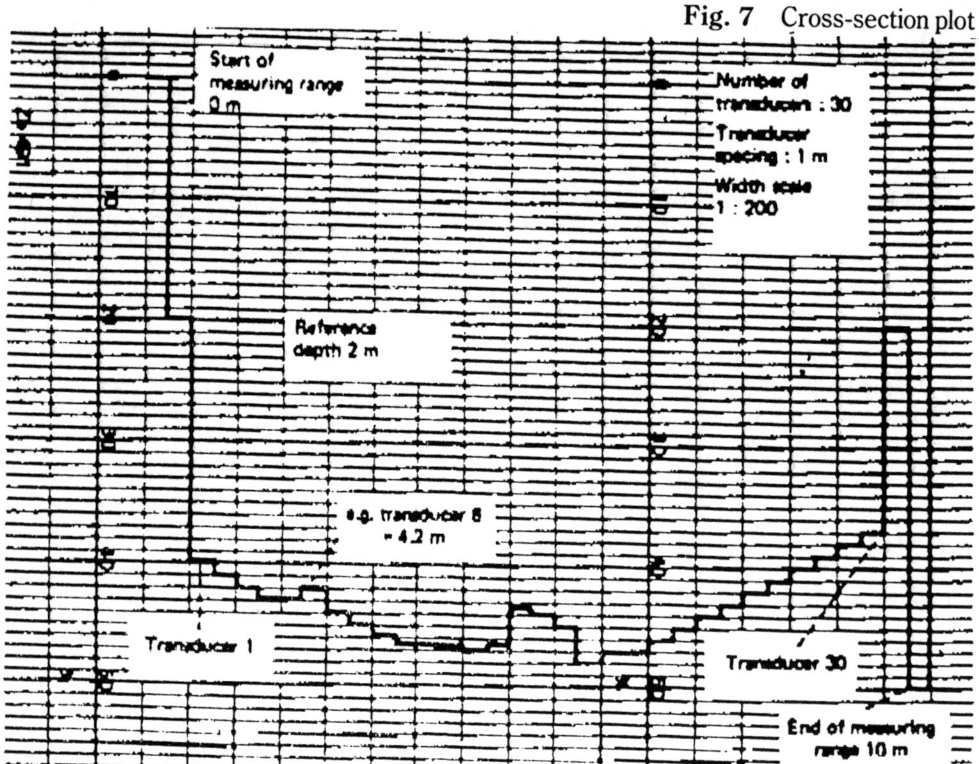

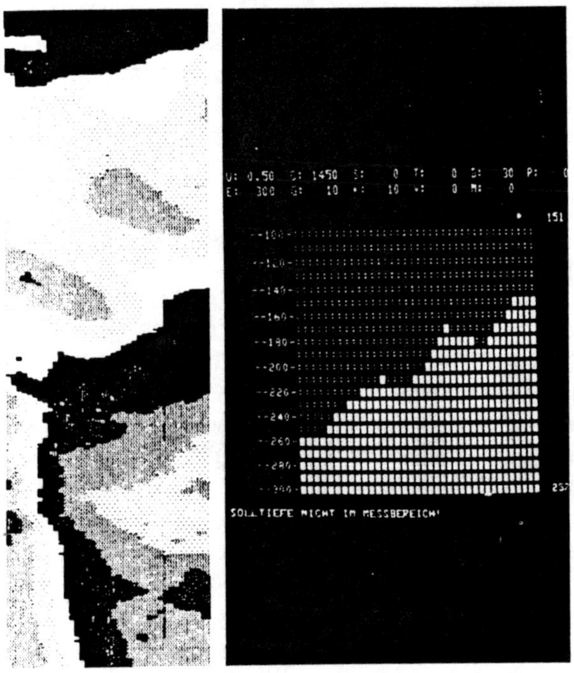

Fig. 8 Bottom morphology map

(c) *Bottom map (Fig. 8).* This is the all important sector. In effect, it presents a bird's eye view of depths using a special dot matrix coding for depth values, where the boundary lines between the dot codes represent the selected isobathic contours. Presentation is twofold: the left record shows all depths deeper, and the right records all those shallower than a selectable reference depth, so that any shoal can be identified quickly at a glance. Depth resolution can be as precise as 5 cm, although for ease of interpretation most users prefer a resolution of 10 cm on the shallow depth record (right side) and of 50 cm on the coarse depth record (left side). If no other sensor input is available, the map is driven with constant speed. Survey parameters are shown numerically at the beginning of a record.

Position determination for safety surveillance (Fig. 9)

With the addition of a sonar doppler speed log, such as the Atlas Dolog 11, the map can be driven true to the ship's speed over ground with an appropriate scale factor in order to secure a true-to-scale record both along and across. Manual events can be generated to update the 'zero distance' by using landmarks as a reference. The distance travelled since the last zero point is numerically plotted beneath the depth record.

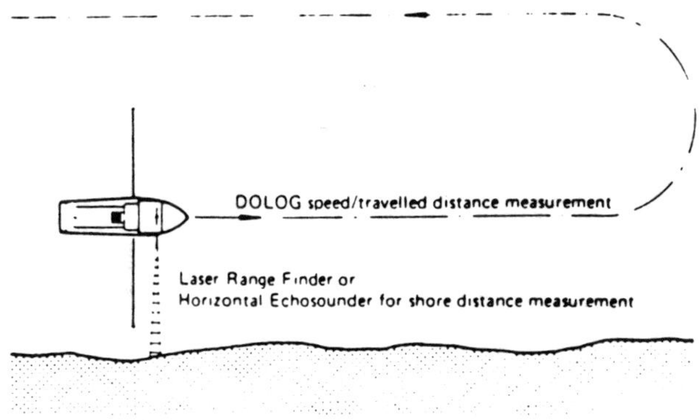

Fig. 9 Course position determination without active reference stations

For safety surveillance, this type of configuration is frequently used in Germany, with most users employing laser range finders, such as the Atlas Lara 90, to obtain information concerning the parallel distance of the survey track with reference to the shore. Some vessels are equipped with a horizontal echosounder which indicates the distance to the shore to the vessel helmsman as an aid to maintaining track. But it should be noted that this method is only truly satisfactory in canals with defined and relatively steep slopes.

These methods of measuring speed and securing distance along track using a doppler log, while additionally measuring distance to the shore by laser or horizontal echosounding, have been shown to be more than adequate for safety surveillance and dredge-guidance management. For the latter task also, an additional software program can be implemented. This calculates the volume of the bottom above the reference depth, formed from single depth values appearing on the shoal depth record, and then multiplied by the horizontal transducer spacing and associated distance increments. The result is numerically plotted at the end of a record.

With these positional methods, the distance accuracy obtained from an Atlas Dolog 11 sonar doppler log is typically of the order of ±1.5 m ±0.5% of distance travelled. Under favourable acoustical conditions, however, the precision can be much better. Accuracy of distance to shore using either method, laser or horizontal echosounder, can be assumed to be within the order of 1-3 m, depending on the cycle of measurements.

Range-azimuth position fixing

Ideally, positional accuracy in rivers and canals for hydrographic purposes should be better than 3 m. For detailed analysis of morphological problems, however, accuracies of better than 1 m may well be required. The system adopted at Hamburg is the Atlas Polarfix, a microprocessor-controlled configuration which uses a laser beam to measure dynamically both range and bearing from a fixed tracking station, usually shore based, to the mast of a vessel equipped with a standard corner-cube prism reflector assembly for return of signal. The pencil-sharp laser beam locks onto the reflector assembly and, by following its movements, range and bearing can be determined. The beam itself continuously tracks the target vessel automatically.

Range and bearing data are likewise processed automatically by the fixed station. But since both are required for use on board the survey vessel, this is facilitated by a supplementary telemetry link between the station and its target. Information is therefore transmitted back to the on-board telemetry survey processing unit for translation into actual position. Thus the data link between the fixed station and target vessel carries all the required operating functions, enabling full control of the system to be exercised on board the vessel itself. Additionally, data protocol is such that a series of self-check and test messages is incorporated in order to maintain quality and reliability of information at all stages.

The telemetric link also permits the system to be set up and calibrated prior to operation. Normally, calibration is carried out using a fixed target in a known position, so that by directing the laser beam at the target, range and bearing can be immediately established. Automatically calibrating the range scale, at the same time the bearing is used as a reference point from which to establish in turn the corresponding relative bearings of the vessel.

For Hamburg, the system also incorporates automatic telemetry for the determination of tidal variations. With a pressure tide gauge directly connected to the fixed station, tide variations during survey operations are relayed to the vessel at regular intervals for data logging or on-line correction of depth measurements. The gauge has been proven to be accurate to within approximately 2 cm over working periods of 6 hours.

As to maximum distance, laser operating range may generally be regarded as being 1.5 times prevailing visibility, and normally this is more than 5 km. Over this sort of range, experience has shown that the system has an accuracy of ±1 m, a tolerance well within the Port of Hamburg's requirements. Conversely, at a range of 1 km, which would normally equate with most port and river dredging operations, accuracy is of the order of ±0.2 m.

As experienced at Hamburg and elsewhere, the operational cost advantages of the system can be considerable. Where, for example, conventional microwave position-fixing methods typically require three remote stations to be deployed at three or more different sites for coverage of a river or channel area, the range-azimuth principle of Polarfix permits coverage of at least 6 km of waterway with significantly improved levels of accuracy, merely by siting the single tracking station on a shoreline.

Computer-Controlled Off-Line Data Processing

Prior to survey operations, pre-plots, and coordinate entries of reference stations and other parameters are carried out as with any automated hydrographic survey system. Depth data from the Boma 20, that is, 50 depth values approximately twice per second, are stored on magnetic tape cartridges together with positional data and with correlation data such as vessel movement (roll, heave, water variations).

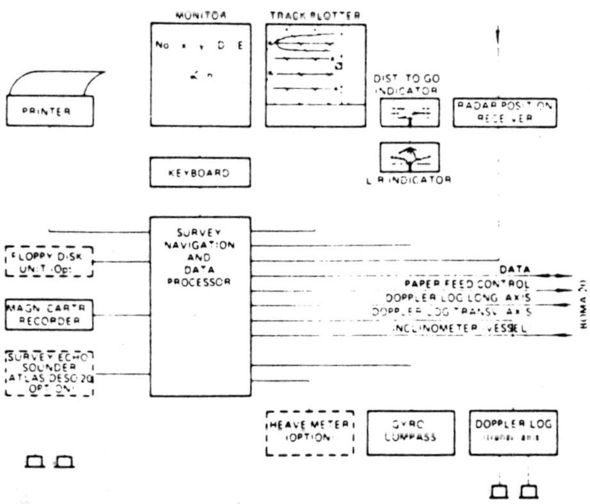

Fig. 10 SUSY 41/Boma 20 survey navigation and processing system

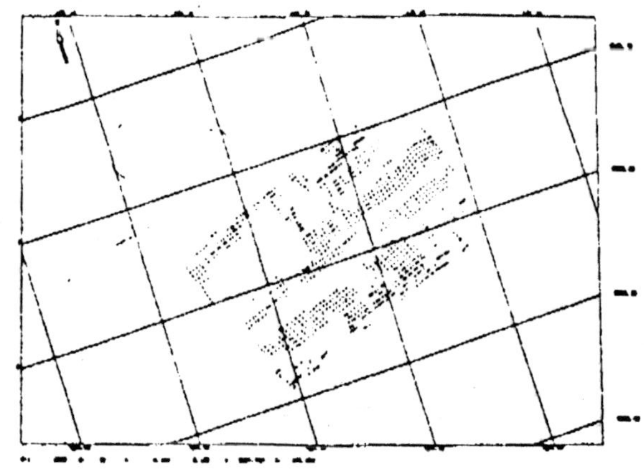

Fig. 11 Numeric area depth point plot

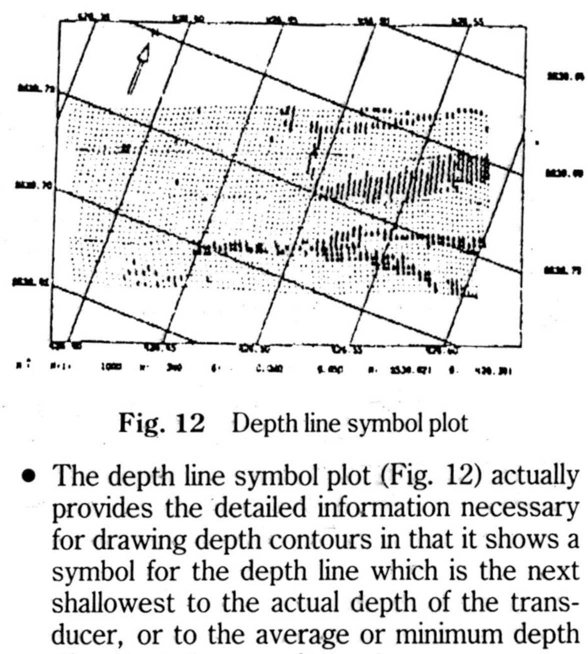

Fig. 12 Depth line symbol plot

In the present configuration at the Port of Hamburg (Fig. 10), only the centre depth of the Boma 20 is recorded on data carrier and numerically plotted onto the plotter. The Authority uses a large-size plotter for fair sheet, with the original vessel track plotted. By means of a special projection table, the bottom map is projected onto the fair sheet with depth contours being manually plotted. The plotted centre depth values are used for reference between the two images.

This manual isolining procedure, however, has proved to be rather time-consuming for areas with difficult currents, that is, when the track of the vessel is not straight. Therefore, in order to improve overall system economy, a shore-processing system is to be installed so that all 50 depth values will be stored on magnetic tape.

Special software programs are necessary to evaluate and output the large volume of depth information in a format suitable for user information. These comprise:

- The numeric area depth point plot (Fig. 11), presenting depth values of each individual transducer or the average or minimum of a group of transducers, depending on the scale of the map. When calculating the position for depth information, the actual gyro course and the position offset of the antenna position are also incorporated in the formula.
- The depth line symbol plot (Fig. 12) actually provides the detailed information necessary for drawing depth contours in that it shows a symbol for the depth line which is the next shallowest to the actual depth of the transducer, or to the average or minimum depth of a selected group of transducers.
- The area-to-profile conversion program provides the means to 'cut' a cross-section out of the real depth data field. The direction of the cross-section can be at any angle with respect to the direction of the actual track of the survey vessel. The program allows various data files to be searched for depth data fulfilling criteria, so that a complete cross-section can be reconstructed from depth data fields of various adjacent survey tracks.
- A volume calculation program provides positive and negative volume calculations within

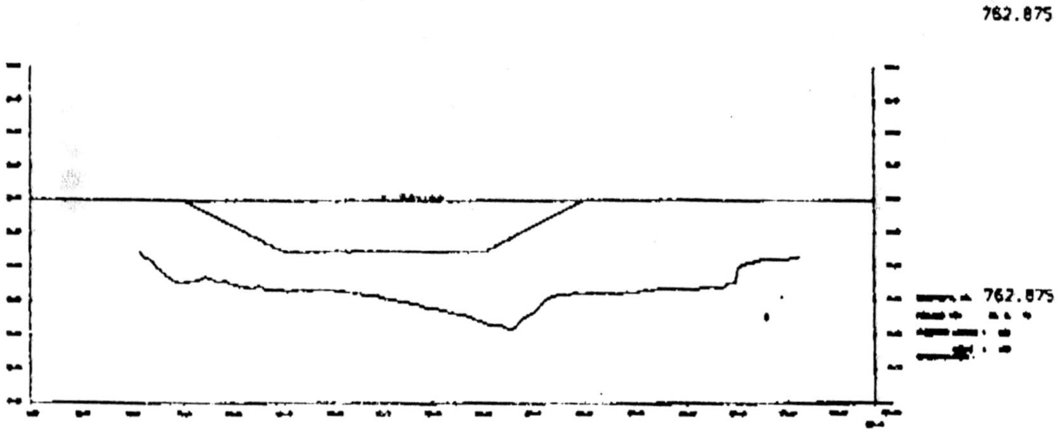

Fig. 13 Graphic cross-section plot

an area defined by reference track line and sweep width, and also with reference to a constant depth or reference depth profile. The program eliminates overlapping areas if the sweep width definitions are properly specified.

- The graphic cross-section plot (Fig. 13) is a standard feature in most automated survey systems. The speciality of this plot, however, is its horizontal accuracy of depth information. As described earlier, this is derived from real depth data in a conversion program and now only those depth values are taken which are found within a specified parallel distance from the profile line.

Thus it is possible with the SUSY 41 system described to store shore profile data on tape prior to activation of the graphic cross-section plot. In this case, the post-plot can combine both hydrographic and geodetic profile data to provide complete information in a ready-to-use format.

As Fig. 14 shows, depth information can also be plotted in a numerical format in a plan view. Both the graphic and cross-section plot can be carried out using data from the Atlas Deso 20 survey echosounder instead of Boma 20 data.

Figure 15 gives a block diagram of the main programs and inputs/outputs of the SUSY 41 system.

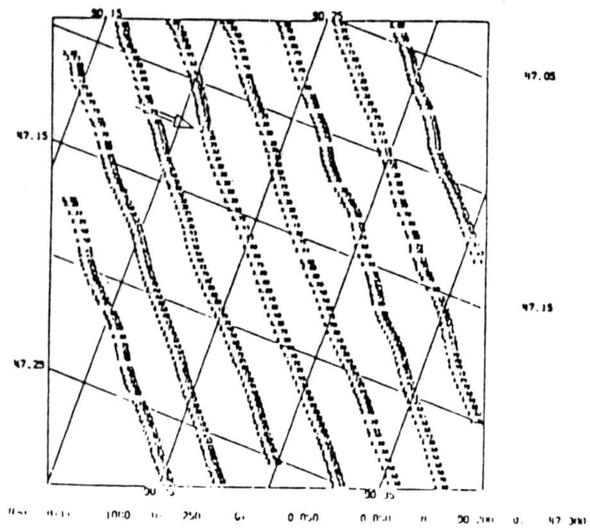

Fig. 14 Numeric cross-section plot

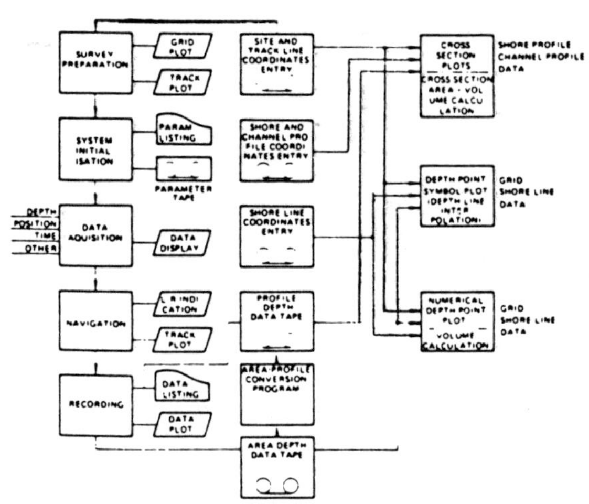

Fig. 15 SUSY 41 system: block diagram of programs and input/output media

NAUTICAL BOTTOM RESEARCH & SURVEY FOR OPTIMISATION OF MAINTENANCE DREDGING IN MUD AREAS

B Malherbe, N V Haecon, P Dewolf, Ostend Ministry of Public Works, Belgium

Presented on **Friday 7th March 1986 at 10.25** in the Clarence Conference Room

EMBARGOED AGAINST DELIVERY

46

Nautical Bottom Research and Survey for Optimization of Maintenance Dredging in Mud Areas

B. Malherbe, Project Manager, Geological Engineer, *N. V. Haecon,*
P. Dewolf, Principal Engineer, Civil Engineer, Ministry of Public Works, Coastal Service, Ostend, Belgium
and *B. Paquot,* Project Engineer, Civil Engineer, Baggerwerken Decloedt & Son

There is a definite need to better define a uniform and exact nautical bottom within loose mud deposits, characterized by low shear strengths in their upper parts. This chapter presents a better approximation of a mud nautical bottom definition, based mainly on the rheological characteristics of the mud considered. From the results it will be clear that a direct application of this theory in the field is possible. Maintenance dredging can be optimized in two ways, by introducing such a definition of the nautical bottom:

- a better definition of the target depth for piloting the dredging works;
- the possiblity of better and more uniform production work.

Additionally, intensive mud survey allows a better understanding of the behaviour of the deposit and may form a programming tool for maintenance dredging; this will be illustrated by some practical examples.

INTRODUCTION

In 1984 the Coastal Service of Ostend, of the Administration of Waterways (Ministry of Public Works, Belgium), ordered a research program wherein, amongst other items, special research is to be carried out on the definition and the detection of the so-called nautical bottom in mud. This chapter presents some of the research results obtained so far.

A nautical bottom as determined by an echo-sounder or a lead wire does not seem to form a basis for discussion, since the hydrographer wants to define the bottom according to safe navigability and manoeuvring capacity, and will thus use for this bottom the level where the medium changes its behaviour drastically, i.e. the water-sediment or water-rock interface. From there on, all calculations can be executed to determine the nautical depth (including keel clearance, sedimentation, tide reduction, etc.), for the programming of maintenance dredging works and for the description of the navigability and manoeuvrability of the access channel or waterway considered.

Loose mud deposits in waterways are characterized by:

- a sharp water-mud interface;
- very low mud shear strengths in the upper parts of the deposit.

These considerations lead to the conclusion that a nautical bottom in mud is not necessarily determined by the water-mud interface, but may be located within the deposit itself. The problem is what level to indicate and what unit to use for the characterization of the nautical bottom, and thus safe navigation also.

Previous work and investigations tend to define the nautical bottom as a density level within the mud deposit. Some relevant work can be summarized here. Extensive investigations have been made in Bangkok and along the coast of Surinam, where ships sail in mud with negative keel-clearance; in Bangkok a volume-mass value of 1.230 t/m^3 of mud is still considered as a safe value in which vessels can sail. On the basis of these investigations and a literature search, it was found that densities up to 1.2 had only a slight influence on manoeuvrability (Kirby and Parker, 1979; Van Bochove, 1979; Nederlof, 1979). Actually in the Rotterdam Europoort, the so-called "nautical bottom" has been set at the 1.2 t/m^3 density limit to be on the safe side (Nederlof, 1981). The question remains whether similar density values are applicable in all cases to define the nautical depth.

From the research results presented in this chapter, an exact and uniform definition of the nautical bottom can only be given when the deformation characteristics (stress - strain relation) of the channel bottom sediments are taken into account. This seems logical, since navigation and manoeuvring capacity are mainly determined by friction, eddy and wave-making resistance, all of these being forces influenced by the viscosity and thus the shear strength of the medium.

In the next section, an attempt will be made to illustrate some typical behaviour features of mud deposits. This mud deposit survey allows better guidance of the research on the nautical bottom and a better understanding of the general behaviour of mud for programming maintenance dredging.

In a further section, the special concept of the nautical bottom definition derived from basic rheological research on mud will be discussed. Also, the universality of this concept will be shown to allow the development of efficient detection techniques.

Finally, the results will be summarized and evaluated with regard to the optimization of maintenance dredging in similar mud deposits.

THE BEHAVIOUR OF "LOOSE MUD" DEPOSITS

The first indication of the presence of "loose mud" deposits is the indication, by a high-frequency acoustic echo-sounder (frequency higher than 10 kHz), while navigating, of a flat sub-horizontal weak reflector (Fig. 1).

Simultaneously, depending upon transducer emission power or coupling of a lower-frequency transducer, deeper penetration is sometimes achieved, revealing a so-called "second reflector or echo", usually interpeted as the hard bottom (see Fig. 1).

The basic questions about detected mud deposits remain:

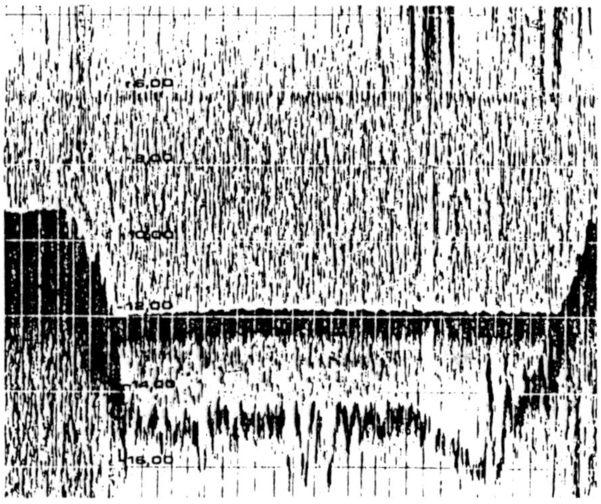

Fig. 1 Cross profile over the access channel to Zeebrugge, showing double echo (top. roof of mud deposit)

- What is the physical significance of these echoes or reflectors, and can they be related in a certain way and with some degree of accuracy to a nautical bottom?
- What is the physical occurrence and behaviour of the deposit, what is his time-dependent variation and what are the factors influencing it?

Physical Significance of Echoes in Mud Deposits

To understand the physical significance of these echoes, one has to realize that reflection of emitted acoustic waves is determined by the reflection coefficient of the two layers encountered by these waves. The reflection coefficient is given by:

$$R = \frac{Z_2 - Z_1}{Z_2 + Z_1}$$

where: R = reflection coefficient; subscript 1 refers to the upper layer; subscript 2 refers to the lower layer; and Z = acoustic impedance, dependent upon wave travelling velocity and medium density.

The greater the difference in acoustic impedance between two layers, the greater R will be, and the better a contrasted reflector will be shown on the echogram. Compared to water, the best reflectors are obtained with rock, sand or compacted sediments and gas.

Loose mud, with typically high water content values, will ovbiously not show a large "acoustic impedance" difference with water and will, in this way, give a weak reflector.

To illustrate these considerations better in Fig. 2 several recorded density profiles in the mud deposits of the access channel to Zeebrugge (Belgium) are compared to the levels of the first and second echo. From this illustration it is obvious that, when mud deposits reach several metres thickness:

- the first echo seems to occur at the water-mud interface, i.e. the first signifi-

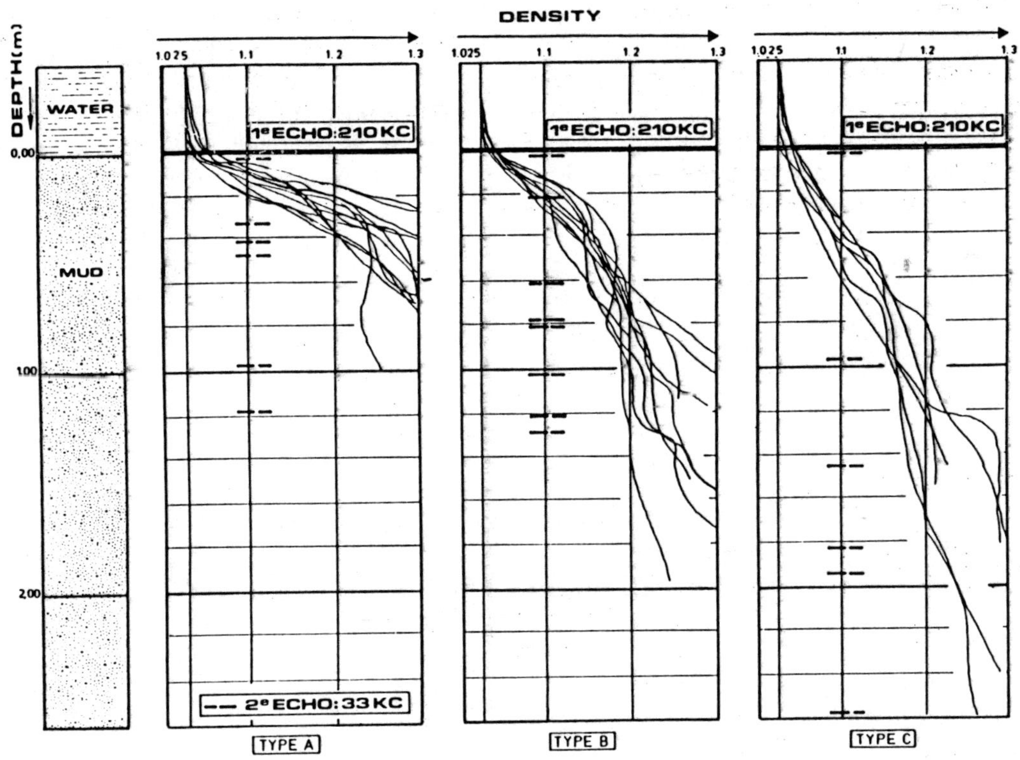

Fig. 2 Comparison of recorded density profiles in Zeebrugge harbour (radioactive back-scatter gauge) with double echo reflections (Atlas Deso 10)

cant "acoustic impedance" changeover;

- the second echo does not seem to correspond to a well defined density value; however, penetration depths of acoustic waves are greater when mud deposits are less consolidated (profile C).

Experience reveals that, when compacted sediments or sand occur at small depths (1–2 m) under the loose mud, this will mostly give a clear reflector. In those cases, the term "hard bottom" can, of course, be considered appropriate for this second echo.

Physical Occurrance and Behaviour of Mud Deposits

To obtain a better knowledge of the physical occurrence and the behaviour of mud deposits, other field-measuring devices become necessary.

With the help of lightweight gravity corers, it is possible to sample several metres of such loose mud deposits at one time. Either freezing in or rapid sampling of the highly loaded fluid core, allow us to distinguish the vertical distribution of physical parameters of the mud deposit, such as:

- water content and concentration;
- sand mixture (proportion of sand in the mud);
- rheological parameters.

Some results of such mud corings are illustrated in Fig. 3. Usually, research pays much attention to the determination of the water content and the granulometric distribution; rheological characteristics are deduced from these two parameters (see next section).

Results of analyses of some mud corings in the access channel to Zeebrugge lead to the following conclusions:

- the sand content (fraction greater than 0.063 mm) may vary within a broad range between 0% and 35% even in this loose mud;
- the sand content within the mud generally increases gradually with depth in the mud;
- the water content or concentration profile in the mud deposit generally shows a rapid

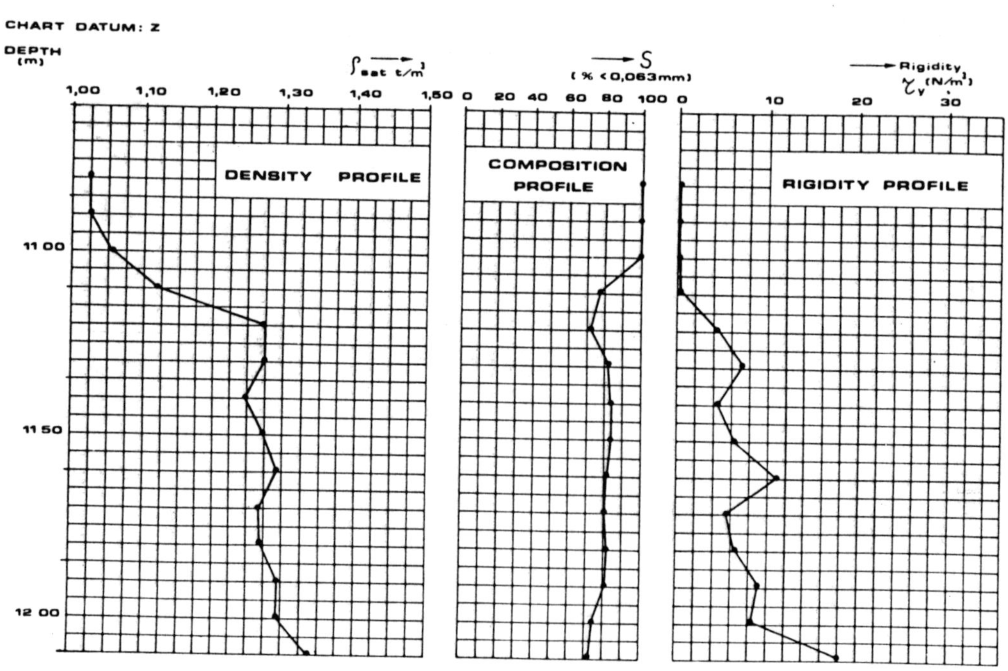

Fig. 3 Mud coring analysis results (mud core no. 8 in access channel to Zeebrugge, Belgium, 11 October 1979).

growth at first (strong density gradient) followed by a weak density gradient.

Radioactive density-measuring devices can be particularly well adapted to the water or marine environment, and allow an operational profiling of the mud density (either horizontal or vertical). Therefore, these kinds of gauges have been developed extensively over the past 10 years (Anguenot, 1972; Kirby and Parker, 1979; Caillot *et al.*, 1984).

Radioactive density gauges may function either by:

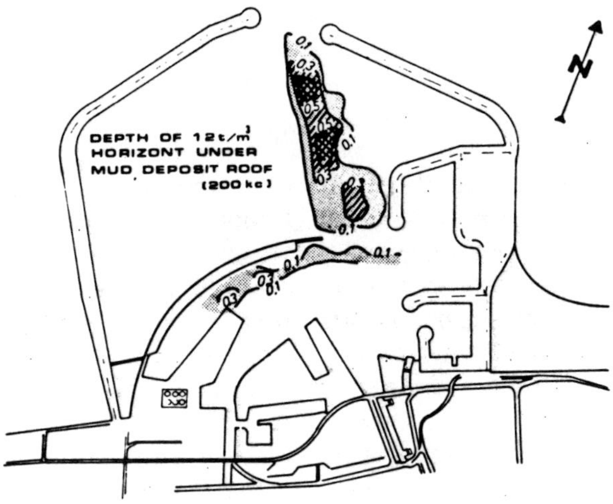

Fig. 4 Example of mud deposit thickness map as recorded in Zeebrugge harbour

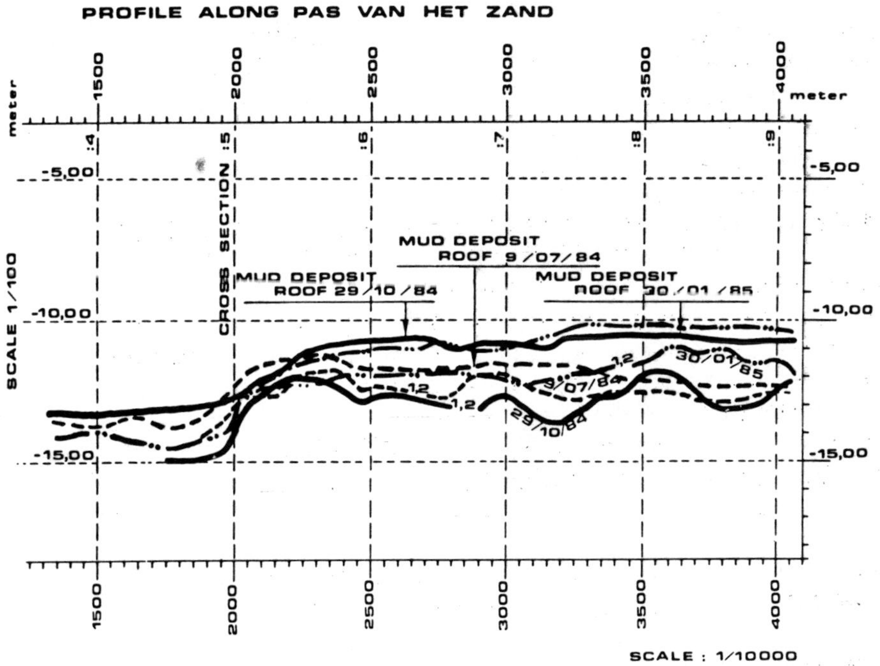

Fig. 5 Profiles along the access channel to Zeebrugge, showing the variation with time of the mud deposit roof and the 1.2 density level

- retro-diffusion or back-scattering of gamma rays;
- transmission of gamma rays.

Both systems have their own measuring characteristics, but can be mounted in hydrodynamically profiled supports to achieve minimum disturbance of the mud.

In this way it is possible to draw so-called "density maps" illustrating either the absolute level of density values, or the layer thickness of mud between the deposit roof and a certain density value horizon (see Fig. 4).

An operational high production survey of the mud deposit density build-up, allows us to recognize the behaviour of the deposit; therefore, an operational fully automatic vertical profiling density gauge was developed.

Figure 5 shows profiles, recorded with this gauge, along the axis of the navigation channel to the harbour of Zeebrugge; the profiles are computations of the results of simultaneous echo-sounding and density profiling measurement campaigns.

It is interesting to observe the time variation of the levels of both the first echo (210 kHz) and of the 1.2 density value in the mud deposit. On the figure these levels are indicated for three distinct campaigns: when the first echo of the top of the mud deposit rises, the level of the 1.2 density goes down, and vice versa. This can be interpreted as a "swelling" and a "shrinking" of the mud deposit, in which the swelling implies an increase of the deposit's volume coupled with a relative dilution (weaker density gradient). Factors influencing this temporary dilution of the mud deposit can be considered to be:

- navigation of deep-draught vessels, and trailing of suction dredgers;
- high wave activity (pore-water pressure build-up);
- gas development due to fermentation.

The direct conclusion of these considerations is that a rise in level of the first echo does not necessarily mean a rise of the nautical bottom (or a decrease in nautical depths).

When measuring the deposit for different tidal coefficients (spring, mean or neap tide) and for approximately the same conditions of maintenance dredging season and wave activity, other features do appear. Figure 6 illustrates two longitudinal density profiles along the access channel within the new harbour of Zeebrugge: one is taken at spring tide while the other is at neap tide; the long profiles are recorded with a time interval of about 1 week. Figure 6 shows a significant difference in the global density build-up of the deposit:

- at spring tide, the deposit seems to be more dilute than at neap tide (stronger density gradient);
- the different density levels and the roof of the mud deposit rises at neap tide.

Both effects seem to indicate a greater net sediment supply in the access channel during neap tide periods.

DETERMINATION OF THE NAUTICAL BOTTOM WITHIN LOOSE MUD DEPOSITS

Introduction

The definition of the nautical depth in mud areas is complicated both by problems of how to define this depth and how to measure it. The nautical depth in mud is not necessarily the water-mud interface, since the upper part of the mud-deposit is characterized by low shear strength. This low shear strength in the upper mud layers is as important for navigation with negative keel-clearance as with restricted keel-clearance (less than 10% above the mud-water interface).

Navigation and manoeuvring with restricted keel-clearance above a hard bottom will be influenced mainly by the underflow; navigating and manoeurving with restricted keel-clearance above loose mud deposits will additionally be influenced by the deformation characteristics of the bed (Sellmeijer and van Oortmerssen, 1983). The mud bed will be deformed according to the applied shear stress and the rheological characteristics of the mud. From model results with ships sailing with restricted keel-

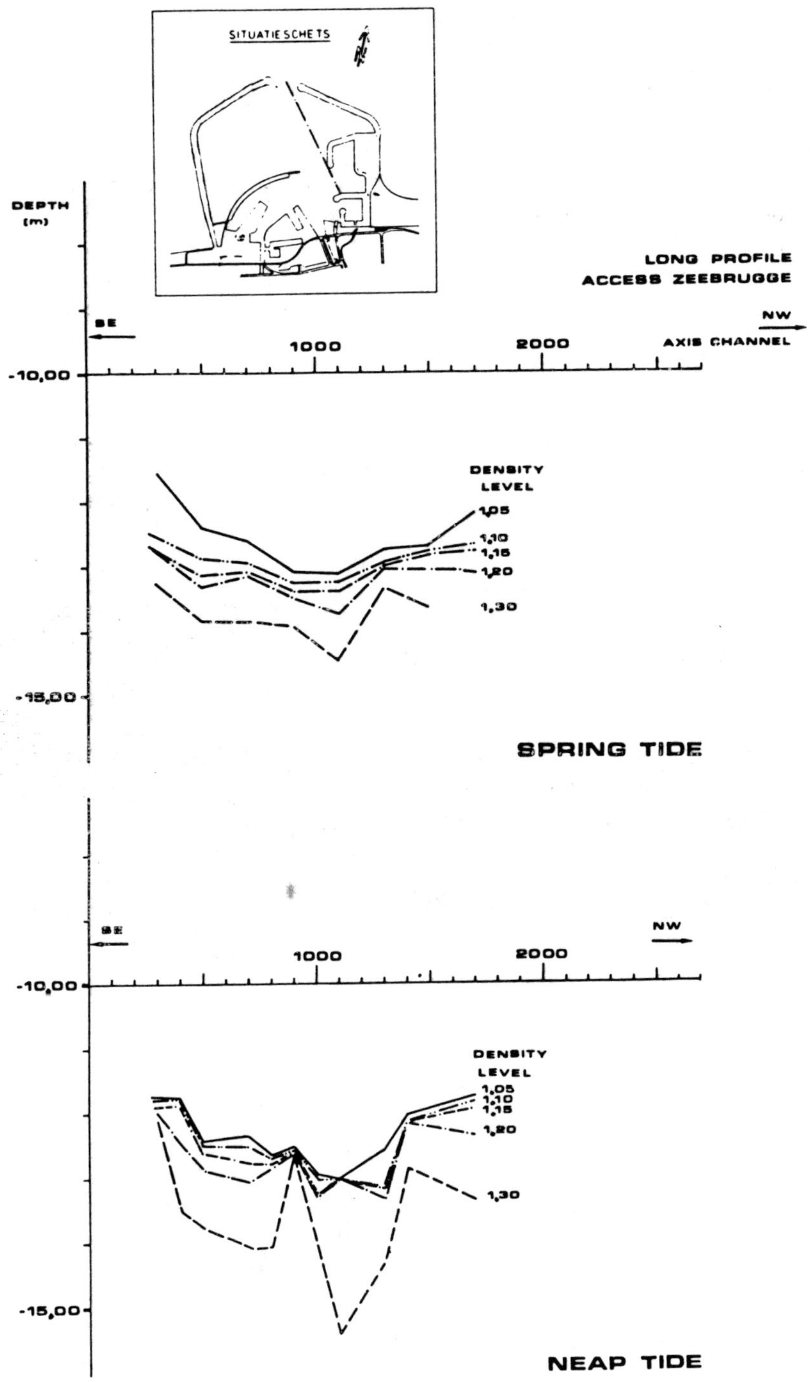

Fig. 6 Tidal variation of density build-up of mud deposits in the harbour of Zeebrugge

clearance, it appeared that in some cases internal waves may form at the water-mud interface (van Bochove and Nederlof, 1979); those internal waves enhance the total wave, causing resistance to the passage of the ship and thus affecting its manoeuvring behaviour. The formation of these waves is, of course, also completely dependent on the deformation characteristics of the mud considered or the mud simulation product.

As mentioned clearly in an earlier publication on physical mud properties (Malherbe, Bastin and De Putter, 1982; PIANC Working Group, 1982–3), the rheological characteristics of mud are mainly determined by:

- the specific gravity or concentration of dry sediment;
- the composition and the sand content of the mud;
- the physico-chemical properties of the mud considered.

Knowing the rheological relationship of the mud, it is possible to deduce theoretically, for instance, the energy absorbed by shear stress when navigating at 0% keel-clearance (to the mud-water interface) above a loose mud bottom (Malherbe *et al.*, 1982); some results of these calculations are illustrated in Fig. 7.

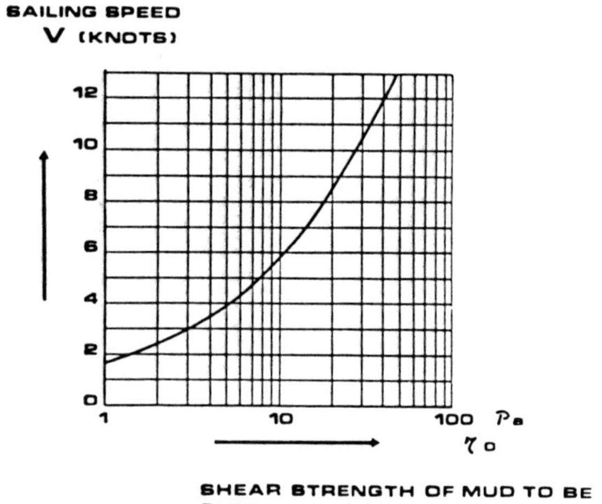

Fig. 7 Navigation above mud at 0% keel-clearance; shear stress exerted by ships' keels at different speeds

Rheological Properties of Mud

The most important rheological characteristics of mud sediments for the definition of the deformation behaviour (stress-strain relation) and the shear strength of the deposit are (see Fig. 7):

- the initial rigidity (or yield strength), τ_y (defined at low shear rate or deformation);
- the dynamic viscosity, η (defined at high shear rate or deformation).

This was first mentioned by C 1. Francis-Boeuf in 1941, in his work upon mud properties; therefore his terminology has been used here.

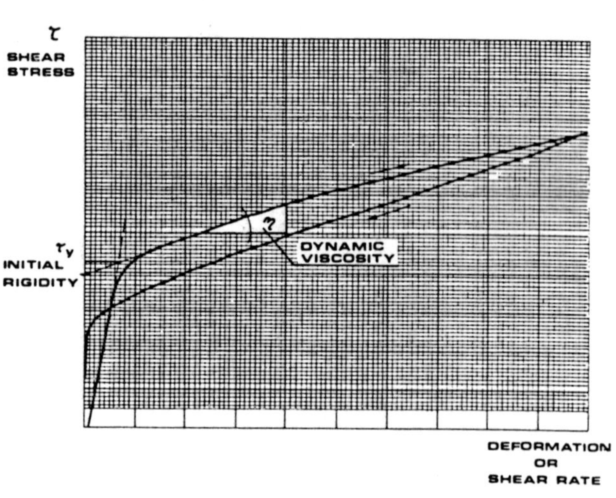

Fig. 8 A mud rheogram

Both characteristics can be measured in a laboratory with an analytical high-precision coaxial viscosity meter. Figures 8 and 9 express the relations between τ_y, η, concentration, T_s and mud content S (% < 63 μm for mud in the harbour at Zeebrugge (Belgium).

Zeebrugge forms a direct access to the North Sea and, due to the high wave activity, mud sediments are mixed in very different amounts with fine marine sand, depending upon the season and the wave activity.

One can deduce from these relationships that two distinct mud behaviours occur:

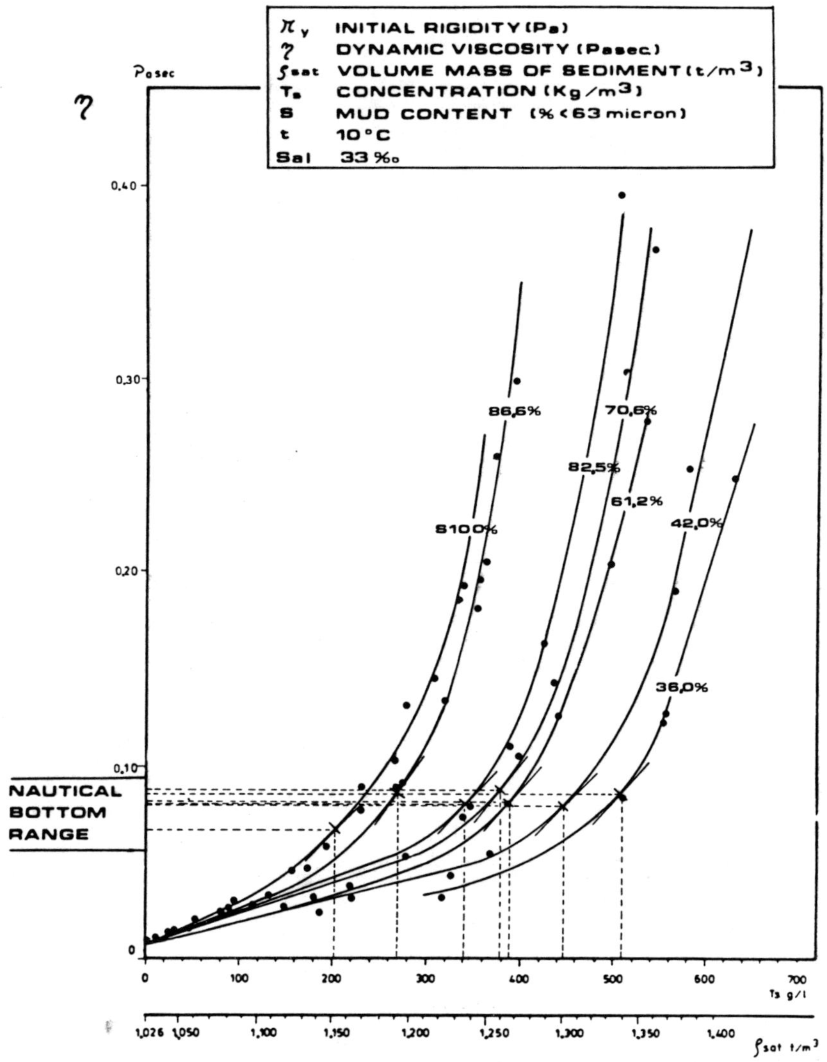

Fig. 9 Initial rigidity τ_y as a function of concentration and mud composition (Zeebrugge mud)

- a first behaviour region where initial rigidity τ_y and dynamic viscosity η are less dependent on the concentration and remain very similar to the values for pure water (first part of curves, small slope);
- a second behaviour region where initial rigidity τ_y and dynamic viscosity η are strongly dependent on the concentration and also very different from the values for pure water (second part of the curves, steep slope).

To illustrate this further, the mud from the Maasmond, Europoort (the Netherlands) was analysed in the same way (see Fig. 11: relation of initial rigidity to concentration); the figure shows a similar kind of relationship as in Fig. 9. The comparison of the relationships of different mud types leads to the following conclusions (Malherbe, 1985):

- From zero up to a certain concentration value, depending mainly on mud type and

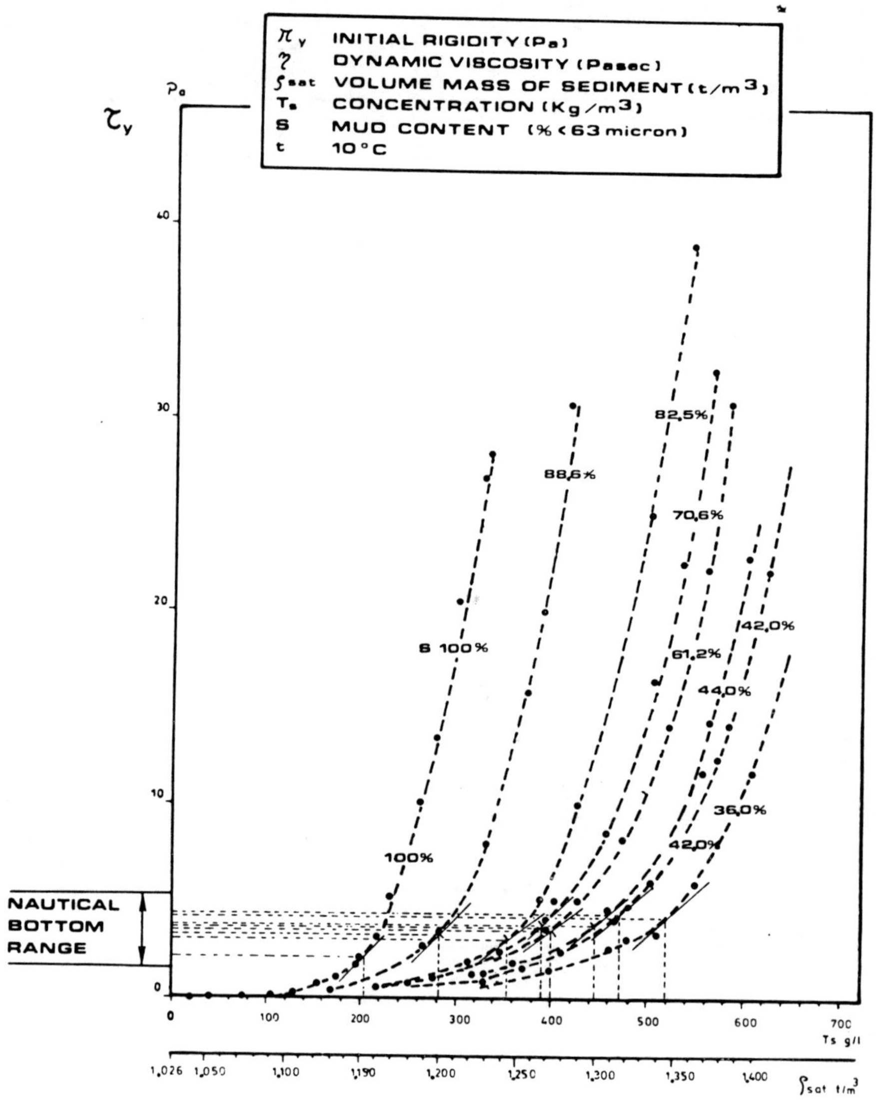

Fig. 10 Dynamic viscosity η as a function of concentration and mud composition (Zeebrugge)

mud composition, the shear strength and the rheology of the mud is not, or is weakly, affected by the mud concentration; for concentrations higher than this certain value, the shear strength and rheological characteristics will increase rapidly with concentration;

- The influence of the composition (sand mixing) is obviously important, but can be very different depending upon the mud type; for mud from the Maasmond, for instance, this influence is stronger than for mud from the harbour of Zeebrugge; also it appears that the effect of the sand content in the mud is most important for low sand content values (0–30%).
- The influence of the mud type particularly affects the rheological characteristic values for a given concentration: depending upon the origin, for a given concentration value, shear strength, initial rigidity or dynamic viscosity can differ by a factor of 10, but the rheological behaviour changeover seems to occur around the same rheological value (for instance, rigidity between 2 and 5 Pa or viscosity between 0.06 and 0.09 Pa s).

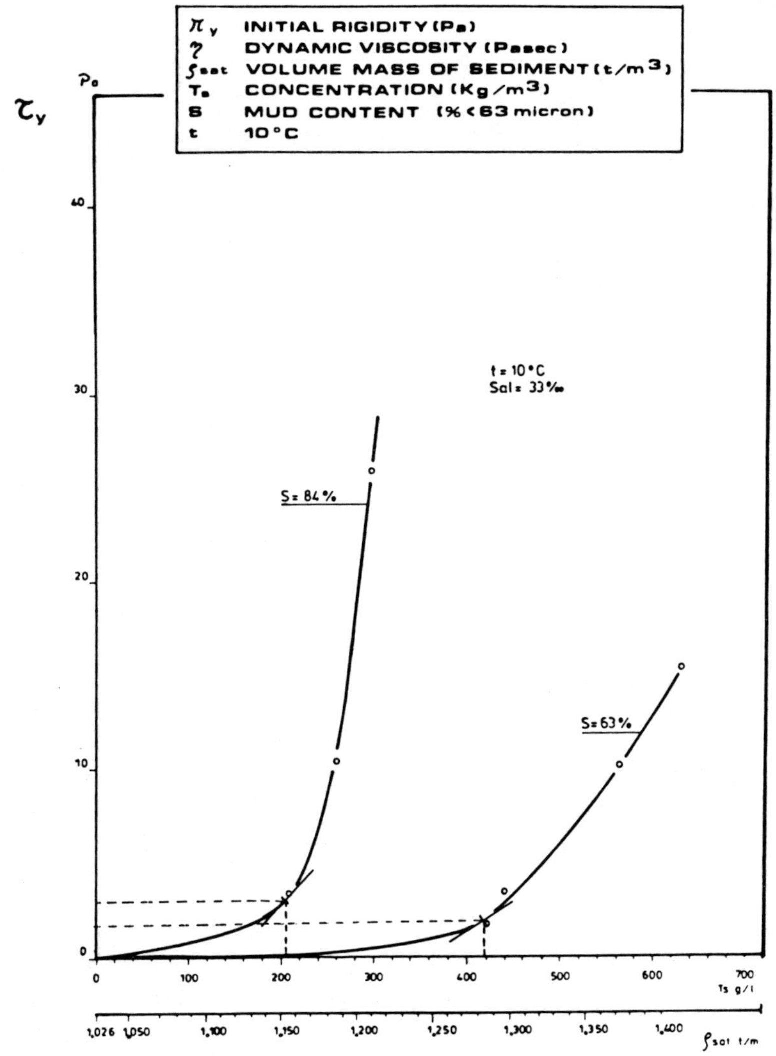

Fig. 11 Initial rigidity as a function of mud concentration (Rotterdam harbour)

The influence of sand content (S = 100%) on the rheological properties seems obvious: the higher the sand content in the mud, the lower the initial rigidity and the dynamic viscosity will be at the same concentration value.

Physically, this can be explained by the fact that rheological properties of cohesive sediments are mainly influenced by the properties of the colloids in it (clay minerals, amorphous clays, biocolloids, etc.). An increase of the proportion of non-colloidal substances (sand, for instance) will decrease the typical values of the mud shear strength.

The Definition of Navigable Depth Based Upon Rheological Properties

Since harbour authorities want to indicate a navigable depth with a certain degree of safety, this depth has to indicate the minimum level at which the medium in which navigation occurs drastically changes its behaviour.

For sandy bottoms it will be clear that the navigable depth will be defined by the water-sand interface.

For muddy bottoms, however, such an interface exists also but a significant change in stress-strain behaviour occurs within the mud deposit itself (see Figs 9, 10, and 11); the level where this significant changeover in shear within the mud deposit occurs will be defined both by:

- the density profile of the deposit; and
- the composition profile (sand content) of the deposit.

From the point of view of navigation and manoeuvring, the presence of two rheological behaviour regions indicates the validity of the following interpretation:

- a first behaviour region where navigation and manoeuvring is not, or is weakly, affected by the concentration of the deposit (navigation rather similar to that in water);
- a second behaviour region where navigation and manoeuvring is strongly affected by the concentration of the deposit (navigation different from that in water).

The rheological behaviour changeover is not very sharply defined. Using an intermediate value between those two behaviour regions to define the nautical bottom will include a certain safety margin, since the second behaviour region of affected navigation will not yet have been reached. Therefore, the determination of the behaviour changeover is done with the help of the point on the curves where the minimum radius of curvature occurs (Kerckaert, Malherbe, and Bastin, 1985).

In this way, it is possible to determine, for Zeebrugge harbour mud, the nautical depth values in terms of volume-mass (abscissa) from the experimental rheological relationships shown in Figs 3 and 4.

The nautical bottom values expressed as volume-masses are summarized in Table 1.

TABLE 1
Deduced Nautical Bottom Volume-mass from Different Mud Compositions

Mud content ($<63\mu m$) S (%)	Nautical bottom volume mass (t/m^3): Determined with initial rigidity	Determined with dynamic viscosity	Mean values
100	1.151	1.153	1.152
86.6	1.199	1.194	1.197
82.5	1.242	1.238	1.240
70.6	1.267	1.260	1.263
61.2	1.271	1.268	1.270
42	1.316	1.301	1.309
36	1.347	1.340	1.344

From the results in Table 1 it appears that the behaviour changeover in mud occurs at approximately the same volume-mass value (for a given sand/mud proportion) independent of low shear rates (initial rigidity) or for high shear rates (dynamic viscosity).

Rheological analyses provide only relative values of shear stresses since the values are dependent upon the system being used. Volume-mass measurements give absolute values and, in this way, provide the link between relative and absolute measurements.

On the other hand, the changeover rheological value (initial rigidity or dynamic viscosity) seems to lie within a small range and is only slightly affected by the mud composition S ($\% < 63\mu m$) and the mud concentration (see Figs 9 and 10).

Table 1 shows that the nautical bottom criterion, expressed as a volume-mass of the sediment, varies in a broad range from 1.151 t/m^3 to 1.347 t/m^3 for the mud compositions tested; using a single volume-mass value in all cases is thus a risky operation and this seems to be valid for all types of mud where sand may mix with the mud.

A test program, using small-scale model experiments and full-scale tests in the field, has been set up to evaluate in which way the inclusion of a mud layer in the keel-clearance of a sailing ship affects the manoeuvring characteristics. These experiments will ascertain the similarity between the changeover in rheological and manoeuvring characteristics. As a part of the whole program, these investigations must lead to the definition of the nautical depth in terms of both mud characteristics and manoeuvring behaviour.

Measuring the Nautical Depth in the Field

From the definition of a nautical bottom as a rheological changeover level and from the considerations mentioned above, it will be clear that a "nautical depth measuring device" should simultaneously measure volume-mass and a rheological property (or a characteristic proportional to one or two rheological properties) to give maximum information about the mud considered.

Volume-mass measuring systems (such as radioactive gauges) can provide all the absolute values needed to draw the density profile within the mud.

On the other hand, rheological measuring devices allow the localization of the nautical depth; the combination of both volume-mass and rheological properties gives all necessury further information about the absolute values to be detected.

Indeed, it seems that measuring rheological properties is always relative: the definition of navigable depth as a mud behaviour changeover is not sensitive to the rheological measuring system as long as these measurements can be calibrated to volume-mass and sand/mud proportions. Of course, measurements of volume-mass and rheological properties have to be reproducible.

OPTIMIZATION OF MAINTENANCE DREDGING

An optimization of maintenance dredging with the aid of rheological knowledge of the mud considered can be performed in two ways, as mentioned in the introduction.

A better Definition of Dredging Level

This allows port authorities to program maintenance dredging according to the changes in the nautical bottom, instead of changes in the top of the mud layer (echo-sounding); the top of the mud layer can vary much more rapidly and strongly than the density levels within the mud deposit.

Intensive mud survey programs reveal a periodical change in the density build-up of the mud deposits in the harbour of Zeebrugge; this results in changes of the mud top layer without a significant need of dredging.

Optimization of Dredging Production

A mud deposit survey based on nautical bottom detection provides a new approach to hydraulic maintenance dredging. It is obvious that a rheological changeover of the mud is also a production changeover; trailing with drag-heads within the mud will give an optimal ratio of gained production/developed power when the head is positioned as close as possible to the *in-situ* detected rheological changeover ("hunting" the rheological changeover).

CONCLUSIONS

1. The detection of loose mud deposits with the aid of echo-sounders is possible, but an exact interpretation of the echograms in terms of density levels remains difficult to achieve. The first echo corresponds mostly to the sharply defined mud-water interface, while the second echo does not seem to correspond to a well defined density level; the reason for this can be found in all the features that can occur within such a mud deposit and that can change the acoustic impedance significantly (gas bubbles, sandy horizons, etc.).

2. The sand content (fraction finer than 0.063 mm) within the profile of these mud deposits may vary within a broad range of values between 0% and 35%, this can be explained by the fact that mud flocs can hold sand grains within their structure.

3. Intensive mud density survey provides a better understanding of the behaviour of this kind of deposit:

- mud deposits are characterized by a weak density gradient versus depth;
- the time variation of this density build-up shows the existence of "swelling" effects (dilution of the deposit) and "shrinking"

(consolidation of the deposit);

- variation of the density build-up of the deposit as a function of the tidal coefficient reveals a net supply of sediment and a general consolidation of the deposit allowed during neap tides.

4. Navigation with restricted keel-clearance in muddy areas requires a better definition of the nautical depth than the obtainable by echo-sounding. The definition of the nautical depth has to take into account both volume-mass and deformation behaviour of the mud.

5. Rheological or deformation behaviour of mud is determined by:

- the volume-mass of the sediment (or the concentration);
- the composition (sand/mud proportion, specially for sand contents between 0% and 30%);
- the physico-chemical properties of the mud (the origin).

Nautical depth within the mud deposit can be defined as the level where the mud changes its rheological behaviour drastically (Kerckaert, Malherbe, and Bastin, 1985). This behaviour changeover occurs at approximately the same volume-mass value for both low and high shear rates (Figs 3 and 4).

6. Nautical depth values defined by rheological properties occur within a small range of initial rigidity or dynamic viscosity, independent of sand/mud proportions, and density or concentration (for the considered mud types, the nautical bottom can be defined with a rigidity ranging from 2 to 5 Pa or a dynamic viscosity ranging from 0.06 to 0.09 Pa s).

7. As the behaviour changeover level is not sharply defined, such a definition of the nautical bottom does include a safety margin: the shear characteristics of the mud are not yet strongly dependent on the concentration value, i.e. the second behaviour region with:

- different rheological concentration relationships;
- different and dangerous navigation characteristics;
- different and strongly varying trailing productions.

8. Recent work has led to the development of a great many instruments capable of measuring either volume-mass or a rheological value. As rheological measurements are dependent upon the system being used, nautical depth sensors should measure both volume-mass (absolute values) and rheological properties (relative values) to get the maximum information about the mud deposit considered.

9. Maintenance dredging in mud can be optimized in two ways:

- better definition of the changes in nautical depth and more efficient steering of maintenance dredging;
- trailing in mud (drag-head within mud deposit) can be done with better ratios of gained production/developed power when "hunting" the in-situ measured rheological changeover.

ACKNOWLEDGEMENTS

As the discussion about the analysis of Zeebrugge mud has already been published in the PIANC Bulletin no. 49 (1985), the contributions of Messrs ir. P. A. Kerckaert (Coastal Service of Ostend, Ministry of Public Works, Belgium) and Dr A Bastin in helping to review this part of the publication are gratefully acknowledged.

REFERENCES

Anguenot, F. 1972. Jauges radioactives pour la mesure des turbidites. Rapport CEA — CEN

Caillot, A., de Meyer, C., Chambeller, D. and Tanguy, J. C. 1984. *A New Nuclear Density Gauge to Measure Directly High Turbidities in Muddy Areas*. Saclay: ORIS-CEA, October.

Francis-Boeuf, C. L. and Bourcart, J. 1942. *La Vase Actualites Scientifiques et Industrielles*. Paris.

Granboulan, J., Fourcassies, C. and Chaumet-Lagrange 1984. Measuring the navigable depth in a muddy channel. Presented at the 2nd International Hydrographic Technical Conference, Plymouth, September.

Kerckaert, P., Melherbe, B. and Bastin A. 1985. Navigation in muddy areas: the Zeebrugge experience. *PIANC Bulletin* 48.

Kirby, R. and Parker, W. 1979. Definition of the seabed in navigation routes through mud areas. *Report IOS*. Taunton.

Malherbe, B. 1985. The determination of the nautical bottom in muddy areas. Presented at MARINTEC China '85 Shanghai, December.

Malherbe, B., Bastin, A. and De Putter, G. 1982. Physical properties of sand and mud sediments. *Symposium on Engineering in the Marine Environment,* Koninklijke Vlaamse Ingenieursvereniging, Brugge (May).

Migniot, C. 1984. Mesures des caracteristiques mecaniques des vases deposees dans les chenaux de navigation. *XVIIIe J. Hydraul. Soc. Hydrotech. France,* Sept.

Nederlof, L. 1981. Sailing through water rich in floating silt: a vessel behaves differently, but remains manageable. Rotterdam Europoort Delta.

Report of PIANC Working Group no. 3a 1982-3. Navigation in mud areas. Report in *PIANC Bulletin* 43.

Sellmeijer, R. and van Oortmerssen, G. 1983. The effect of mud on tanker maneouvres. Presented at The Royal Institute of Naval Architects spring meeting.

van Bochove, G. and Nederlof, J. 1979. Vaargedrag van diepstekende schepen in slibrijke gebieden. *De Ingenieur* 30/31 (July).

A VESSEL-BORNE GAMMA-RAY SILT DENSITY GAUGE FOR THE 'IN-SITU' EXAMINATION OF SEDIMENTS

H-U Fanger, E Bossow, K Grisard, Dr H Kuhn,
GKSS-Forschungszentrum Geesthacht, Federal
Republic of Germany

Presented on **Friday 7th March 1986 at 10.50**
in the Clarence Conference Room

EMBARGOED AGAINST DELIVERY

47

A Vessel-borne Gamma-ray Silt-density Gauge for the "in-situ" *Examination of Sediments*

H.-U. Fanger, E. Bössow and *H. Kuhn,* Institut für Physik, GKSS-Forschungszentrum Geesthacht, D-2054 Geesthacht, Federal Republic of Germany
and *K. Grisard,* ME Meerestechnik-Elektronik, D-2351 Trappenkamp, Federal Republic of Germany

A radiometric density gauge using 662 keV gamma rays has been developed at the GKSS research centre, for the determination of local silt distributions on the beds of waterways. Features of the techniques utilized are transmission geometry, data telemetry via a single conductor cable, a high data-sampling rate, and computer-based on-line data evaluation. The probe was subjected to various *in-situ* tests and successfully applied for extensive measurements in Hamburg harbour and in the tidal region of the River Weser.

INTRODUCTION

The sedimentation of suspended particulate matter and sand in harbours and shipping channels of tidal rivers or estuaries causes a high maintenance dredging cost. Beyond that, removal or deposition of the dredged material raises serious ecological problems because of heavy metal pollution and organic contamination. Sediment in waters has thus attracted considerable research interest.

In particular, the density of silt on the beds of waterways has been recognized to be an important parameter describing not only the actual mass distribution, but also indicating the so-called navigational depth. Since the body causing echo-sounding horizons at depth may vary from consolidated mud (or sand) to "dirty water", the latter being no hindrance to navigation, the vertical density profile might be a decision criterion for the necessity of expensive dredging operations.

Therefore, a vessel-borne silt-density gauge has been developed, which may penetrate the ground to depths of up to 1.60 g/cm^3, by gravity alone. It differs from other commercially available gauges by the application of transmission geometry which is superior to backscattering arrangements. Also with respect to signal processing, data transfer and on-line evaluation, a state-of-the-art concept has been realized. Details of construction and operation, results of tests and field measurements will be given in the following sections.

IN-SITU SEDIMENT INVESTIGATION METHODS

Since the capabilities of methods other than gamma-ray absorptiometry appear to be widely overestimated with respect to the quantitative analysis of bottom silt, it is necessary to point out a few features of competing methods such as echo-sounding and radiometry.

The well known echo-sounding technique utilizes the time elapsed between emission of an acoustic signal and receipt of the echo to measure the water depth; the strength of the reflected signal is proportional to the gradient of the product of density and the propagation velocity of sound in the reflecting layer. Evidently, for soft transitions between water and the bottom, as is the case with silt, the echo horizon is not well defined. In any case, the reflection is determined by the local change of density with depth, rather than by the absolute value of the density.

This fact was quite well demonstrated by a tank experiment performed in 1967 (Hellema, 1979), the results of which are given in Fig. 1. In that experiment, starting with a well mixed homogenous mixture, the sedimentation was studied for 32 days by daily sampling and echo-sounding using different frequencies. As may be seen in the figure, the acoustic 30 kHz horizon crosses several density levels between 1.20 and 1.27 g/cm^3, whereas higher frequency horizons seem to follow the water-silt borderline in the 10–20 cm depth region.

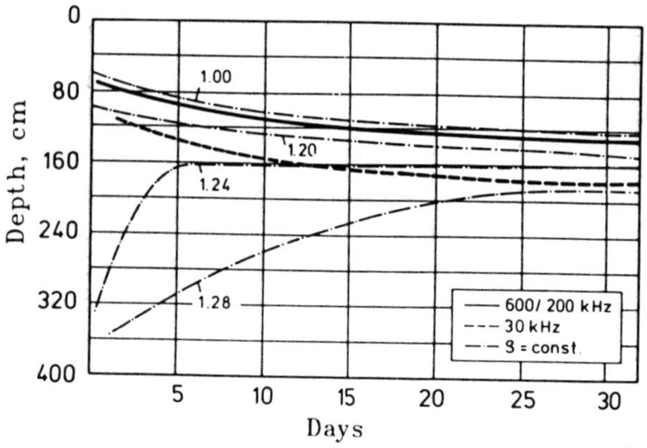

Fig. 1 Temporal trends of acoustic (echo-sounding) and sampling density horizons in a silt-tank experiment (after Hellema, 1979)

Recently, another radiometric method using the natural radioactivity of ^{40}K, which is mainly bound or absorbed in fine-grained fractions of clay (Führböter *et al.*, 1984), has been discussed for *in-situ* sediment investigations. Although this is an interesting approach to a mineral classification of the bottom, it does not yield direct information on the existence or density of silt; in addition, care has to be taken not to confuse readings with fluctuations of the natural radiation background (uranium decay products) or anthropogenic influences (fall-out, radioactive waste waters).

Active radiometric methods apply gamma-ray or neutron sources and use scattering, absorption or the radiation of nuclear reaction products. Some element concentrations on the river-bed surface have been determined quantitatively by the last of these techniques (Moxham *et al.*, 1976), but not in such a way that the silty character of the bottom can be deduced.

RADIOMETRIC DENSITY DETERMINATION

In an energy range between roughly 0.5 and 1.5 MeV, the scattering probability for gamma-ray quanta is proportional to the number of electrons per unit volume and is therefore coupled to the material density ρ by a factor Z/A, where Z is the atomic number, and A is the atomic mass of the chemical elements. From this follows the well known exponential attenuation law for the ratio of intensities with (I) and without (I_0) scattering material in the beam path:

$$I/I_0 = e^{-\bar{\mu}\rho L}$$

where L is the transmission length, and $\bar{\mu}$ the mass attenuation coefficient.

In practice, the interaction process quoted above is used in transmission and backscattering geometry. Whereas scattering contributions can be almost eliminated in transmission geometry by careful collimation and energy discrimination, one has to trade in transmission effects when using backscattering arrange-

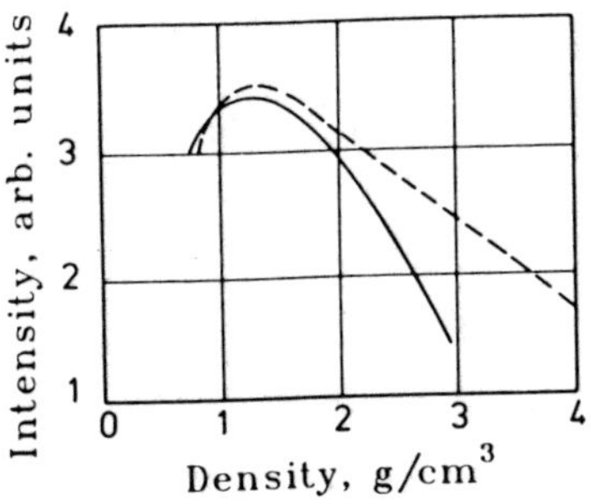

Fig. 2 Intensity of scattered gamma-rays vs. density of scattering material (after Hartmann, 1969). ____, Radium source; ...., 60°C

ments. This point is well illustrated in Fig. 2 (Hartmann, 1969), where the scattering intensity, after first rising with growing density, finally declines due to the prevalence of attenuation (i.e. transmission effects).

For *in-situ* silt-density measurements it is obvious that the backscattering geometry (b.s.) offers the advantage of simplicity in probe construction and handling, since the source and detector with adjacent electronics may be placed in a single tube. On the other hand, the ambiguity of signal-density relationships, the rather poor local resolution in the vertical source-detector axis, the preponderance of the surrounding material contributing to the density information together with the fact that the sensitivity to chemical variations is more pronounced with the b.s. method, were considered to be strong arguments in favour of transmission geometry.

THE GAMMA-RAY DENSITY GAUGE

Construction and Components

Figure 3 shows the construction features of the transmission probe. The total length is more than 3 m, and the clear space between the bifurcated tube which corresponds to the relevant transmission length is 25 cm.

The gamma-ray source is contained in a tungsten cylinder at the right-hand lower end of the probe; the tungsten block serves simultaneously as an efficient radiation shield and

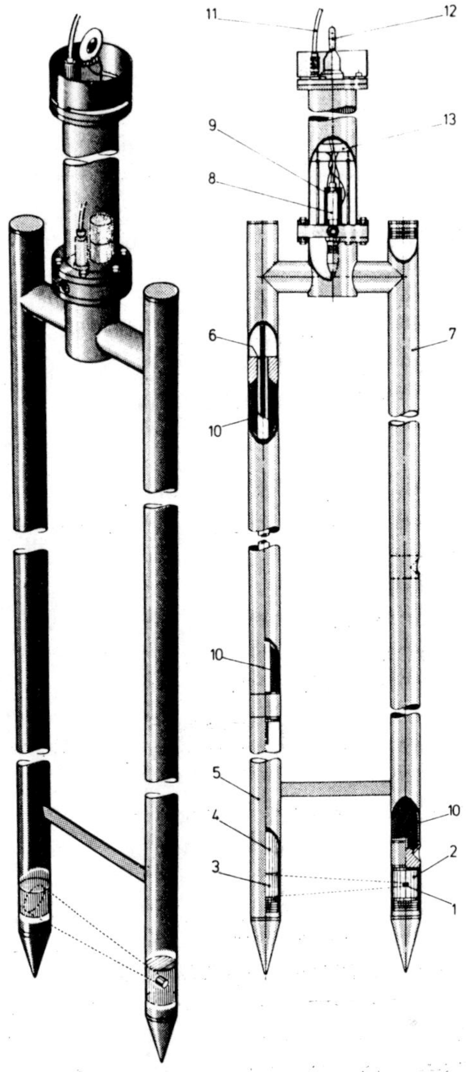

1 Source
2 Shielding
3 Detector
4 Preamplifier
5 High voltage generator
6 Cable duct
7 Measuring and control electronics
8 Pressure sensor
9 Inclinometer
10 Granulated lead
11 Single conductor cable
12 Cable eye
13 Telemetry

Fig. 3 View of the submersible body of the gamma-ray density gauge

beam collimator. An activity of 0.37 GBq ^{137}Cs has been chosen which yields, with the geometry and shielding given, equivalent dose rates in air of 25 μSv/h in 10 cm distance from the source tube and 100 μSv/h at the rear of the detector (i.e. in the direction of the beam). Although there is no hazard implied in handling, for transport in air an easily removable stainless steel shield has been provided around the source tube, while a small water tank gives sufficient radiation protection for long-term storage. The radioactive half-life of the 662 keV gamma-ray emitter is 30.2 years: this guarantees constant conditions over long measuring times.

On the opposite side of the source, the radiation is received by a reinforced scintillation detector. In order to obtain well reproducible results even under stress conditions, high mechanical stiffness has to be guaranteed. The flat metal bridge about 20 cm above the source–detector axis contributes to the rigidity with respect to the measuring line between the twin tubes.

Electronic components belonging to the gamma detector, e.g. the high-voltage power package for the photomultiplier, amplifier, discriminators, and stabilizing circuits, are arranged in a compartment above the scintillation crystal. Granulated lead has been packed into the empty spaces of the twin tubes to obtain a sufficient probe weight for penetration into the bottom and to lower the centre of mass. The weight is determined by the requirement that gravity alone should be sufficient to drive the probe into silty ground, i.e. without additional kinetic energy from lowering.

The cylindrical pressure case above the bifurcation contains sensors for pressure and tilt, and the data telemetry system. This part of the electronics is essentially adopted from the "Kiel Multisonde" (Kroebel, 1973) — an extended CTD probe developed by "Meerestechnik Elektronik (ME)" in Trappenkamp (Federal Republic of Germany) — and therefore well proven in oceanographic applications. The telemetry system represents a considerable improvement over the pre-prototype probe described in Fanger *et al.*, 1985a. The main advantages come from the use of a single conductor cable for data transmission and power supply, but the compatibility with the CTD system also permits the addition of sensors, e.g. for conductivity, temperature, etc.

The pressure sensor yielding the depth reference of the density measurement has been arranged 2.80 m above the source-detector axis to avoid the formation of silt plugs in the sensor outlet. The measuring range is 5 bar, and the overall accuracy corresponds to 5 cm head of water. The inclination determined with a two-axis sensor has to be known for the cosine correction of depth (due to the long distance between the transmission line and the pressure sensor), and for the general judgement of the quality of measured density profiles.

As already indicated, the probe is tethered to a single conductor cable. A winch equipped with slip-ring contacts allows comfortable handling.

Signal and Data Processing

The gross features of the electronic system are displayed in the block diagram of Fig. 4. Each scintillation event causes a current pulse in the photomultiplier, proportional to the energy quanta deposited in the crystal. After amplification the pulses are sorted by three single-channel discriminators. The balance of two side channels is taken for multiplier gain control and the central main channel is fed into a 16-bit counter. The counting rate is a measure of the transmission intensity. The pressure sensor signal is converted into a 16-bit digital value, and the inclinometers are connected to 12-bit A/D converters. The Gamma counter and the A/D converters are read out simultaneously every 1/10 s. After parallel/serial conversion, the data are transmitted in the frequency shift keyed (FSK) format, via the single conductor cable, to the shipboard unit.

The on-board system consists of the power supply and telemetry unit, and a powerful free programmable 16-bit microcomputer system for on-line data evaluation and presentation. The FSK signals coming from the underwater unit are demodulated and interfaced to the computer. Raw pressure and inclination data are calibrated to engineering units, and silt-

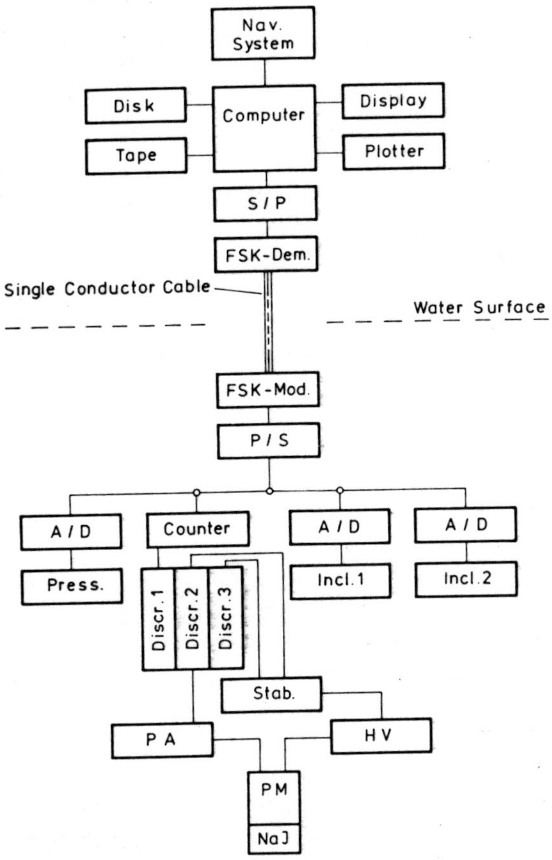

Fig. 4 Block diagram of the system components. A/D, analog-to-digital converter; Discr., window discriminator; HV, high-voltage generator; PA, preamplifier; PM, photomultipier; S/P, serial-to-parallel converter; Stab., stabilizer

density values are obtained from the count rates by evaluation of the equation given above. The depth values are calculated from the pressure, inclination, and atmospheric pressure; corrections for water density (derived from CTD profiles) are optional. A selectable option permits depth-normalization of the data (i.e. to suppress heave motions). All calculations are done in real time 10 times per second. Also, the profile data are displayed in real time on a CRT unit in digital form, or as a density (inclination) vs. depth diagram, and hard copies can be obtained on a plotter. The software is capable of communicating with a navigational system to obtain position data automatically when the probe penetrates the ground. All data can be stored on disk (or floppy), cartridge tape, or 9-track reel-to-reel tape. An off-line program can be used to view or plot stored profiles.

A typical example for the presentation of results is given in Fig. 5. This shows another useful feature of the operational software: one or more density thresholds can be freely selected, and the depths corresponding to the density values are calculated on line. In the example shown, a depth appertaining to a density of 1.2 g/cm^3 is marked and — with some arbitrariness - denoted as "navigational depth" (NDPT).

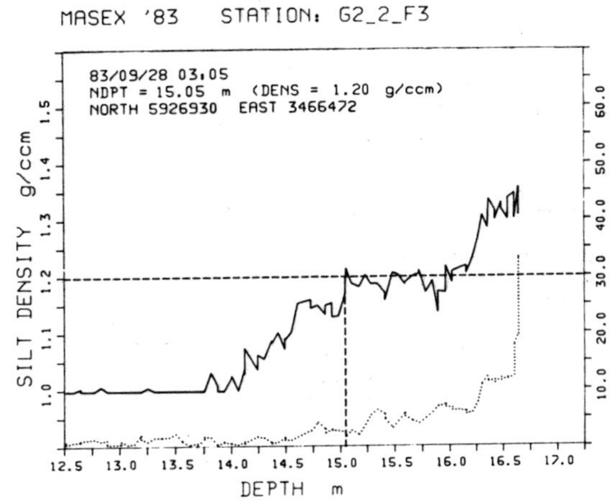

Fig. 5 On-line computer plot of a measured vertical density profile. Dark line = density (see left-hand axis); dotted line = inclination (see right-hand axis); NDPT, navigational depth (for pre-selectable density values). East/north co-ordinates refer to the Gauss–Krueger system

Calibration and Accuracy

For physically well understood reasons — as indicated in a preceding section — the radiometric density determination is influenced by the chemical composition. In brief, the already mentioned Z/A factor connecting electron and material density is not constant for all elements (decreasing with growing Z and having irregularities, as does hydrogen, for example) and there are small admixtures of strongly Z-dependent effects (photoelectric absorption).

Therefore, the range of validity of an empirical function describing the density-transmission relationship is limited to certain chemical elements. In other words, it must not be assumed *a priori* that, for any material with variable chemical composition, a satisfying calibration of a density gauge might be possible.

In the case of silt-density measurement, theoretical considerations have anticipated only very weak fluctuations in radiometric density values due to chemical variations, according to the literature (Christiansen et al., 1982). Thus, it was not too surprising to find that there is an excellent linear relationship between the natural logarithm of transmission (intensity with reference to that of pure water) and density, and it is the same for silt samples from different sources, such as the Elbe or Weser rivers, as shown in Fig. 6. The experiments were conducted with large amounts of original material in tanks at the laboratories of GKSS.

In another laboratory experiment, the density resolution of the gauge has been studied. For that purpose, layers of different density were simulated by an aluminium step profile in a water tank. The thickness of a step, related to the total transmission length (25 cm), presents the density weighting factor (2.70 g/cm^3) by which aluminium contributes to the mean density. The results of the study are given in Fig. 7. The dotted line represents the nominal values, and the full line the measured densities. Although the experiment gives a rough impression of the resolution, it may not be transferred directly to natural conditions, since the beam geometry for extended layers differs considerably. This fact is indicated by systematically lower measuring values for small steps due to small-angle scattering effects.

For the vertical resolution achieved by the probe, the speed of the vertical movement also has to be considered. As the measuring cycle is 1/10 s, the vertical resolution for recommended lowering velocities of 30–50 cm/s is 3–5 cm. The cycle period is a compromise between the requirement of high resolution at reasonable lowering speed and the just tolerable statistical fluctutations of the counting rate. In principle, the fluctuations could be depressed by using higher radiation activities; this would imply an alteration of the existing electronic and data-processing concept and might require more effort in shielding, as well as more care in handling the probe onboard ship. With the

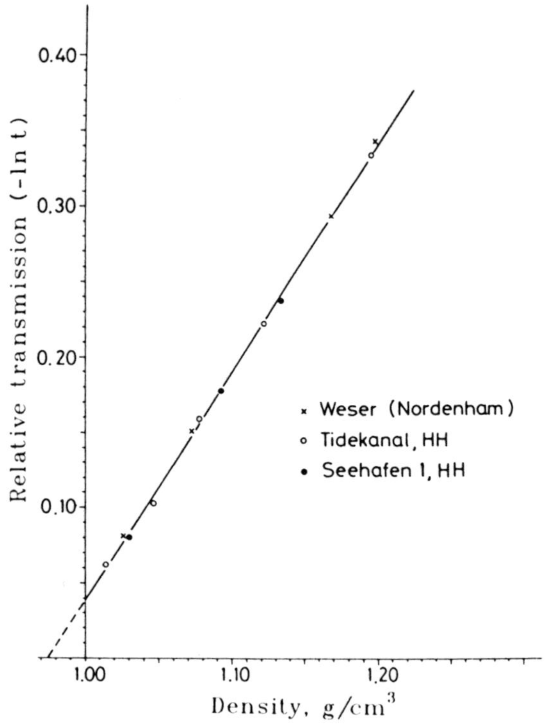

Fig. 6 Calibration curve for the radiometric silt-density gauge

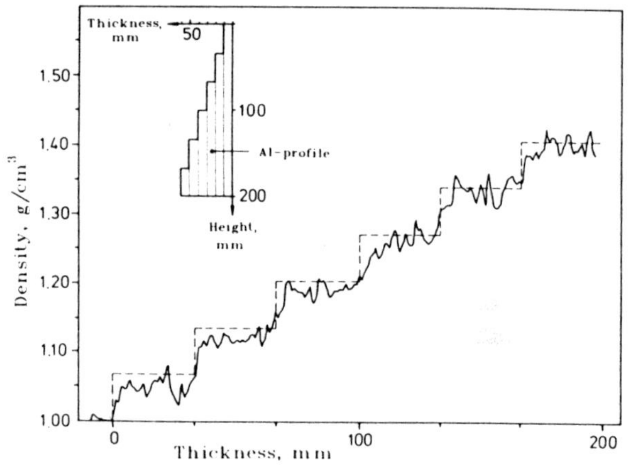

Fig. 7 Density resolution test using an aluminium profile: for details see text

actual layout of the gauge, the statistical fluctuations involve a standard deviation of 0.015 g/cm^3 for a single measured density value. The precision of density profiles is improved further by graphical smoothing or digital sliding averaging.

Many attempts have been made to assess the accuracy of the radiometric method. A comparison with echo-sounding failed because this technique is not quantitative in itself, as explained before. Another test together with sediment coring suffered from the fact that the two techniques could not be applied simultaneously at exactly the same position, so that tidal variations and local inhomogeneities prevented an exact comparison. In addition, it turned out that coring techniques seem to disturb soft sediment layers, or depress the horizons of hard layers. Some of the results are discussed in Fanger et al. (1985a, b). The conclusion is that radiometric density profiles have to be considered to be valid at least as long as rapid tilt movements are avoided.

Figure 8 shows that one prerequisite for good accuracy is fulfilled by reproduceability of profiles: the measurements were taken in a small harbour at Bremerhaven. The curves of the three consecutive profiles also confirm the predicted statistical fluctuations.

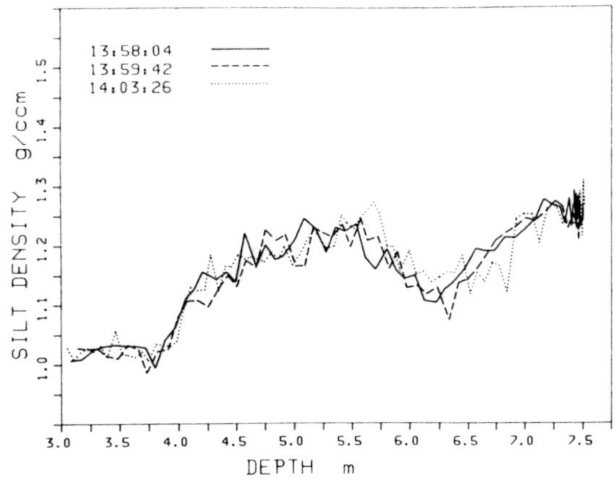

Fig. 8 Consecutive density profiling at identical positions in Bremerhaven harbour

IN SITU MEASUREMENTS OF SILT DENSITY

The first long-term operation of the radiometric density gauge occurred during a joint experiment (MASEX '83) on RV Victor Hensen, described in Fanger (1985b). Amongst many other activities, the vertical density profiles at three locations at equivalent distances of about 500 m along the river axis were measured every 2 hours for five consecutive tidal periods, day and night. The operational area was close to the small city of Nordenham, on the Lower Weser River, and the locations were chosen to coincide with an interesting small groove of 1.5 km length and 2–4 m depth, filled with silt.

The groove width of only 50 m and the inclination of the hard base below it proved to be a major problem in evaluating the data obtained. Therefore, the time-series of depth horizons of equal density reflected the navigational skill to hold position rather than revealing tidal changes caused by sediment transport directly. Nevertheless, careful analysis, taking into account the exact positional data of the Krupp-Atlas RALGO system, appeared to indicate an increase of the silt-thickness by about 20 cm during the total measuring time.

Although the pre-prototype of the probe had been used, with much less comfort in data acquisition and processing, no problems arose, and it worked continuously throughout the experiment.

In the course of over two years, numerous tests were performed in various basins of Hamburg harbour. Most of these experiments were initiated by the Hamburg administrative authority and were supported by the "Strom und Hafenbau" Office. More and more, the tests graded into systematic measurement sessions.

The first example of results from an early measurement in the Hamburg Vorhafen is given in Fig. 9. The horizons of equal density with levels of 1.02 (water-silt borderline), 1.10 and 1.20 g/cm^3 reveal the typical shape of a lens, which is the subject of many investigations nowadays. It seems that this kind of deposition at the entrance of harbour basins is produced by centripetal ground currents of flood-induced vortices (Christiansen and Radtke, 1985).

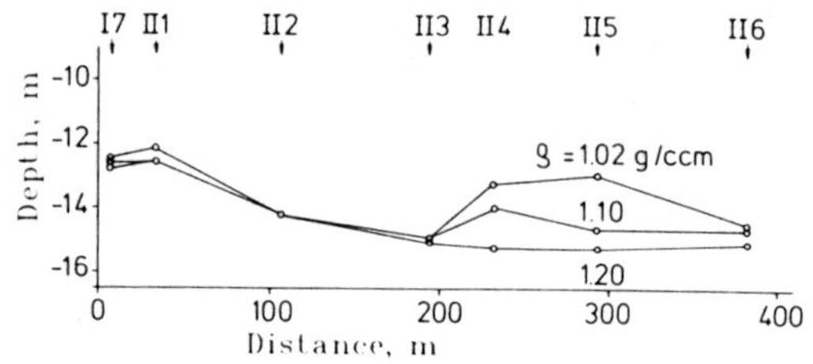

Fig. 9 Local silt distribution in a basin of Hamburg harbour

The second example shown refers to a scanning measurement carried out in the large Hamburg harbour area from Kohlfleethafen to Dradenauhafen over several days in November 1984. The mesh width of gamma-ray gauge profiling was about 50 m; all the measuring locations are plotted in the schematic topographic sketch of Fig. 10 (upper part). The lower part of the figure shows a patch of the total area, with lines of equal silt thickness, derived from the maximum gauge penetration depths. The isolines are drawn in steps of 0.5 m, and were shown to correspond very well with sedimentation rates determined otherwise. Again, the phenomenon of lens-shaped depositions at the entrance of the basin (upper edge) was observed.

The second version of the gauge (the one following the pre-prototype) has been given to the Wasser-und Schiffahrtsamt Bremerhaven for further investigations in the Lower Weser River, and a third one is just being completed for continuing research work in the Elbe River. Future gauges will be manufactured under licence by "Meerestechnik Elektronik (ME)" in Trappenkamp, Federal Republic of Germany.

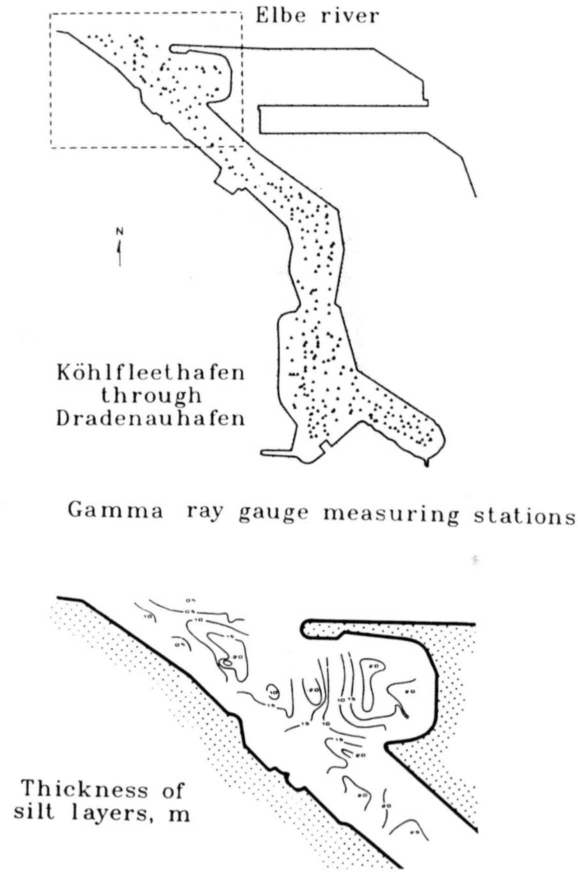

Fig. 10 Profiling scan in part of Hamburg harbour. Top, gauge measuring locations; bottom; lines of equal silt thickness

REFERENCES

Christiansen, H., Ohlmann, G. and Tent, L. 1982. Probleme im Zusammenhang mit dem Anfall von Baggergut im Hamburger Hafen. *Wasserwirtschaft* **72**(11), 385–9.

Christiansen, H. and Radtks, G. 1985. Strömungsverhältnisse und Sedimentation in ausgewählten Hamburger Hafeneinfahrten. *Gewässerkundliche Studie* nr. 4. Hamburg: Strom- und Hafenbau, pp. 1–20.

Fanger, H.-U., Bössow, E. and Kuhn, H. 1985a. Eine Gammasonde zur Schlickdichte-Bestimmung: ihre Entwicklung, Erprobung und Bewahrung im Feldeinsatz. *Die Küste* **42**, 189–207.

Fanger, H.-U., Neumann, L., Ohm, K. and Riethmuller, R. 1985b. MASEX '83, eine Untersuchung uber die Trubungszone der Unterweser. *Die Kuste* **42**, 171–87.

Fuhrboter, A. and Mittelstadt, M. 1984. Strahlungsmessungen zur Klassifizierung von Baggergut aus dem Hamburger Hafen. *Leichtweiss-Inst. für Wasserbau der TU Braunschweig, Bericht nr.* 575, May.

Hartmann, W. 1969. Messverfahren unter Anwendung ionisierender Strahlen. In: *Handbuch der Messtechnik in der Betriebskontrolle,* vol. V. Leipzig: AVG, p. 995.

Hellema, J. A. 1979. *Density Measurements* in situ. Report of the Lower Reaches Directorate of the Netherlands' Public Works Department, Dordrecht, (Holland), November.

Kroebel, W. 1973. Die Kieler Multimeeressonde. *"Meteor"-Forschungsergebnisse (A), no. 12,* p. 53.

Moxham, R. M., Tanner, A. B. and Senftle, F. E. 1976. *In situ* neutron activation analysis of bottom sediments, Anacostia River, Washington, D.C. *Mar. Technol. Soc.* **11**(1), 14.

HOVERCRAFT IN SURVEYING

B J Russell, A R E, Haslar, UK

Presented on **Friday 7th March 1986 at 11.35**
in the Clarence Conference Room

EMBARGOED AGAINST DELIVERY

48

Hovercraft in Surveying

Brian J. Russell, Admiralty Research Establishment,
Haslar, Gosport, Hants, PO12 2AG, UK

The advantages of the hovercraft principle of high speed and low footprint pressure have been exploited in a number of applications, particularly in marginal and difficult terrains. Since the late 1960s, hovercraft have been employed for both hydrographic and seismographic surveying. Initially, medium-sized hovercraft were employed, but more recent applications have involved smaller, more cost-effective craft. Techniques and equipment have been developed that enable hydrographic surveying to be completed at up to three times the speed achievable with more traditional methods. Similar benefits accrue in seismographic operations. It is conjectured that, as the advantages become more widely known, the demand for hovercraft in surveying is likely to increase significantly.

INTRODUCTION

Recent years have seen increasing pressures to exploit oil or other resources, no matter where they exist. This has resulted in the extraction of oil and natural gas from the North Sea and the exploration of the Irish Sea and other areas. Although the depth of water and severity of the environment created difficulties, objectives have been and are being satisfactorily completed. Similarly, overland explorations have presented no major obstacles.

However, there are other areas where surveying needs to be carried out, which present their own particular problems. Shallow-water areas, surf zones, tidal areas — particularly those that dry out — and "tender areas" such as the Arctic tundra, are typical examples. Surveying by traditional means is either difficult, or impossible, so alternative methods were needed. An alternative method becoming increasingly popular is to use medium-sized hovercraft, both for seismographic and hydrographic surveying.

In fact, the first usage took place in the late 1960s, with later developments leading to more efficient surveying. Surveys have taken place in a large number of locations, scattered around the globe. More recent applications have seen the use of smaller hovercraft, which have proved capable of completing a worthwhile job both efficiently and cost-effectively, often with the employment of ethnic personnel.

Wide ranging though past and current uses

are, the employment of hovercraft for surveying has only begun to scratch the surface of potential applications. The object of this chapter is to bring the advantages to the notice of a wider audience than has previously been attempted.

HYDROGRAPHIC SURVEYING

Background and Early Operations

Hovercraft confer two major advantages to hydrographic surveying: the ability to carry out surveys in shallow or tidal areas, and an increase in surveying speed from around 8 knots to nearly 30 knots.

The first major entry to this field occurred in October 1968, when Hovermarine (now Vosper Hovermarine) and the Decca Navigator Company (now Racal), signed an agreement with the East Pakistan Inland Water Authority (EPIWA) to produce a high-speed survey system. The resulting Surveymarine craft was launched in January 1969, and is based on the HM2 side-wall hovercraft, but having a much larger bridge to accommodate the surveying equipment. Additional fuel stowage increased the range from 140 to 400 nautical miles and, with the fitting of air conditioning and domestic facilities, the craft can operate away from base for up to five days.

Seafix, Hifix or Decca Main Chain were used to fix the craft position, with outputs to the data logger, a left/right indicator, used when manual steering was employed, and an autopilot for automatic operation, with the surveyor presetting requirements. The echo sounder employed was an Atlas Survey type AN6014, which makes simultaneous soundings on 30 Hz and 210 kHz. The transducer was mounted on a retractable "fish" hung below the hull amidships. Accuracy of sounding data is within 1% and soundings are printed on a paper-roll display and digitized for data logging.

The data logger recorded administrative data, time (from an internal electronic clock) and soundings, in alphanumeric form and as a punched tape. The former enabled onboard progress to be monitored and data checked, and the latter was used for computer processing to produce an X-Y plot of soundings.

Pakistan

Surveying speed was important in Pakistan, for the trade routes to Dacca, and area communications in general depend on good navigation of the many hundreds of miles of waterways in the Ganges delta. The delta area is notorious for constantly shifting shoals and sandbanks, and channels filling up.

The speed of surveying offers great potential to this sort of problem, because it allows a picture to be built up of shoal movements etc. This might then enable long-term solutions to be arrived at, with a significant cost saving if regular dredging can be avoided.

The Surveymarine operated successfully in Pakistan until that country's internal upheaval. At this juncture, the craft is understood to have been taken to the Singapore area, where it continues to operate.

Severn Estuary Survey

The Institute of Geological Sciences (now the British Geological Survey) required to carry out a bottom survey of the area around the proposed Severn Barrage in late 1980. This survey involved the collection of mud and sand samples within a 250 square mile (~650 km^2) area, some of which was only exposed for short periods at low tide.

At Air Vehicles AV Tiger was chosen for the survey, the area stretching from Penarth to Sudbrooke on the Welsh shore, and from south-west of Weston-Super-Mare to Avonmouth on the English coast. Sample sites were based on a 1 km^2 grid, position fixing being by triangulation, using a Motorola system.

The hovercraft used two bases, one near Weston-Super-Mare and the other at Portishead. An area was selected for the day's survey, the craft working out from the shore, taking surface or core samples as appropriate. The weather during the survey period was far from ideal, being typified by strong to gale-force winds, but operations were completed well within the allotted time.

The crew consisted of the pilot, hovercraft engineer, navigator and two geologists. The craft proved extremely reliable during this demanding operation, which often involved travel over shingle, sandbanks and mud-flats and journeys of 45 miles (72.4 km) when operating near the Welsh shore. In fact, no mechanical or operating problems were encountered.

The survey team were favourably impressed with the hovercraft's capabilities, particularly as some samples had to be collected from areas of soft mud unable to support a person's weight. In these areas sampling was carried out from the craft fore-deck.

Solway Firth

A similar exercise was conducted for three weeks in 1981, when another AV Tiger was employed in the Solway Firth. The object was to check the effects on the environment of any radioactivity that might be released from Windscale (now Sellafield).

Skima 12

Work has also been conducted with Pindair in the hydrographic survey role. Wemmel were interested in surveying a coastal area in Belgium and considered a hovercraft the best vessel for the task. The survey stemmed from a requirement similar to that in the Ganges Delta, to cover an estuary with silting problems in the shortest possible time.

Trials were completed with a German transducer, either carried on or under the hovercraft, with a recorder mounted in the hovercraft cabin. The sonar transducer was streamlined and could be raised when approaching shallow water. Test runs in up to 20 knots of wind showed that the system worked satisfactorily.

SRN6 Survey Hovercraft

Eurosense have been contracted by the Belgian Government to carry out hydrographic surveys in the coastal regions of that country. Both deep- and shallow-water capabilities were required, with speedy and efficient surveying being of paramount importance. These requirements led to the consideration and eventual adoption of an SRN6 MK1S amphibious hovercraft, modified for hydrographic surveying.

The craft had been previously used as a passenger ferry on the Solent route and extensive modifications were completed, in preparation for use as a survey craft. The trim and floor covering have been chosen for their hard-wearing properties, and tinted windows have been fitted, to give protection from glare when the sun is near the horizon. The main cabin has a large table for surveying tasks and houses an air-conditioned computer room and domestic facilities (see Fig. 1).

Fig. 1 The Eurosense SRN6 hovercraft, currently employed on hydrographic surveys in Belgium

Surveying Equipment

Identical auxiliary power supply units and fish deployment systems are fitted on raised side decks, with safety handrails at the periphery around the bow (see Fig. 2).

Each power supply unit consists of an Isuzu diesel engine driving a 12½ kVA Ansaldo generator, via the main power take-off. A hydraulic pump is driven from the front of the engine and provides the hydraulic power to lower and raise the fish deployment gear. The hydraulic systems are cross-connected, enabling both deployment systems to be operated from each power unit. The container is lined with stainless steel and is fitted with a fire wire detector and remotely operated fire exting-

Fig. 2 Details of the fish deployment gear on the Belgian SRN6 hovercraft

uisher. Mounted outside the diesel generator containers are the diesel fuel tanks, the diesel engine starting batteries and the hydraulic oil tanks.

Two depth transducers are mounted in a fish made from welded stainless steel sheet. The fish is streamlined to reduce drag and is fitted with a single vertical tail fin, for stability and to assist with alignment with the water flow. The fish is supported by a broad streamlined sword, fitted with two peripheral fences. A strong support bracket made from stainless plate attaches the sword to the main jib attachment bracket, also fabricated in stainless steel, via a main pivot bearing. The mechanical connection is completed via a weak link at the front of the mounting brackets. In the event of the fish or sword hitting an obstruction, or grounding, the weak link breaks, allowing the sword and fish to pivot back and clear the obstruction with minimal damage.

The fish depth is controlled by the onboard computer, the actual depth required being determined, to some extent, by craft speed. Electrical supplies and signals are supplied to and taken from the transducer by an electrical cable that runs down the inside of the sword. In the event of power or hydraulic failure, the deployment structure can be manually retracted with a turfer winch. The design speed for surveying is 27 knots, and it is likely that higher speeds will be possible in practice.

Operations

After the hovercraft was fitted out, it went to Belgium for the computer and other equipment to be fitted and set to work. The contract runs for 5 years, the initial period leading to the satisfactory demonstration of the fully automatic surveying system. Surveying is completed at 40 knots, this along with craft heading being displayed on the pilot's TV presentation. The required track is set up by the surveyor and appears as a dotted line on the screen. The hovercraft's actual track appears as a full line, which the pilot tries to superimpose on the required track.

Air Vehicles Tiger 12

The SRN6 is a relatively old hovercraft design, powered by a gas turbine, propelled by an open-air propeller, and being some 17 m long. Not all applications require such a large craft and, indeed, some potential areas of operation would not tolerate the noise generated by the propeller. One recent development has seen the use of an 8 m long Air Vehicles Tiger 12, which has an automotive engine and a ducted propeller, configured for hydrographic surveying.

The Tiger has been fitted out for an American customer, Tetra Tech Inc., and will be used for shallow-water hydrographic surveying in the Middle East. The depth transducer is mounted at the bow, this position being chosen to avoid the craft trim problems that would have been experienced with alternatives. The fish is suspended under an A frame, which pivots in two large triangular brackets. The frame is raised and lowered by two hydraulic rams, and can be raised by hand in the event of a hydraulic failure. The A frame and transducer support sword weigh 60 kg and deployment is usually undertaken at speeds of about 6–7 knots. The transducer 2 ft (0.6 m) immersion depth is set by lining up pre-positioned marks on the frame and craft. A weak link is fitted at the front of the sword support bracket, the sword being flipped back if this link is broken by impact with debris etc.

The port side of the cabin is fitted with a large

table, with a swivelling chair for the surveyor. A TV monitor is provided on the pilot's left and is used to provide track and other relevant information. Adjacent to the monitor are the two large push buttons which control the fish position. The electrical power requirements are met by two 12 V 96 Amp hr batteries, and a Honda APU driven 2.5 kW generator. This is mounted on a support structure on the port side of the craft: a similar arrangement on the starboard side carries another APU and the air-conditioning compressor and evaporator.

This craft is about to complete its first contract for surveying in shallow waters, where boats cannot be used. Tetra Tech are completely satisfied with the hovercraft's capabilities.

SEISMOGRAPHIC/GEOPHYSICAL SURVEYS

Background

Geophysical surveys are carried out in order to detect and locate subsurface geological features, and if possible, measure their size and some physical properties. This information may be required for engineering or prospecting purposes, such as when searching for economical deposits of minerals and oil.

During surveys, a set of measurements are taken, either in a horizontal line, or vertically, in a borehole. These measurements are of changes in a particular force field with space or time, which yield an indication of the nature and structure of the subsurface. A number of survey methods are available, the one that is chosen depending on the information required. Magnetic and gravitational surveys make use of natural force fields. Seismographic surveys make use of the elastic properties of rocks, and require the introduction of external energy, the distance between the source and detector being variable, leading to more precise techniques.

Seismic surveying is the most advanced method and yields the most information. It involves the generation of an elastic pulse near the surface of the ground, and the measurement of resultant ground motion by small seismometers. Measurement of the time intervals between the generation of the pulse and its reception at seismometers at various distances gives the velocity of pulse propagation in the ground. Because the ground is not elastically homogeneous, the velocity will vary with depth, velocity boundaries usually coinciding with natural geological boundaries. This type of surveying has been extensively used in the search for oil, although it detects the structure likely to contain oil rather than the oil itself.

Method

The most usual method of pulse generation is to fire a shot of high explosives in a hole (up to 100 ft (30.5 m) deep), tamped with water. If the survey is in shallow conditions, smaller charges and depths are employed. Instead of limited-duration pulses, vibrators can be used to generate a continuous wave to a limited surface area. Both techniques result in complex ground motions, due to reflection and refraction at subsurface geological features.

The second method of applied seismology is known as the reflection method; the spacing between the shot and the seismometer is small compared with the depth of investigation. In principle, information down to several thousand feet can be recorded, and requires shots of a few tens of pounds of high explosives. The method lends itself to routine techniques in which the ground is covered by a series of profiles, the rate of coverage usually being dependent on the time required to drill shot holes.

In conjunction with Seismographic Services Ltd (SSL), Hovertravel/Hoverwork have executed a number of surveys, including such areas as the Arabian Gulf, the Mediterranean coast of Africa, Morecambe Bay, the North Sea and the Waddenzee (Holland).

Early Operations — Marine Laying

Most early operations involved the use of two hovercraft (see Fig. 3). The first craft towed the marine cable, and contained the shot-firing and recording equipment, and the navigation equipment. The crew carried to operate the surveying equipment usually consisted of an

Fig. 3 A typical two-hovercraft seismographic operation

observer in overall charge and who operated the shot-firing and recording gear, a surveyor who covered navigational requirements, and two back-up personnel.

The second craft carried the explosives and was fitted with the shot-laying equipment. It was also used for general duties, carrying the additional personnel, such as labourers, necessary for the operation. The shot laying was done by two people, the explosives expert charging the shot prior to laying, and the other laying or streaming the shot.

The towing craft would pick up the towing cable and proceed at about 3–5 knots to the first shot point, where it was brought to rest.

While the towing craft was doing this, the shot laying craft would maintain a position slightly astern and around 20 m to port or starboard. Just before the shot point, the laying craft would overtake the towing craft and be brought to a halt over the shot point. The shot would be laid and fired, provided the towing craft was in the required position.

Once the shot was fired, the towing craft would lift off and proceed to the next shot point, 100 m away. The shot-laying craft would execute the most appropriate pattern to arrive in position for the next shot. A two-minute shot rate could be satisfactorily achieved.

Some operations were undertaken using a single hovercraft, which performed all duties, including shot laying and firing. However, difficulties were experienced — particularly on crosswind and downwind headings — and the hovercraft currently operates with workboats performing the shot laying and firing function.

Hovercraft Deployment - Land Laying

Land laying is a laborious process, involving the separate operations of: selection and marking of line of shoot, cable laying, borehole slushing and laying of charges, shot firing and, finally, cable lifting.

Two craft are employed, one housing the recording, shot-firing and navigation equipment, together with the facilities for cable deployment and retrieval. The second craft carries the explosives and bore flushing equipment, and can also handle cables.

The line of shoot is marked by a survey team, the 100 m shot points and 600 m cable lengths being indicated. The cable is then laid along the line, each 600 m length being connected to its neighbour(s), a minimum of two lengths being used before shooting commences.

The shot-laying craft is then positioned in turn at each shot point, some 20 ft (6 m) from the line, where two 15–20 ft (4.5–6 m) deep holes are sunk. Charges are laid, complete with detonators, and the shot wires terminated at a marker peg. Once the shots are laid, the craft proceeds to the middle of the line. Each shot is fired and recordings taken. At the same time, the second craft is preparing the next set of boreholes and the surveying team is marking further positions.

Surveying Equipment and Hovercraft Modifications

The standard SRN6 employed on seismographic surveys were fitted with side decks, to facilitate equipment carriage and to enable the crew to work over the edge and bow of the craft.

The shot-laying hovercraft originally used had two diesel-powered water pumps on either side deck for borehole flushing. Water was carried in four 45 gallon (204 l) water barrels. An echo sounder was installed, with the probe on the starboard bow. Radio shot-firing equipment, clothing and equipment lockers, a plotting table and an echo-sounding operator's position were fitted in the cabin.

The towing craft was fitted out with the seismographic equipment, which included the position-fixing equipment and track plotter. In addition, the 48-channel digital recording equipment, shot-firing equipment and observer's and surveyor's positions were installed. On the side decks were installed the explosive storage and deployment gear and a diesel-powered 250 V a.c. generator (enabling the seismographic equipment to be independent of craft electrical supplies). The original 48-channel, 1200 m long cable was towed from a bridle attached to the craft's rear strong points.

Over the years, apart from the adoption of the single hovercraft and workboat technique, there have been other changes. Lighter and smaller equipment availability has meant that recorders with more channels could be used. Also, longer cables are being employed, as hydroplanes are no longer contained in a streamer, but are contained in reels suspended from the towed cables. The current cable employed is 1.25 miles (2 km) long, with 48 hydrophones at 50 m spacing. The next cable, to enter service shortly, will have the same length, but 96 hydrophones will be fitted at 25 m spacing. Other changes include the distance between shots, generally reduced to 50 m, and the increased weather limitations for craft operations.

Recent Activities

Hovercraft have been used for surveys on a continuing basis since the first operation in the Waddenzee in 1969.

Operations in the 1980s have been centred on the Middle East, with seismic surveys being conducted in Abu Dhabi (in very shallow water and coral reefs), Iraq (in inland lakes with mudbanks and small reeds), and Egypt (in shallow water and coral reefs). For the past four years an SRN6 has been based at Gemsa in the Gulf of Suez, where annual overhauls are undertaken.

The latest contract for surveying in Abu Qir Bay, to the east of Alexandria, clearly illustrates two major advantages of hovercraft in surveying; easy and rapid transits to alternative areas; and the completion of operations without the need for extensive support facilities. This latest deployment involved the first transit of the Suez Canal by a commercial hovercraft.

Leaving the main base at Gemsa, in calm conditions, on 21 July, the craft reached Suez Bay, 175 nautical miles away, in just under 9¼ hours underway. The next day was taken up by the paperwork needed before the canal could be entered, so departure for Ismailia did not occur until 23 July. Speed during the canal transit was limited to a maximum of 10 knots, but once the Great Bitter Lake was reached, the local pilot on board allowed this to be exceeded, and the 95 nautical miles to Port Said were completed in just under 7 hours.

The final 115 nautical mile leg to the new base at Rosetta in Abu Qir Bay was completed on 24 July in just over 6 hours, the prevailing northwesterly winds at the mouth of the River Nile causing conditions to deteriorate during the final 90 minutes.

Operating 8 hours a day, 5 hours at the survey site and 3 hours on positioning trips, the crews finished this latest contract during the third week in August. Conditions in the area were generally good, and the hovercraft was the only vehicle that could conduct surveying close to the shore in the surf zone. On occasions, the craft had to wait until conditions reduced to those acceptable to the Zodiacs used to carry the explosives. The craft can operate in 1.25 m of surf, and the cable was taken to the operating area by road, snaked out on the beach, and connected to the hovercraft which then pulled it into the water. After completion of the contract, the craft returned to the Red Sea base at Gemsa for a minor refurbishment programme and the installation of more sophisticated navigation and recording gear by SSL. This will double the rate of data collection capability to produce better pictorial records of the survey location. The first contract using the new equipment should be underway soon.

Support of Chinese Exploration

Geophysical Service Inc. (a subsidiary of Texas Instruments) has been awarded a number of contracts by the People's Republic of China, to jointly operate seismic crews and processing centres. One of the objectives is for the Chinese to take over the operations once the contracts have been completed.

One of the survey areas is in the Yellow River Delta, with the environment ranging from low scrub and bush, through reeds, to mud and water. The distance between high- and low-water marks can be as much as 5 miles (8 km), crossed by occasional sea-defence dykes and tidal gullies up to 3 ft (0.9 m) deep. The surveying activities are undertaken from very slow-moving large tracked vehicles called "buggies". Hovercraft were chosen for the fast transport of personnel and equipment from surfaced roads to the buggies.

Three Griffon Hovercraft 1000TD, 8.5 m long craft are used, a 190 hp Deutz diesel engine driving the lift fan and ducted propeller (see Fig. 4). These craft have undertaken other specific tasks: the movement of survey parties and their equipment rapidly over the site, as the seismic line progresses to move geophones and related equipment from the back to the front of the line; to convey bulk liquids, water, diesel fuel, etc.; to convey mechanics and equipment to broken-down vehicles; as a mobile platform from which to detonate charges; and to lay out and recover geophone strings and related equipment.

The personnel provided by the Chinese for the hovercraft operation were local peasants and so added to the fundamental language problem was the fact that none had operated a machine in their lives before. The fact that, in 6 months, 40% of the original intake were reasonable pilots and mechanics, says a lot for the design and the personnel responsible for the hovercraft operation and training. The Chinese currently operate the craft only in winds below 25 knots, in daylight hours and in good visibility. Outside these conditions the GSI personnel take over, and all operations are limited to less than 35 knots of wind.

The alternatives to the use of a hovercraft all have disadvantages: flotation tyred vehicles are very slow and difficult to manoeuvre over water; tracked vehicles are too slow, and limited by water depth and excessive seismic noise; helicopters are too expensive and their maintenance too involved; and airboats have no overland capability. The hovercraft have had minor problems, and greater manoeuvrability, better fendering, stronger side bodies and a 2 tonne payload capability would be advantageous. These could all be achieved by design modifications, and do not detract from the basic hovercraft capability of achieving task objectives that are virtually impossible or very difficult and/or costly by other means.

Fig. 4 A Griffon 1000TD hovercraft, three of which are supporting seismographic surveying in China

CONCLUDING REMARKS

This chapter set out to describe the advantages that hovercraft can bring to the various surveying tasks that need to be undertaken. These advantages are particularly significant in marginal terrains, where to use alternative vehicles is virtually impossible, or at best very costly. Examples of typical hydrographic and seismographic operations have been described, in order to illustrate the hovercraft's advantages.

Basic commercial designs have been used, with minimal modifications to suit them for the specific roles involved. Increased utilization of hovercraft will no doubt lead to suggestions for design improvements. However, at last there is a true amphibian available, whose reliability and capabilities make the exploration of shallow water and marginal terrains practical.

HYDROGRAPHIC SURVEYING EDUCATION IN NIGERIA

Dr F A O Egberongbe, University of
Lagos, Nigeria

Presented on **Friday 7th March 1986 at 12.00**
in the Clarence Room

EMBARGOED AGAINST DELIVERY

49

Hydrographic Surveying Education in Nigeria

F. O. A. Egberongbe, University of Lagos, Akoka, Nigeria

INTRODUCTION

Since the arrival of seamen in the Bight of Benin, continuous survey and sounding of the bars rivers and ports of Nigeria have been carried out by master mariners, shipping agents, the Royal Navy, oil companies, dredging companies and the Nigerian Ports Authority. The Ports Authority is responsible for surveying the approaches and the ports proper, and also for the production of charts.

With the discovery of oil in the mid-1950s, it became necessary to establish facilities for the export of crude oil. Since the early discovery was mainly in the riverine area, the need to maintain a routine but efficient hydrographic survey department became even more significant.

Prior to the establishment of various schools, training in hydrography was either an overseas programme or an 'on-the-job' process. For a long time, and until quite recently, the apprenticeship system was combined with formal training. The Nigerian Navy only recently came into existence, and hence formal training in hydrography could not commence until very late. Prior to the establishment of the sea school, most companies, and the Nigerian Port Authority, participated in short training courses for hydrographic surveyors, mainly in the United Kingdom. Those who benefited from this type of training were graduates in physics, mathematics or geography who had worked in a marine environment for some time. They were mostly land surveyors who then became hydrographic surveyors.

Independence brought with it an awareness of our position and the need to catch up in every aspect of development, including maritime activities. The acquisition of more ships and the improvement and building of new ports were added incentives for the training of hydrographic surveyors. The crisis of 1966, though it affected the country's economic life, brought to light the importance of maritime power and the necessity for channel surveillance. The advent of larger ships and supertankers has made great demands on port management since such vessels require deeper and wider channels which, in order to reduce downtime, must be accessible throughout the day and night, irrespective of tidal conditions. As a consequence, following an investigation carried out by NEDECCO in 1957, new ports were created, especially in the oil mining areas of the country, while old ports had to undertake major dredging of critical navigation areas.

Moreover, the dredging of ports and river channels increased so dramatically in the late 1960s and early 1970s that hydrographic surveying for dredging control formed a substantial part of the outlay on development both for the Nigerian Ports Authority and for the oil companies.

TRAINING OF HYDROGRAPHIC SURVEYORS

Hydrographic surveying assistants are trained through either the apprenticeship system on board merchant ships or through graduate courses of training in naval school.

According to the general requirements for the qualification of hydrographic surveyors (NP135), an assistant is defined as having a 3rd class Assistant Surveyor or lower qualification, while courses for technologists who may take charge terminate with the award of 1st class certificate. Full details are given in (GIHS), Section 2.

Because of the vast investment involved, it became mandatory for the Ports Authority, as well as some oil companies, notably Shell Petroleum Development Nigeria Limited, to embark on a limited hydrographic surveying training programme for their Nigerian senior staff, who were mostly land surveyors. This training took place overseas.

The Nigerian Navy, realizing the huge investment involved in the training of men abroad, soon embarked on local training by taking over the privately owned School of Nautical Science in Oron, while opening its own School of Maritime Science in Port Harcourt some few years later (1981). Both schools cater for technical/technologist (ratings) recorder level and 1st class assistant surveyor level. Both grades can carry out limited hydrographic surveys without responsibility for the accuracy of the survey.

Because of the high demand for hydrographic surveyors, higher institutions of learning, notably the University of Nigeria (Enugu Campus) and the University of Lagos, introduced courses in hydrographic surveying in their land surveying curriculum, while Rivers State University of Science and Technology did so at postgraduate level. As to be expected, courses on nautical science, naval sciences and shipbuilding do not yet form part of their curriculum.

Unlike its counterpart on land, hydrographic surveying did not involve training in the higher institutions until recently, and even now only very few such institutions offering training in this subject exist throughout the world. The majority of hydrographic surveyors are retired non-commissioned officers or master mariners who have had several years of sea experience. The percentage of those with formal training is negligible. Under these circumstances and coupled with the new Law of the Sea, as well as the demands on port management for deeper and wider channels to accommodate large capacity ships, including huge oil tankers, most Third World countries have to embark on massive training of their personnel in the few schools available in the advanced countries while simultaneously trying to establish their own sea schools for the training of technicians and technologists. Additionally, some encourage higher institutions such as universities to include elements of hydrographic surveying in their land surveying curriculum.

In addition to government-established schools, the Aiyetoro community in Ondo State established a sort of sea school which serves the community's fishing industry. Here no formal training in hydrographic surveying is given, except for the interpretation of echo from a school of fish noted on the plot from an echosounder. The Aiyetoro Community School basically specializes in shipbuilding and nautical science. Similarly, the fishermen rarely go beyond 20 km from shore and, even then, the area is familiar to them.

The School of Nautical Science in Oron offers various courses at the technician and technologist level and is open to members of the public. Equivalent to the technician levels are the 4th class (probationary grade) and 2nd class (intermediate) levels. The final examinations, or 1st class hydrographic surveyor (technologist)

Basic	*Intermediate*	*Final*
Simple survey instruments, including distance-measuring equipment; simple map projection systems; basic survey techniques and marking and coast lining; fixing position afloat; Astro and navigation seamanship; tides and tidal streams; sweeping; elementary oceanography; geophysics; Draughtsmanship and plotting	All subjects in 1st year but in more detail with more emphasis on navigation, seamanship, astronomy at sea. Also elaborated upon are tidal streams, current heights, levelling, position fixing afloat, triangulation and control	Subjects are: instrument geodesy, air photography, soundings, oceanography, projections, triangulation and control, and geophysics A Hydrographic Surveyor, 1st class, is expected to have a thorough knowledge of the whole of the syllabus for Hydrographic Surveyor 2nd class, but with wider experience

examinations are partially taken overseas. This is because the school has yet to gain international recognition, having only been founded less than ten years ago.

The courses offered in each year are listed above.

These courses are designed in such a way that one leads to another. Admission into the six-month 4th class course is given to School Certificate holders with credit level passes in mathematics and English and any other two subjects, preferably in the sciences. On successful completion of this course the students go back to their sponsors for a period of not less than two years' sea experience. The intermediate courses are offered to candidates recommended by merchant ship captains of several years' experience or, alternatively to candidates who can provide a folio of projects carried out on their own and attested by their sponsors. The final course can only be considered after a further 18 months of experience at sea, subject to good conduct and the ability to command a ship.

As would be expected, these courses emphasize practical applications. Consequently, students are examined in both theory and seamanship, including navigation. At the moment there is no provision for final courses at this school.

In addition to the School of Nautical Science, the Nigerian Navy has recently established its own school, which is similar to the one described above. The Navy School is fashioned according to the General Instructions for Hydrographic Surveyors (GIHS). The final course, however, will take place at the Naval School for Cadets in Lagos. This course is equivalent to the Higher National Diploma in Hydrographic Surveying.

TRAINING OF PROFESSIONAL HYDROGRAPHIC SURVEYORS

The training of professional hydrographic surveyors in Nigeria at present takes place in three of the country's universities as part of the programme in land surveying. Most of these programmes were fashioned to meet environmental requirements. Emphasis is placed on port management and on inland waterways.

Seamanship and navigation do not at present form part of their curriculum. Moreover, experienced naval officers who could impart this knowledge successfully are not available to do so as the Nigerian Navy itself is relatively new. Even then, due to the poor financial situation, intensive seamanship training will have to be suspended until such time as adequate arrangements can be made with the Nigerian Navy and Nigerian Ports Authority.

The three universities involved in the training of hydrographic surveyors are the University of Nigeria (Enugu Campus), the Rivers State University of Science and Technology, and the University of Lagos.

University of Nigeria, Enugu Campus

The Department of Surveying, Photogrammetry and Geodesy of the University of Nigeria started a degree programme in surveying in 1962. At present, the department offers a five-year programme of study in surveying leading to a BSc (Hons) degree. According to the departmental brochure, 'the courses are designed to produce surveyors with sound knowledge in theory and practice of the engineering, scientific, technological and professional aspects of Geodesy, Photogrammetry and Hydrography'. Normal entry into the programme is through the joint matriculation examination for holders of a secondary school certificate with credits in five subjects, including mathematics and physics. However, holders of an A level certificate in mathematics and physics or the equivalent are considered for a four-year programme; such students join the 2nd year of the five-year programme.

The subjects offered in the first year of the five-year programme consists mainly of mathematics and physics courses. The four-year programme is as follows:

Second year	*Third Year*
Mathematics	Electronics and radio systems
Computer programming	Large-scale surveying I
Statistics	Topographic surveying
Social science	Survey computations
Fundamental surveying	Fundamental photogrammetry
Cartographic interpretation	Practical photogrammetry I
Draughtsmanship	Hydraulics for surveyors
Practical surveying I	Swimming and water safety
Use of English	Practical surveying II
	Elective

Seamanship and navigation do not currently form part of the curriculum. Moreover, experienced naval officers who could impart this knowledge successfully are not available to do so as the Nigerian Navy itself is relatively new.

Fourth year	*Fifth year*
Land registration	Large-scale surveying II
Map reproduction	Electronic surveying II
Electronic surveying I	Geodesy II
Practical surveying III	Geodetic astronomy II
Geodesy I	Practical geodesy/astronomy
Geodetic astronomy I	Map projections
Practical geodetic astronomy I	Photogrammetry II
Adjustment computations	Project
Photogrammetry I	
Hydrographic surveying I	
Practical photogrammetry II	
Practical hydrography I	

Elective courses in the third year include French, German, computer science, town planning, mathematics and statistics. In the fifth year, electives include hydrographic survey II, physical geodesy, engineering photogrammetry and analytical photogrammetry. Although organized industrial training does not at present form part of the curriculum, efforts are made to place students in places where they can gain useful practical experience during the long vacations.

The hydrographic surveying programme includes the following: tides and tidal theory planning and chart preparation, positioning, soundings, automated surveying, acoustic wave propagation and instrumentation.

University of Lagos, Lagos

Undergraduate training began at the University of Lagos in 1974. At that time an integrated course programme led to a BSc (Surveying) degree in four years. In October 1976 the undergraduate curriculum was changed to a five-year sandwich course, so as to allow for more practical training of students under actual working conditions.

The basic subjects of mathematics, physics, plane and geodetic surveying, photogrammetry and cartography are taught during the first 3½ years. The student spends the second half of the fourth year in industry in addition to the three months already spent at the end of his third year, so as to enable him to gain practical experience while he is studying. The final year is devoted to higher surveying methods, quality control in surveying, geodesy, hydrographic surveying, photogrammetry, cadastral surveying (survey laws and regulations) and applied mathematics methods. In addition, to obtain a degree each student must undertake a small-scale research project, and a high premium is placed on his performance in practical work during the course. The students are also expected to perform well during the industrial attachment periods.

The entry requirements for the degree programme require that candidates must possess a secondary school certificate with credit passes in at least five subjects, including mathematics and physics, in addition to a good pass (at least 50% and above) in the Joint Matriculation examination set by the Joint Matriculation Board on Examinations. However, candidates may be admitted into the second year of the five-year programme provided that they possess GCE A level passes in mathematics and physics in addition to satisfying the minimum matriculation requirement of the University of Lagos.

Year I
General engineering
Preliminary courses in mathematics and physics (including mechanics) as well as history of surveying
Introduction to surveying and technical drawing

Year II
Mathematics (including programming)
Fundamentals of electrical engineering
Classical mechanics I
Optics
Practical physics
Large-scale and engineering surveying
Cartography and map projections
Elements of environmental studies
French
General African studies

Year III
Numerical analysis
Statistical analysis
Classical mechanics II
Astrophysics
Topographic and geodetic surveying
Photogrammetry I

Year IV
Potential theory and spherical harmonics
Applied maths methods I
Photogrammetry II
Mining, surveying and town planning
Cadastral surveying I
Field and geodetic astronomy
Adjustment computations and electronic surveying
Geology
Geophysics and hydrology

Year V
Applied maths methods II
Photogrammetry IV
Hydrographic surveying
Cadastral surveying II
Geodesy
Project
Economics, law and industrial management
Technical communications

The hydrographic surveying programme has two other options in Year V. These are near shore surveying and offshore and ocean surveying. The basic hydrographic surveying includes such topics as oceanography, which involves basic concepts, marine circulation dynamics and meteorology, tides and tidal streams, position fixing afloat, tract control, sounding surveys, sweeping surveys, airborne and satellite techniques, coastal topography, sedimentology (interpretation and processes) physical oceanography and Law of the Sea.

Rivers State University of Science and Technology

This university is currently offering hydrographic surveying under hydrographic engineering at Master's degree level, and it is open to graduates with a surveying engineering or physical science background. Such candidates must possess an honours degree at 2nd class (lower) in his area of specialization. The course lasts two years and four semesters.

The subjects offered in the two-year programme are as follows:

Year 1/Semester 1
Survey and sea laws
Principles of computer science

Year 1/Semester 2
Geotechnical engineering
Numerical analysis

Engineering surveying
Geodesy and map projections
Photogrammetry and remote sensing
Fluid mechanics and hydraulics
Electronic surveying
Adjustment computation
Environmental hazard management
Hydrographic surveying
Practical hydrographic surveying

Year 2/Semester 1
Mathematical analysis of data
Advanced hydrographic surveying
Electronic surveying and navigation systems
Satellite and marine geodesy
Applied hydraulics and river engineering

Year 2/Semester 2
Thesis
Physical oceanography and coastal processes
Practical hydrographic surveying

Academic Recognition

From the courses listed above, hydrography could be considered as an element of a larger unit – land surveying. The addition of nautical science and seamanship as well as six months' practical experience at sea to the programmes will be of advantage to a category-A Hydrographic Surveyor of the International Advisory Board on the Standards of Competence for Hydrographic Surveyors. Nevertheless, there is need for academic recognition at the Board's level. If this had been the case, the additional requirement would be to include a test of practical competence on the completion of practical experience as is usual with other professional organizations.

Moreover, the concept advocated above in the three institutions is that of a system of essential courses combined with modular optional courses, the essential basics being mathematics and statistics, general physics, land surveying geodesy and photogrammetry, tides, precise positioning, sounding, sonar and sweeping and such other courses as cartography, oceanography, astronomy, geology and geophysics, and Law of the Sea. There is of course no doubt a need for a more practical content to the entire system.

CONTRACT HYDROGRAPHIC SURVEYING

Nigeria is a developing country and is endowed with a lot of minerals including oil and gas. This is perhaps the main reason why heavy traffic is a daily sight at her harbours and seaports. Independence, coupled with the experience of the last Civil War, have brought about an awareness of need to catch up in every aspect of development including maritime activities. This and the advent of larger ships and supertankers contributed to the practice of contracting surveys in order to reduce the time needed to construct new but modern seaports and harbours as well as reduce downtime of vessels at such ports.

While the country pays over and above the international rate for such ventures, experience has shown that the country is yet to get its money's worth. The training of personnel to keep the system in operation after work has been completed has not been achieved. The country's hopes of industrial work experience for students from its institutions of higher learning in the form of attachment both locally and abroad (parent company) are yet to be realized. Most companies see their presence in the country as an opportunity to make a quick profit without any investment in the country's education.

Surveys under contract in our country have been carried out on a turnkey basis, and often these have not been of the highest standard. The rapid growth in the numbers of firms claiming competency to undertake hydrographic work has not been matched by any increase in better training facilities. Furthermore, the number of professionally qualified sea surveyors is relatively small and, even then, nearly all of them are employed in managerial positions and not in the field. As a result the work carried out has not always been of the highest standard due to insufficiently experienced field staff.

I believe that to accelerate the transfer of technology and, more important still, to increase the growth of the profession, the recipient survey companies need to invest directly in education in the host country by sharing their experience with the local personnel as well as encouraging the local industrial work experience schemes. Suggestions as to their requirements as well as observations on such students as may come their way, should flow from industry to institutions of higher learning. These institutions would benefit from the wealth of experience of the companies and in turn would gain by recruiting competent staff. Through their interactions with companies the survey teachers are made aware of problems facing the profession, while the companies benefit from the latest ideas on instrumentation and methodology.

CONCLUSION

In this paper an attempt has been made to review the development of hydrographic surveying education in Nigeria and to describe the various training programmes in the country. I have also shown that hydrographic education is based on a modular system with emphasis on relevant mathematics, statistics and physics courses, including computer programming and adjustment computations. The essential specialization includes tides and tidal analysis, precise positioning, sounding, sonar and sweeping, geodesy, terrestrial surveys, photogrammetry, oceanography (physical), cartography, astronomy and measurement science, while optional courses include such areas as coastal engineering, offshore surveys, geology and geophysics. It is also noted that, though practical training is presently restricted, this could be improved through cooperation between the industry and institutions of higher learning.

It is obvious that if the hydrographic surveying profession is to grow in outlook and widen its horizons, more trained personnel are needed. In turn this means that steps must be taken to increase the production of trained indigenous manpower in the country through the judicious use of existing facilities. Cooperation between well-established institutions and the multinational companies operating within the country is likely to produce some satisfactory results. Finally, external help in the form of assistance from older establishments outside the country would also go a long way to solving our mutual problem, that of building a dynamic professional group throughout the world.

ACKNOWLEDGEMENTS

I wish to thank the Hydrographer of the Navy, the heads of the departments of surveying and registrars of institutions cited in this report for making their course programmes available to me for the purpose of preparing this paper.

REFERENCES

Adebekun, O. 1980. Training of surveyors in Nigeria. *Paper presented at the Departmental Seminar Series*, Department of Surveying, University of Lagos, Lagos, Nigeria.

Ayeni, O. O., Egberongbe, F. O. A. and Fajemirokun, F. A. 1983. Surveying and national security. In *Proceedings of the Nigerian Institution of Surveyors*, Owerri, Map Maker.

Campbell, R. J. 1980. Hydrographic surveying for ports and approaches. *International Hydrographic Review, Monaco*, **LVII** (2).

Ekblom, R. 1982. Hydrographic Surveying Requirements for the Third World Ports. **International Dredging and Port Construction**, I(2).

Ewing, G. N. 1977. Report of IHO/FIG Working Group. In *XI International Hydrographic Conference on Education and Standards of Competence for Hydrographic Surveyors.*

Fajemirokun, F. A. 1981. Surveying Education in Nigeria. In *Proceedings of the International Federation of Surveyors; Commission 2 (27.3), XVI International Congress*, Montreux, Switzerland.

Sathaye, N. N. Cmdr 1980. The requirements for hydrographic surveys in ports and anchorages. International Hydrographic Review, Monaco, **LVII** (2).

DOUBTFUL HYDROGRAPHIC DATA

G L Haskins, International Hydrographic
Bureau, Monaco

Presented on **Friday 7th March 1986 at 12.25**
in the Clarence Conference Room

EMBARGOED AGAINST DELIVERY

50

Doubtful Hydrographic Data

Geoffrey L. Haskins, Professional Assistant (Hydrography) to the Directing Committee, International Hydrographic Bureau, Principality of Monaco

Over the years, mariners have reported finding shoals and reefs, islands, and many other features during the course of their voyages. Unless these have been verified as actually existing and their position fixed with the least water depth over the associated shoals, they are depicted on published navigation charts as "doubtful data". Existence Doubtful (ED), Sounding Doubtful (SD), and Position Doubtful (PD) are marked alongside such features on the charts.

The documentation supporting many of these historical data is very sparse. Their verification by searching is an expensive and time-consuming matter during which every effort must be made to prove (or disprove) the reported feature conclusively; otherwise a great deal of money and resources will have been expended to no purpose.

The International Hydrographic Organization has recently made a study of how best to ensure that positive charting action can be taken after searches for doubtful features. The result is the publication of Standard Criteria by which it may be judged whether or not a search can lead to the doubt being eliminated. Recent developments in instrumentation and methodology for positioning and remote sensing by sonar systems and satellite imagery will enable some of the thousands of such features to be removed or confirmed on the charts.

A VOICE FROM THE PAST

> "There is little chance of finding a reported shoal in the open ocean by directing the course of the searching vessel to the geographical position assigned in the report and endeavouring to re-discover the shoal by passing in its vicinity with look-outs posted aloft or with a sounding line suspended in the depths. Unsuccessful proceedings of this kind have sometimes been cited in disproof, and it is, therefore, of interest to arrive at an estimate of the expectation that would thus be justified ... it appears that there would not be one chance in hundreds of finding the shoal by thus performing the search for it."

Thus wrote Littlehales in 1933.

A Mariner's Tale

Thirteen years after Littlehales wrote the above in the *International Hydrographic Review,* early in the New Year of 1946, His Britannic Majesty's Landing Ship (Tank) 324 passed, westbound and with a strong tidal stream under her stern, through the Balabak Strait on passage from the Pacific towards Labaun off the West coast of Borneo. It was dusk; as the tropical night fell under an overcast sky, it was decided to stand-on towards the west into deeper water before heading southward to avoid the congested and shoal-littered inshore passage down the coast of what is now Sabah.

The navigator had an uncomfortable feeling that his dead reckoning was being influenced, either ahead or astern, by the strong streams that flow through the Balabak Strait. If the westing was too great the ship might find herself in the vast "not surveyed" area in the middle of the South China Sea. There were no radar or modern navigation aids other than the magnetic and (unreliable) gyrocompasses, chronometer, sextant and an antique Kelvin hand sounding machine; to make matters worse, the only rotor for the patent log had parted company with the log-line a thousand miles astern. He felt underequipped and isolated. At the age of nineteen, he was the only professional seaman officer on board. The captain, in civilian life the stage manager of a London theatre, and other officers were, albeit competent seamen, "hostilities only" and virtually devoid of recent intimate acquaintance with the art of astronomical navigation out of sight of land.

As the ship continued westward, the navigator expressed his doubts to an uneasy commanding officer and suggested a sounding be taken with the dormant, never-before-used sounding machine. There is always a first time — and this was it! All was prepared: the lead streamed, the "feeler" placed on the wire, and . . . CLUNK: there was an indication of the bottom. The pressure tube was recovered and read off against the box-wood scale to give a recorded depth of nine fathoms. Engines were stopped forthwith while the situation was assessed.

Some time later, after several more fruitless attempts to obtain a vertical sounding in what was in all probability the Palawan Trough, the sky cleared, a full moon lit the scene, which gave an approximate longitude by chronometer, and the ship steamed thankfully and without further incident to her destination. The subsequent report to the senior officer in Singapore (if it was ever made) could well have given rise to yet another of the many "doubtful data" charted in this area today by Japan on International chart INT508. The navigator, on the basis of his reputation as a shoal finder, subsequently dedicated his working life to hydrographic surveying.

So — a "sounding of 9 fathoms was reported . . . by HMLST 324 in 1946". With hindsight, and 40 years on, the erstwhile navigator is not too sure about that clunk in the night via the "feeler". Most likely it was his imagination.

This true (and unpurged) account is related to illustrate the possibly nebulous origins of many reports of doubtful data which are, alas, perpetuated on the world's charts until they can be disproved. By the end of the fourth decade of this century, echo-sounders were almost universally available on board vessels, and became a regular source of reports from mariners of anomalous depths in the oceans. Many were correctly identified and reported, but conversely, some reports stemmed from unskilled use of the phasing mechanism or "second-time-around" reflections which appeared in close proximity to the transmission line on the echo-trace.

Yet other reports were reflections from the deep-scattering layer rather than the sea bed itself. Ritchie (1953) discussed this possibility in the light of the improvements being made in merchant ships' echo-sounders during the 1950s. Similar false echoes in shallower waters arise from isolated patches of plankton and shoals of fish.

THE BACKGROUND

National hydrographers have always been conscious that, as with any map, their charts are only as good as the information provided for compiling them. The sources of such information do not consist solely of hydrographic surveys; indeed, most of the details shown on small scale "crossing" charts have been gleaned by vessels on passage, such vessels not being under the control of the hydrographers themselves. Thus a vast number of reports on features encountered, soundings taken on passage and investigations by vessels either not under their control, or perhaps only indirectly, are compiled into an ocean chart. Some of these features are of uncertain position or existence.

At the 2nd International Hydrographic Conference (IHC) in 1926 (IHB, 1926), the United States of America proposed that the IHB "carry into execution a plan for the dissemination of information concerning dangers shown on charts as 'position doubtful' and 'existence doubtful' ". The US had already published a list of such dangers in the Pacific, each of which carried a short note describing the origin and any research pertaining to it. The conference agreed that such information would be useful, but voted against the Bureau being involved—apparently because of the size of the task and the limited resources of the Bureau. The vote was not unanimous.

Despite this disagreement, in 1927, under Circular Letter No. 28-H, a provisional version of a list was sent to member states for review and was subsequently issued as Special Publication No. 20 (SP20) in the following January. In 1929, at the 1st Supplementary IHC (and only one to date), the matter seems to have been formalized (IHB, 1929). Resolution No. 18 was passed, and it reads:

> "The International Hydrographic Bureau is requested to draw up a list of dangers marked "P.D" and "E.D" on charts, each being accompanied by a short historical note. All States are invited to provide the Bureau with the information necessary for the preparation of this list.

This publication, updated and re-edited a number of times over the intervening years, remained a standard item of the Bureau's work until 1982 (IHB, 1973). The 12th IHC in 1982, by Decision No. 16 (IHB, 1982), discontinued it's publication, and the last edition of 1973 thereby became an historical document (see Fig. 1). The Bureau itself had proposed to shed its role, since all such data was by then being shown on the small-scale International (INT) charts on scales 1: 3500000 and 1:10000000. Interested enquirers could now approach the producer nations of the INT charts concerned as to the history of the features shown.

In fact, by 1962, the elimination of doubtful data had come to be considered as far more important than the maintenance of the lists. The ratio of the amount of doubtful data to the world's resources available to remove such doubt is very large indeed; clearly the work has to be tackled on a regular priority basis to remove those data which are a hindrance to passage. A United Nations study (1985), prepared by the Bureau on its behalf, states:

> "Large areas of continental shelf and EEZ waters are inadequately surveyed or not surveyed at all. The majority of coastal states have little or no capability for carrying out even vital ... surveys; for publishing charts ... and have little hope of remedying the situation in the near future...".

Doubtful data in international waters must exacerbate the problem.

THE EFFECT OF DOUBTFUL DATA

The obvious reaction by the prudent mariner to a doubtful feature on or near his planned track is to give it a wide berth. The question then arises as to how far off is a "wide berth" to ensure that he does not find the obstruction inadvertently.

The date may give some clue. A recent report would probably not be of doubtful position unless the reporting vessel's SATNAV receiver was out of action. Less recently, but still very valid, OMEGA would ensure a reasonably reliable position. But older reports would depend upon dead reckoning between perhaps infrequent astronomical observations or extrapolated from the last land fix. Even older

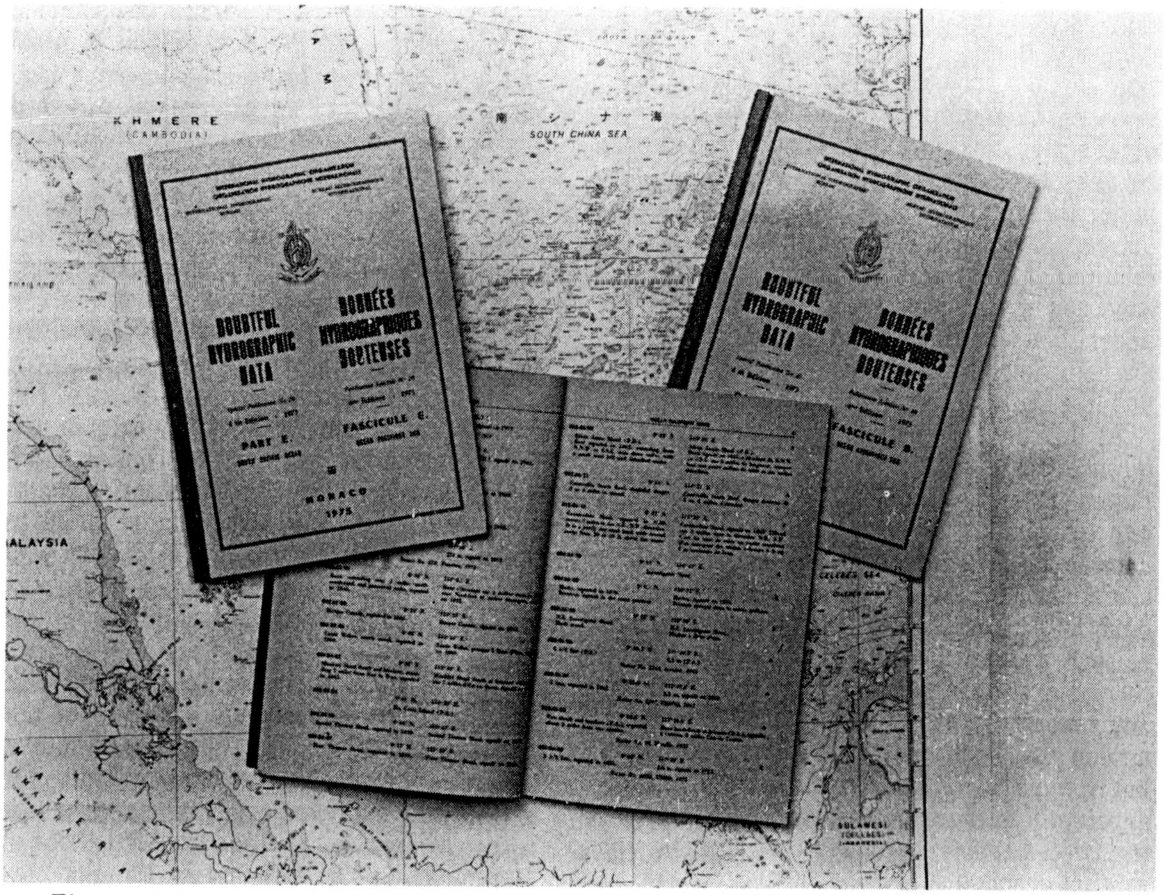

Fig. 1 An historic document: SP20 for the South China Sea, with INT chart 508 in the background

reports may have a large longitude error due to the uncertainty of the ship's chronometer. To leave the clearance too fine runs the risk of finding the feature with the ship's keel (a not unknown occurrence). A wide berth increases the steaming time, and hence fuel consumption, on passage.

The *China Sea Pilot,* issued by the British Admiralty (1978), describes, with notable restraint, the South China Sea where there is "an incompletely surveyed region encumbered with numerous coral reefs and shoals". It continues with a brief description of the two principle routes between Singapore and Hong Kong, one of 1450 miles and the other 1935 miles, which avoid this horrific reef-studded no-go area. Even so, there are many features adjacent to these routes which are classified as doubtful, and caution must be exercized by the mariner. Today the trade patterns are changing, but there is no route described in the pilot which crosses the South China Sea in an east–west direction; to do so is a very circuitous voyage.

Thus, one of the effects of doubtful data is economy. If vessels can navigate in safety, with complete confidence that the chart is a true statement of the facts, then two immediate benefits could arise: the cost of the deviations will be reduced to an essential minimum, and insurance rates will drop. Result: improved trading conditions.

The problem of costs associated with the management of doubtful data presents another effect. In waters over which coastal states have a clear jurisdiction, the states are responsible for maintaining up-to-date navigation information. Each doubtful feature will have a history attached to it: when, and by whom, it was first reported; subsequent sightings; formal efforts to locate and survey it; new reported features nearby that may be the same one (or may not); the decisions as to its classification; and, finally, the decision to take appropriate charting action. This information must be kept meticulously.

In international waters, however, there is no clear line of charting responsibility. The small-scale INT charts covering the whole world and allocated to producer nations by international agreement, contain thousands of doubtful features about which it is no-one's responsibility to do anything. The only possible exception could be the requirement by a nation for a Deep Water (DW) route or Traffic Separation Scheme (TSS) in such waters; in which case the onus must logically fall upon the requesting state to carry out the hydrographic survey to make sure the area is safe for navigation.

It follows that the tedious task of dealing with doubtful data, not all of which are admittedly a danger to navigation, is costly if it is to be done properly. The problem is made worse by the limited resources available for executing hydrographic surveys, as highlighted by the United Nations (1985).

THE ELIMINATION OF DOUBTFUL DATA

Technical Resolution No. A1.11 of the IHO (IHB, 1985) had its origins in the 8th IHC held in 1962. It makes strong recommendations that member states allocate resources annually to the task of elimination of doubtful data. It tasks the IHB with investigating what resources are available and which states would be willing to act co-operatively. Provision is made for searches in another nation's waters by the co-operating states; and there are a number of recommendations and resolutions about scheduling the work and reporting the results.

Over the subsequent five years, several countries accepted responsibility in terms of this Resolution and it is to be presumed that the work still progresses. Indeed, Section 2 of Booklet D of SP20 (1973) (IHB, 1973), lists "Data for which doubt seems (sic) to have been removed", and indicates that work completed between 1967 and 1973 in the North Pacific area was as follows:

United Kingdom	37 reports
USA (DMA)	8
Philippines	2
Japan	1
Canada	1 report

Since 1973, and the demise of SP20, reports of this work are no longer notified directly to the Bureau, nor are the Notices to Mariners carefully perused to abstract them. In 1983, with the agreement of member states, the Bureau began preparing minimum standards of surveying and reporting by which features could be confirmed or disproved. The work did not have a high priority, but in 1985 the so-called criteria were approved as an internationally agreed standard.

THE DEVELOPMENT OF THE CRITERIA

The Bureau followed its standard procedures in preparing the criteria. A Circular Letter was sent to all the member states requesting details as to their own approach to the task. Replies came mainly from those states with well developed hydrographic offices, active in such investigations on a regular basis. They indicated that there are two main factors be be considered, as follows.

The Dimensions of the Area to be Searched

The area to be searched is dependent upon the accuracy of the reported position. In many cases the navigation accuracy is stated by the reporting vessel, and the size of the search area is then simply a matter of the probability of indecision.

If the report is unqualified, the navigation accuracy has to be inferred, taking into account the sources of any possible errors. The questions then posed are:

- distance from land;
- suitability of the charted land features for

providing a reliable set of cross-bearings and/or radar ranges;

- likely methods of navigation available at the time of the report — very old reports could be many miles in error of sight of a land fix;
- a reasonable factor to apply to the estimated accuracy in order to allow for theoretical statistical probability.

Some countries use empirical values based upon distance from land, others upon the age of the report, and some a combination of both.

The Density of the Search Pattern

It was immediately apparent that the density of the sounding lines in waters of less than 200 m water depth was amply catered for by the existing IHO Standards for Hydrographic Surveys, SP44 Section A.1 (IHB, n.d.), which provides for a recommended scale of 1:50 000 for the survey, and a minimum scale of 1:100 000. The interval between lines is to be not more than 10 mm at the scale of survey (A.2.2) and should be reduced where the sea floor is irregular (A.2.3). The inference, then, is that line spacing for general hydrographic surveys in water depths from 30 m to 200 m should normally be 500 m apart, this interval being reduced under certain stated circumstances.

It is important to realize, however, that obstructions which could be a danger to navigation can rise vertically from water depths of about 100 m. Wrecks, basaltic cores of extinct volcanoes, pinnacle rocks and the like can all rise with little or even no slope; therefore total ensonification by echo-sounder or sonar, or clearance by means of mechanical sweeping, is the only sure way of detecting or disproving their existence, and the cone of ensonification of the search equipment plays an important part in deciding line spacing.

In waters that are deeper than 200 m, certain assumptions have to be made as to the probable shape of the geological feature. In deep waters, this is generally accepted to approximate as follows:

- a maximum slope of 60° to a depth of 2000 m;
- deeper than 3000 m, the maximum slope is unlikely to exceed 15° from the horizontal;
- between 2000 and 3000 m, the change in slope occurs uniformly and a mean figure of 27° is usually inferred.

Littlehales (1933) provides an interesting description of the theory underlying these assumptions, and deduces an expression for calculating the shape.

The interval between lines must be selected to ensure that an indication of the existence of a feature is likely to be found if it exists. Four variables have to be considered:

- general surrounding water depth;
- the changing slope of the feature as a function of water depth;
- the area covered by the cone of ensonification of the searching system (echo-sounder, multi-beam array, low frequency side-scan sonar);
- the relative accuracy of the available positioning system(s).

The member states of the IHO responding to the initial Bureau approach had differing practical ways of solving the problem, but in general their fundamental views were almost identical. Using these responses, and a few of its own ideas, the Bureau prepared a draft set of Criteria for consideration by all the member states. A form of report of the survey was also designed to assist the hydrographic offices to decide whether or not charting action was justified on the basis of the results of the search. They duly commented and these were taken into account in preparing a final document, which was approved by a large majority.

It cannot be emphasized too strongly that the objective is to locate the feature. The determination of the least depth and extent once it has been found is quite a separate matter and well covered by normal hydrographic practices.

THE IHO CRITERIA FOR THE ELIMINATION OF DOUBTFUL DATA

Search Area

When the accuracy of the positioning of the reported anomaly can be determined with confidence from the original report, the minimum recommended radius of the search area shall include a circle of radius R such that

$$R = 2.5PE$$

where PE = probable error of the position

When the accuracy of the positioning of the anomaly cannot be determined with confidence from the original report, the minimum recommended radius of the search area shall be assumed to be as noted as below:

(a) *Within sight of clearly defined land features,* the radius (R) of the area to be searched, regardless of the date of report of the anomaly, shall be:

$$R = 1.5 \text{ nautical miles}$$

(b) *Out of sight of clearly defined land features,* the radius (R) of the area to be searched shall be the radius R_1 based upon whether or not the reported position of the questioned anomaly lies on the continental shelf, plus a factor based upon the date the anomaly was reported, such that

$$R = R_1 + T/5$$

where T = (1950 − date of report) to a maximum value of $T = 60$ and a minimum value of $T = -15$ (resulting values for R as computed below should not be less than R = 3 nautical miles on the continental shelf, and R = 5 nautical miles off the continental shelf).

- On the continental shelf

  $R_1 = 6$ nautical miles

  and thus $R = 6 + T/5$.

- Off the continental shelf

  $R_1 = 8$ nautical miles

  and thus $R = 8 + T/5$.

Where the search area optionally chosen is a square, the sides should be tangential to the circle defined above.

Search Procedures

In water depths less than 100 m

Equipment. Equipment must include vertical echo-sounder plus side-scan sonar or its equivalent. An equivalent to side-scan sonar could be multi-beam, or sector-scanning sonar. Wire-drag sweeping could also be employed for the search, but may only verify that the area is clear down to such depths that sweeping can be deemed to be effective (possibly limited to 35 m). Magnetometer is also recommended for searches involving ferrous objects.

Line spacing. Spacing will depend upon the range of the sonar coverage, and the accuracy of the positioning system, allowing for an overlap of twice their combined probable errors. When side-scan sonar is used, the overlap should normally ensure coverage of the sea bed beneath the keel of the search vessel on adjacent lines.

The speed of the vessel should be such as to ensure that sufficient sonic echoes are received by all systems from any possible pinnacle features for registration on the analogue records (accepted as being 5 to 6 echoes).

When wire-drag sweeping, the overlaps between adjacent lines must compensate for all possible errors in positioning and variations in the width of the sweep.

In water depths greater than 100 m

Equipment required. A wide beam echo-sounder, but side-scan sonar or its equivalent is strongly recommended near the continental shelf and is required in depths under 200 m. A magnetometer is also recommended if searching for a reported ferrous hazard. A multi-beam array echo-sounder or long-range hull-mounted sonar is a valuable aid for searches; when either is used, its angle of coverage should be used for α in the expressions below.

Line spacing. Spacing of the lines of search (l) is a function of (a) the potential slope of the sea bed; (b) the general depth of water being sounded; (c) the coverage of the echo-sounder or sonar beam; and (d) the relative accuracy of the search vessel's positioning system(s), where:

I = line spacing (m)
H = general water depth, or least depth found on adjacent line (m),
α = apex angle of cone of ensonification of the echo-sounder, and
E_s = the probable error ("repeatability") of the search vessel's positioning systems(s).

In the first case considered, 100–2000 m:

$I = 2H(\tan \alpha/3 + 0.29) - 2E_s$

and in the second case, over 2000 m:

$I = 2H(\tan \alpha/3 + 0.0002H) - 440 - 2E_s$.

Line spacing may be increased moderately where the geology of the sea bed is well known and does not support the probability of the existence of the reported anomaly, or where dense ship traffic similarly makes the reported anomaly's existence improbable. However, any increase in value of I determined above should not exceed 33%. For ease of reference, graphs can be constructed for various values of H and α.

Cross-check lines should be run at intervals not greater than 10 times the expected line spacing.

The search operation report should be in the approved format.

THE NEW TOOLS

The classic wide beam, low-frequency oceanographic echo-sounder is still used for ocean bathymetry. An example of new narrower beam sounders, one of several more recently developed, is the Krupp Atlas DES020. Using a 4000 W transmission at 33 kHz, giving a beam width of 8° by 6°, 5000 m water depth is claimed as achievable, depending upon acoustic conditions. This kind of development is really designed for higher-resolution bathymetry than was normally obtainable by the older deep-water sounders with their wide beams, such as the Edo Western Model 185 and variants.

For searches in deep water, the picture has altered radically. General Instrument Corporation's SEA BEAM multiplexes an array of narrow-beam transducers to survey a swathe 3 km wide in 4000 m depths, processing the acoustic data in association with roll, heave, pitch, speed, etc., sensors to plot, in real time, a contoured strip chart.

The UK Institute of Oceanographic Sciences has been successfully operating various marks of GLORIA, its long-range, low-frequency, side-scan sonar which may be towed at 8 knots to map swathes of seabed reflective data 60 km across.

Döppler Satellite navigation will, hopefully, be replaced by the Global Positioning System (GPS) which could virtually eliminate the need for narrowing the search lines to allow for positioning errors within the next five years.

For searches in shallow water, side-scan sonar is already almost universally available. Developments in data processing and integration to position and speed sensors now permit on-line seabed mapping on the more sophisticated versions. Other swathe systems, such as interferometric techniques, are in the pipeline.

The Hydrographer, Royal Australian Navy (1985), reports on the use of satellite imagery to compile hydrographic data on new charts — AUS376, "Torres Strait", and AUS836, "Northern Part of the Great Barrier Reef" are mentioned. These techniques have also been used to identify shoal areas off the coasts of Tanzania and Papua New Guinea, providing data where information was doubtful or non-existent.

The means for eliminating doubtful data are thus being much enhanced by the aid of electronics and space or airborne sensor techniques. If the opportunity is now taken to concentrate the advantages on the task of removing doubtful data from the world's charts, the movements of shipping along newly opened routes, safely and without the prospect of having to avoid some, perhaps nebulous, obstructions could lead to better trading conditions.

CONCLUSION

The removal of doubt about hydrographic features which could be a danger to safe navigation is clearly of much importance to world shipping interests. It has taken the International Hydrographic Organization 60 years to produce specifications for the survey effort involved in clearing up that doubt.

This is, of course, only the beginning. However, a note of optimism can be sounded, because remote sensing from aircraft and satellites can already play an increasingly effective role in finding shoal areas, determining the water depth over them, and bringing the charts up to date. It may not be too long before they are free of the notes PA, SD, or ED.

By international co-operation, and the free exchange of technology and data, together with some badly needed funding, maritime safety will be greatly improved. The International Bureau will continue its traditional efforts to support this objective.

ACKNOWLEDGEMENTS

The author would like to thank the Directing Committee of the Bureau and its President for permission to produce and present this paper. The contribution made by all the staff, was essential to the outcome, and the work of Captain James E. Ayres, US Navy (Ret.) in developing the criteria jointly with the author is much appreciated — we had many long and interesting discussions, particularly about the performance of deep-water hydrographic tools.

REFERENCES

Hydrographer of the Navy 1988. *China Sea Pilot,* vol. 1, 4th edition, p. 51.

Hydrographer, RAN 1985. *Report of the Hydrographic Service of the Royal Australian Navy for the year ended 30 June 1984.* Canberra: Royal Australian Navy.

IHB (International Hydrographic Bureau) 1926. Report of the Proceedings of the 2nd I.H. Conference, pp. 422–4.

IHB 1929. *Report of the Proceedings of the 1st Supplementary I.H. Conference,* pp. 25, 76, 162, 243 and 325.

IHB 1973. *Doubtful Hydrographic Data,* 5th edn. SP20. Monaco: IHB.

IHB 1982. *Report of the Proceedings of the 12th I.H. Conference,* pp. 47–108.

IHB 1985. *IHO Resolutions of the International Hydrographic Organization.* Monaco: IHB.

IHB n.d. *IHO Standards for Hydrographic Surveys and Classification for Deep Sea Soundings.* SP44. Book1, p.4.

Littlehales, G. W. 1933. Measure for settlement of report of doubtful dangers in the ocean. *I.H. Review,* 96–105.

Ritchie, G. S. 1953. The deep scattering layer. *I.H. Review* **XXXI** (2), 121–8.

United Nations 1985. *Status of Hydrographic Surveying and National Charting World Wide.* UN Economic and Social Council, E/CONF.77/L.12.

51

A Seismic Cone Penetrometer for Offshore Applications

R. G. Campanella, P. K. Robertson and D. Gillespie,

A new test, called the seismic cone penetration test (SCPT) is described. A small rugged velocity seismometer has been incorporated into an electronic cone penetrometer. The combination of the seismic downhole method and the CPT logging provides an extremely rapid, reliable and economic means of determining stratigraphic, strength and modulus information in one sounding. Results using the seismic cone penetration test are presented and compared to conventional onshore *in-situ* techniques. Results are also presented from seismic cone testing performed offshore in the shallow Canadian Beaufort Sea.

INTRODUCTION

In the offshore environment, the dynamic behaviour of soil can be of great importance. Rapid development has taken place in the past decade on new analytical and dynamic testing techniques.

For onshore investigations, the crosshole and downhole methods have become the standard techniques for dynamic testing to determine the *in-situ* shear wave velocity. A polarized shear wave is generated in one borehole (or at the surface), and the time is measured for the shear wave to travel a known distance to the geophone in the borehole. Elastic theory relates the shear modulus, G, soil density, ρ, and shear wave velocity, V_s, as follows:

$$G = \rho V_s^2 \quad (1)$$

Hence, the shear modulus can be determined using *in-situ* seismic methods for the determination of the shear wave velocity. The shear modulus is largest at low strains and decreases with increasing shear strain (Seed and Idris, 1970). The shear strain amplitude in *in-situ* seismic tests is usually low and of the order of 10^{-4} %. Thus, the very low strain level dynamic shear modulus, G_{max}, is usually obtained. The cost of such a test is usually high because of the requirement to have one or more boreholes. This has generally made the technique difficult or impossible for offshore use.

A new test, called the seismic cone penetration test (SCPT), which can dramatically reduce the cost associated with the *in-situ* determination of shear wave velocity, will be described in this chapter.

SEISMIC CPT

The cone penetration test (CPT) is already used extensively offshore for geotechnical investigations. A cone of 10 cm^2 base area with an apex angle of 60° is generally accepted as standard, and has been specified in the European and American Standards. A friction sleeve, located above the conical tip, has a standard area of 150 cm^2. A pore-pressure transducer has recently been added to measure the dynamic pore pressures during penetration. The cone penetrometer is pushed at the standard rate of 2.0 cm/s. A small slope sensor is usually incorporated to monitor sounding verticality.

Full details of the design of an electronic cone are given by Campanella and Robertson (1981) and Schaap and Zuidberg (1982). The piezometer-CPT is regarded as the premier test for the continuous logging of soil stratigraphy and shear strength. An example of the extensive data obtained from a piezometer-CPT is shown in Fig. 1. The cone data can be interpreted to give a good continuous prediction of soil type and shear strength (Robertson and Campanella, 1983). Predictions of soil stiffness (modulus) from the cone resistance can be rather poor, especially for overconsolidated soils, with a large potential error. The introduction of seismic measurements into the cone penetration test procedures enables the specific determination of the dynamic shear modulus (G_{max}). Thus, the combined measurement of soil strength and modulus provides a means to identify the complete stress–strain relationship of soil units which is required for many complex finite-element analyses.

To obtain the measurement of dynamic shear modulus, a small rugged velocity seismometer has been incorporated into the cone penetrometer. The miniature seismometer is a Geospace GSC-14-L3 (1.7 cm dia) with a nominal natural frequency of 28 Hz. The seismometer is

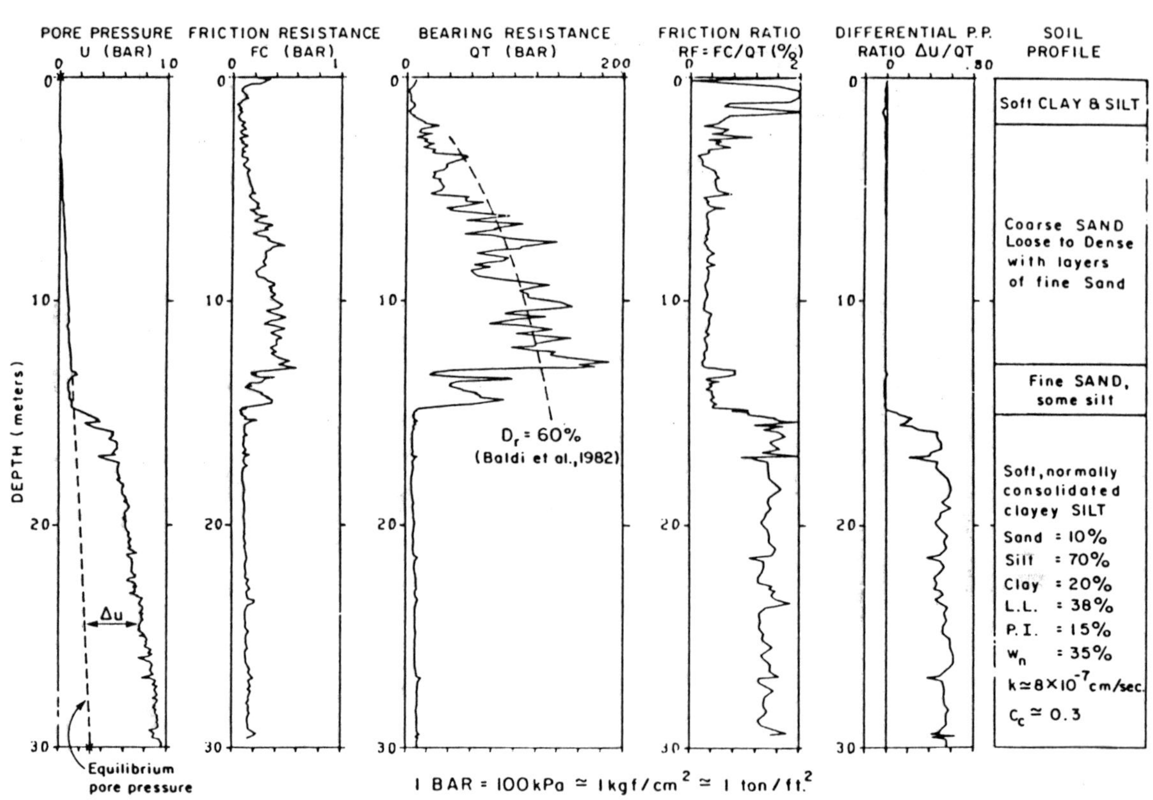

Fig. 1 Soil profile for research site at McDonald's Farm site (after Campanella *et al.* 1983)

placed in the horizontal direction and orientated transverse to the signal source to detect the horizontal component of the shear wave arrivals. A schematic diagram showing the layout of the standard downhole technique is shown in Fig. 2.

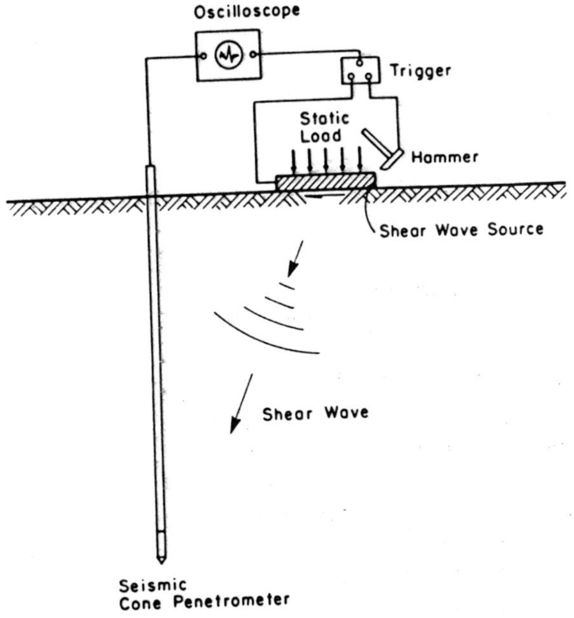

Fig. 2 Schematic layout of downhole seismic cone penetration test

A suitable seismic signal source should preferentially generate large-amplitude shear waves with little or no compressional wave component. The shear waves travel through the soil skeleton and are thus related to the soil shear modulus. Results indicatc that an cxccllent onshore downhole seismic shear wave source consists of a rigid beam or platform, steel jacketed and weighted to the ground. It may be struck with a sledge-hammer as shown in Fig. 2. If the cone is being pushed by a drill rig, the beam can be weighted down by the rear pads of the drill rig. If the cone is being pushed by a cone penetration vehicle, the beam can be weighted down by the pads of the vehicle or incorporated into the stabilizing pads for the truck. The beam-type signal source is usually placed with ends equidistant, within about 3 m of the cone hole. The beam should be rigidly placed on the ground so that no energy losses should occur due to plastic shearing of the soil beneath the beam.

For the offshore environment, the generation of an effective shear source is more complex. For deep offshore investigations (depth > 30 m), cone penetration testing is usually carried out using either downhole devices or seabed supported devices. It should be possible to incorporate an automatic shear wave source, similar to that used onshore, on existing seabed devices. The downhole devices could also incorporate a similar seabed device to generate shear waves. For the shallow offshore environment, it should be possible to manufacture a simple shear source to be lowered to the sea bed. It is possible to generate shear waves in the seabed soils by detonation of an energy source in the water. Further details of this method are given later in the chapter. The generation of an effective shear source is currently an area of intense research.

The design and construction of the seismometer carrier provides a snug seating for the seismometer package. The method of advancing the cone penetrometer provides continuous firm mechanical contact between the seismometer carrier and the surrounding soil. This allows excellent signal response. In addition, seismometer orientation can be controlled and accurate depth measurements obtained.

The seismic wave traces detected by the seismometer are recorded on a Nicolet 4094 digital oscilloscope with floppy disk capability. This unit has a 15-bit analog-to-digital signal resolution, very accurate timing capability, and trigger delay capacity. The high-resolution oscilloscope is capable of recording clean shear wave traces from forward and reverse single hammer impulses to depths of over 40 m, as shown in Fig. 3. Figure 3 provides a quantitative comparison of the geophone response amplitude and relative shear wave travel times with depth. The geophone output voltage is directly related to the particle oscillation velocity, as shown on the inset scales.

The strain level caused by the shear waves can be estimated at any depth during the CPT downhole seismic survey. The relationship

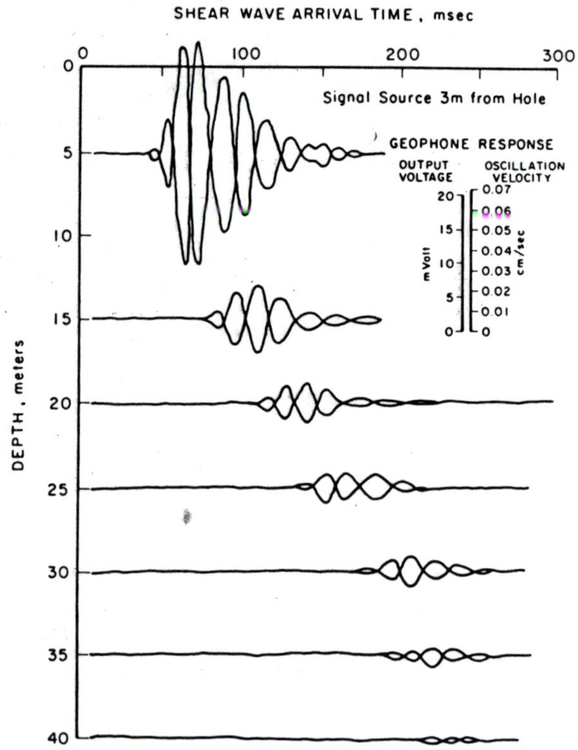

Fig. 3 Quantitative comparison of geophone response amplitude and relative shear wave travel times from seismic cone penetration test

between shear strain, γ, shear wave velocity V_s, and peak oscillation velocity, u, is given by White (1965):

$$\gamma \times u/V_s \qquad (2)$$

Analysis of the existing field data shows that the strain amplitudes caused by the hammer-beam source are generally less than $10^{-4}\%$ and decrease with depth.

It has been found that the time for the first crossover point (shear wave changes sign) is easily identified from the polarized waves (forward and reverse) and provides the most repeatable reference arrival time. The arrival time from source to detector is converted vectorially to a vertical travel path. The difference between successive 1 m depth measurements of vertical travel path time is used to determine the shear wave travel time over the 1 m depth interval. Because of the short distances and small travel times involved, the oscilloscope must have very high resolution, fast sample times and a very fast, repeatable trigger. The trigger used is similar to that suggested by Hoar and Stokoe (1978). The trigger incorporates a MC1455 linear integrated circuit with a rise time of less than 1 μs. Since the shear wave velocity is squared to calculate G_{max}, a high priority must be given to the accuracy of travel-time measurements.

To assess the variability of the arrival time measurements and the accuracy and reliability of the trigger system, a second geophone system was placed 1 m vertically above the first geophone (Rice, 1984). This equipment modification allowed true interval surveys to be carried out and compared to the pseudo-time-interval method described above. The arrival time data was then analyzed assuming a normal statistical distribution at each depth interval. The results showed that there is very little error using a single geophone and the pseudo-time-interval method, as compared to the true time-interval method with a pair of geophones.

Onshore Seismic CPT Results

Downhole seismic shear wave velocity measurements have been made at several sites and, in some cases, compared to results obtained by others using the conventional crosshole techniques.

A comparison between shear wave velocities measured using the seismic cone penetration test and conventional crosshole techniques is shown in Figs. 4 and 5. Seismic CPT tests were performed at several sites in Norway, with the co-operation of the Norwegian Geotechnical Institute (NGI) (Eidsmoen *et al.*, 1984). These are well documented sites with extensive field and laboratory data.

The Holmen site (Fig. 4) consists of loose, medium to coarse sand to a depth of 25 m. On average, the two seismic shear wave velocity profiles are almost identical at this site showing little, if any, discrepancy between the SCPT downhole and conventional crosshole.

The Museumsparken site (Fig. 5) consists of the well documented Drammen Clay, to a depth of 15 m. Again, the two seismic profiles compare very favourably.

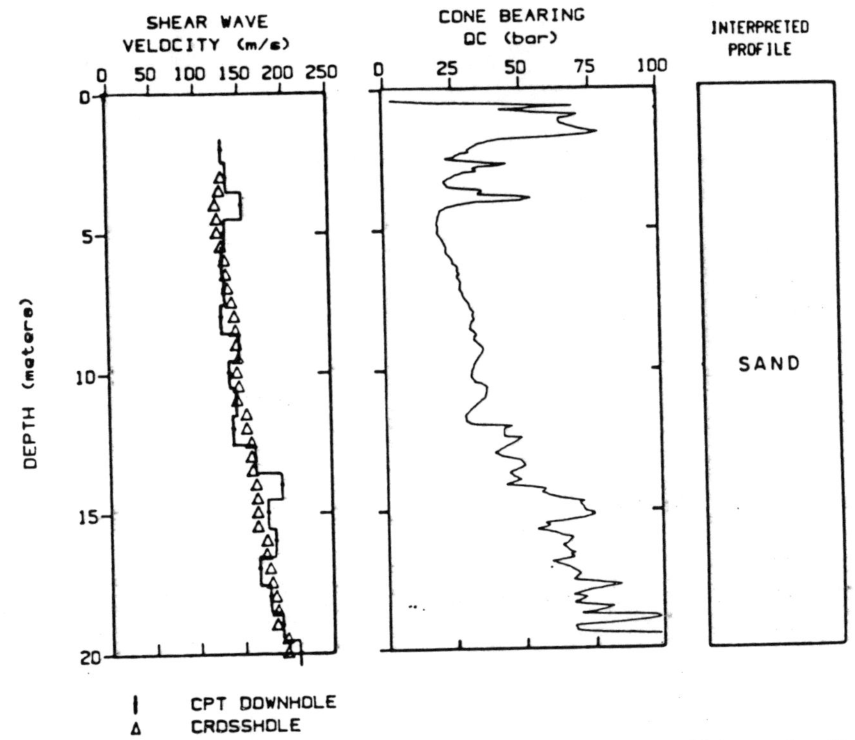

Fig. 4 Comparison of seismic CPT downhole and crosshole data at Holmen site, Norway

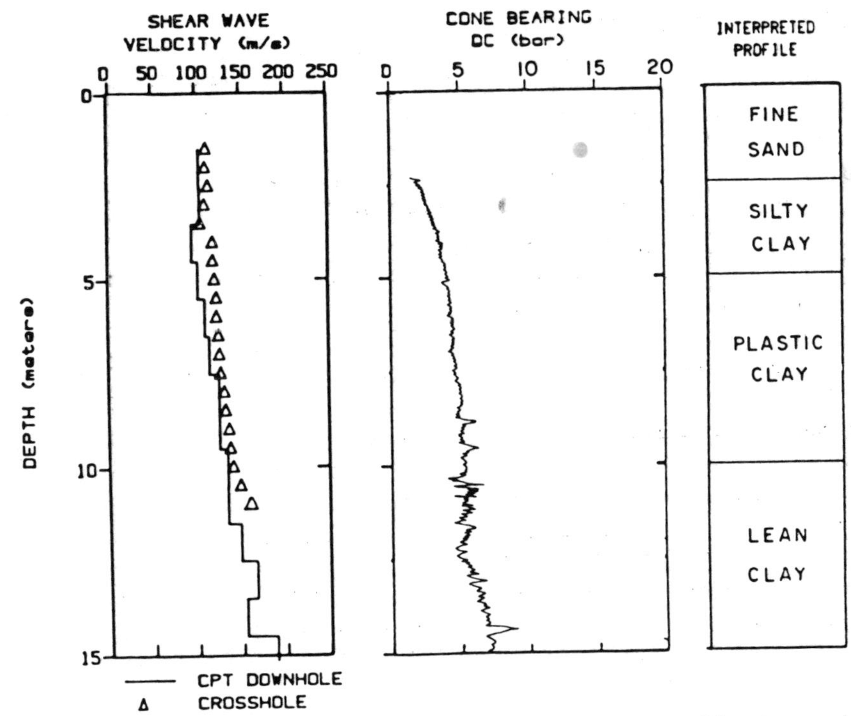

Fig. 5 Comparison of seismic CPT downhole and crosshole data at Museumsparken site, Drammen, Norway

Offshore Seismic CPT Results

Seismic cone penetration tests were performed in the shallow offshore area of the Canadian Beaufort Sea. Testing was performed during the winter of 1985 using the approximately 2 m thick sea-ice as a stable platform. The depth of water beneath the ice was generally less than 8 m.

Several seismic sources were tested to evaluate their effectiveness in the generation of compressional (P) and shear (S) waves. Explosive caps were the predominant seismic source. The explosive caps were generally placed in the water beneath the ice. When the sea-ice extended to the sea floor (i.e. no water beneath ice) the seismic signal was generated by striking the rear support pads of the cone vehicle with a 7 kg hammer, as shown schematically in Fig. 2. Using the hammer technique, it was possible to generate polarized shear (S) waves with very little compressional (P) waves.

Seismic velocities were calculated using the pseudo-time-interval technique. Generally, seismic measurements were made at 1 m intervals starting approximately 2 m below mudline.

The average P-wave velocity for most of the tests was close to 1500 m/s. However, calculations using the pseudo-time-interval method for high P-wave velocities are sensitive to small variations in seismic cap location. In most saturated soils, the compression wave velocity is dominated by the compression wave velocity of water (i.e. 1500 m/s).

Initially, difficulty was experienced in identifying the shear wave arrivals using the seismic cap explosive source. Figure 6 illustrates the geophone response from a series of tests over the depth range of 4–18 m below mudline for a typical sounding. The fast, high-frequency P-wave arrivals are also visible in Fig. 6. The average shear (S) and compression wave (P) velocities shown in Fig. 6 are 220 m/s and 1600 m/s, respectively. It is interesting to note the reflected P-wave arrivals in Fig. 6. The depth of water beneath the sea-ice at that location was less than 1 m. It appears the P-wave is reflected off the permafrost that was at a depth of 18.4 m below mudline, and then reflected again off the sea-ice. There also appears to be a reflected S-wave produced as the P-wave reaches the permafrost.

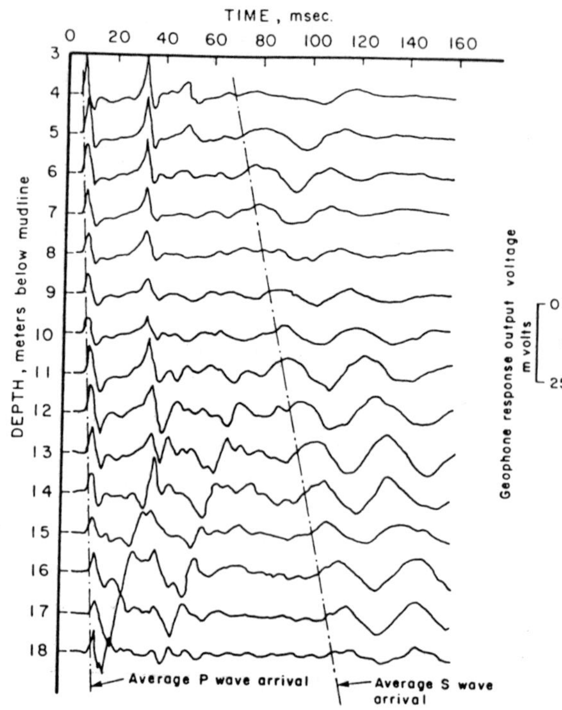

Fig. 6 Geophone response from a series of seismic tests during a single seismic CPT profile in the shallow Canadian Beaufort Sea

At one location, the depth to mudline increased rapidly over a short distance. This enabled the hammer technique to be used where the sea-ice extended down to and into the sea bed and also to use the explosive cap source nearby where water existed beneath the sea-ice. A comparison of the shear wave velocities measured using the hammer source and the seismic cap source is shown in Fig. 7. Good comparison was achieved between the hammer source data and the seismic cap data. This provides some confidence in the selection of shear wave arrivals using the seismic cap source.

SUMMARY

A new test, called the seismic cone penetration test (SCPT) has been described. The cone

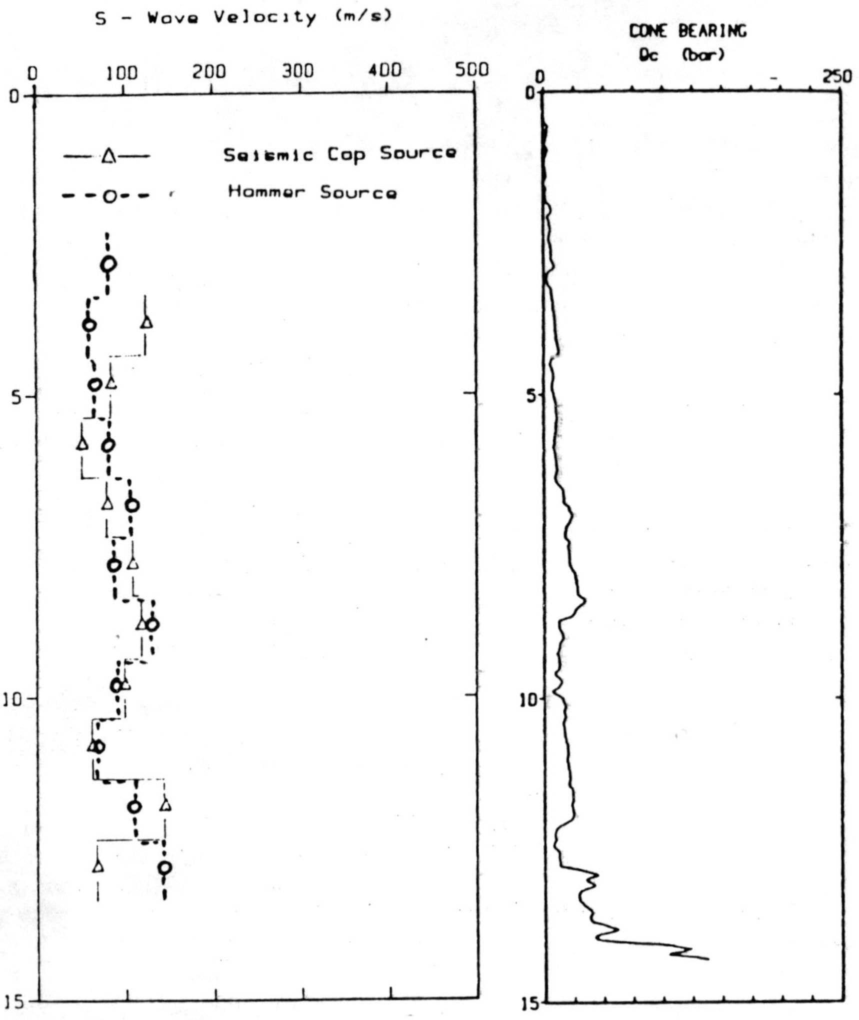

Fig. 7 Comparison of shear wave(s) velocities measured using hammer source and seismic cap source

bearing, friction sleeve stress and cone pore-pressure data can be used to provide a fast and reliable determination of soil type and shear strength. Downhole seismic shear wave velocity measurements can be made during brief pauses in the cone penetration. The shear wave velocity data can be used to provide reliable determination of the maximum dynamic shear modulus. Accurate depth determination is made by measuring the rod length, and seismometer orientation is easily maintained throughout the sounding. Hole verticality is monitored throughout the sounding with a small slope sensor installed in the cone. The combination of the seismic downhole method and the CPT logging provide and extremely rapid, reliable and economic means of determining stratigraphic, strength and modulus information in one sounding.

A current area of research is to investigate the full application of SCPT data. Comparisons are presently underway between SCPT data and self-bored pressuremeter, crosshole, flat plate dilatometer, screw plate and laboratory test results from several onshore sites.

Comparison of onshore seismic CPT downhole shear wave velocity measurements with those obtained by conventional crosshole tech-

niques show excellent agreement. The seismic CPT is, however, considerably less expensive and a more rapid procedure than the crosshole technique. The seismic CPT shows particular promise for use offshore, where CPT equipment is already used extensively and where shear wave velocity measurements would be of value in design of large offshore platforms.

Preliminary test results using the seismic CPT in the shallow offshore environment show that the simple procedure of detonating explosive seismic caps in the water can provide adequate data, provided that care and experience are used in the data interpretation. However, it should be possible to incorporate a simple automatic hammer type seismic source onto existing seabed cone penetration devices. Further research is presently under way to investigate various sources and receivers for offshore work.

ACKNOWLEDGEMENTS

The assistance of the Natural Sciences and Engineering Research Council; T. Lunne and T. Eidsmoen of the Norwegian Geotechnical Institute; and the technical staff of the Civil Engineering Department, University of British Columbia is gratefully acknowledged. The work of N. Laing and J. Greig is also appreciated. Acknowledgement is also given to the Geological Survey of Canada for providing the opportunity and financial assistance for the offshore testing.

REFERENCES

Campanella, R. G. and Robertson P. K. 1981. Applied cone research. In: *Symposium on Cone Penetration Testing and Experience,* pp. 343–362. Geotechnical Engineering Division, ASCE, October.

Eidsmoen, T., Gillespie, D., Lunne, T. and Campanella, R. G. 1984. Evaluation of the seismic cone penetration test. *Joint Report.* Norwegian Geotechnical Institute and University of British Columbia.

Hoar, R. J. and Stokoe, K. H. 1978. Generation and measurement of shear waves in-situ. In: *Dynamic Geotechnical Testing,* pp. 3–29. ASTM STP, American Society for Testing Materials.

Rice, A. 1984. The seismic cone penetrometer. hesis submitted in partial fulfilment for Master of Applied Science, Department of Civil Engineering, University of British Columbia.

Robertson, P. K. and Campanella, R. G. 1983. Interpretation of cone penetration tests; parts I and II. *Can. Geotech. J.* **20** (4), 718–745.

Schaap, L. H. J. and Zuidberg, H. M. 1982. Mechanical and electrical aspects of the electric cone penetration tip. *Proceedings of the Second European Symposium on Penetration Testing,* ESOPT II, Amsterdam, Vol. 2, pp. 841–851.

Seed, H. B. and Idris, I. M. 1970. Soil moduli and damping factors for dynamic response analyses. *Report to EERC 70–10.* Earthquake Engineering Research Center, University of California, Berkeley.

White, J. E. 1965. *Seismic Waves: Radiation, Transmission and Attenuation.* New York: McGraw Hill.